MW01165792

Microwave Circuit Modeling Using Electromagnetic Field Simulation

For a complete listing of the *Artech House Microwave Library*,
turn to the back of this book.

To Mike
With Best Wishes
Dan Swanson

Microwave Circuit Modeling Using Electromagnetic Field Simulation

Daniel G. Swanson, Jr.
Wolfgang J. R. Hoefer

Artech House
Boston • London
www.artechhouse.com

Library of Congress Cataloging-in-Publication Data

Library of Congress CIP information available on request.

British Library Cataloguing in Publication Data

British CIP information available on request.

Cover design by Yekaterina Ratner

685 Canton Street
Norwood, MA 02062

International Standard Book Number: 1-58053-308-6
Library of Congress Catalog Card Number:

10 9 8 7 6 5 4 3 2 1

To Mom and Dad, my first and best teachers.

DGS

Contents

Preface

This book is about modeling microwave circuits using commercial electromagnetic field-solvers. But before we can model a circuit we have to understand how the tools work. All the field-solvers we will discuss are based on well-established numerical methods for solving Maxwell's equations. We have tried to gather just enough background material on the major numerical methods to help the reader appreciate what is going on behind the interface. We will spend a lot of effort outlining the strengths and weaknesses of each numerical method in a fair and balanced way. This knowledge helps us choose the right software tool for a specific task and set up the problem more intelligently.

I have included some, but not a lot of information on simulation times. I am not interested in benchmarking various tools against each other because that borders on marketing. When I do quote times it is mostly for historical reasons and to point out how far we have come in only a decade. I may also quote simulation times to emphasize the difference between a lossless and a lossy analysis. Given the right problem and an intelligently constructed model, all of the software packages will give a usable answer in a reasonable amount of time. All the factors we have to consider when constructing that model is what this book is about.

Design case studies make up about half the material in this book. The examples are not intended to be a complete design procedure for any particular component. Rather, they are intended to demonstrate the trade-offs and compromises that must be made to get an efficient solution. I have also tried to document some cases where the modeling process did not work correctly the first time and what was needed to correct the model. In the cases where a bad solution was the result of a bug in the software I hope the vendors will forgive me. But these are large, complicated codes and being critical of results and looking for bugs should be a part of the modeling process.

I have avoided the temptation of using example files from the various software vendors or from colleagues. It would be nice to have a very broad set of examples that cover many disciplines, but I feel uncomfortable presenting an example where I am not personally aware of all the details and background material. Unfortunately, this also limits the range of examples that I can present. It would be nice to have some active circuit examples, some antenna examples, and maybe some EMC/EMI-related projects. But my fundamental approach to using these tools should be universal and easily applied to other areas.

Still, when I started this project my goal was to have a balanced number of examples from each of the major software packages. This was perhaps a worthy but not very practical goal. The reality is that I have used design examples that span more than a decade in time and date from the first introduction of commercial electromagnetic field-solvers. So the tools that entered the market first, namely Sonnet ***em*** and Ansoft HFSS, are perhaps overrepresented simply because I have been using them the longest.

I have also avoided the temptation of showing plot after plot of near-perfect agreement between measured and predicted results, as this would be somewhat dishonest. We don't get perfect results every time in the lab and we often learn more from failures than from successes. I also tend to favor small projects rather than an end-to-end analysis of a large, complicated geometry. Small projects fit the capabilities of the tool better. Small projects run faster and tend to encourage some "what if" experimentation with the geometry. And with a small project there is always a chance that we will gain some valuable insight into how a particular structure really behaves. Big projects take a long time to compute and tend to stifle "what if" experiments. A big project can only give you numbers, which may be right or wrong, and without measured data or previous experience it is difficult to judge the quality of the solution.

I am thrilled that Wolfgang Hoefer could join me on this project. Over the years he has been one of the experts who has very patiently explained to me some of the inner workings of numerical electromagnetics. Wolfgang is by nature a teacher and his enthusiasm for the subject comes through. He and I have taught a 1-day tutorial based on just some of the material in this book several times now. It is always fun and I always learn something new.

There are many other friends and colleagues in both the academic and industrial communities that I could recognize. But one person in particular has stimulated my thinking on how to apply these tools more creatively and that was Dr. John Bandler. Our progress in optimization using field-solvers is largely due to the motivation of his ideas and those of his students. I should also recognize the generous support of all the software vendors that made this work possible by giving me access to their tools. And all the staff members at the various software providers that patiently answered my many questions. I also owe a debt to the students in my classes who challenged me to come up with new ways of presenting this material.

Finally, I would like to thank my wife Ibis and my daughter Melissa for their love, patience, and support during the writing of this book.

Dan Swanson
Westford, MA

§

I have greatly enjoyed the collaboration and exchange with Dan Swanson that eventually led to this book. The project evolved over several years through individual and joint workshop presentations, tutorials, and lectures. Dan has become well-known in the microwave community as an enthusiastic and expert user of electromagnetic simulators from the early days of their commercial availability, and he has been instrumental in promoting their acceptance as effective, reliable engineering tools by microwave designers. This book is thus unique in the way it broaches the subject of electromagnetic simulators, not "from the inside out," beginning perhaps with an extensive theoretical development and culminating in a algorithmic implementation. Rather, the reader is invited to discover and experience an extensive arsenal of modeling and simulation features from the perspective of microwave practitioners, building upon their traditional design experience, their knowledge of laboratory practice, and their intuitive understanding of microwave components and systems. The study of the field-theoretical foundations of commercial software tools thus becomes more than a mere academic pursuit: it empowers the user to apply them more effectively, more intelligently, and with greater confidence. What type of simulator is best suited for what kind of technology? What is the expected margin of error? What is the best trade-off between accuracy and computational burden? What are the strengths and weaknesses of the different numerical techniques that underlie the various software tools? These are the questions that guide our approach and emphasis throughout this text.

I share Dan's conviction that the key to successful electromagnetic field simulation is to begin with simple, easily manageable problems for which the solution is known in advance. This enables the user to build a sound technical judgment and an appreciation for the sensitivity of the solution to various critical simulation parameters, such as meshing, frequency or time resolution, definition of geometrical detail, and the configuration of field excitation and sensing elements. Techniques for error checking and assessment of convergence can thus be systematically articulated and refined. This, in turn, motivates the user to explore the underlying theoretical foundations of a tool, a process that is considerably helped by the dynamic field and data display capabilities of most simulators. Interactive computer graphics allow us to observe electromagnetic field behavior which we could previously only imagine, enriching our physical perception to an extent rarely achieved by any other tool in science or engineering. Graphical dynamic representation reveals most electromagnetic processes in their full complexity and allows us to perceive the relationship between field behavior and specifications of microwave components more clearly than equations or diagrams. It is not only extremely satisfying to see one's theoretical projection confirmed by a simulation, but the involvement of our intuitive abilities through visualization effectively complements our analytical skills, enhances creative projection, and spawns innovation.

The extensive use of case studies reflects Dan's background and expertise as a microwave designer and reveals the primary target audience of this book, namely designers and practicing engineers. However, the focus on practical design applications will also be invaluable to students, researchers, and educators who use electromagnetic simulators mainly for demonstration, analysis, and physical insight. Last, but not least, it will provide input to those who develop software tools for electromagnetic modeling and simulation.

Wolfgang J.R. Hoefer
Victoria, BC

Chapter 1

Introduction

The history of microwave engineering is relatively short, beginning with the development of RADAR during World War II. Computer aided design (CAD) from a strictly circuit theory point of view gained momentum in the 1970s with the wide availability of mainframe computers that could be time shared. With easier access to computer power, numerical electromagnetics began to emerge at about the same time in the academic community. Only 20 years later, in the 1990s, the UNIX workstation and the personal computer (PC) made commercial field-solvers a practical reality.

Today, electromagnetic (EM) field-solvers have given the radio frequency (RF) or high-speed digital design engineer new tools to attack his or her more difficult design problems. Used often in conjunction with circuit-theory-based CAD, these new tools generate solutions derived directly from Maxwell's equations. Generally we are most interested in finding scattering parameters (*S*-parameters) or an equivalent circuit model for a given structure. But with the field-solver, we also have the capability to look inside the structure and display surface currents, various types of electric-field and magnetic-field plots, or other quantities derived from the fields. The visualization capabilities built into most field-solvers can lead to startling new insights into how RF and high-speed digital components actually behave. Perhaps you have had a colleague who could look at a complex structure and "see the fields." These rare individuals are highly regarded for their grasp of especially challenging design problems. Those engineers not blessed with this gift can use the visualization tools in today's field-solvers to develop some of these skills and see their design work in an entirely new way.

Long solution times limited early users of field-solvers to an analysis of relatively small, fixed geometries. These discontinuity size problems were quite valuable on their own or as sets of solutions that could be used to generate faster, circuit-theory-based models. By the mid-1990s, faster computers and more efficient software made it possible to optimize planar and three-dimensional (3D) RF structures using direct driven electromagnetic simulation. Although practical problem size is still limited, field-solver tools can now be more fully integrated into the

Table 1.1

A List of General Field-Solver Applications

Solenoids, transformers, rotating machines	Antennas
Magnetic recording heads	Active devices
Computer backplanes	RF and microwave circuits
Board level and chip level interconnect	Electromagnetic compatibility (EMC)
Packaging of high-speed devices	Electromagnetic interference (EMI)
Radar cross-section (RCS)	

design environment. Today, many field-solver vendors offer a "design environment" that manages any number of smaller field-solver solutions and integrates them into a higher level solution. At some point, practical problem size can also be cast as a trade-off between raw numbers and insight. Large problems may only give you numbers; small problems often lead to a deeper understanding of fundamentals.

In this book, we will start with a summary of CAD for RF and microwave circuits followed by very brief review of the more popular numerical methods. Some understanding of the method underneath the interface is needed to more fully grasp the strengths and weaknesses of each field-solver. Next we explore several issues that are common to all work with these tools. These special issues include meshing, convergence, de-embedding, and visualization. Part of this discussion focuses on validation structures and some simple "calibration elements" that stimulate our thinking and make us confident that we are using the tool correctly.

Half of this book is devoted to actual design case histories developed by the author. Some of these examples are filter structures. A filter is actually an excellent test case; there is an exact answer that makes comparisons between measured and modeled results quite easy. A filter is also a very sensitive structure; it is a collection of resonators that must be synchronously tuned. When we use active circuits as test cases, the uncertainty in the active device parameters can sometimes make comparisons between measured and modeled results difficult. In any case, the type of problem we present is less important than the fundamental concepts we are trying to demonstrate. The examples we present not only demonstrate the accuracy of the field-solver but also develop a design philosophy that has been very successful.

1.1 GENERAL FIELD-SOLVER APPLICATIONS

Numerical methods have been applied to any number of interesting electrical engineering problems over the years (Table 1.1). At low frequencies solenoids, transformers, and rotating machines have been popular topics. One popular demonstration of the early finite element method (FEM) tools was an analysis of an automobile alternator. The cost of tooling a new design more than justified the effort put into the analysis. The study of magnetic recording heads has been very important in the computer industry.

Mainframe computer manufacturers spent much time and effort understanding high-speed backplane problems. These were mostly internal efforts that resulted in custom codes that were not published widely. Workstation and personal computer designers have continued these efforts. Today, board level and chip level interconnect problems are receiving additional attention. Packaging of high-speed devices is another interesting topic. Multilayer boards using various construction techniques are of interest to both the RF and digital communities.

In the RF/microwave arena, radar cross-section problems (RCS) have received a great deal of funding over the years; stealth technology is the culmination of this work. The study of antennas has generated much interesting work as well. Today planar antennas for various wireless applications are attracting considerable attention. Simulating active microwave devices has also been a popular topic. Models based on the physics of the active device may soon appear in commercial microwave circuit simulators. However, it is only recently that much attention has been focused on RF and microwave circuits. And now, electromagnetic compatibility (EMC) and electromagnetic interference (EMI) will receive more attention. EMC is actually a very challenging application because, in general, we do not know exactly where the electromagnetic sources are.

1.2 A NOTE ON COLOR PLOTS

One of the unique features of this book is the large number of false color current plots and field plots. The most desirable method of presentation would include a scale for each color plot. Unfortunately, time and space do not always permit this. Most of the field-solver software vendors initially adopted a colors of the rainbow spectrum (red, orange, yellow, green, blue, indigo, violet, or ROYGBIV) for their false color current and field plots (Figure 1.1(a)). Red generally indicates high values, and dark blue or violet indicates low values. While the color red is easily associated with "hot" values and the colors blue or violet with "cold," the intermediate colors of the rainbow have no values intuitively associated with them. The viewer is forced to adapt to a relatively nonintuitive display format [1].

Later, the various commercial software vendors began to offer alternative color schemes, including a "temperature" scheme that runs from black or blue "cold," through shades of red, shades of yellow-orange, and finally white "hot." While this scale may be generally more intuitive, at least to those who have ever witnessed metals heated to various temperatures, the white values tend to get lost on a white page (Figure 1.1(b)). One color scheme that seems somewhat intuitive to this author uses shades of red for magnitudes with positive phase and shades of blue for magnitudes with negative phase [2]. However, this particular scheme has not been

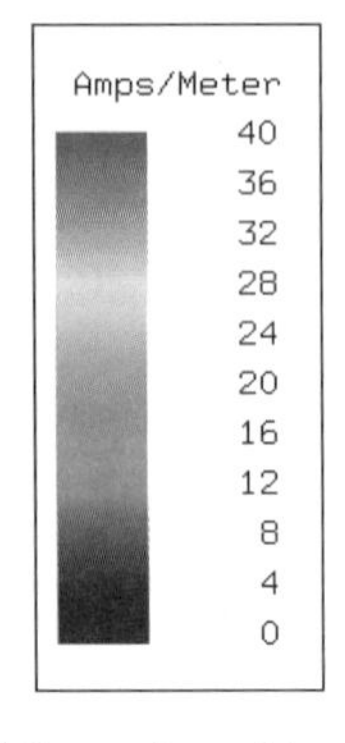

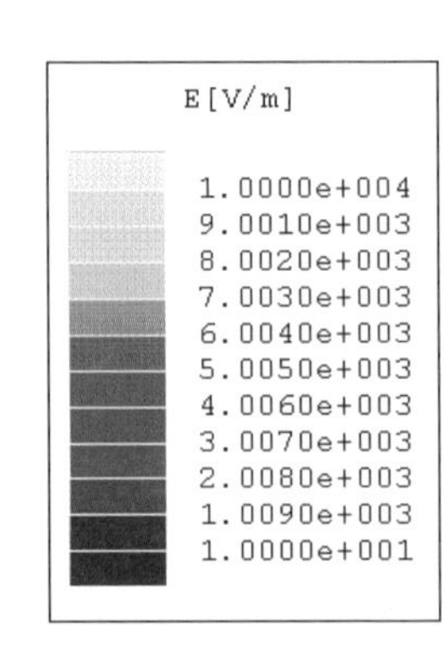

(a) Sonnet ***em*** Ver 8.0 (b) Ansoft HFSS Ver 8.5

Figure 1.1 Typical false color mappings: (a) conduction current magnitude using colors of the rainbow (ROYGBIV); and (b) E-field magnitude using a temperature mapping.

widely adopted. Now that the field-solver codes are more mature, perhaps it is time to re-think data display options and come up with some alternative approaches [3, 4].

In this book, the scale for each color plot will be stated in the text whenever possible. Dynamic range is also a problem with these plots. The quantities we are trying to display easily cover five to six orders of magnitude or more. It is difficult to display the full range of the variable of interest with only eight to 16 colors. In many cases the scale of the plot has been compressed at the high or low end to highlight the desired feature. Fine mesh resolution is also needed to produce a pleasing color picture. However, we can often compute accurate *S*-parameters with much coarser mesh resolution.

1.3 A NOTE ON 3D WIREFRAME VIEWS

When we begin to discuss various 3D geometries and the field-solvers that we use to solve those problems, we will show many 3D wireframe views. In the case of the 3D finite element method solvers the assumed background material is perfectly conducting metal. Or in other words, our model starts with a solid block of metal and we remove material and add interior details to build the model.

For example, if we wish to model a simple, air-filled coaxial transmission line, we "remove" a cylinder of air from the metal background material and then draw the metal inner conductor (Figure 1.2(a)). The boundary of the air-filled cylinder is perfectly conducting metal by default. To model a Teflon-filled coax we would simply change the material properties of the larger cylinder to Teflon. For clarity, we can explicitly draw a cylindrical outer metal boundary (Figure 1.2(b)). While this is also a perfectly valid model, the extra detail in the outer conductor is not needed

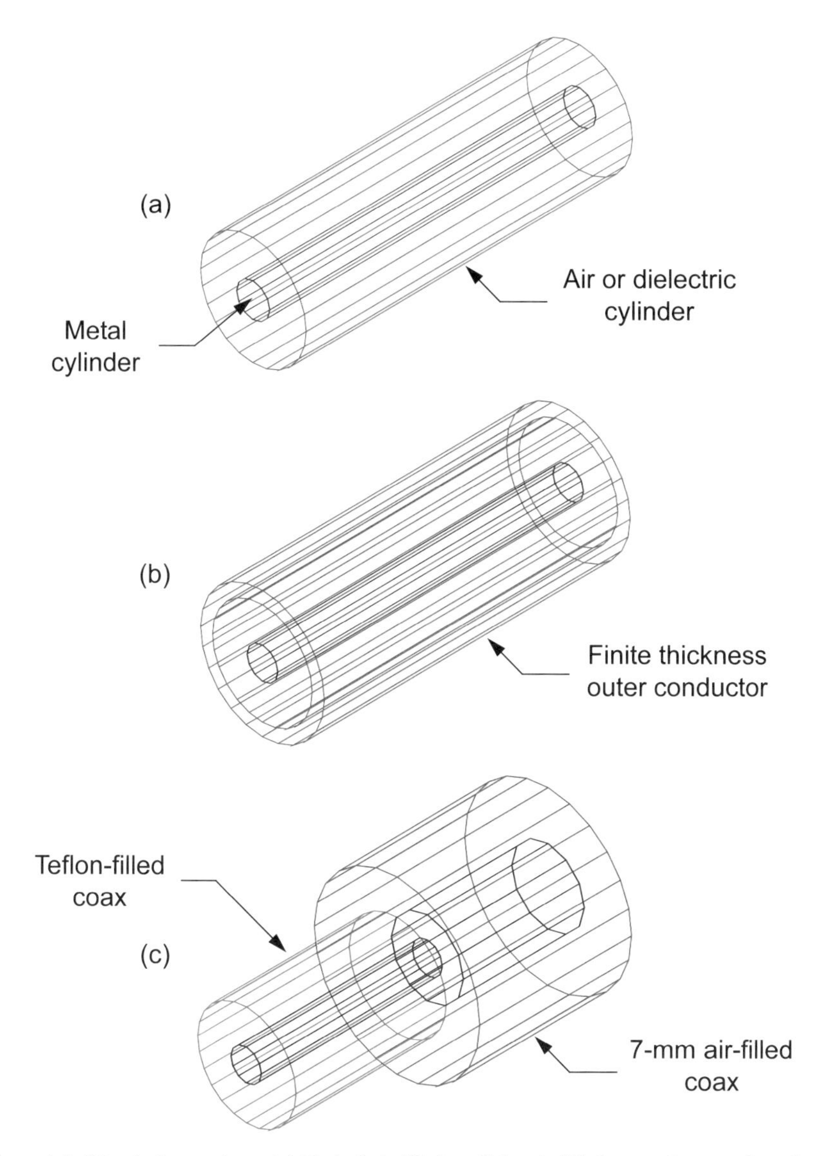

Figure 1.2 3D wireframe views. (a) Typical air-filled or dielectric-filled coax; the outer boundary is metal by default. (b) Outer conductor with finite thickness; the interior of the outer conductor is ignored. (c) Transition from Teflon-filled coax (SMA connector) to 7-mm air-filled coax. (Ansoft HFSS Ver. 5.6.)

and adds nothing to the electromagnetic treatment of the problem. The field-solver, by default, will ignore the interior of the coaxial outer conductor (and the interior of the center conductor). Figure 1.2(c) shows a smaller diameter Teflon- filled coax that transitions to a larger diameter air-filled coax. This is typically all the detail we need to represent a transition from a subminiature A (SMA) connector to a 7-mm air-filled coax.

Many of the 3D finite difference time domain (FDTD) and transmission line matrix (TLM) solvers also start with an assumed solid metal background. The default background can also be defined as air or to a configuration that absorbs electromagnetic energy almost perfectly.

1.4 A BRIEF HISTORICAL VIEW

In Table 1.2 we have created a very brief historical summary of the development of commercial numerical electromagnetics and its relationship to developments in the computer industry. It is not intended to be an exhaustive history of numerical electromagnetics. Rather, we would just like to note a few major events and put them in perspective relative to developments in computer hardware.

Numerical electromagnetics got its start in the days of the mainframe computer. Operating systems and compilers were unique to each vendor's hardware and options for high-resolution graphics were nonexistent or very expensive. It was the development of the microprocessor and the UNIX workstation that made commercial field-solver software economically viable. In the early years of microprocessor development we can track clock speed improvements on a yearly time scale. By the late 1990s, we need a monthly time scale to track improvements. The acquisitions and mergers among the software vendors starting in the mid-1990s is another indication of maturity in the market.

Table 1.2

A Brief Historical Summary

1966	– Yee proposes the FDTD method
1968	– Method of moments concept introduced by Harrington
1969	– Finite elements introduced in electrical engineering by Silvester
1971	– First formulation of 2D TLM method by Johns and Beurle
1975	– Simple FORTRAN TLM code published in Akhtarzad's thesis
1978	– Intel releases the 8086 microprocessor
1979	– Motorola releases the 68000 microprocessor
1980	– Apollo introduces a line of workstations using the Motorola 68000
1981	– IBM announces the personal computer
1982	– Sun Microsystems is founded
1987	– Sun introduces its first SPARC-based system with 10-MIPs performance
	– Symmetrical condensed TLM node introduced by Peter Johns
1989	– EMSim introduced
	– Sonnet ***em*** introduced
	– Sun introduces 20-MHz SPARCstation 1 with 12.5-MIPs performance
	– Intel announces the i486 at 25 MHz
1990	– High Frequency Structure Simulator (HFSS) introduced
	– EMAS introduced
	– Sun announces the SPARCstation 2 series
1991	– First TLM simulator for the PC is introduced
	– Intel introduces the 60-MHz Pentium processor
	– Gateway 2000 ships its 1 millionth PC
1992	– IE3D introduced
	– OSA demonstrates optimization with Empipe and Sonnet ***em***
1993	– EEsof acquired by Hewlett-Packard
1994	– Intel ships 90 and 100-MHz versions of the Pentium processors
1995	– Movement away from workstations towards Pentiums/Windows NT
1996	– OSA demonstrates optimization with Empipe3D and HFSS
	– MicroWaveLab acquired by Ansoft
1997	– Hewlett-Packard version of HFSS introduced
	– Boulder Microwave Technology (Ensemble) acquired by Ansoft
	– Intel ships 233-MHz Pentium II
1998	– OSA acquired by Hewlett-Packard
1999	– Intel ships 500-MHz Pentium III (May)
	– AMD ships 600-MHz Athlon (Aug.)
2000	– PC processor clocks hit 1 GHz
	– Support for multithreading and multiprocessors begins to appear
	– KCC Ltd. merges with Flomerics
2001	– Ansoft purchases Agilent HFSS
	– PC processor clocks hit 2 GHz (Sept.)
2002	– 64-bit hardware and software becomes available

References

[1] Tufte, E. R., *Envisioning Information*, Cheshire, CT: Graphics Press, 1990, p. 92.

[2] Li, K., et al., "Simulation of EM Phenomena Using a Finite Difference-Time Domain Technique," in *NSF/IEEE Center for Computer Applications in Electromagnetics Education Software Book*, Volume 1, Chapter 16, M. Iskander (ed.), Salt Lake City, UT: CAEME, 1991.

[3] Lefkowitz, H., and G. T. Herman, "Color Scales for Image Data," *IEEE Computer Graphics and Applications*, Vol. 12, No. 1, 1992, pp. 72–80.

[4] Rogowitz, B. E., and L. A. Treinish, "Data Visualization: The End of the Rainbow," *IEEE Spectrum*, Vol. 35, No. 12, 1998, pp. 52–59.

Chapter 2

CAD of Passive Components

Computer-aided design of passive RF and microwave components has advanced slowly but steadily over the past four decades. The 1960s and 1970s were the decades of the mainframe computer. In the early years, CAD tools were proprietary, in-house efforts running on text-only terminals. The few graphics terminals available were large, expensive, and required a short, direct connection to the mainframe. Later in this period, commercial tools became available for use on in-house machines or through time-sharing services. A simulation of a RF or microwave network was based on a combination of lumped and distributed elements. The elements were connected in cascade using *ABCD* parameters or in a nodal network using admittance- or *Y*-parameters. The connection between elements and the control parameters for the simulation were stored in a text file called a netlist. The netlist syntax was similar but unique for each software tool. The mathematical foundations for a more sophisticated analysis based on Maxwell's equations were developed in this same time period [1–5]. However, the computer technology of the day could not support effective commercial implementation of these more advanced codes.

The 1980s brought the development of the microprocessor and UNIX workstations. The UNIX workstation played a large role in the development of more sophisticated CAD tools. For the first time there was a common operating system and computer language (the C language) to support the development of cross-platform applications. UNIX workstations also featured large, bit-mapped graphics displays for interaction with the user. The same microprocessor technology that launched the workstation also made the personal computer possible. Although the workstation architecture was initially more sophisticated, personal computer hardware and software has grown steadily more elaborate. Today, the choice between a workstation and a PC is largely a personal one. CAD tools in this time period were still based on lumped and distributed concepts. The innovations brought about by the cheaper, graphics-based hardware had largely to do with schematic capture and layout. Schematic capture replaced the netlist on the input side of the analysis, and automatic or semiautomatic layout provided a quicker path to the finished circuit after analysis and optimization.

The greatest innovation in the 1990s was the emergence of CAD tools based on the direct solution of Maxwell's equations. Finally, there was enough computer horsepower to support commercial versions of the codes that had been in development since the late 1960s and early 1970s. These codes are in general labeled electromagnetic field-solvers although any one code may be based on one of several different numerical methods. One of the earliest commercial codes was EMSim [6–8], a method of moments (MoM) code written by Chow and marketed by EEsof. However, EMSim was optimized for electrically thin substrates and was limited to a small number of dielectric and metal layers. Despite these limitations, some excellent results were achieved, including the complete analysis of a two stage MMIC amplifier [9]. Sonnet ***em*** [10, 11], also based on the method of moments, was the first commercially viable tool designed for RF and microwave engineers. Only a few months later, Hewlett-Packard HFSS [12], an FEM code co-developed with Ansoft Corp., was released to the design community. Among the time domain codes MAFIA [13], using the finite integral technique, and a PC-based TLM code by Hoefer and So [14] were the earliest contributions. Because they have been available for over a decade, many of the examples in this book were developed using Sonnet ***em*** and Ansoft HFSS.

All of these tools approximate the true fields or currents in the problem space by subdividing the problem into basic "cells" or "elements" that are roughly 1/10 to 1/20 of a guide wavelength in size. For any guided electromagnetic wave, the guide wavelength is the distance spanned by one full cycle of the electric or magnetic field. The problem is to find the magnitude of the assumed current, field or potential on each cell or the field at the junction of elements. The final solution is then just the sum of each small contribution from each basic unit. Most of these codes first appeared on UNIX workstations and then migrated to the personal computer, as that hardware became more powerful. In the later years of this decade, field-solver codes appeared that were developed on and for the personal computer. In the early years, the typical field-solver problem was a single discontinuity or some other structure that was small in terms of wavelengths. Today, groups of discontinuities, complete matching networks, or small parts of a multilayer printed circuit board (PCB) are all suitable problems for a field-solver. Field-solver data in *S*-parameter form is typically imported into a circuit simulator and combined with lumped and distributed models to complete the analysis of the structure.

2.1 CIRCUIT-THEORY-BASED CAD

CAD of low-frequency circuits is at least 30 years old, and microwave circuits have been analyzed by computer for at least 20 years. At very low frequencies, we can connect inductors, capacitors, resistors, and active devices in a very arbitrary way. The lumped lowpass filter shown in Figure 2.1(a) is a simple example. This very

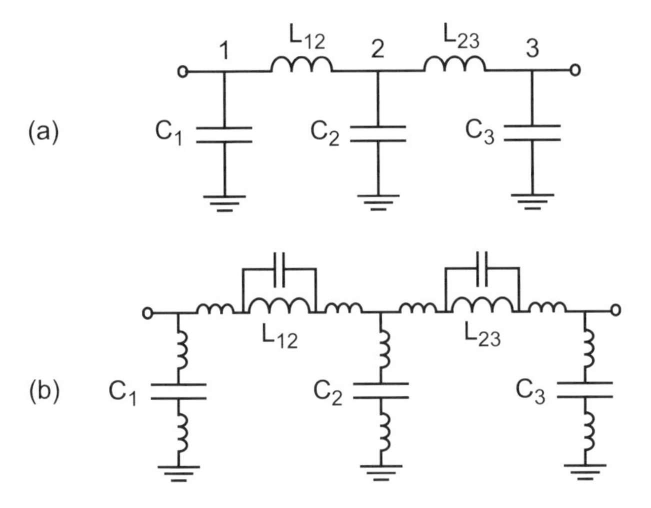

Figure 2.1 (a) Ideal lumped element lowpass filter or matching network and (b) the same network with parasitic elements due to lead inductance and inter-turn capacitance.

simple circuit has only three nodes. Most network analysis programs will form an admittance matrix (*Y*-matrix) internally and invert the matrix to find a solution. The *Y*-matrix (2.1) for the ideal three-node circuit is filled using some fairly simple rules. A shunt element connected to node one generates an entry at y_{11}. A series element connected between nodes one and two generates entries at y_{11}, y_{12}, y_{21}, and y_{22}. A large ladder network with sequential node numbering results in a large, tridiagonal matrix with many zeros off axis:

$$\mathbf{Y} = \begin{bmatrix} j\omega C_1 - j\dfrac{1}{\omega L_{12}} & j\dfrac{1}{\omega L_{12}} & 0 \\ j\dfrac{1}{\omega L_{12}} & j\omega C_2 - j\dfrac{1}{\omega L_{12}} - j\dfrac{1}{\omega L_{23}} & j\dfrac{1}{\omega L_{23}} \\ 0 & j\dfrac{1}{\omega L_{23}} & j\omega C_3 - j\dfrac{1}{\omega L_{23}} \end{bmatrix} \tag{2.1}$$

The *Y*-matrix links the known source currents to the unknown node voltages. **I** is a vector of source currents. Typically the input node is excited with a one-amp source and the rest of the nodes are set to zero. **V** is the vector of unknown node voltages. To find **V**, we invert the matrix **Y** and multiply by the known source currents:

$$\mathbf{I} = \mathbf{Y}\mathbf{V} \tag{2.2}$$

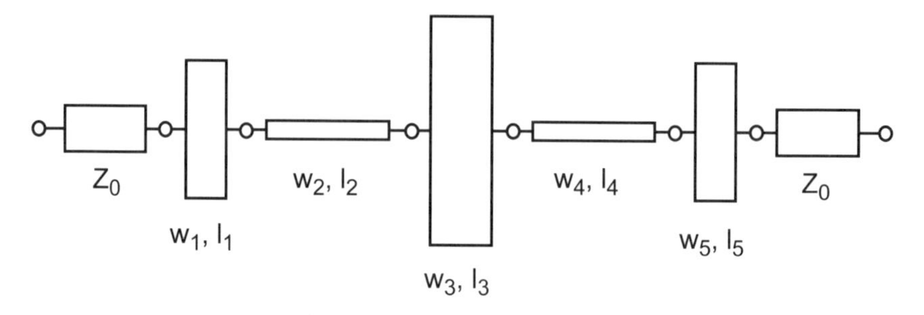

Figure 2.2 Distributed lowpass filter modeled using transmission lines. Step discontinuities are ignored. © 2000 CRC Press [15].

$$\mathbf{V} = \mathbf{Y}^{-1}\mathbf{I} \tag{2.3}$$

The time needed to invert an N by N matrix is roughly proportional to N^3. Filling and inverting the Y-matrix for each frequency of interest will be very fast, in this case, so fast it will be difficult to measure the computation time unless we specify a very large number of frequencies. This very simple approach might be good up to 1 MHz or so.

In our low-frequency model there is no concept of wavelength or even physical size. Any phase shift we compute is strictly due to the reactance of the component, not its physical size. There is also no concept of radiation; power can only be dissipated in resistive components. As we move into the HF frequency range (1–30 MHz) the real components we buy will have significant parasitics (Figure 2.1(b)). Lead lengths and proximity to the ground plane become very important and our physical construction techniques will have a big impact on the results achieved. Even leadless components based on surface mount technology (SMT) will have significant parasitics as we move higher in frequency.

By the time we reach VHF frequencies (50–150 MHz) we are forced to adopt distributed concepts in the physical construction and analysis of our circuits. The connections between components become transmission lines and many components themselves are based on transmission line models. Our simple lowpass circuit might become a cascade of low and high impedance transmission lines (Figure 2.2).

If this were a microstrip circuit, we would typically specify the substrate parameters and the width and length of each transmission line. We have ignored the step discontinuities due to changes in line width in this simplified example. Internally, the software would use analytical equations to convert our physical dimensions to impedances and electrical lengths. The software might use a Y-matrix, a cascade of *ABCD* parameter blocks, or a cascade of S-parameter blocks for the actual analysis. At the ports, we typically ask for S-parameters referenced to the system impedance.

Table 2.1

Boundary Between Lumped and Distributed Behavior

Less than $\lambda/10$	*Grey area*	$\lambda/8$ *or greater*
Lumped		Distributed
L, C, R, G		Transmission lines
Voltage, current		[S], [Z], [Y]
No radiation		Radiation possible
Only reactance can shift phase of V or I		Physical distance can shift phase of V or I
Fields rise and fall at same time all through the structure		There is phase shift in the fields across the structure

Notice that we still have a small number of nodes to consider. Our circuit is clearly distributed but the solution time does not depend on its size in terms of wavelengths. The evaluation time for the analytical transmission line models is not a function of their electrical length. Any phase shift we compute is directly related to the physical size of the network. Although we can include conductor and substrate losses, there is still no radiation loss mechanism. It is also difficult to include enclosure effects; there may be box resonances or waveguide modes in our physical implementation. There is also no mechanism for parasitic coupling between our various circuit models.

The boundary between a lumped circuit point of view and a distributed point of view can be somewhat fuzzy. A quick review of some rules of thumb and terminology might be helpful. One common rule of thumb says that the boundary between lumped and distributed behavior is somewhere between a tenth and an eighth of a guide wavelength. Remember that wavelength in inches is defined by

$$\lambda = \frac{c}{f\sqrt{\varepsilon_{eff}}} = \frac{11.803}{f\sqrt{\varepsilon_{eff}}} \text{ inches} \tag{2.4}$$

where ε_{eff} is the effective dielectric constant of the medium and f is in GHz. At 1 GHz, $\lambda = 11.803$ inches in air and $\lambda = 6.465$ inches for a 50-ohm line on 0.014-inch thick FR4. In Table 2.1 we can relate the physical size of our structure to the concept of wavelength and to some common terminology.

2.2 FIELD-THEORY-BASED CAD

A field-theory-based solution is an alternative to the previous distributed, circuit-theory-based approach. The field-solver takes a more microscopic view of any dis-

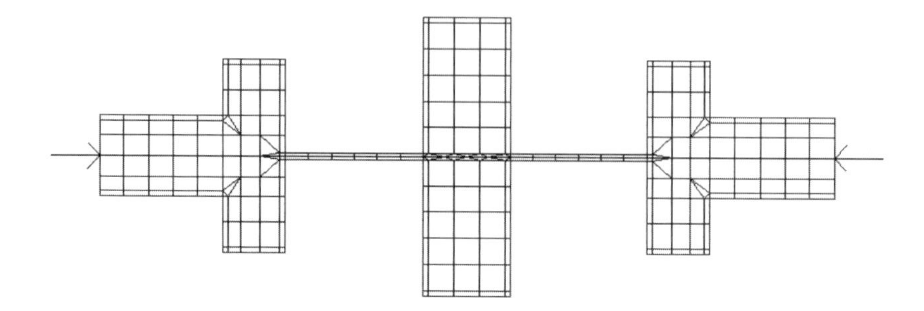

Figure 2.3 A typical MoM mesh for the distributed lowpass filter circuit. The number of unknowns, N is 474 (Agilent Momentum, ADS Ver. 1.3). © 2000 CRC Press [15].

tributed geometry. Most field-solvers we might employ must subdivide the geometry based on guide wavelength. Typically we need 10 to 30 elements or cells per guide wavelength to capture the fields or currents in our structure. Figure 2.3 shows a typical mesh generated by Agilent Momentum [16] for our microstrip lowpass filter example. Narrow cells are used on the edges of the strip to capture the spatial wavelength, or highly nonuniform current distribution across the width of the strips. This MoM code has subdivided the microstrip metal and will solve for the current on each small rectangular or triangular patch. The default settings for mesh generation were used.

For this type of field-solver there is a strong analogy between the Y-matrix description we discussed for our lumped element circuit and what the field-solver must do internally. Imagine a lumped capacitor to ground at the center of each "cell" in our field-solver description. Series inductors connect these capacitors to each other. Coupling between nonadjacent cells can be represented by mutual inductances. So we have to fill and invert a matrix, but this matrix is now large and dense compared to our simple, lumped element circuit Y-matrix. For the mesh above, $N = 474$ and we must fill and invert an N by N matrix.

One reason we turn to the field-solver is because it can potentially include all electromagnetic effects from first principles. We can include all loss mechanisms including surface waves and radiation. We can also include parasitic coupling between elements and the effects of compacting a circuit into a small space. The effects of the package or housing on our circuit performance can also be included in the field-solver analysis. However, the size of the numerical problem is now proportional to the structure size in wavelengths. The details of how enclosures are included in our analysis will vary from solver to solver. In some tools an enclosure is part of the basic formulation. In other tools, the analysis environment is "laterally open"; there are no sidewalls although there may be a cover. One of the exciting aspects of field-solvers is the ability to observe fields and currents in the circuit,

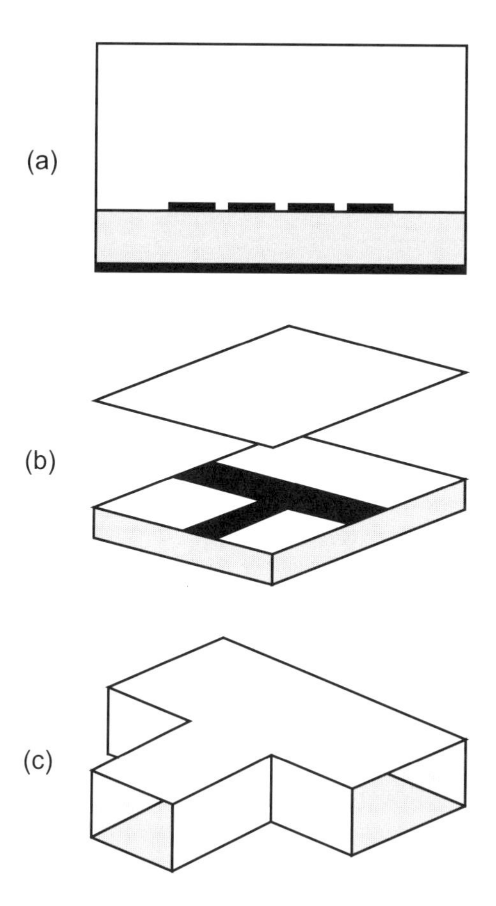

Figure 2.4 Field-solvers classified by geometrical dimensionality: (a) 2D cross-section, (b) 2.5D mostly planar, and (c) 3D fully arbitrary.

which sometimes leads to a deeper understanding of how the circuit actually operates. However, the size of the numerical problem will also be greater using a field-solver versus circuit theory, so we must choose carefully which pieces of global problem we will attack with the field-solver.

Although our discussion so far has focused on planar, distributed circuits, there are actually three broad classes of field-solver codes. The 2D cross-section codes (Figure 2.4(a)) solve for the transverse field distributions, yielding the modal impedance and phase velocity of 1 to N strips with a uniform cross-section. This class of problem includes coupled microstrips, coupled slots and conductors of arbitrary cross-section buried in a multilayer PC board, and waveguides with arbitrary cross-section. These tools use a variety of numerical methods including MoM, FEM, and the spectral domain method (SDM). Field-solver engines that solve for multiple strips in a layered environment are built into several linear and nonlinear simulators. A multistrip model of this type is a building block for more complicated geometries like Lange couplers, spiral inductors, baluns, and many distributed filters. The advantage of this approach is speed; only the 2D cross-section must be discretized and solved.

The second general class of codes mesh or subdivide the surfaces of planar metals (Figure 2.4(b)). The assumed environment for these surface meshing codes is a set of homogeneous dielectric layers with patterned metal conductors at the layer interfaces. Vertical vias are available to form connections between metal layers. This is where the half dimension comes from in the 2.5D description; we are somewhere in between a strictly planar structure and a completely arbitrary 3D structure. There are two fundamental formulations for these codes, closed box and laterally open. In the closed box formulation the boundaries of the problem space are perfectly conducting walls. In the laterally open formulation, the dielectric layers extend to infinity. The numerical method for this class of tool is generally MoM. Surface meshing codes can solve a broad range of strip and slot-based planar circuits and antennas. Compared to the 2D cross-section solvers, the numerical effort is considerably higher.

The third general class of codes meshes or subdivides a 3D volume. These volume meshing codes (Figure 2.4(c)) can handle virtually any three-dimensional object, with some restrictions on where ports can be located. Typical problems are waveguide discontinuities, various coaxial junctions, and transitions between different guiding systems, such as transitions from coax to waveguide. These codes can also be quite efficient for computing transitions between layers in multilayer PC boards and connector transitions between boards or off the board. The more popular volume meshing codes employ FEM, FDTD, and the transmission line matrix (TLM) method. Although the volume meshing codes can solve a very broad range of problems, the penalty for this generality is total solution time. It typically takes longer to set up and run a 3D problem compared to a surface meshing or cross-section problem. Sadiku [17] has compiled a very thorough introduction to many of these numerical methods.

2.3 SOLUTION TIME FOR CIRCUIT THEORY AND FIELD THEORY

When we use circuit theory to analyze a RF or microwave network, we are building a Y-matrix of dimension N, where N is the number of nodes. A typical amplifier or oscillator design may have only a couple of dozen nodes. Depending on the solution method, the solution time is proportional to a factor between N^2 and N^3. When we talk about a "solution" we really mean matrix inversion. In Figure 2.5 we have plotted solution time as a function of matrix size N. The vertical time scale is somewhat arbitrary but should be typical of workstations and personal computers today.

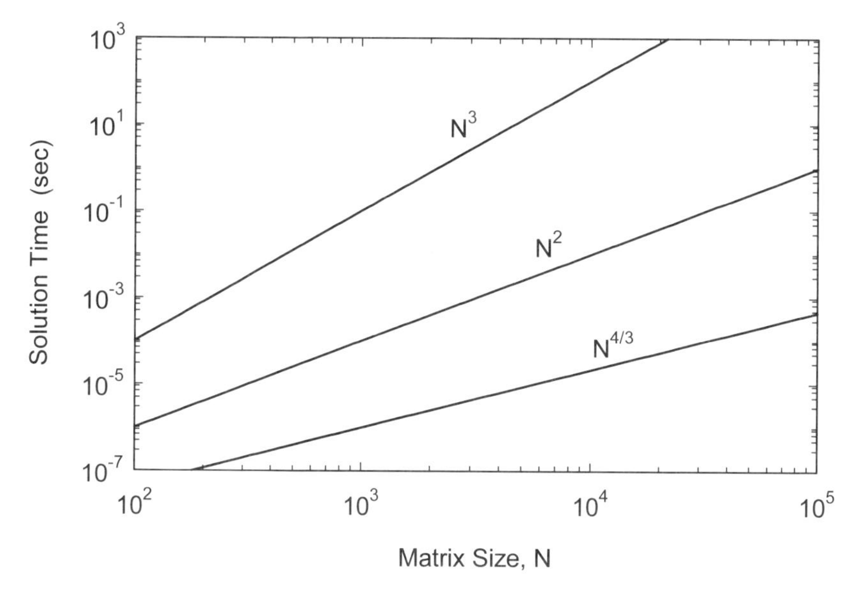

Figure 2.5 Solution time as a function of matrix size, N, circa 2002. Solution times for circuit simulators, MoM field-solvers, and FEM field-solvers fall between the N^2 and N^3 curves. Solution times for FDTD and TLM simulators fall between the $N^{4/3}$ and N^2 curves.

When we use a MoM field-solver, a "small" problem may have a matrix dimension of $N = 500$–1,000. Medium size problems may be around $N = 5{,}000$ and large problems quickly get into the $N = 10{,}000$–15,000 range. Because of the N^2/N^3 effect, the solution time is impacted dramatically as the problem size grows. In this case we can identify two processes, filling the matrix with all the couplings between cells and inverting or solving that matrix. So we are motivated to keep our problem size as small as possible. The FEM codes also must fill and invert a matrix. Compared to MoM, the matrix tends to be larger but more sparse. As we move into 64-bit computing and break the 2 GB memory limit on the PC the definition of a "large" problem will certainly take a dramatic shift upwards.

The time domain solvers using FDTD or TLM are exceptions to the N^2/N^3 rule. The solution process for these codes is iterative; there is no matrix to fill or invert with these solvers. Thus, the memory required and the solution time grow more linearly with problem size in terms of the number or cells or unknowns. This is one reason these tools have been very popular for RCS analysis of ships and airplanes. However, because these are time stepping codes, we must perform a Fourier transform on the time domain solution to get the frequency domain solution. Closely spaced resonances in the frequency domain require a large number of time samples in the time domain. Therefore, time stepping codes may not be the most efficient choice for structures like filters, although there are techniques available to

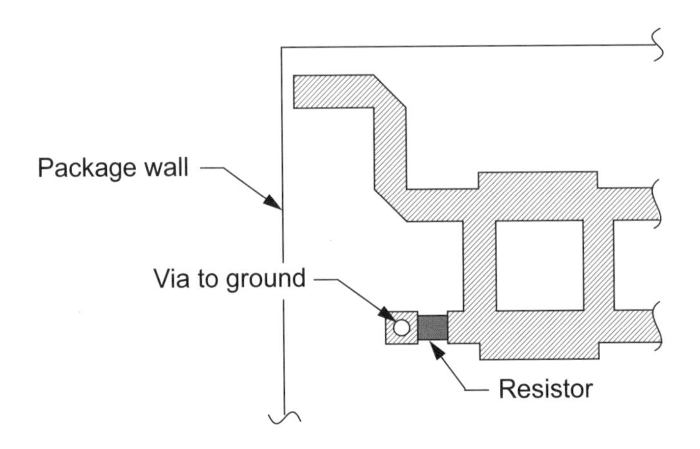

Figure 2.6 Part of an RF printed circuit board that includes a branchline coupler, a resistive termination to ground, and several mitered bends. © 2000 CRC Press [15].

speed up convergence. Veidt [18] presents a good summary of how solution time scales for various numerical methods.

2.4 A "HYBRID" APPROACH TO CIRCUIT ANALYSIS

If long solution times prevent us from analyzing complete circuits with a field-solver, what is the best strategy for integrating these tools into the design process? The best approach is to identify the key pieces of the problem that need the field-solver, and to do the rest with circuit theory. Thus, the final result is a "hybrid solution" using different techniques, and even different tools from different vendors. It is also possible to solve a single field-solver project using a "hybrid" of two different numerical methods, but we will not discuss that option here. As computer power grows and software techniques improve, we can do larger and larger pieces of the problem with a field-solver. A simple example will help to demonstrate this approach. The circuit in Figure 2.6 is part of a larger RF printed circuit board. In one corner of the board we have a branchline coupler, a resistive termination, and several mitered bends.

Using the library of elements in our favorite linear simulator, there are several possible ways to subdivide this network for analysis (Figure 2.7). In this case we get about 21 nodes in our circuit. Solution time is roughly proportional to N^3, so if we ignore the overhead of computing any of the individual models, we would expect the solution to come back very quickly. But we have clearly neglected several things in our analysis. Parasitic coupling between the arms of the coupler, interaction between the discontinuities, and any potential interaction with the package have all been ignored. Some of our analytical models may not be as accurate as

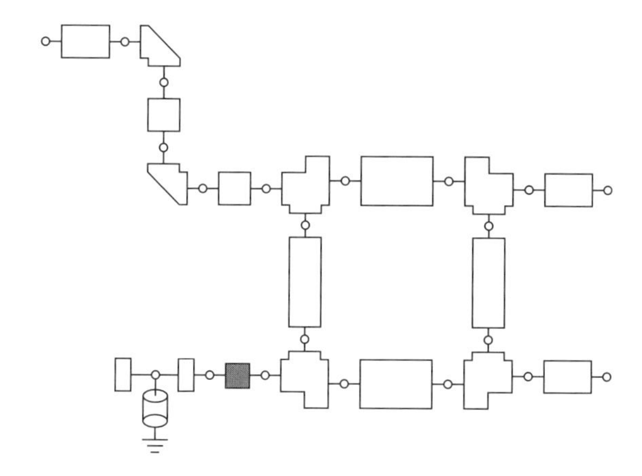

Figure 2.7 The layout in Figure 2.6 has been subdivided for analysis using the standard library elements found in many circuit-theory-based simulators. © 2000 CRC Press [15].

we would like, and in some cases a combination of models may not accurately describe our actual circuit. If this circuit were compacted into a much denser layout, all of the effects mentioned above would become more pronounced.

Each of the circuit elements in our schematic has some kind of analytic model inside the software. For a transmission line, the model would relate physical width and length to impedance and electrical length through a set of closed form equations. For a discontinuity like the mitered bend, the physical parameters might be mapped to an equivalent lumped element circuit (Figure 2.8), again through a set of closed form equations. The field-solver will take a more microscopic view of the same mitered bend discontinuity. Any tool we use will subdivide the metal pattern using 10 to 30 elements per guide wavelength. The sharp inside corner where current changes direction rapidly will force an even finer subdivision. If we want to solve the bend discontinuity individually, we must also connect a short length of series line to each port. Agilent Momentum generated the mesh in Figure 2.9. The

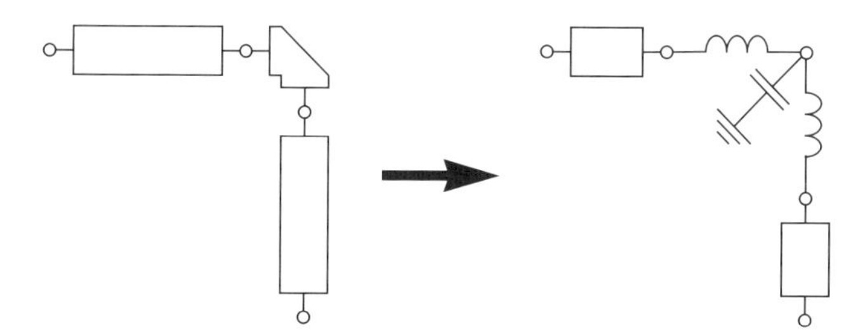

Figure 2.8 The equivalent circuit of a microstrip mitered bend. The physical dimensions are mapped to an equivalent lumped element circuit. © 2000 CRC Press [15].

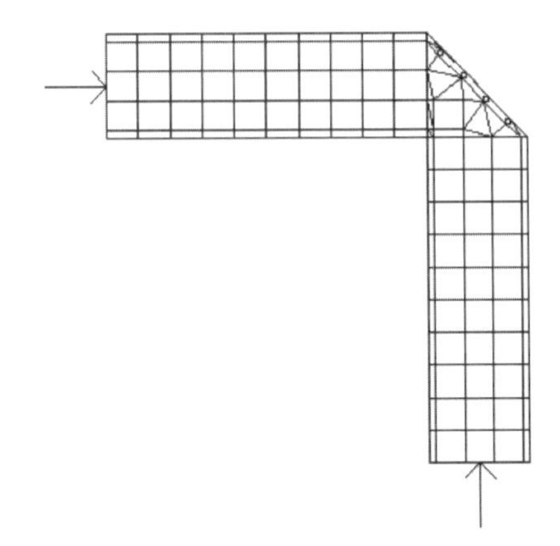

Figure 2.9 A typical MoM mesh for the microstrip mitered bend. The solution space is "laterally open," with no box walls. The number of unknowns, N is 221 (Agilent Momentum, ADS Ver. 1.3). © 2000 CRC Press [15].

number of unknowns is 221. If the line widths are not variable in our design, we could compute this bend once, and use it over and over again in our circuit design. Or, we might do a validation experiment to convince ourselves that an existing analytical model is accurate, given our particular substrate parameters and frequency range.

Another potential field-solver problem is in the corner of the package near the input trace (Figure 2.10). You might be able include the box wall effect on the series line, but wall effects are generally not included in discontinuity models. However, it is quite easy to set up a field-solver problem that would include the

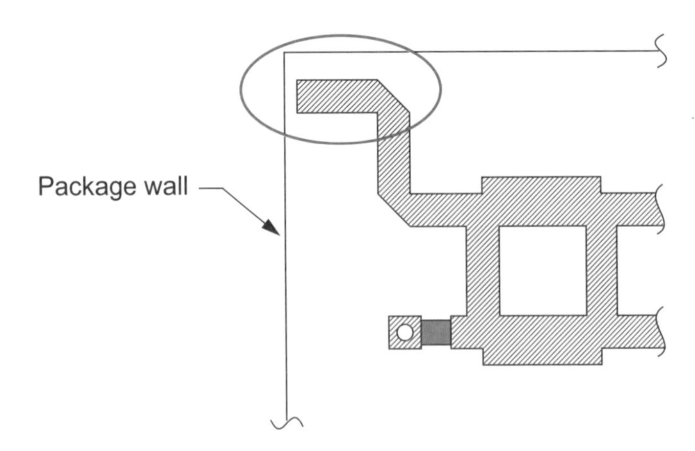

Figure 2.10 In our original problem, one part of the circuit is very close to the box walls. © 2000 CRC Press [15].

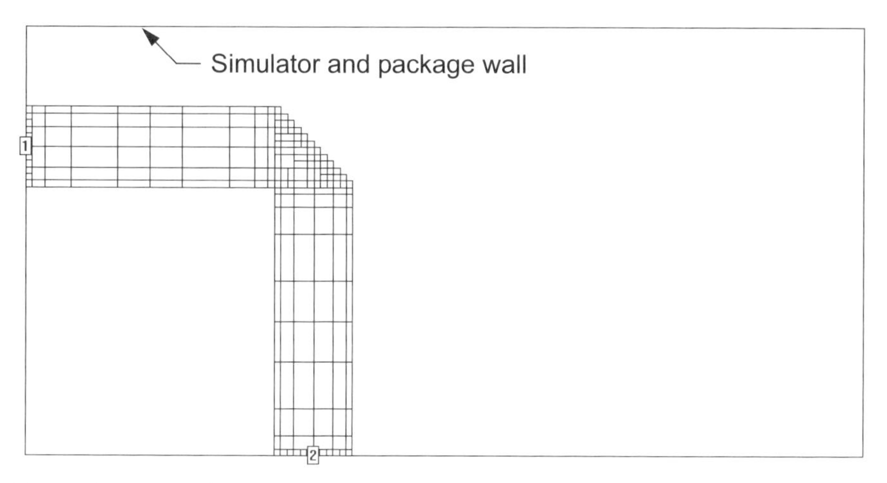

Figure 2.11 An analysis of the input line and mitered bend in the presence of the package walls. The number of unknowns, N is 360 (Sonnet ***em*** Ver. 6.0). © 2000 CRC Press [15].

microstrip line, the mitered bend and the influence of the walls. The project in Figure 2.11 was drawn using Sonnet ***em***. The box walls to the left and top in the electromagnetic simulation mimic the true location of the package walls in the real hardware. There are 360 unknowns in this simulation.

One of the more interesting ways to use a field-solver is to analyze groups of discontinuities rather than single discontinuities. A good example of this is the termination resistor and via [19, 20] in our example circuit. A field-solver analysis of this group may be much more accurate than a combination of individual analytical models. We could also optimize the termination, then use the analysis data and the

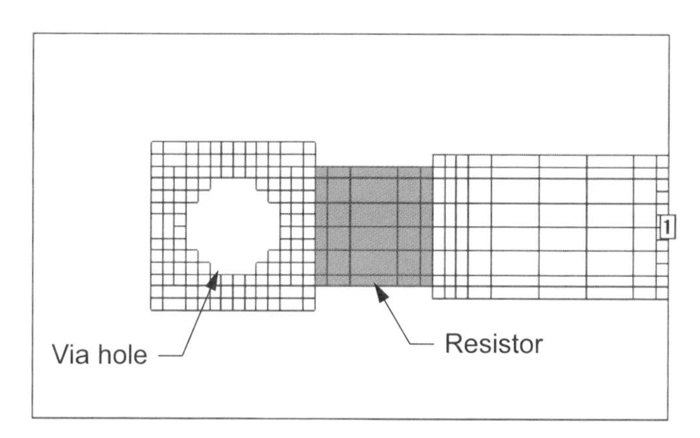

Figure 2.12 A MoM analysis of a group of discontinuities including a thin-film resistor, two steps in width, and a via hole to ground. The number of unknowns, N is 452 (Sonnet ***em*** Ver. 6.0). © 2000 CRC Press [15].

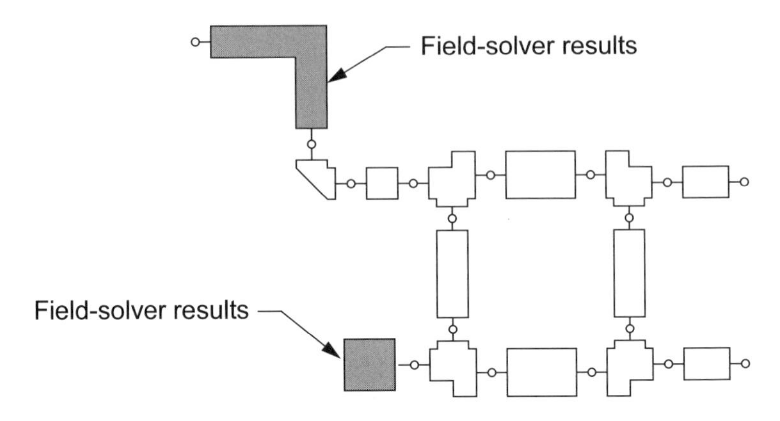

Figure 2.13 Substituting field-solver results into the original solution scheme mixes field-theory and circuit-theory in a cost-effective way. © 2000 CRC Press [15].

optimized geometry over and over again in this project or other projects. The mesh for the resistor via combination (Figure 2.12) was generated using Sonnet ***em*** and represents a problem with 452 unknowns.

Our original analysis scheme based on circuit theory models alone was shown in Figure 2.6. Although this will give us the fastest analysis, there may be room for improvement. We can substitute in our field-solver results for the elements near the package walls and for the resistor/via combination (Figure 2.13). The data from the field-solver would typically be *S*-parameter files. This "hybrid" solution mixes field theory and circuit theory in a cost-effective way [21]. The challenge to the

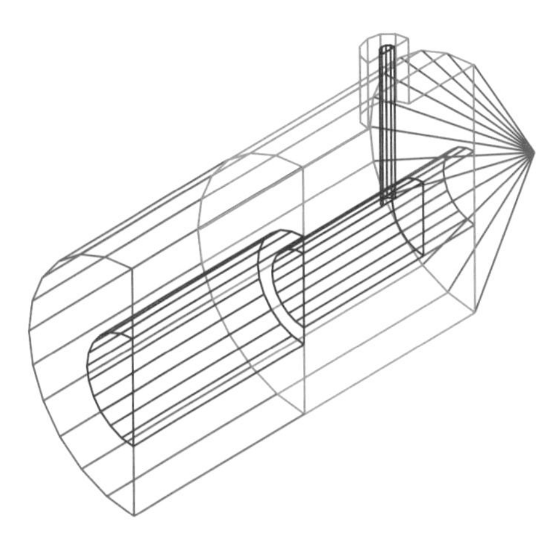

Figure 2.14 A right angle, coax-to-coax transition that was optimized for return loss; one half of the complete geometry is shown. The number of unknowns, *N* is 8,172 (Ansoft HFSS Ver. 7.0). © 2000 CRC Press [15].

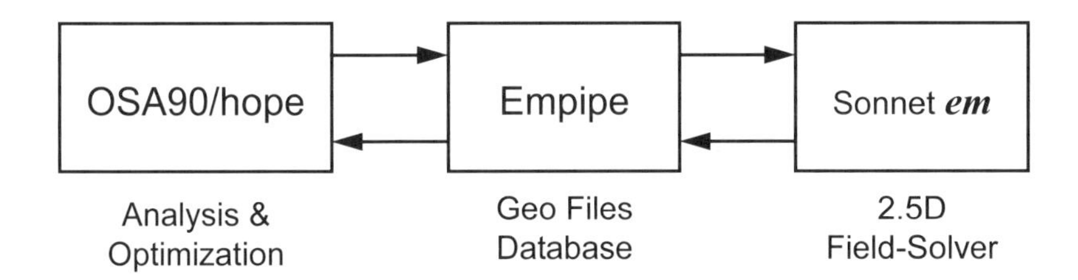

Figure 2.15 The first commercially successful optimization scheme that included a field-solver inside the optimization loop. © 2000 CRC Press [15].

design engineer is to identify the critical components that should be addressed using the field-solver.

The hybrid solution philosophy is not limited to planar components; three-dimensional problems can be solved and cascaded as well. The right angle coax bend shown in Figure 2.14 is one example of a 3D component that was analyzed and optimized using Ansoft HFSS [22]. In this case we have taken advantage of a symmetry plane down the center of the problem in order to reduce solution time. This component includes a large step in inner conductor diameter and a Teflon sleeve to support the larger inner conductor. After optimizing two dimensions, the computed return loss is greater than –30 dB. The coax bend is only one of several problems taken from a larger assembly, which included a lowpass filter, coupler, amplifier, and bandpass filter.

2.5 OPTIMIZATION

Optimization is a key component of modern linear and nonlinear circuit design. Many optimization schemes require gradient information, which is often computed by taking simple forward or central differences. The extra computations required to find gradients become very costly if there is a field-solver inside the optimization loop. So it is important to minimize the number of field-solver analysis runs. It is also necessary to capture the desired changes in the geometry and pass this information to the field-solver. Bandler, et al., [23, 24] developed an elegant solution to both of these problems in 1993. The key concept was a "data pipe" program sitting between the simulator and the field-solver (Figure 2.15). When the linear simulator calls for a field-solver analysis the data pipe generates a new geometry file and passes it to the field-solver. In the reverse direction, the data pipe stores the analysis results and interpolates between data sets if possible. The final iterations of the optimization operate entirely on interpolated data without requiring any new field-solver runs. This concept was applied quite successfully to both surface meshing [25] and volume meshing solvers [26]. The same basic rules that lead to

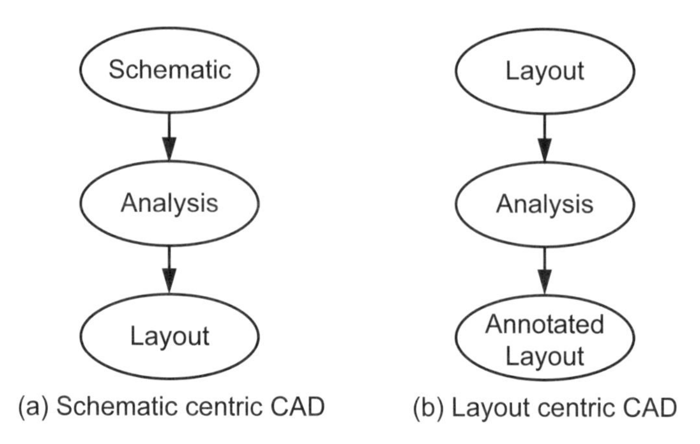

Figure 2.16 Two possible approaches to RF/microwave CAD: (a) a schematic centric approach, and (b) a layout centric approach.

successful circuit-theory-based optimization are even more important when a field-solver is in the loop as well. First, a good starting point leads to more rapid and consistent convergence. And second, it is important to limit the number of variables.

2.6 MODERN MICROWAVE CAD—WHAT'S MISSING?

Most microwave CAD tools have abandoned their original netlist orientation in favor of a schematic driven approach (Figure 2.16(a)). While a schematic is very useful for documentation, it has some limitations for RF and microwave circuits. A schematic, like a netlist, is a collection of individual, noninteracting discontinuities. This point of view leads to decomposition of the desired circuit into basic building blocks based on visual clues rather than actual current patterns, as in Figure 2.7. The schematic has no fundamental description of the layout, except for the most simple circuits, like the examples earlier in this chapter. Real circuits tend to be heavily compacted; the schematic needs a lot of patches and fixes to force the desired layout. Finally, a schematic driven design built from a library of available elements tends to limit creativity. The average user sees only what is available on the toolbar of the simulator.

If, however, a simulator was designed to be layout centric, it would have several advantages for RF and high-speed digital circuits (Figure 2.16(b)). A layout-based tool could easily capture groups of discontinuities for EM analysis, without regard for their individual behavior. A layout-based tool would also maintain the desired and parasitic physical relationships between various parts of the circuit. The low temperature co-fired ceramic (LTCC) module shown in Figure 2.17 contains a simple lowpass/highpass diplexer. From a strictly schematic point of view, there

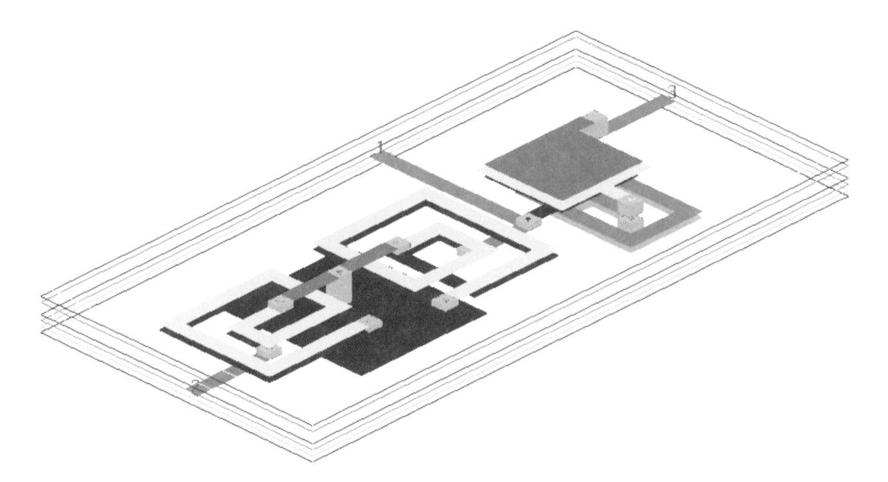

Figure 2.17 A pseudo-lumped element lowpass/highpass diplexer realized using LTCC technology. Figure courtesy of Muelhous Consulting, used with permission.

are three capacitors and three inductors in this circuit. However, it is the layout that captures the complex behavior of the individual multilayer components and the interactions between components due to their proximity. A library of distributed inductor and capacitor models would also fail to capture the interactions between components in a highly compacted layout.

In a layout centric simulator some form of "nodes" would still have to be inserted into the layout to mark the boundaries between EM-based models and analytical-based models. Nodes would also be needed to connect active and passive components to the traces. This implies that we need a flexible and accurate internal port scheme to allow these connections between planar traces and SMT type components. As for documentation, an annotated layout may be just as descriptive or more so when compared to a conventional schematic.

The artwork for early microwave circuits was generated at 20X to 50X scale by laying down strips of red tape on clear acetate. Artwork could also be produced by removing rubylithe from a solid sheet on a clear backing. Straight edges, compasses with knife blades, circle templates, ellipse templates, and even French curves could be used to bend and flow the circuit layout.

This design freedom and the rather primitive CAD of the day resulted in some quite remarkable, intuitive circuits. Figure 2.18 shows a reproduction from memory of an input matching circuit for an 8- to 18-GHz, broadband, GaAs FET amplifier. The matching network must present a conjugate match to the input of the FET and provide a path to ground for the gate to source bias. This circuit was designed at Narda Microwave West, circa 1978, and is the result of some very careful tuning of prototype circuits on the bench and a large measure of intuition. This circuit liter-

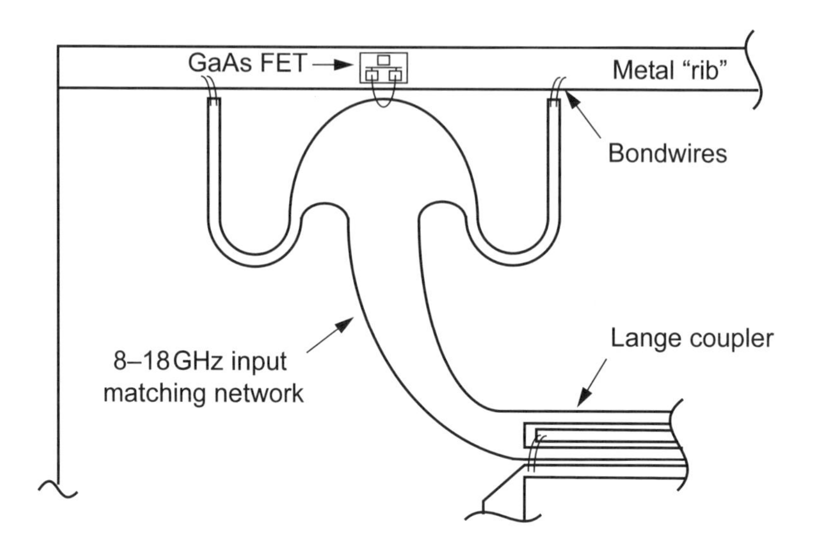

Figure 2.18 Broadband, input matching network for a balanced GaAs FET amplifier, circa 1978. The source bypass capacitors and bias resistor are not shown.

ally flows from the Lange coupler to the FET gate bondwires with a minimum number of discontinuities. It is hard to imagine such a circuit being built from a palette of predefined shapes.

One of the dangers for students and new engineers in schematic driven, library oriented CAD systems is a loss of creativity. At various times, some of the material in this book has been presented to graduate students at universities. When asked to review a student project we have drawn a sketch for the student and suggested that he or she "try something like this." Almost invariably, their response is "but there's no picture or icon that looks like that on the CAD system." Too much reliance on a standard library of elements puts an unnecessary restriction on the creativity of the circuit designer. The field-solver lets the designer create almost any "picture" they would like and get an accurate solution for that geometry. Unfortunately, these custom field-solver solutions tend to break the forward/backward flow between schematic and layout in today's schematic centric design tools.

2.7 THE NEXT DECADE

The need for inexpensive wireless systems has forced the RF community to rapidly adopt low-cost, multilayer PC board technology. In the simpler examples, most circuitry and components are mounted on the top layer while inner layers are used for routing of RF signals and dc bias. However, more complex examples can be found where printed passive components and discontinuities are located in one or more buried layers. Given the large number of variables in PC board construction it will be difficult for vendors of linear and nonlinear circuit simulators to support large libraries of passive models that cover all possible scenarios. However, a field-solver can be used to generate new models as needed for any novel layer stackup. Of course, the user is also free to use the field-solver data to develop custom, proprietary models for his or her particular technology.

The traditional hierarchy of construction for RF systems has been chip device, mounted to leaded package, mounted to printed circuit board located in system cabinet or housing. Today however, the "package" may be a multilayer LTCC substrate or a multilayer PCB using ball grid array (BGA) interconnects. Thus, the boundary between package and PC board has blurred somewhat. No matter what the technology details are, the problem remains to transfer a signal from the outside world into the system, onto the main system board, through the package, and into the chip. And, of course, there is an analogous connection from the chip back to the outside world. From this point of view, the problem becomes a complex, multilevel passive interconnect that must support not only the signal currents but also the ground currents in the return path. It is often the ground return path that limits package isolation or causes unexpected oscillations in active circuits [27]. The high-speed digital community is faced with very similar passive interconnect challenges at similar, if not higher frequencies and typically much higher signal densities. Again, there is ample opportunity to apply field-solver technology to these problems although practical problem size is still somewhat limited. The challenge to the design engineer is to identify and correct problems at multiple points in the signal path.

References

[1] Yee, K. S., "Numerical Solution of Initial Boundary-Value Problems Involving Maxwell's Equations in Isotropic Media," *IEEE Trans. Ant. Prop.*, Vol. 14, No. 5, 1966, pp. 302–207.

[2] Wexler, A., "Solution of Waveguide Discontinuities by Modal Analysis," *IEEE Trans. Microwave Theory Tech.*, Vol. 15, No. 9, 1967, pp. 508–517.

[3] Harrington, R. F., *Field Computation by Moment Methods*, New York, NY: Macmillan, 1968.

[4] Johns, P. B., and R. L. Beurle, "Numerical Solution of 2-Dimensional Scattering Problems Using a Transmission-Line Matrix," *Proc. Inst. Electr. Eng.*, Vol. 118, No. 9, 1971, pp. 1203–1208.

[5] Silvester, P., "Finite Element Analysis of Planar Microwave Networks," *IEEE Trans. Microwave Theory Tech.*, Vol. 21, No. 2, 1973, pp. 104–108.

[6] Chow, Y. L., "An Approximate Dynamic Spatial Green's Function in Three Dimensions for Finite Length Microstrip Lines," *IEEE Trans. Microwave Theory Tech.*, Vol. 28, No. 4, 1980, pp. 393–397.

[7] Chow, Y. L., et al., "A Modified Moment Method for the Computation of Complex MMIC Circuits," *Proc. 16th European Microwave Conference*, Dublin, Sept. 1986, pp. 625–630.

[8] Stubbs, M., L. Chow, and G. Howard, "Simulation Tool Accurately Models MMIC Passive Elements," *Microwaves and RF*, Vol. 26, No. 1, 1988, pp. 75–79.

[9] Draxler, P. J., G. E. Howard, and Y. L. Chow, "Mixed Spectral/Spatial Domain Moment Method Simulation of Components and Circuits," *Proc. 21st European Microwave Conference*, Stuttgart, 1991, pp. 1284–1289.

[10] ***em***™, Sonnet Software, Liverpool, NY.

[11] Rautio, J. C., and R. F. Harrington, "An Electromagnetic Time-Harmonic Analysis of Shielded Microstrip Circuits," *IEEE Trans. Microwave Theory Tech.*, Vol. 35, No. 8, 1987, pp. 726–730.

[12] HFSS, Hewlett-Packard, Santa Rosa, CA and Ansoft, Pittsburgh, PA.

[13] MAFIA, Computer Simulation Technology (CST), Darmstadt, Germany.

[14] Hoefer, W. J. R., and P. So, *The Electromagnetic Wave Simulator*, Chichester, UK: John Wiley & Sons, Inc., 1991.

[15] Swanson, Jr., D. G., "Computer Aided Design of Passive Components," in *The RF and Microwave Handbook*, pp. (8-34)–(8-44), M. Golio (ed.), Boca Raton, FL: CRC Press, 2000.

[16] Momentum, Agilent EEsof EDA, Santa Rosa, CA.

[17] Sadiku, M., *Numerical Techniques in Electromagnetics*, Boca Raton, FL: CRC Press, 1992.

[18] Veidt, B., "Selecting 3D Electromagnetic Software," *Microwave Journal*, Vol. 41, No. 9, 1998, pp. 126–137.

[19] Goldfarb, M., and R. Pucel, "Modeling Via Hole Grounds in Microstrip," *IEEE Microwave and Guided Wave Letters*, Vol. 1, No. 6, 1991, pp. 135–137.

[20] Swanson, Jr., D. G., "Grounding Microstrip Lines with Via Holes," *IEEE Trans. Microwave Theory Tech.*, Vol. 40, No. 8, 1992, pp. 1719–1721.

[21] Swanson, Jr., D. G., "Using a Microstrip Bandpass Filter to Compare Different Circuit Analysis Techniques," *Int. J. MIMICAE*, Vol. 5, No. 1, 1995, pp. 4–12.

[22] HFSS™, Ansoft Corp., Pittsburgh, PA.

[23] Bandler, J. W., et al., "Minimax Microstrip Filter Design Using Direct EM Field Simulation," *IEEE MTT-S Int. Microwave Symposium Digest*, 1993, pp. 889–892.

[24] Bandler, J. W., et al., "Microstrip Filter Design Using Direct EM Field Simulation," *IEEE Trans. Microwave Theory Tech.*, Vol. 42, No. 7, 1994, pp. 1353–1359.

[25] Swanson, Jr., D. G., "Optimizing a Microstrip Bandpass Filter Using Electromagnetics," *Int. J. MIMICAE*, Vol. 5, No. 9, 1995, pp. 344–351.

[26] So, P. P. M., et al., "Hybrid Frequency/Time Domain Field Theory Based CAD of Microwave Circuits," *Proc. 23rd European Microwave Conference*, Madrid, Spain, 1993, pp. 218–219.

[27] Swanson, Jr., D. G., D. Baker, and M. O'Mahoney, "Connecting MMIC Chips to Ground in a Microstrip Environment," *Microwave Journal*, Vol. 34, No. 12, 1993, pp. 58–64.

Chapter 3

Numerical Electromagnetics

Numerical electromagnetics is the theory and practice of solving electromagnetic field problems on digital computers. It reflects the general trend in science and engineering to formulate the laws of nature as computer algorithms and to simulate physical processes on digital computers. While theory and experiment remain the two traditional pillars of science and engineering, numerical modeling and simulation represent a third pillar that supports, complements, and sometimes replaces them. Numerical modeling and simulation have revolutionized all aspects of engineering design to the extent that the concepts of computer-aided engineering (CAE) and (CAD) have become synonymous with progressive, state-of-the-art engineering practice.

To the microwave and high-speed electronics engineer, numerical electromagnetics offers the key to comprehensive solutions of Maxwell's equations by means of electromagnetic simulators or field-solvers. Field-based solutions have become necessary due to the evolution of analog and digital systems towards higher clock rates, higher frequencies, larger bandwidths, higher packaging density, and higher complexity. While field solutions require more computation time and memory than circuit-based simulations, they can account for all parasitic interactions, packaging effects, and the distributed nature of the fields in a structure. However, a field simulation can also be used to generate realistic equivalent network models of electromagnetic structures that include these parasitic and distributed effects, and thus yield accurate results with minimum computational expenditure, a major asset when optimization is to be used.

In this chapter, the predominant methods used in *computational electromagnetics* will be discussed. They determine the properties of the numerical "engines" of the various commercial simulators and define their respective characteristics, strengths, and limitations. Developers of field solvers are making considerable efforts to ensure that users can solve electromagnetic problems without expert knowledge in the numerical method used in their tools. However, a user who knows the fundamental properties and characteristics of the method implemented in a simulation tool will be better prepared to exploit its full capabilities, achieve reliable results more quickly, and avoid errors and pitfalls that occur when the limitations of a particular numerical method are ignored. Thus, we have included this chapter with the aim to introduce present and future users of electromagnetic simulators to the theoretical and computational foundations of numerical methods, the fundamentals of computational electromagnetics. Most of these methods can be interpreted mathematically as *projective approximations* and are backed by an extensive mathematical framework developed since the early 1900s. The massive amount of literature on the subject is staggering, and we can summarize here only the most elementary concepts and ideas without presenting the extensive mathematical formalism that a rigorous account of the theory would require. We thus refer the reader to the list of specialized works on the various numerical techniques at the end of this chapter.

3.1 MICROWAVE ANALYSIS AND DESIGN

Microwave technology typically involves components with dimensions of the order of the operating wavelength. The electrical characteristics of microwave components thus depend strongly on their geometry or topology, as well as on the properties of the surrounding space and packaging in the case of open structures. This "distributed" nature of microwave circuits has a number of fundamental implications: the traditional definitions of voltage, current, and impedance are no longer adequate. Instead, the electric and magnetic field vectors governed by Maxwell's equations are the primary unknown quantities, while voltages and currents are secondary quantities, derived from the fields by integration along specific paths upon which they depend. Furthermore, since the field solutions of interest here are predominantly of a wave nature, specifications are often given in terms of scattering rather than impedance or admittance parameters prevalent in classical circuit and network theory. The microwave engineer's task may thus be generalized as follows: establish a relationship between the geometry or topology of a component and its functionality.

The topology of an electromagnetic structure can be very complex and can have many degrees of freedom (many variables describing it), particularly when it is discretized into small elements, while its functionality is usually less complex and can be described by fewer independent variables or degrees of freedom, as suggested in Figure 3.1.

If we know the topology of a structure, we can obtain its functionality by performing an electromagnetic analysis (or a measurement). The result of analysis is unique and should be the same, irrespective of the method used, and within its error margin. Electromagnetic synthesis, however, which is the reverse process, does usually not yield unique results because the functionality is described by fewer degrees of freedom than the topology. In other words, the same functionality may be realized by several different topologies. A waveguide bandpass filter, for

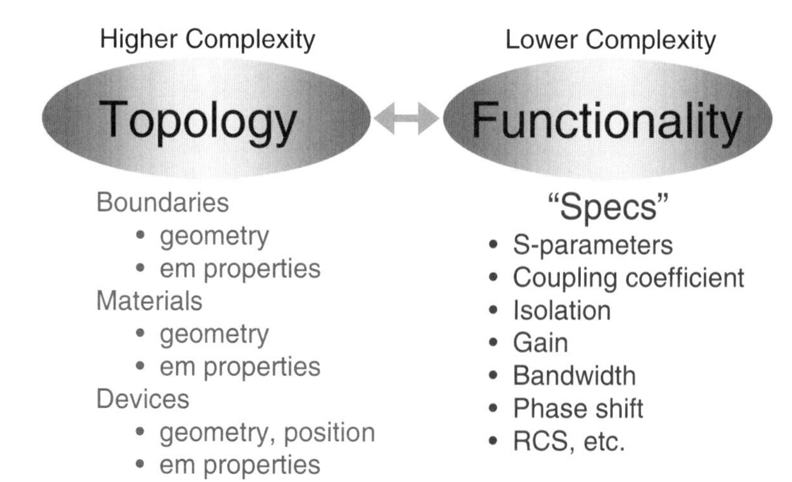

Figure 3.1 The microwave engineer's task is to establish a relationship between the topology and the functionality of a component.

instance, may consist of either inductive irises or posts of various cross-sections, and the designer must introduce additional constraints and characteristics in order to obtain a particular topology. This is probably the reason why design is considered more of an art and analysis more of a science; see Figure 3.2.

Traditionally, microwave circuits and components have been modeled by equivalent lumped and transmission line element networks, resulting in very efficient analysis tools. However, as mentioned at the beginning of this chapter, these

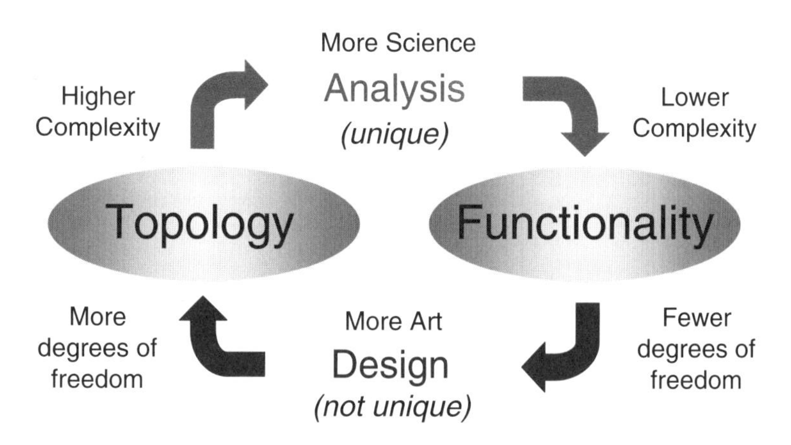

Figure 3.2 Analysis yields the functionality of a structure while synthesis creates a topology from specifications.

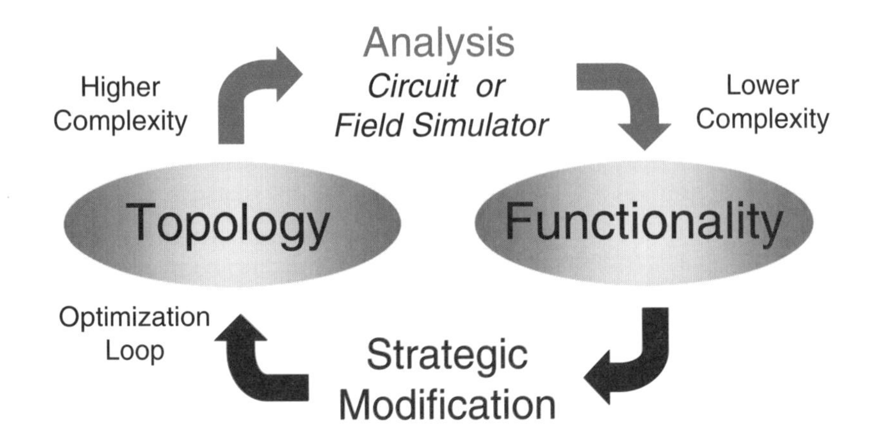

Figure 3.3 Design by analysis and optimization. The analysis is either performed by solving an equivalent circuit or a suitable surrogate model, or by a full electromagnetic field analysis. The topology is strategically modified and re-analyzed until specifications are met.

circuit models do not account for parasitic interactions between elements. One way to include these effects is to perform an electromagnetic analysis of a critical subdomain and extract a more comprehensive equivalent circuit and import it into the circuit-based simulator.

The alternative is to perform a full electromagnetic analysis. Design would usually call for the inversion of analysis, namely synthesis, which is more challenging since it does not always yield unique results. A very successful alternative is analysis with optimization, in which a starting topology (first guess) is repeatedly analyzed and strategically modified until specifications are met, as shown in Figure 3.3. In fact, analysis with optimization has become the prevalent paradigm in computer-aided microwave design, while direct field-based synthesis is still in the research stage. For this reason, this book focuses exclusively on electromagnetic analysis.

3.2 METHODS OF ELECTROMAGNETIC ANALYSIS

Electromagnetic analysis amounts to finding field and/or source functions that:

- Obey Maxwell's equations;
- Satisfy all boundary conditions;
- Satisfy all interface and material conditions;
- Satisfy all excitation conditions.

Figure 3.4 A dazzling and sometimes confusing array of analysis methods for electromagnetic field problems that form the heart of modern field simulators and solvers.

Electromagnetic analysis methods can be classified as follows:

(a) Analytical Methods – Closed-form solutions in terms of analytical functions can only be found for a few special geometries (for example in rectangular, elliptical, and spherical waveguides and resonators). In spite of their limited practical applicability, analytical solutions are extremely useful for the purpose of validating numerical methods since they provide error-free reference solutions.

(b) Semianalytical Methods – Semianalytical methods were developed before the advent of powerful computers. They involve extensive analytical processing of a field problem resulting in a complicated integral, an infinite series, a variational formula, an asymptotic approximation, in short, an expression that requires a final computational treatment to yield a quantitative solution. The analytical preprocessing often leads to rather fast and efficient computer algorithms, but the resulting programs are necessarily specialized since specific types of boundary and material conditions have been incorporated in the formulation.

(c) Numerical Methods – Numerical methods transform the continuous integral or differential equations of Maxwell into an approximate discrete formulation that requires either the inversion of a large matrix or an iterative procedure. There exist many ways to discretize an electromagnetic problem, ranging from very problem-specific to very general purpose approaches. The large number of available numerical methods can be quite daunting and confusing to anyone who wants to study them, as symbolized in Figure 3.4.

It is also not easy for the newcomer to get a balanced and impartial account of the strengths and weaknesses of these methods since researchers, vendors, and users tend to become quite ideological about their favorite approach or the type of simulator they have adopted and mastered. It will thus be helpful to discuss first what all these numerical methods have in common, and what distinguishes them.

3.3 THE FEATURES COMMON TO ALL NUMERICAL METHODS

The purpose of all numerical methods in electromagnetics is to find approximate solutions to Maxwell's equations (or of equations derived from them) that satisfy given boundary and initial conditions. Formulating an electromagnetic problem amounts to specifying the *properties* that a solution must have in order to qualify. These properties can be specified as local (differential) or global (integral) properties, both in the field space and at its boundaries. In other words, we must solve a differential or an integral equation subject to specific conditions.

If we had enough experience we could try to guess a solution and then verify if it has indeed the required properties. If it does, the problem is solved, but that outcome is highly unlikely. Next, we could try to improve our guess until its properties meet the specifications, at least approximately. In other words, we would try to optimize our guess in some sense.

To implement this approach on a computer, we must formalize it in such a way that it converges accurately, quickly, and reliably in a wide variety of electromagnetic scenarios. The basis for such a computer solution is the classical mathematical technique of approximating a function $f(x)$ (our unknown solution) by a sum of *known* functions $f_n(x)$, also called *expansion functions* or *basis functions*. (We will use the term *expansion functions* rather than *basis functions f*or reasons that will become clear later.)

$$f(x) \approx \sum_n \alpha_n \cdot f_n(x) \tag{3.1}$$

The *weight* or *coefficient* α_n of each expansion function must be determined such that the sum approaches the function as closely as possible. A typical example is the representation of a function by a Fourier series; here, the expansion functions f_n are *sine* and *cosine* functions, and the αn are the Fourier coefficients or amplitudes of the expansion functions. In fact, all numerical methods in electromagnetics employ this common strategy: the unknown solution is expanded in terms of known expansion functions with unknown coefficients. The coefficients are then determined such that the sum in (3.1) meets, as closely as possible, all the criteria stated in the formulation of the problem. Note that we can formulate and interpret all method of moments, finite element, finite difference, and transmission line matrix methods in these terms, even though they are usually derived and stated differently.

3.4 THE DIFFERENCES BETWEEN NUMERICAL METHODS

The differences between various numerical techniques reside essentially in the following aspects:

- The electromagnetic quantity that is being approximated;
- The expansion functions that are used to approximate the unknown solution;
- The strategy employed to determine the coefficients of the expansion functions.

The solution of an electromagnetic problem may require finding the electric or magnetic field, a potential function, or a distribution of charges and/or currents. While these quantities are related, they have different properties; hence, problem formulations for field, potential, and charge or current solutions are different. Finding fields or potentials will require expansion functions in the field space *(domain methods)*, while unknown charge or current distributions are expanded into functions defined mostly on boundaries *(boundary methods)*. Finally, there exists a variety of strategies for computing the unknown coefficients, which involve the inversion of large matrices, implicit and explicit iteration schemes, evolutionary algorithms, or random walks. The various existing numerical methods employ different combinations of these aspects. In the following, we will attempt to classify them, and highlight those that have become prominent as numerical engines of mainstream commercial electromagnetic simulators.

3.5 CATEGORIES OF NUMERICAL METHODS

Numerical methods fall into two broad categories: *frequency domain* and *time domain* methods.[1] This distinction reflects the difference in our perception of space and time. In physics and mathematics, space and time are treated as dimensions of the same manifold. However, in the physical world, and at the human scale, space and time present very different properties. While space appears stationary and can be crossed in all directions, time flows continuously and only in positive direction. The perceived differences between these two categories are better captured by the terms *time-harmonic* and *transient* methods. However, in the formal sense, frequency domain formulations are time domain formulations in which the time dimension has been subject to a Fourier transform, thus reducing the number of independent variables by one. Expressed in a simplistic way, frequency domain formulations are obtained by replacing the time differential operator d/dt by $j\omega$, and the time integration operator by $-j/\omega$, thus effectively transforming a time differentiation into a multiplication, and a time integration into a division by $j\omega$. This not only lightens the computational burden, particularly in narrow and moderate band-

1 We interpret a static method either as a frequency domain method for zero frequency, or a time domain method in which $d/dt = 0$.

width applications, but it also permits the use of complex notation with all its advantages. However, it causes complications in nonlinear situations and raises causality issues. The transient or time domain methods are thus the most general and comprehensive formulations, suitable for the widest possible range of applications, while time-harmonic or frequency domain methods emphasize the spectral view of electromagnetics and microwaves. The latter is the more traditional, albeit more abstract, paradigm characterized by the complex formalism and the representation of fields by phasors. It is prevalent in analog microwave and millimeter-wave engineering.

Another way of categorizing both the numerical techniques and the computer tools based on them relies on the number of independent *space* variables upon which the field and source functions depend. In all categories we can again distinguish between frequency domain and time domain formulations.

(a) 1D Methods – These are methods for solving problems where the field and source functions depend on *one* space dimension only. Typical applications are transmission line problems, uniform plane wave propagation, and spherically or cylindrically symmetrical problems with only radial dependence. Transmission-line circuit solvers and the SPICE program are well-known examples of 1D solvers.

(b) 2D Methods – These are methods for solving problems where the field and source functions depend on *two* space dimensions. Typical applications are cross-section problems in transmission lines and waveguides, TE_{n0} propagation in rectangular waveguide structures, coaxial TEM problems, and spherical problems depending only on radius and azimuth or radius and elevation.

(c) 2.5D Methods – These are methods for solving problems where the fields depend on *three* space dimensions, while their sources (the currents) are mainly confined planes with two space dimensions. Typical examples are planar structures such as microstrip circuits, co-planar circuits, patch antennas, and general multilayer structures that contain planar conductor pattern. The predominant solution method for such structures is the method of moments in the space and spectral domains; however, the method of lines is also suitable for planar and quasi-planar structures.

(d) 3D Methods – These are methods for solving problems where both the field and source functions depend on *three* space dimensions. This category comprises all volumic full-wave general-purpose formulations. The most prominent 3D frequency domain methods are finite element, finite difference, and method of moments formulations. Among the 3D time domain methods, the FDTD, FIT, and TLM formulations dominate. Many other 3D methods have been developed over the years and translated into computer simulation tools for more specialized applications, including the mode-matching technique, the coupled integral equation technique, the spectral domain technique, the general multipole technique, the method of lines, and the boundary element method. Hybrid formulations combining two or more different numerical techniques have also been developed and implemented for particular applications.

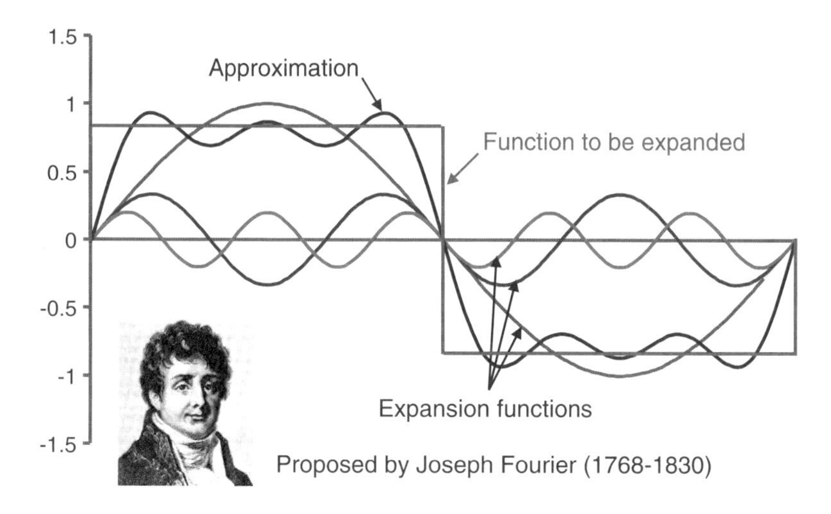

Figure 3.5 Full Domain Expansion Functions extend over the entire domain in which the solution is to be calculated. A finite number of expansion functions results in an imperfect approximation.

3.6 EXPANSION FUNCTIONS

We mentioned earlier that different numerical methods employ different types of *expansion functions* to approximate the unknown solution. In some methods the expansion functions extend over the entire domain in which the solution must be found. Figure 3.5 shows the well-known example of sine and cosine expansion functions proposed by Fourier, which are typically found in mode matching and spectral domain formulations for rectangular waveguides. Note that in order to have a perfect representation of a function, we need a *complete set* of expansion functions (in this case an infinite but discrete number of harmonics). However, in a practical application we must truncate the number of expansion functions and accept a tolerable approximation error. Obviously, this will always involve a compromise between accuracy and computational burden. Other typical full domain expansion functions are power series of the type

$$f_n(x) = x - x^{n+1} \tag{3.2}$$

or, for cylindrical or spherical problems, Bessel functions or Legendre polynomials. These are obviously specialized structure functions that have the characteristics of the unknown solution and satisfy the same boundary conditions. We can think of them as *problem-specific*. However, when the geometry and material properties are very complicated, more flexibility is provided by the so-called *subsectional expansion functions*, and it is not surprising that all general-purpose field simulators

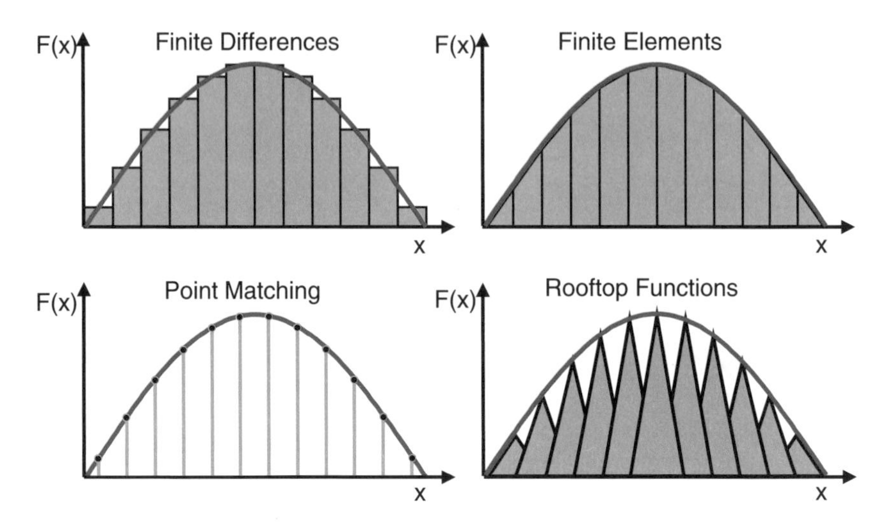

Figure 3.6 Subsectional expansion functions are defined only over a small part of the computational domain. They provide the greatest flexibility in the approximation of solutions.

employ this type of approach. Some typical examples are shown in Figure 3.6 for the simple one-dimensional case. Naturally, the expansion functions must depend on as many independent variables as the solution we want to approximate.

A finite difference formulation amounts to approximating the solution (red curve) by a series of pulse functions, each being defined over a small subsection of the computational domain, resulting in a step approximation of the solution. Note that the height of each pulse is the coefficient of that expansion function; it exclusively determines the value of $f(x)$ in the center of its subdomain. Linear finite element formulations represent the unknown solution by piecewise linear functions, and the coefficients are their values at the break points between the straight sections. Point-matching formulations yield samples of the solution at discrete points.[2] Rooftop functions are also very flexible and well suited for general situations, and even though they overlap with their neighbors, their peak nevertheless determines the solution at the center of each subdomain.

The size of the subdomains, also called the *support*, must not necessarily be equal for all expansion functions. In regions where the solution is fairly uniform, only a few expansion functions with large support may be necessary, while in regions of high nonuniformity, more expansion functions with compact support will be required. This irregular subdivision of the domain is synonymous with

2 Point matching or collocation has been presented by Harrington [1] as a variant of the Method of Moments with Dirac delta functions as testing functions, but strictly speaking, it is not a projective approximation since the Dirac delta functions are not square integrable and do not belong to a Hilbert space with a suitable inner product definition.

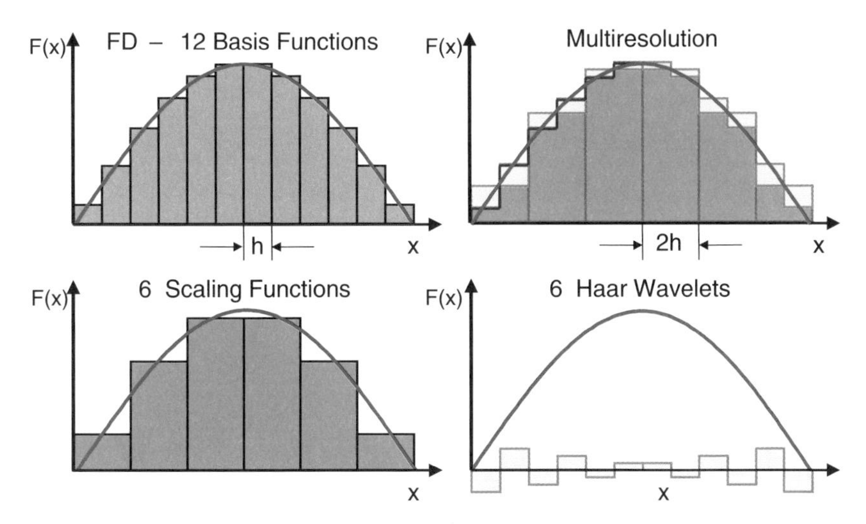

Figure 3.7 Comparison of pulse and wavelet expansion functions. Approximation of the solution with six Haar scaling and six Haar wavelet functions of support $2h$ (top right) is identical to the approximation with 12 pulse functions of support h (top left).

irregular or *graded meshing*. Usually, this subdivision is determined before the computation starts, and thus requires an a priori knowledge of the characteristics of the solution. *Adaptive meshing* is a technique that automatically modifies the subdomain discretization as the solution evolves. One way to realize such an adaptive resolution is to employ *wavelets* as expansion functions. This approach, also called multiresolution, involves families of functions with increasing resolution, such as the Haar wavelets, as demonstrated in Figure 3.7. Here, a sinusoidal field distribution is approximated by finite differences yielding 12 pulse functions of support h.

On the other hand, the Haar family of expansion functions consists of six scaling functions and six wavelets, both of support $2h$. The scaling functions represent a coarse approximation which can be refined by adding wavelets of increasing periodicity. The main difference in this approach is that the wavelets can be omitted in regions where the solution is rather uniform, and can be selectively added wherever a higher resolution is required. The decision whether to add wavelets or not is made by *thresholding*, where the relative or absolute change in the solution due to the addition of a wavelet is tested in each subdomain. This leads to adaptive refinement of the discretization at the cost of additional computational overhead.

In all cases, the approximation of the problem solution by a finite set of known expansion functions with unknown coefficients amounts to a discretization of the problem. The coefficients represent its *degrees of freedom*. We will now discuss the various strategies for finding these coefficients. Like the choice of the expansion functions, the solution strategy allows us to distinguish between the different numerical techniques employed in computational electromagnetics.

3.7 STRATEGIES FOR FINDING THE UNKNOWN COEFFICIENTS

As stated at the beginning of Section 3.3, our objective is to find solutions to Maxwell's equations (or of equations derived from them) in differential or integral form. In symbolic notation, the differential formulation can be stated as follows:

$$Lf(\boldsymbol{r}) = g(\boldsymbol{r}) \tag{3.3}$$

where L is an operator, g is a known *source* or *excitation* function, and f is the unknown field or response to be determined. r is the vector of coordinates.

A practical example is the well-known electrostatic problem stated by the Poisson equation:

$$-\nabla^2\phi(\boldsymbol{r}) = \rho(\boldsymbol{r})/\varepsilon \tag{3.4}$$

where the operator L is defined in Cartesian coordinates as

$$L = -\nabla^2 = -\left(\frac{d^2}{dx^2} + \frac{d^2}{dy^2} + \frac{d^2}{dz^2}\right) \tag{3.5}$$

the unknown function f is the electrostatic potential ϕ, and the source function g is the electric charge density ρ, divided by the permittivity ε. Note that in order to obtain a unique solution, we must specify appropriate boundary conditions (i.e., values of the function f or of its derivatives at the boundaries of the problem domain).

The same problem can also be stated in integral form using the inverse operator L^{-1}:

$$L^{-1}g(\boldsymbol{r}') = f(\boldsymbol{r}) \tag{3.6}$$

where L^{-1} is defined using a suitable Green's function $G(\boldsymbol{r}, \boldsymbol{r}')$ as follows:

$$L^{-1}(g(\boldsymbol{r}')) = \int G(\boldsymbol{r}, \boldsymbol{r}')g(\boldsymbol{r}')d^3r' \tag{3.7}$$

$\boldsymbol{r}$ and $\boldsymbol{r}'$ are the field point and source point vectors, respectively, and the integration must be carried out over the entire source volume. For our electrostatic example and in unbounded space, this is the integral of Coulomb's law:

$$\phi = \int \frac{1}{4\pi|\boldsymbol{r} - \boldsymbol{r}'|}\frac{\rho}{\varepsilon}d^3r' \tag{3.8}$$

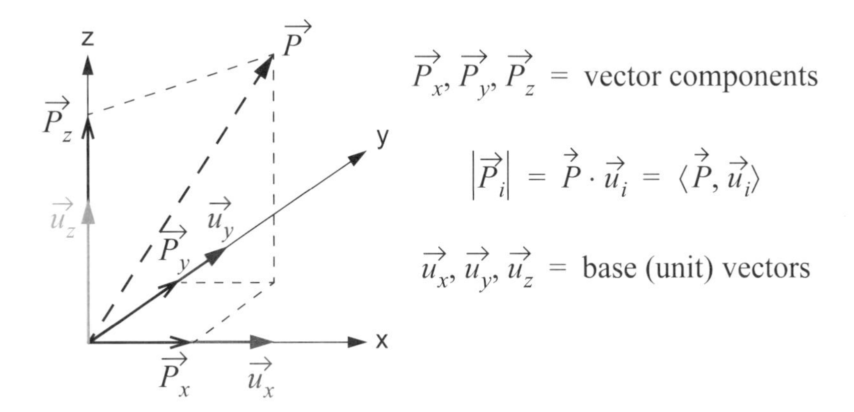

Figure 3.8 The components of a vector are obtained by projecting it unto the basis or unit vectors that span the vector space. In a function space, vectors are replaced by functions.

which means that the Green's function for this situation is

$$G(\boldsymbol{r}, \boldsymbol{r}') = \frac{1}{4\pi|\boldsymbol{r} - \boldsymbol{r}'|} \tag{3.9}$$

It can be interpreted as the potential produced by a unit point source. In the presence of boundaries the Green's function will be different; hence, the integral operator in (3.6) includes the boundary conditions, and it is not necessary to state them since they are implicit in the inverse operator L^{-1}.

Regardless of whether the problem is formulated in differential or integral form, the most universal strategy for finding the unknown coefficients is the so-called method of *projective approximation.* The terms *Petrov-Galerkin* method, *method of weighted residuals*, or *method of moments* all refer to variations of this approach. It is best described in terms of linear function spaces, an abstract concept that may be understood by analogy with vector spaces.

Consider first a three-dimensional Euclidian space described by a Cartesian coordinate system. The three *unit vectors* that point in the three coordinate directions are also called the *basis vectors* of that space (see Figure 3.8). The *components* of any vector $\boldsymbol{P}$ are the *projections* of that vector onto the basis vectors, obtained by forming the *dot product* or *inner product* of the vector $\boldsymbol{P}$ with the basis vectors. Furthermore, we can affirm that two vectors $\boldsymbol{P}_1$ and $\boldsymbol{P}_2$ are identical when they have identical *x*-, *y*-, and *z*-components (projections or inner products with the basis vectors). That statement can be generalized to n-dimensional spaces.

By analogy, we can imagine an abstract *function space* that is "spanned" by *basis functions* instead of basis vectors. Any function within that space can then be decomposed into "components" that are obtained by projecting the function onto the basis functions (i.e., by forming its inner product with the basis functions). The inner product of two functions g_1 and g_2 of m variables is usually defined as

$$\langle g_1, g_2 \rangle = \langle g_2, g_1 \rangle = \int_a^b g_1(\boldsymbol{r}) g_2(\boldsymbol{r}) dR^m \tag{3.10}$$

where a and b bound the domain of interest. Expanding a function into basis functions is thus similar to decomposing a vector into components along coordinate axes. We can carry the analogy further by affirming that two functions f_1 and f_2 are identical if their respective projections unto each basis function are the same.

We will now return to the description of the *projective approximation* procedure. Assuming the case of a problem statement in integral form, as described in (3.6), we decompose the unknown source function g into a finite number n of known expansion functions $g_n(\boldsymbol{r}')$ with unknown coefficients α_n,

$$g(\boldsymbol{r}') \approx \sum_n \alpha_n \cdot g_n(\boldsymbol{r}') \tag{3.11}$$

enter this representation into (3.6), and make use of the linearity of the operator L^{-1}.

$$\sum_n \alpha_n \cdot L^{-1} g_n(\boldsymbol{r}') \approx f(\boldsymbol{r}) \tag{3.12}$$

Let us assume that both sides of (3.12) are equal (keeping in mind that, strictly speaking, we need an infinite number of expansion functions to justify an equal sign). We will now test the validity of (3.12) in a *function space* spanned by m functions b_m. Since the term *basis functions* is often used for the expansion functions of the unknown solution, we call from now on b_m the *testing* or *weighting* functions. For both sides of (3.12) to be equal, their projections unto each of the m testing functions must be identical. Since these projections are their inner products with the testing functions, we can write:

$$\sum_n \alpha_n \cdot \langle b_m, L^{-1} g_n \rangle = \langle b_m, f \rangle \quad \textit{for } m = 1, 2, 3... \tag{3.13}$$

yielding m algebraic equations. Using as many testing functions as expansion functions ($m = n$) we obtain n equations with the n unknown coefficients α_n, which we can solve with a variety of available numerical procedures for matrix inversion.

(We are free to use more testing functions than expansion functions ($m>n$), but not less.)

A special choice of testing functions, attributed to *Galerkin*, is to use the expansion functions themselves as testing functions, resulting in the following system:

$$\sum_n \alpha_n \cdot \langle g_m, L^{-1} g_n \rangle = \langle g_m, f \rangle \quad \text{for } m = 1, 2, 3 \ldots \tag{3.14}$$

We can write both equations (3.13) and (3.14) in matrix form:

$$[I_{mn}][\alpha_n] = [f_m] \tag{3.15}$$

where $[I_{mn}]$ is an $m \times n$ matrix of inner products, $[\alpha n]$ the vector of unknown coefficients, and $[f_m]$ the vector of projections of f onto the testing functions. To find the coefficients αn we must invert the matrix $[I_{mn}]$ using available numerical algorithms, since

$$[\alpha_n] = [I_{mn}]^{-1}[f_m] \tag{3.16}$$

Once we have determined the coefficients α_n, we can construct the solution using (3.11).

Note that the projection of the initial operator equation on the manifold of test functions results in a *discretization* of the problem. Without going into specific numerical methods, we can already draw some general conclusions from the above.

(1) The more expansion functions we use, the larger a matrix we must invert.

(2) The computational effort required to compute the inner products (filling the matrix) strongly affects the solution time.

(3) The matrix will be much faster to compute and to invert if we choose testing and expansion functions such that all inner products are zero except for $m=n$. That leads to diagonal matrix which is very easy to invert.

We are now ready to take a closer look at specific numerical techniques. While discussing their particular features and properties, we will be able to relate them to the general principles and solution strategy outlined above.

3.8 THE METHOD OF MOMENTS

The term *Method of Moments* got its name from earlier applications in mechanical and civil engineering. The original MoM employs pulse expansion functions and Dirac testing functions (collocation). Harrington [1] has extended this concept by describing it in terms such that it is essentially identical to the general method of *projective approximations*, including *collocation* as a special case. In the light of Harrington's unifying representation it could rightly be called the *Mother of all Methods* (MOM). However, the Method of Moments is usually understood in a narrower sense, and viewed as distinct from finite element and finite difference methods.[3]

In the narrower sense, MoM is the method of choice for solving problems stated in the form of an electric field integral equation (EFIE) or a magnetic field integral equation (MFIE) of the following form:

$$\text{EFIE:} \quad L_e^{-1}\boldsymbol{J} = \boldsymbol{E} \tag{3.17}$$

$$\text{MFIE:} \quad L_m^{-1}\boldsymbol{J} = \boldsymbol{H} \tag{3.18}$$

$\boldsymbol{E}$ and $\boldsymbol{H}$ are field functions, and $\boldsymbol{J}$ is a source function (current density). These expressions are of the type given in (3.6), and the inverse operators thus involve Green's functions that depend on the boundaries and material distributions of the problem. In most cases these integral equations are formulated in the frequency domain although time domain applications exist. Instead of the fields $\boldsymbol{E}$ and $\boldsymbol{H}$ we may also find scalar or vector potentials.

The solution strategy closely follows the steps outlined in Section 3.7, and we will give another example here to clarify the methodology. Figure 3.9 shows a section of a metal strip on a dielectric sheet. The problem consists of finding the current density $\boldsymbol{J}$ on the strip that is induced by an incident electric field $\boldsymbol{E}^i(\boldsymbol{r})$. The physics of the problem may be described as follows. The incident field induces a current distribution on the strip. This induced current, in turn, will radiate a *scattered* electric field $\boldsymbol{E}_s(\boldsymbol{r})$ that will be superimposed on the incident field, resulting in the total field $\boldsymbol{E}_t(\boldsymbol{r})$. We know neither the induced current density nor the scattered field. The only thing we know is that the tangential component of the *total* field must vanish everywhere on the strip.

The relationship between current density and field functions is given by the following electric field integral equation:

$$\int_{S'} G(\boldsymbol{r}, \boldsymbol{r}')\boldsymbol{J}(\boldsymbol{r}')ds' = \boldsymbol{E}_t(\boldsymbol{r}) - \boldsymbol{E}^i(\boldsymbol{r}) = \boldsymbol{E}_s(\boldsymbol{r}) \tag{3.19}$$

where the integration must be performed over the surface S' of the strip. Since at the surface of the strip, the total tangential field $\boldsymbol{E}_t(\boldsymbol{r})$ must vanish, the right-hand side of (3.19) is equal to $-\boldsymbol{E}^i(\boldsymbol{r})$ everywhere on the surface, so that we have

3 The reader may be aware of the fact that discussions of relationships between numerical methods can quickly shift from methodology to ideology.

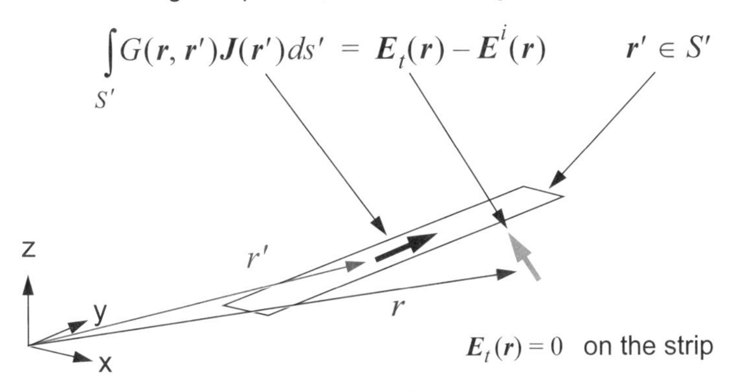

Figure 3.9 A metal strip problem formulated as an electric field integral equation.

$$\int_{S'} G(\boldsymbol{r}, \boldsymbol{r}')\boldsymbol{J}(\boldsymbol{r}')ds' = -\boldsymbol{E}^i(\boldsymbol{r}) \qquad \boldsymbol{r}, \boldsymbol{r}' \in S' \tag{3.20}$$

which is an expression of the form given in (3.6). Before we can attempt a solution we must derive the Green's function $G(\boldsymbol{r}, \boldsymbol{r}')$ so that we know the inverse operator L^{-1}. Figure 3.10 shows how we can derive a relationship between the surface current density $\boldsymbol{J}$ and the scattered field $\boldsymbol{E}_s$ via the vector potential $\boldsymbol{A}$. For a time-harmonic current density on the strip we first determine the (retarded) vector potential $\boldsymbol{A}(\boldsymbol{r})$ by adding the contributions of all current elements on the strip as shown in Figure 3.10. For infinitesimal current elements the vector potential becomes a vector integral taken over the surface of the strip:

$$\boldsymbol{A}(\boldsymbol{r}) = \iint_{S'} \frac{e^{-jk|\boldsymbol{r}-\boldsymbol{r}'|}}{4\pi|\boldsymbol{r}-\boldsymbol{r}'|}\boldsymbol{J}(\boldsymbol{r}')dS' \tag{3.21}$$

k is the free space propagation constant. The kernel of the integral is the free space Green's function for the vector potential $\boldsymbol{A}$. In a second step we find the electric field $\boldsymbol{E}_s$ from the vector potential $\boldsymbol{A}$:

$$\boldsymbol{E}_s(\boldsymbol{r}) = -j\omega\mu\boldsymbol{A}(\boldsymbol{r}) + \frac{1}{j\omega\varepsilon}\nabla(\nabla \cdot \boldsymbol{A}(\boldsymbol{r})) \tag{3.22}$$

Combining (3.21) and (3.22) yields after some manipulation

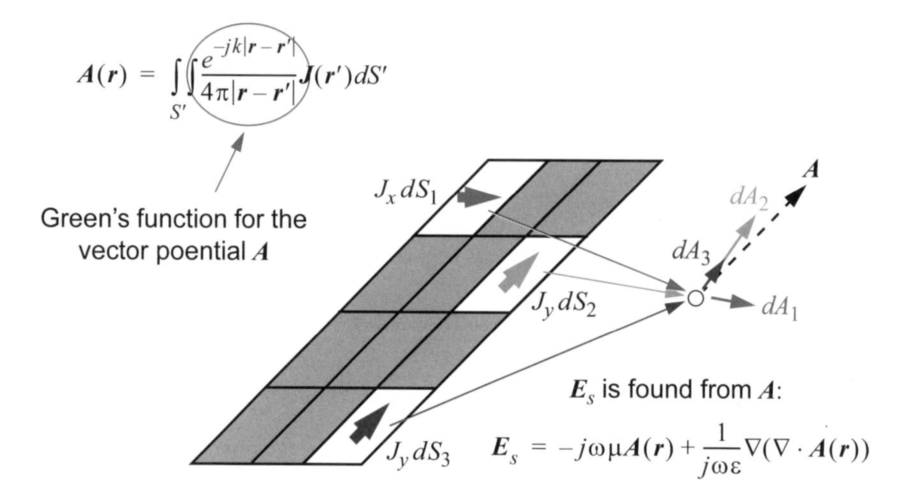

Figure 3.10 Calculation of the vector potential and the electric field produced by a current on a strip.

$$E_s(r) = \frac{1}{j4\pi\omega\varepsilon} \iint_{S'} (k^2 + \nabla\nabla) \frac{e^{-jk|r-r'|}}{|r-r'|} J(r')dS' \qquad (3.23)$$

This integral equation is not only valid for a simple strip, but also for any arbitrarily shaped conducting object in free space, such as an airplane. In fact, it is the general EFIE scattering problem formulation for conducting objects in free space. Naturally, for complex 3D shapes, the current density function $\boldsymbol{J}$ will depend on three rather than two space variables, and the vector integral must be evaluated over a more complex surface shape.

Once we have formulated the integral equation, specified the surface S' of the scattering object and the incident field (for example, a uniform plane wave of known frequency $\omega/2\pi$ and propagation constant k), we can proceed with the solution by following the general strategy outlined in Section 3.7. In the following, we will focus on some specific characteristics of MoM solutions for the EFIE scattering problem formulation.

- **Solution Step 1: Expand $\boldsymbol{J}$ into known expansion functions with unknown coefficients.**

The first step is to express the unknown surface current density $\boldsymbol{J}(\boldsymbol{r}')$ as a sum of known expansion functions with unknown coefficients. At this stage we recognize already one of the strong points of the method of moments: the expansion functions are confined to the surface of the scatterer! If we want to use full-domain expansion functions, we need to define them only on the surface S'. If we want to use subsectional expansion functions, we need to discretize only the surface S' into subsections (*surface* or *boundary meshing*) and define the expansion functions on them. This will require far less expansion functions than a differential problem formulation for which the entire 3D field space must be discretized, as we shall see later.

The choice of the expansion functions depends on the geometry of the scatterer. If the object has a simple regular shape (strip, cylinder, needle, ellipsoid) the surface current density can be approximated with a few simple full domain functions, hence, the number of unknown coefficients will be small, requiring the inversion of a small matrix only. However, the inner products (the elements of the matrix) may be difficult to evaluate. If the shape is more complex, a discretization of the surface into rectangular and/or triangular subsections will provide maximum flexibility, and the expansion functions could be pulse functions, rooftop functions, piecewise sinusoidal functions, or wavelets. Clearly, the number of unknown coefficients will be much larger in this case, yielding a large matrix. Another approach employed in some MoM solvers consists of replacing the continuous surface of the scatterer by a conformal mesh of conducting wires [2]. In this case, the expansion functions are linear currents in the wire sections, and the surface integral in (3.23) becomes a summation over discrete current elements. In the infinitesimal limit, such a discrete wire mesh melts into a continuous surface. Note that in all cases, the expansion functions are vector functions, since they have both a magnitude and a direction in space.

- **Step 2: Select a set of appropriate testing or weighting functions.**

Here we have three general choices:

Option 1: Collocation or Point Matching – If we select Dirac delta functions as testing functions, their inner products with both sides of the integral equation are very easy to evaluate. In fact, they are just the samples of both sides at the location of the Dirac pulses. This approach represents the original method of moments and works well with most types of subsectional expansion functions and object shapes. We would normally locate one Dirac testing function at, or close to, the center of each subdomain of the surface S'. The sampling of both sides of the integral equation amounts to enforcing it at discrete points, hence the term *point matching*. Obviously, the larger the number of subsection samples, the better will be the approximation, but the price to pay is the inversion of a larger matrix. Another shortcoming of collocation is the lack of information on the approximation error at all points other than the sampling points.

Option 2: Galerkin's Method – This is a special case of projective approximation in which the testing functions are identical to the expansion functions. This choice always results in a square system matrix and has the advantage that the results of the approximation are variational, thus requiring considerably less expansion functions for a given accuracy of approximation and hence, smaller matrices than point matching. However, the evaluation of the inner products require the computation of a double integral for each matrix element, and it takes more time to "fill the matrix."

Option 3: General Projective Approximations – Here the testing functions are different from the expansion functions, but unlike the Dirac delta functions, they are usually square integrable and belong to a *Hilbert* space with a suitably defined inner product. Since the testing functions are different from the expansion functions, one can actually choose more testing functions than expansion functions, which allows the construction of least-square solutions for the unknown coefficients, yielding useful information as to the precision of the solution.

- **Step 3: Form the inner products and generate the matrix equation.**

This step closely follows the procedure outlined in Section 3.7. Note that the computation of the inner products may be quite involved and represent a large percentage of the total solution time. It depends on the ease with which the expansion and testing functions can be integrated, so that their choice critically affects the computational burden.

- **Step 4: Invert the matrix and determine the coefficients α_n.**

Techniques and procedures for matrix inversions are well known and very mature and will not be discussed here. Naturally, care must be taken to ensure that none of the inner products that form the elements of the matrix becomes singular. Once the coefficients have been determined, the approximate solution can be obtained using the ansatz in (3.11). This solution can then be further processed to extract engineering parameters, such as the radar cross-section or scattering parameters.

Coming back to our scattering example and contemplating the inverse operator that involves a convolution integral with a complicated Green's function in the kernel, (3.23), we appreciate that it is not straightforward to formulate general complex electromagnetic topologies in this form. One of the difficulties is to determine a suitable Green's function for a general topology because it is problem dependent. For example, in the presence of a metal plane close to the scattering object, the vector potential will not only be due to the primary induced current elements on the strip, but also to their images in that metal plane. Furthermore, if the surrounding medium is not homogeneous, partial reflections at the material interfaces must be included in the Green's function. However, if we can impose some constraints on the topologies to be solved, the integral formulation can be implemented quite effectively in practical MoM solvers. One option is to admit only homogeneous media surrounding the scatterer, while remaining flexible as to its shape. This allows us to use the same integral equation (3.23) in all cases and build a MoM field-solver that computes the current distribution and hence, the scattered field, for conducting bodies of arbitrary shape in homogeneous space. The emphasis here would be to implement a capability for handling complex three-dimensional current distributions. Such realizations are indeed suitable for field simulators capable of handling a wide range of scattering and antenna problems.

On the other hand, we could restrict the scattering surfaces to planar topologies, and admit inhomogeneous, albeit layered piece-wise uniform media, typically found in multilayered printed circuit boards. This would require a more complicated Green's function, but we would only need to discretize (mesh) plane parallel surfaces. Such implementations have indeed been realized in several mainstream electromagnetic solvers for planar circuit simulation and are commonly called *2.5D planar MoM solvers.* They are widely used in both analog microwave and high-speed digital circuit design and will, therefore, be discussed in some detail.

3.8.1 2.5D Planar MoM Solvers

Planar and quasi-planar circuit topologies are either completely open, laterally open, or fully enclosed by conducting planes (see Figures 4.2 and 4.3). In the laterally open case, the homogeneous dielectric layers are assumed to extend to infinity; while in the laterally closed case (closed box), the metallic sidewalls act like perfect mirrors for the electromagnetic fields and currents, thus creating a periodic series of images in the lateral directions. Naturally, the Green's function is quite different in these two cases.

In a multilayered, laterally open environment, the preferred formulation of the Green's function is in terms of Sommerfeld integrals which have no analytical solution, and must, therefore, be computed numerically. Only in the near-field and the far-field regions of the scatterer can we use approximate analytical expressions for the Sommerfeld integrals. The method of moments is used to compute the current distribution on the metal surfaces as described in Section 3.8, using rectangular and/or triangular subsections with pulse or rooftop expansion functions. Since the Green's function is not available in analytical form, the inner products (matrix elements) must be computed numerically in the laterally open configurations.

Closed-box formulations take advantage of the lateral periodicity introduced by the conducting sidewalls. By subjecting all field and current functions to a discrete spatial Fourier transform in both transverse directions, one obtains a set of "box modes" that propagate in a direction normal to the dielectric and metallic layers. Each of these modes can be expressed analytically as a rectangular waveguide mode that resonates between the top and bottom wall of the box, obeying a transcendental resonance condition in a piecewise homogeneous waveguide section. The Fourier transforms of the current distributions appear as modal current sources in the planes of metallization. This approach is known as the *spectral domain technique*. There are two ways to combine the spectral domain formulation with the method of moments [3].

(1) The spectral components of the Green's function are computed in the spectral domain and recombined in the space domain by inverse spatial Fourier transform. The method of moments is then applied in the space domain as usual. In this approach, no spatial Fourier transform of the current distributions on the metal planes is needed.

(2) The integral equation is written and solved in the spectral domain. This means that both the fields and the current distributions are Fourier transformed, and

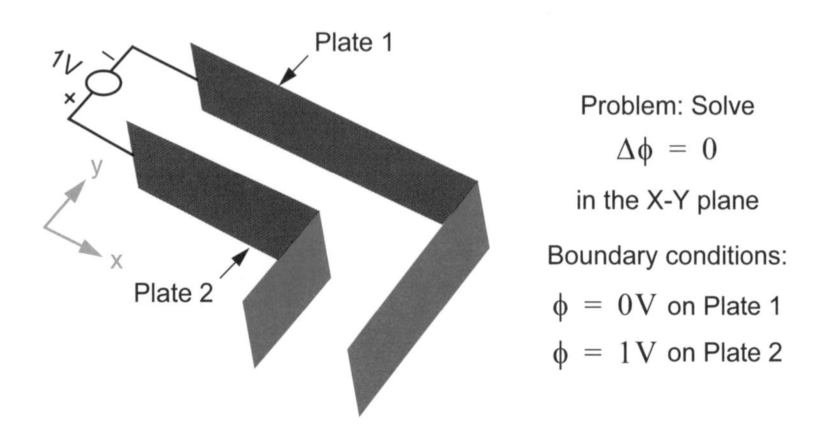

Figure 3.11 A simple electrostatic problem to be solved with finite elements.

the method of moments is applied in the spectral domain. The space domain current distribution is then recovered by inverse Fourier transform of its spectral terms. This approach is particularly effective when the conducting strips have relatively simple shapes so that the current on these can be expanded into simple functions that have analytic Fourier transforms.

In both cases, the Green's functions are analytical expressions in terms of trigonometric functions. The resolution accuracy in the transverse directions depends on the number of terms that are carried in the spatial Fourier transform. The higher the order of the term, the higher is its spatial wavelength and hence, its ability to resolve small geometric detail.

The manner in which the electromagnetic problem is formulated and implemented, has a direct bearing on the characteristics of the MoM simulators based on these techniques. These characteristics are discussed in more practical terms in Chapter 5.

3.9 THE FINITE ELEMENT METHOD

The finite element method originated as a method for modeling stress in structural mechanics applications, but was adapted to electromagnetics by Silvester and coworkers [4] in the 1970s. Today, FEM has become a mainstay numerical method and is well known in the electromagnetics and microwaves community as the "engine" that drives Ansoft HFSS and can solve electromagnetic fields in structures with arbitrary boundary shape (i.e., [5]).

The primary unknown quantity in finite element analysis is usually a field or a potential. To express it as a sum of known expansion functions with unknown coef-

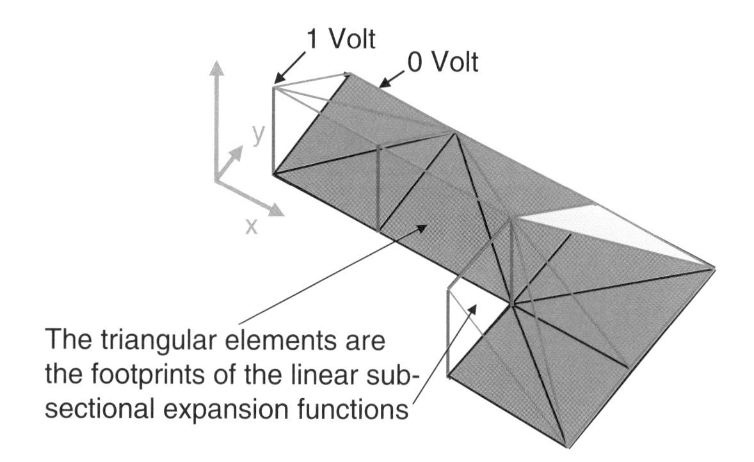

Figure 3.12 A very coarse approximation of the potential between the two plates. All nodes are situated on one of the boundaries, and no unknown coefficients exist.

ficients, we must discretize the *field domain* rather than the *boundary surfaces*. As the name of the method implies, the expansion functions are always subsectional. The finite subsections (finite elements) over which they are defined, are contiguous and of the simplest possible shape. In a 1D problem they are line elements; in 2D they are triangular (and sometimes rectangular) surface elements; and in 3D they are tetrahedral volume elements. This choice of simplest possible subdivisions provides maximum flexibility in the discretization of arbitrary geometries, which is a major strength of the method.

3.9.1 Linear Expansion Functions and Unknown Coefficients

The expansion functions are usually *linear* functions of the coordinates (see Figure 3.6), but can be *higher order* polynomials. For a demonstration of the linear expansion functions in FEM, consider the simple L-shaped parallel plate capacitor shown in Figure 3.11. The problem is two-dimensional in the X-Y plane, and the electrostatic field is essentially confined between the two plates; we can discretize the solution domain into triangles as shown in Figure 3.12. The finite elements are the orange triangular floor tiles in the X-Y plane, and the linear expansion functions are the yellow triangular panes that approximate the potential function between the two plates. Note the following:

(a) The corners of all triangles are either at $\phi = 0\text{V}$ or at $\phi = 1\text{V}$.

(b) The triangular expansion functions are all connected at the seams. Continuity from one element to the other is thus ensured.

There are no free corners or *nodes* in this expansion because they are all at a fixed potential determined by the boundary condition. This system has no degrees

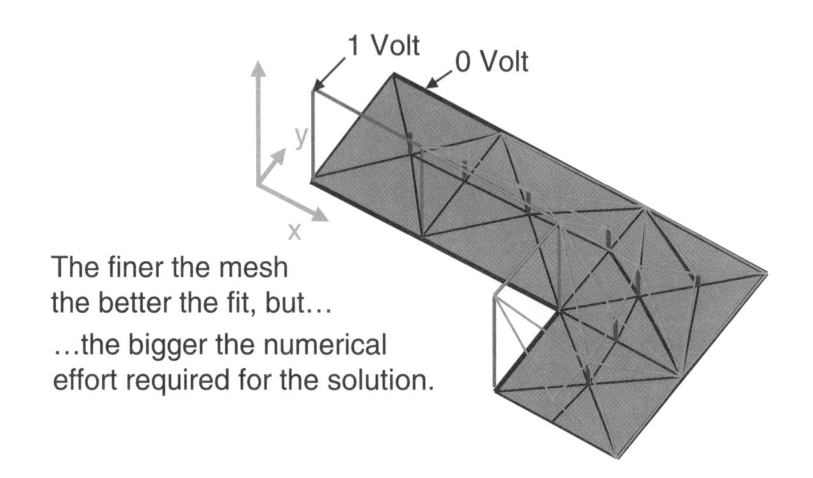

Figure 3.13 A finer mesh results in several free nodes halfway between the plates. The potential values at these nodes represent the unknown coefficients of the expansion.

of freedom. If we want to refine the approximation, we must use a finer discretization or meshing, such as in Figure 3.13. Some free nodes now exist halfway between the plates, and our task is to determine the potential values at these nodes such that the approximation is optimal. In fact, the free node potentials represent the unknown coefficients of the expansion. To derive the FEM formalism, consider a triangular element in the X-Y plane, with corners situated at (x_p, y_p), $p=1, 2, 3$. The potential ϕ^e within this subdomain element is assumed to be a linear function of x and y of the form:

$$\phi^e = a + bx + cy \tag{3.24}$$

We can express the constants a,b,c by the potentials and coordinates of the corners:

$$\begin{bmatrix} a \\ b \\ c \end{bmatrix} = \begin{bmatrix} 1 & x_1 & y_1 \\ 1 & x_2 & y_2 \\ 1 & x_3 & y_3 \end{bmatrix}^{-1} \begin{bmatrix} \phi_1^e \\ \phi_2^e \\ \phi_3^e \end{bmatrix} \tag{3.25}$$

Substituting these values into (3.23) yields

$$\phi^e = \sum_p N_p^e \phi_p^e \tag{3.26}$$

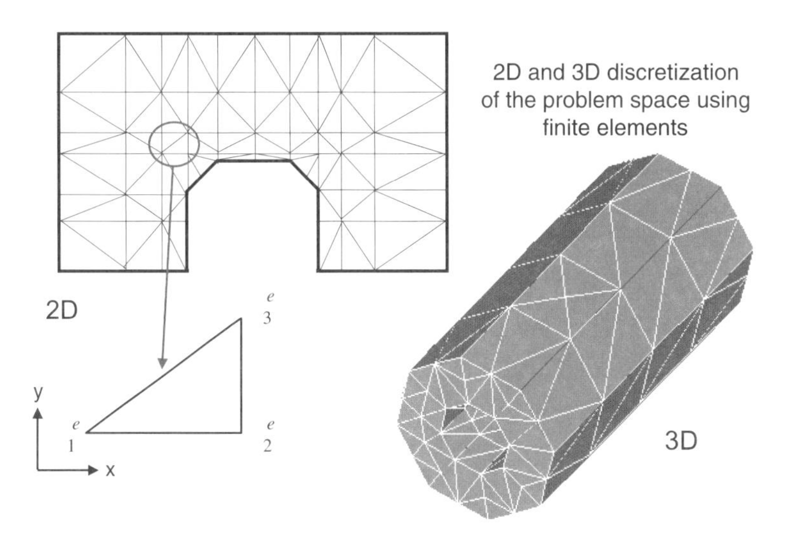

Figure 3.14 Discretization into triangular (2D) and tetrahedral (3D) finite elements affords maximum flexibility in the approximation of arbitrary shapes.

where N_p^e are the so-called shape functions. The potential function ϕ^e within a triangle is thus a linear combination of the shape functions and the three nodal potentials (expansion coefficients) of the triangle. In the three-dimensional case, the subdomains are tetrahedra. Figure 3.14 shows typical 2D and 3D meshes.

3.9.2 Strategy for Determining the Unknown Expansion Coefficients

The strategy for finding the unknown coefficients exploits the fact that electromagnetic energy is always minimized by the correct solution. In other words, we determine the unknown coefficients in such a way that the total energy contained in the approximation is minimized. Energy is computed by integrating the potentials or fields. Typical energy integrals are for the 2D electrostatic case

$$F_1 = \int_S \frac{\varepsilon|E|^2}{2} dv = \int_V \frac{\varepsilon|\nabla\phi|^2}{2} ds \tag{3.27}$$

or for 3D time-periodic lossy case

$$F_2 = \int_V \left\{ \frac{\mu|H|^2}{2} + \frac{\varepsilon|E|^2}{2} - \frac{JE}{2j\omega} \right\} dv \tag{3.28}$$

The total energy in the solution domain is simply the sum of all energy contents of the subdomains. In the language of the calculus of variations, the above energy integrals are called *functionals* that are made *stationary* by the correct potential or field function. The functionals have an associated *Euler-Lagrange* differential equation such that the solution of that differential equation makes the functional stationary. For the functional in (3.23), the associated Euler-Lagrange equation is the Laplace equation. In other words, finding the potential ϕ that minimizes the functional F_1 is equivalent to solving the Laplace equation.

The so-called *Rayleigh-Ritz* procedure allows us to find the stationary solution in the following manner. For example, if we substitute the expression (3.23) into F_1 for each element and add them all up, we obtain an approximate expression for the total energy in terms of the unknown potentials (coefficients) at the nodes.

$$F_1 \approx \sum_i \int_{S_i} \frac{\varepsilon_i |\nabla \phi_i^e|^2}{2} ds \tag{3.29}$$

Differentiating this expression with respect to each coefficient and setting this derivative equal to zero

$$\frac{dF_1}{d\phi_i^e} = 0 \tag{3.30}$$

yields i equations that allow us to compute the i unknown coefficients. This system of equations is in matrix form:

$$[C][\phi^e] = 0 \tag{3.31}$$

$[C]$ is a sparse, symmetric and banded matrix of size $I \times I$ (I = total number of nodes). Fine discretization of a large structure leads to very big matrices. Fortunately, special computational procedures are available for inverting such sparse matrices efficiently. Note that (3.23) closely resembles the integral equations that we solved with the method of moments, in which the expansion and basis functions are the same. In fact, the Rayleigh-Ritz procedure is equivalent to the Galerkin method of projection.

In early finite element solvers the discretization into triangles or tetrahedra was done by hand and constituted a tedious task. In a modern simulator the discretization is done automatically using tesselation algorithms. These algorithms have some built-in intelligence that allows them to predict the required mesh density for particular boundary topologies; for instance, at sharp edges or in tight subsections, more elements are generated than in relatively uniform field regions. In the vicinity of a singularity of the solution, the energy density becomes very high, and the value of the functional is very sensitive to the coefficients in that area. This requires a finer discretization for better accuracy, and also an appropriate functional that does not involve quantities that become singular at corners and edges.

In summary, the finite element method is one of the most flexible numerical modeling approaches since it can be applied to almost arbitrarily shaped boundaries. In contrast to the MoM applications discussed earlier, which require discretization of boundaries, FEM requires discretization of the field space or computational domain which must be entirely bounded to be finite. However, so-called *infinite elements* have been developed for open problems, and hybrid combinations of FEM with other methods have been developed which are more suitable for handling open boundary problems. While the vast majority of FEM applications are time-harmonic, time domain formulations exist as well, but they are still in a state of evolution.

3.10 FINITE DIFFERENCE AND FINITE INTEGRATION METHODS

The name of these methods indicates that they approximate the differential or integral operators by finite differences or finite integrals over small subdomains of the problem space. It is possible to derive these formulations by solving the differential or integral equations with the method of moments. For example, as Harrington [1] has pointed out, finite difference approximations of differential equations can be obtained by using pulse functions as expansion and testing functions. Krumpholz, Huber, and Russer [6] have demonstrated that the FDTD formulation by Yee [7] can be derived by using pulse functions and a Galerkin procedure for both the electric and the magnetic field components in space and time. While this underscores the common roots of the various numerical techniques in electromagnetics, finite difference and finite integration methods can be derived more directly from the differential or integral form of Maxwell's equations or the wave equation without applying the method of moments.

3.10.1 Finite Difference Formulations

The simplest geometrical interpretation of finite differences is shown in Figures 3.15 and 3.16 where the first and second derivatives of a function are represented as the slopes of the tangents to the function and to its derivative. It is immediately obvious that the central difference is more accurate than the forward and backward differences. By formulating the three approximations in terms of Taylor series (see, for example, Booton [8]), one can confirm analytically that the error of the central difference approximation is of second order (decreasing as the square of the interval h), while the two others are only first-order accurate (the error decreases linearly with the interval h). It is thus important to use central differencing whenever possible in finite difference formulations of electromagnetic problems.

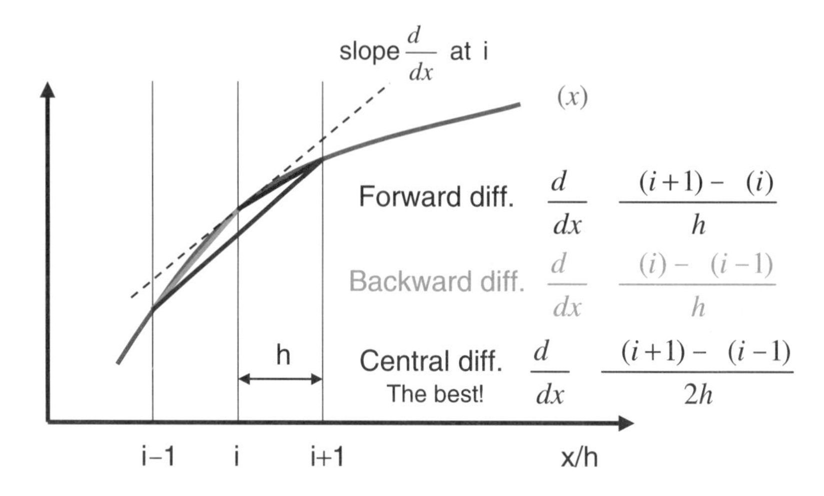

Figure 3.15 Geometrical interpretation of the approximation of the first derivative of a function by a forward, backward, and central finite difference. The superior accuracy of the central difference is obvious.

In static and time-harmonic formulations, the finite differences are defined in space by first establishing a grid of discrete points and then writing for each point the central difference approximation of the unknown solution. Consider again the example of the Laplace equation for the electrostatic potential in source-free 3D space

$$\nabla^2\phi = \frac{d^2\phi}{dx^2} + \frac{d^2\phi}{dy^2} + \frac{d^2\phi}{dz^2} = 0 \tag{3.32}$$

and use the stencil in Figure 3.17 to formulate the central difference operator in the three coordinate directions. The central difference operators can be written by inspection:

$$\frac{d\phi}{dx} \approx \frac{\phi(l+1, m, n) - \phi(l-1, m, n)}{2h} \tag{3.33}$$

Analogous expressions approximate the first derivatives with respect to y and z. The second derivatives are approximated by central difference operators as follows:

$$\frac{d^2\phi}{dx^2} \approx \frac{\phi(l+1, m, n) - 2\phi(l, m, n) + \phi(l-1, m, n)}{h^2} \tag{3.34}$$

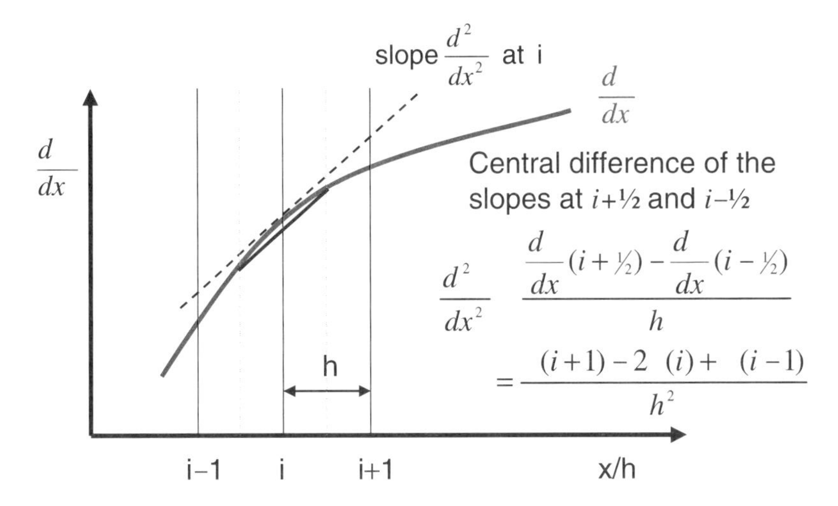

Figure 3.16 Geometrical interpretation of the approximation of the second derivative by a central finite difference of first derivatives.

Analogous expressions are found for the two other partial second derivatives. The Laplace equation thus becomes in finite difference form:

$$\begin{aligned} &\phi(l+1, m, n) + \phi(l-1, m, n) + \phi(l, m+1, n) + \phi(l, m-1, n) + \dots \\ &\qquad \phi(l, m, n+1) + \phi(l, m, n-1) \ = \ 6\phi(l, m, n) \end{aligned} \tag{3.35}$$

This formula can be interpreted as follows: the potential at any point in the computational domain is the arithmetic average of the potentials at the six closest neighboring points. This represents a discrete formulation of the general property of the electrostatic potential; namely, that its average value taken over a spherical surface equals the potential at its center.

The same finite difference approximation can be applied to solve eigenvalue problems involving, for example, the time-harmonic homogeneous wave equation of the form

$$\nabla^2 E_z + k^2 E_z \ = \ 0 \tag{3.36}$$

which describes the *TM-to-z* modes in a waveguide cavity. The longitudinal component E_z must satisfy all boundary conditions and has nontrivial solutions only for discrete values of the (yet unknown) eigenvalues $k = \omega(\mu\varepsilon)^{1/2}$. Hence, we must determine both the field values and the eigenvalues that simultaneously satisfy (3.36). Again, the differential Laplace operator is replaced by its corresponding

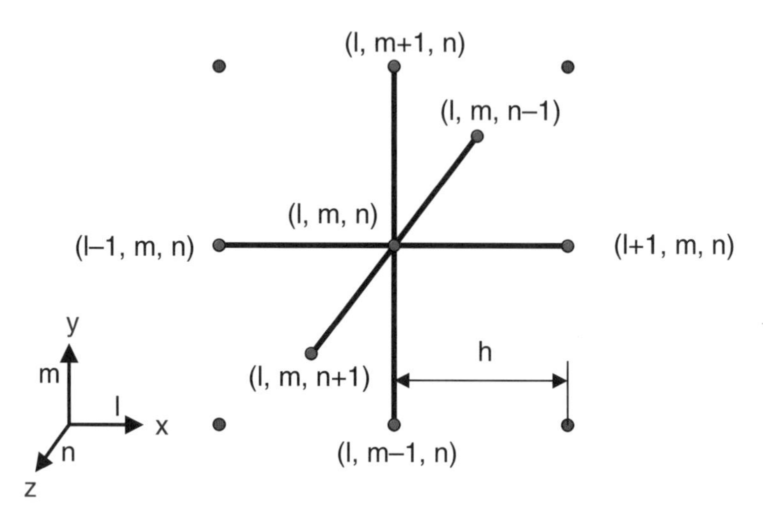

Figure 3.17 Three-dimensional stencil of the discrete finite difference operator.

central difference operator (as in (3.34) and (3.35), in which ϕ is to be replaced by E_z), resulting in a finite difference equation for the E_z-values at discrete mesh points. The manner in which the eigenvalues k are determined depends on the solution strategy that is discussed below.

3.10.2 Finite Integration Formulation

Returning to the electrostatic problem, we can derive the same update equation for ϕ by computing the net flux of the electrostatic electric field through the surface of the unit cube of side h, shown in Figure 3.18. We assume that over each face of the cube, the normal electric field is constant and equal to its value at the face center. (Note that this is equivalent to approximating the components of E by 2D pulse expansion function.) The electric field can be expressed in terms of the values of ϕ. For example, the x-components of the electric field that are normal to the faces at $(l+1/2)$ and $(l-1/2)$ are, respectively:

$$E_x(l+\tfrac{1}{2}) = \frac{\phi(l,m,n)-\phi(l+1,m,n)}{h} \tag{3.37}$$

$$E_x(l-\tfrac{1}{2}) = \frac{\phi(l-1,m,n)-\phi(l,m,n)}{h} \tag{3.38}$$

Hence, the net flux of E_x through the cube becomes

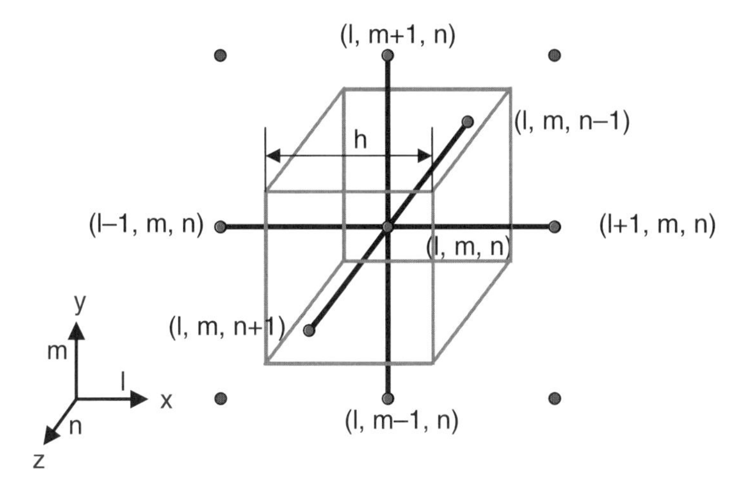

Figure 3.18 The electric field normal to the faces of a unit cube centered at *(l, m, n)* is expressed in terms of the potential values ϕ. The net flux of E through the cube surface is then computed according to Gauss' law by finite integration over the cube.

$$h^2[E_x(l + \tfrac{1}{2}) - E_x(l - \tfrac{1}{2})] = h[2\phi(l, m, n) - \phi(l + 1, m, n) - \phi(l - 1, m, n)] \quad (3.39)$$

Analogous expressions yield the net flux of E_y and E_z through the cube. The total flux must be zero in the source-free case governed by Laplace's equation.

The integral representing the total flux of the discretized electric field through the six faces of the finite-size cube thus yields the following difference equation:

$$\begin{aligned} 6\phi(l, m, n) - \phi(l + 1, m, n) - \phi(l - 1, m, n) - \phi(l, m + 1, n) \\ -\phi(l, m - 1, n) - \phi(l, m, n + 1) - \phi(l, m, n - 1) = 0 \end{aligned} \quad (3.40)$$

which is identical to (3.35). Note that the finite integration formulation can be quite flexible since the Gaussian integration surface must not necessarily be a cube but can have any shape, even though this will result in a more complex difference equation.

3.10.3 Solution Strategies

Whether we use a finite difference or a finite integration approximation, we always obtain a system of finite difference equations, one for each mesh point in the computational space, which we can solve for the unknown values of ϕ by matrix inversion. Since each equation involves only the ϕ-values at the closest neighboring

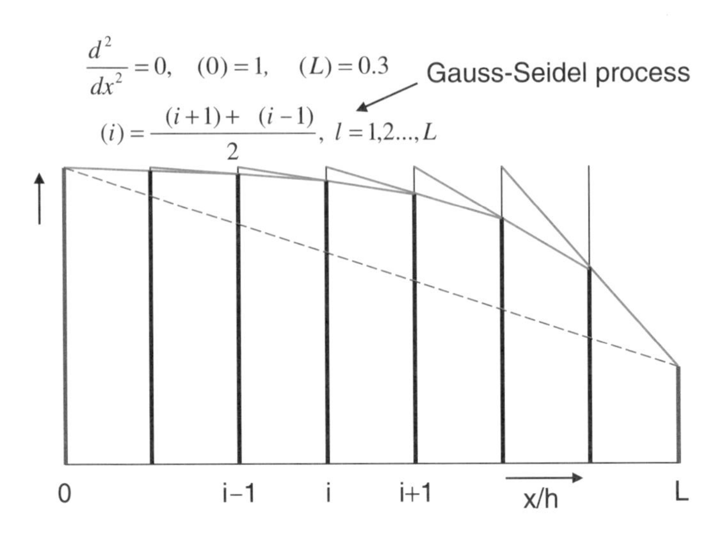

Figure 3.19 Iterative solution of the one-dimensional Laplace equation by a Gauss-Seidel process. Several iterations will be necessary for the solution to converge (dashed line).

mesh points, the coefficient matrix contains a large number of zero elements (banded sparse matrix), and only the diagonal and nearby elements are filled. Alternatively, we can consider (3.35) and (3.40) as update equations to be sequentially executed for each point at the discrete coordinates (l, m, n). The potential of the points on the boundaries is fixed by the boundary conditions. Furthermore, the starting values at the inner mesh points must be selected first by educated guess, and then replaced iteratively by updated values until all ϕ converge within a predetermined tolerance. This so-called *Gauss-Seidel* iterative process is illustrated for the one-dimensional case in Figure 3.19 where all values of ϕ have initially been set to unity, and then iteratively adjusted from the right to the left to be the average of the neighboring values. Several iterations will be necessary to converge to the solution, which, in this case, is a linear function (dashed red line).

The process is also known as *relaxation* by analogy with successively relaxing the tension in an elastic band stretched over the initial values. The convergence can be accelerated by *overrelaxation* (i.e., overcorrecting by a certain percentage the value predicted by the update formula). However, one must be careful not to overrelax too much because this makes the process unstable. For solutions of the Laplace equation, overrelaxation by 150% converges quickly, while 200% causes instability.

The same solution strategies, namely matrix inversion and relaxation, can be applied to solve eigenvalue problems such as (3.36). If the matrix inversion strategy is used, the eigenvalues are first determined such that the determinant of the system matrix is zero. (The eigenvalues are the roots of the determinant.) If the relaxation strategy is used, a separate iterative procedure for approximating the eigenvalues must be employed. To this end, both sides of (3.36) are multiplied by E and integrated over the entire cavity volume, resulting in a variational expression for k^2 that is stationary with respect to E:

$$k^2 = -\frac{\iiint E\nabla^2 E dxdydz}{\iiint E^2 dxdydz} \tag{3.41}$$

Use of the finite difference approximation for E transforms this expression into a ratio of double summations that is evaluated after each update of the electric field values in the cavity (see, for example, Booton [8]). The choice of the initial values for E determines the eigenvalue and eigenmode towards which the solution converges. Choosing E at all interior points to have the same starting value usually causes the solution to converge to the lowest eigenmode.

3.11 FINITE DIFFERENCE TIME DOMAIN FORMULATIONS

We have mentioned earlier that the time dimension can be treated mathematically in the same way as the space dimensions. Thus, in order to solve transient electromagnetic phenomena in the time domain, we could discretize the homogeneous time domain wave equation in Cartesian coordinates,

$$\nabla^2 E_z - \frac{1}{c^2}\frac{\partial^2 E_z}{\partial t^2} = 0 \tag{3.42}$$

by approximating both the Laplacian operator and the second time derivative by central differences. The latter term becomes

$$\frac{d^2 E_z}{dt^2} \approx \frac{{}_{k+1}E_z(l,m,n) - 2{}_kE_z(l,m,n) + {}_{k-1}E_z(l-1,m,n)}{(\Delta t)^2} \tag{3.43}$$

where the prescript k is the discrete time index, and Δt the time step $(t = k\Delta t)$. We can then use the same solution strategies that we described for static and time-harmonic problems, provided that the values of E_z are specified for the first two time steps. Alternatively, the starting field values and their first time derivatives may be specified as initial conditions.

However, most time domain simulators based on finite differences do not solve the discretized time domain wave equation for E or H, but employ the method proposed by Yee [7] in 1966 and subsequently developed further by Taflove and Brod-

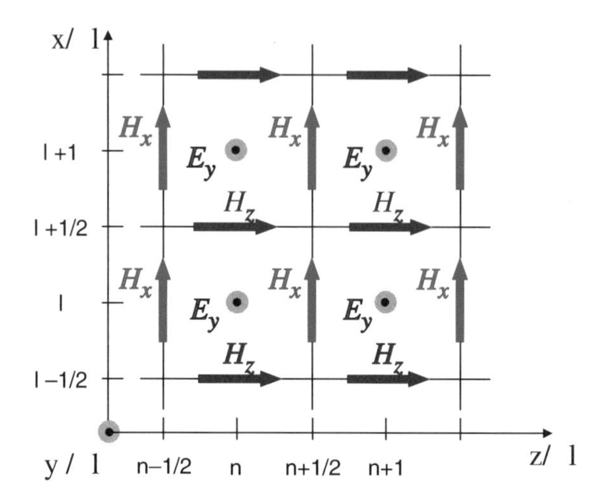

Figure 3.20 Two-dimensional FDTD grid (Yee cells) for the *TM-to-y* case. The sampling positions for the electric and magnetic field components are staggered in space and time.

win [9]. Yee simply replaced the partial derivatives in Maxwell's curl equations by central finite differences. Weiland [10] derived an equivalent discretization approach using finite integration of Maxwell's equations in 1977.

The FDTD approach is shown for the two-dimensional *TM-to-y* case in Figure 3.20. In Cartesian coordinates the curl equations reduce in this case to the following three scalar differential equations:

$$\frac{dE_y}{dx} = -\mu\frac{dH_z}{dt} \tag{3.44}$$

$$\frac{dE_y}{dz} = \mu\frac{dH_x}{dt} \tag{3.45}$$

$$\frac{dH_x}{dz} - \frac{dH_z}{dx} = \varepsilon\frac{dE_y}{dt} \tag{3.46}$$

To obtain central difference approximations of these expressions, the discrete samples of the electric and magnetic field components are staggered in both space and time. This means that the instances and positions of the electric field samples are defined half way between those of the magnetic field samples. If the electric field component E_y is sampled at the discrete time points k and the discrete positions (l, n), then the magnetic field components H_x and H_z are sampled at time points $k + 1/2$ and at positions $(l, n + 1/2)$ and $(l + 1/2, n)$, respectively. Hence,

$$\frac{{}_kE_y(l+1,n) - {}_kE_y(l,n)}{\Delta x} = -\mu \frac{{}_{k+1/2}H_z(l+1/2,n) - {}_{k-1/2}H_z(l+1/2,n)}{\Delta t} \tag{3.47}$$

$$\frac{{}_kE_y(l,n+1) - {}_kE_y(l,n)}{\Delta z} = \mu \frac{{}_{k+1/2}H_x(l,n+1/2) - {}_{k-1/2}H_z(l,n+1/2)}{\Delta t} \tag{3.48}$$

$$\begin{aligned} &\frac{{}_{k+1/2}H_x(l,n+1/2) - {}_{k+1/2}H_x(l,n-1/2)}{\Delta z} + \ldots \\ &\ldots + \frac{{}_{k+1/2}H_z(l+1/2,n) - {}_{k+1/2}H_z(l-1/2,n)}{\Delta x} \\ &= \varepsilon \frac{{}_{k+1}E_y(l,n) - {}_kE_y(l,n)}{\Delta t} \end{aligned} \tag{3.49}$$

These expressions are also referred to as *field update equations* since they allow us to explicitly compute future values of the H-field components from their previous values and the present spatial variations of the E-field, and vice versa. Appropriate initial and boundary conditions must be defined before the update process begins. The fully three-dimensional version of the FDTD approximation of Maxwell's curl equations involves six update equations, one for each field component.

Figure 3.21 shows a unit FDTD cell (Yee cell) of a Cartesian space grid. Continuous space and time coordinates (x, y, z, t) are replaced by discrete coordinates $l\Delta x$, $m\Delta y$, $n\Delta z$, $k\Delta t$, where l, m, n, k are integers and Δx, Δy, Δz, and Δt are the space and time steps. Note that the three electric field components are defined along the cell edges, while the magnetic field components are normal to the cell faces. The staggering of the field components by one-half of the cell dimensions allows for the central difference approximation of the differential operators. For the same reason, electric and magnetic field components are also staggered in time, the electric field components being defined at time points $k\Delta t$, and the magnetic field components at $(k+1/2)\Delta t$. While a cubic Yee cell yields the simplest FDTD algorithm, the three cell dimensions can, in general, be different. If we assume that $\Delta x = p\Delta l$, $\Delta y = q\Delta l$, $\Delta z = r\Delta l$, where Δl is the unit reference length, and the scaling coefficients p, q, and r are all smaller or equal to unity, then the finite difference update equations for the electric and magnetic field components in each cell are given by

$$\begin{aligned} {}_{k+1}E_x(l+½,m,n) &= {}_kE_x(l+½,m,n) \\ +sx\{[{}_{k+½}H_z(l+½,m+½,n) &- {}_{k+½}H_z(l+½,m-½,n)]/q \\ +[{}_{k+½}H_y(l+½,m,n-½) &- {}_{k+½}H_y(l+½,m,n+½)]/r\} \end{aligned} \tag{3.50}$$

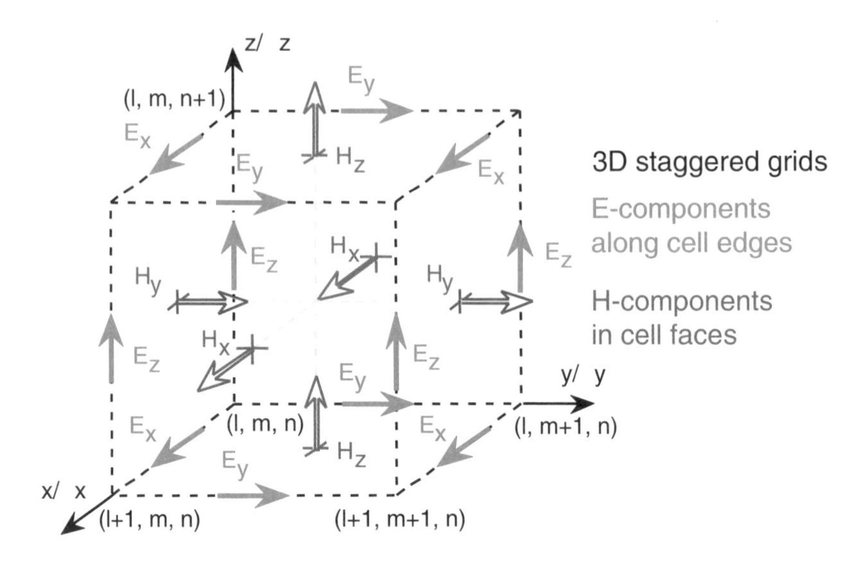

Figure 3.21 Three-dimensional Yee cell showing the staggered positions of the field component samples.

$$\begin{aligned} {}_{k+1}E_y(l, m+½, n) &= {}_kE_y(l, m+½, n) \\ &+sy\{[{}_{k+½}H_x(l, m+½, n+½) - {}_{k+½}H_x(l, m+½, n-½)]/r \\ &+[{}_{k+½}H_z(l-½, m+½, n) - {}_{k+½}H_z(l+½, m+½, n)]/p\} \end{aligned} \tag{3.51}$$

$$\begin{aligned} {}_{k+1}E_z(l, m, n+½) &= {}_kE_z(l, m, n+½) \\ &+sz\{[{}_{k+½}H_x(l, m-½, n+½) - {}_{k+½}H_x(l, m+½, n+½)]/q \\ &+[{}_{k+½}H_y(l, m, n+½) - {}_{k+½}H_y(l-½, m, n+½)]/p\} \end{aligned} \tag{3.52}$$

$$\begin{aligned} {}_{k+½}H_x(l, m+½, n+½) &= {}_{k-½}H_x(l, m+½, n+½) \\ &+sx'\{[{}_kE_y(l, m+½, n+1) - {}_kE_y(l, m+½, n)]/p \\ &+[{}_kE_z(l, m, n+½) - {}_kE_z(l, m+1, n+½)]/q\} \end{aligned} \tag{3.53}$$

$$\begin{aligned} {}_{k+½}H_y(l+½, m, n+½) &= {}_{k-½}H_y(l+½, m, n+½) \\ &+sy'\{[{}_kE_x(l+½, m, n) - {}_kE_x(l+½, m, n+1)]/r \\ &+[{}_kE_z(l+1, m, n+½) - {}_kE_z(l, m, n+½)]/p\} \end{aligned} \tag{3.54}$$

$$
\begin{aligned}
{}_{k+\frac{1}{2}}H_z(l+\tfrac{1}{2}, m+\tfrac{1}{2}, n) &= {}_{k-\frac{1}{2}}H_z(l+\tfrac{1}{2}, m+\tfrac{1}{2}, n) \\
&+sz'\{[{}_kE_x(l+\tfrac{1}{2}, m+1, n) - {}_kE_x(l+\tfrac{1}{2}, m, n)]/q \\
&+[{}_kE_y(l, m+\tfrac{1}{2}, n) - {}_kE_y(l+1, m+\tfrac{1}{2}, n)]/r\}
\end{aligned}
\tag{3.55}
$$

where

$$
\begin{aligned}
sx &= Z_0 c\Delta t/(\varepsilon_{rx}\Delta l) & sx' &= c\Delta t/(\mu_{rx}Z_0\Delta l) \\
sy &= Z_0 c\Delta t/(\varepsilon_{ry}\Delta l) & sy' &= c\Delta t/(\mu_{ry}Z_0\Delta l) \\
sz &= Z_0 c\Delta t/(\varepsilon_{rz}\Delta l) & sz' &= c\Delta t/(\mu_{rz}Z_0\Delta l)
\end{aligned}
\tag{3.56}
$$

In these expressions, c and Z_0 are the velocity of light and the wave impedance in vacuo, and ε_{rx}, ε_{ry}, ε_{rz} and μ_{rx}, μ_{ry}, μ_{rz} are the diagonal elements of the relative permittivity and permeability tensors of the medium, respectively. This algorithm explicitly updates each field component in a leapfrog time-stepping process. The future value of each E-field component is computed from its previous value and from the four H-field components circulating around it, and vice versa. The permittivity and permeability can be different in each cell, thus allowing the representation of inhomogeneous media. Losses can be included as well by carrying and discretizing the loss terms in Maxwell's curl equations. Details can be found in the FDTD literature (see, for example, [11]).

3.11.1 Stability

The explicit field updating process is stable as long as the time step is smaller than a maximum value known as the so-called *Courant stability limit*. For electrically and magnetically isotropic media characterized by ε_r and μ_r, the stability criterion is

$$
\Delta t \le \frac{\Delta l\sqrt{\mu_r\varepsilon_r}}{c\sqrt{\dfrac{1}{p^2}+\dfrac{1}{q^2}+\dfrac{1}{r^2}}}
\tag{3.57}
$$

Since in anisotropic media the wave velocity depends on the (generally unknown) polarization, it is prudent to enter the smallest of the three μ- and ε-values of the diagonal tensors into the stability condition. For free space discretized into cubic cells $(\mu_r = \varepsilon_r = p = q = r = 1)$ it becomes

$$
\Delta t \le \Delta l/(c\sqrt{3})
\tag{3.58}
$$

For 2D FDTD with square cells, the stability limit is $\Delta t \leq \Delta l/(c\sqrt{2})$, and for the 1D case it is $\Delta t \leq \Delta l/c$. The physical interpretation of the stability criterion is that the time step cannot be larger than the time required for the field to travel across the largest diagonal dimension of a cell. This has important consequences for the meshing of the computational domain. First, a reduction in the cell size also requires a corresponding reduction in the time step, so that more updates are necessary to cover the same absolute time interval. Second, if a graded mesh with varying cell size is used, the smallest cell in the mesh dictates the time step. This explains why mesh grading in FDTD considerably increases the computational burden by imposing the smallest stable time step on the entire mesh. This is also true for other explicit space-time-discrete methods such as the transmission line matrix method.

3.11.2 Initial and Boundary Conditions

At the start of a computation, the initial values of all electric or/and magnetic field components in the computational domain must be specified before the updating process can begin. By imposing or adding field values in certain regions at each time step, source functions with arbitrary time and space dependence can be modeled.

Boundary conditions must be enforced at each time step as well. Electric and magnetic walls can be modeled by imposing appropriate condition on either the electric or the magnetic field only. For example, an electric wall (perfectly conducting boundary) can be imposed by forcing the tangential electric field to be zero at its location at all times (Dirichlet type). Alternatively, we can force the tangential magnetic field to be identical on either side of the boundary location, or the tangential electric field to be equal and opposite in sign on either side (Neumann type). Similarly, the tangential electric field must be identical on either side of a magnetic wall (ideal open circuit), or the tangential magnetic field must be either zero on the boundary, or equal and opposite on either side of the boundary. Lossy resistive boundary conditions call for a fixed ratio between the tangential electric and magnetic field components at the boundary. Since electric and magnetic fields are staggered in space and time, the imposition of general impedance conditions requires additional space- and time-averaging operations at the boundary. More complex boundary conditions such as wideband absorbing walls or frequency-dispersive boundaries call for special algorithms such as one-way absorbing boundary conditions [12, 13] or Berenger's perfectly matched layer [14]. Similar approaches are required for the modeling of complex materials and devices. Such boundary conditions are available in most commercial FDTD simulators.

3.11.3 Output from FDTD Simulators

FDTD simulators generate massive amounts of data that must be processed further to yield meaningful engineering information. Since these output data are similar to those generated by TLM simulators, they will be discussed jointly in Section 3.13.

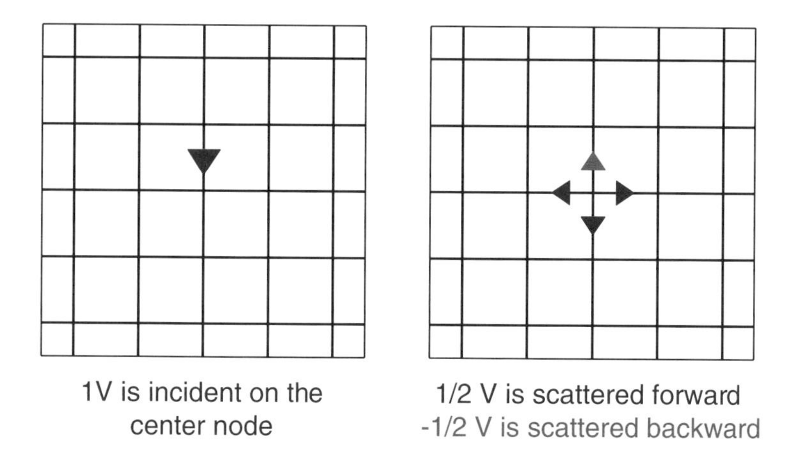

Figure 3.22 Scattering of a voltage impulse at a shunt node in a 2D TLM transmission line network.

3.12 TRANSMISSION LINE MATRIX METHODS

The TLM method was first proposed in 1971 by Johns and Beurle [15] who described a novel numerical technique for solving two-dimensional scattering problems. Inspired by earlier network simulation techniques (i.e., [16]), Johns and Beurle modeled the two-dimensional propagation space by a Cartesian mesh (or matrix) of shunt-connected TEM transmission lines, with the nodes of this mesh acting as scattering centers for short voltage impulses that are propagating in it. We will first discuss the basic features of TLM using the original 2D shunt implementation and then proceed to the 3D TLM schemes.

3.12.1 TLM Basics and the Two-Dimensional TLM Shunt Mesh

TLM is essentially a discretized network representation of Maxwell's equations as opposed to FDTD, which is a discretized mathematical field equation model. Furthermore, TLM employs a scattering formulation involving incident and reflected wave impulses in the TLM mesh, while FDTD models the differential relationships between the total electric and magnetic field quantities.

Figure 3.22 shows the top view of a small subsection of a 2D mesh of shunt-connected transmission lines of characteristic impedance Z_l. A narrow voltage impulse of 1V is incident on the center node. This impulse is scattered at the node into a reflected voltage impulse of $-1/2$ V and three transmitted voltage impulses of $+1/2$ V, satisfying the requirements of continuity of voltage and conservation of energy. The impulse voltage represents the electric field, and the impulse current on the transmission line $(i = v/Z_l)$ represents the magnetic field. The scattered impulses then travel to the neighboring nodes where they become incident impulses

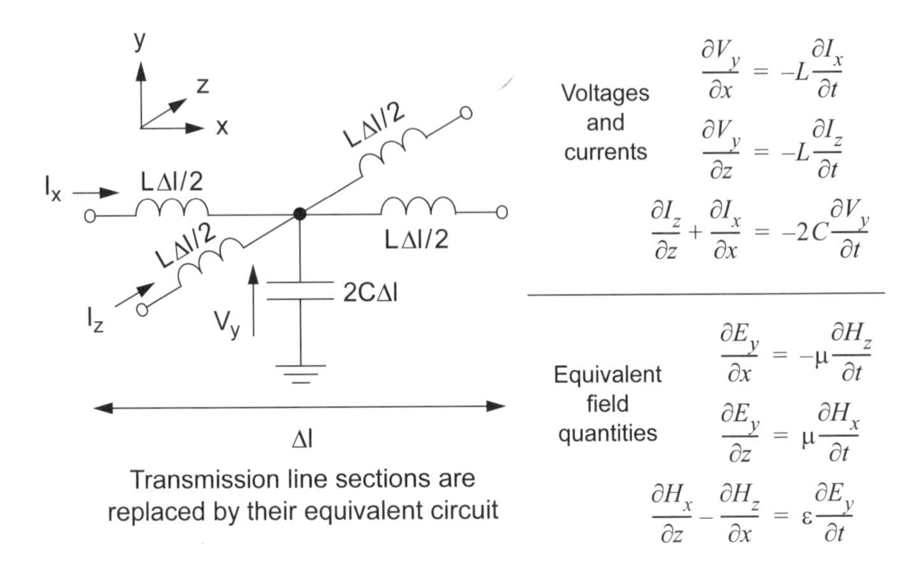

Figure 3.23 The equivalent lumped element network of a 2D TLM cell for cell dimensions much smaller than the wavelength. The differential equations governing the voltages and currents in that network are isomorphic with the 2D Maxwell equations for the *TM-to-y* case.

at the next time step and are scattered again according to the same rule. Boundaries reflect the impulses back into the TLM mesh with an appropriate reflection coefficient. The TLM algorithm thus consists of two alternating steps:

(1) The scattering of the voltage impulses incident upon a node on its connected transmission lines (link lines);

(2) The transmission of the scattered impulses to the neighboring nodes where they become incident impulses at the next time step, and the reflection of voltage impulses by the boundaries.

This series of events can be described in symbolic form as follows:

$$\left[{}_{k}\boldsymbol{v}^{r}\right] = [\boldsymbol{S}] \cdot \left[{}_{k}\boldsymbol{v}^{i}\right] \qquad \left[{}_{k+1}\boldsymbol{v}^{i}\right] = [\boldsymbol{C}] \cdot \left[{}_{k}\boldsymbol{v}^{r}\right] \tag{3.59}$$

where $[{}_{k}\boldsymbol{v}^{r}]$ and $[{}_{k}\boldsymbol{v}^{i}]$ are the vectors of reflected and incident impulses at the *k th* time step, $[\boldsymbol{S}]$ is the impulse scattering matrix of the node, and $[\boldsymbol{C}]$ is a connection matrix describing the topology of the TLM mesh. It governs the transfer of the reflected pulses to the connected ports of the neighboring cells and/or the reflection from boundaries. The subscripts k and $k+1$ denote the discrete time points at which the pulses are scattered at the nodes.

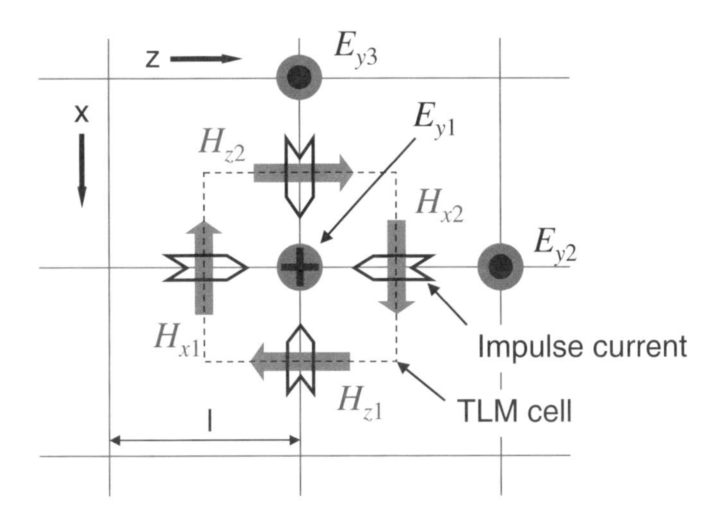

Figure 3.24 Relationship between 2D TLM and 2D FDTD grids. The E_y field samples of the FDTD scheme are defined at the TLM nodes at the instant of scattering, while the H_x and H_z samples are defined half-way between TLM nodes and scattering events; they are associated with the impulse currents flowing in the TLM lines.

One way to demonstrate that the TLM mesh effectively models Maxwell's equations is to derive the differential equations that govern the voltages and currents in the TLM cells in which the transmission line sections have been replaced by equivalent lumped inductances and capacitances. For infinitesimal mesh size Δl, these equations have exactly the same mathematical form as Maxwell's equations for the 2D *TM-to-y* case, as shown in Figure 3.23 (isomorphism) and are thus suitable for modeling electromagnetic field propagation.

Like the FDTD algorithm, the TLM scattering and connecting equations can also be derived directly from Maxwell's equations via the method of moments as Krumpholz, *et al.,* [6] have shown. This confirms our initial statement that all numerical solution methods in computational electromagnetics approximate the unknown field by an expansion into suitable basis functions and then use a particular strategy for determining the unknown expansion coefficients.

The relationship between the 2D FDTD and 2D shunt node TLM grids is shown in Figure 3.24. The TLM mesh and the FDTD mesh have identical cell size. The electric field component E_y is sampled at the TLM nodes, and the magnetic field components H_x and H_z are sampled at the cell boundaries. The H-field components are associated with the impulse currents traveling in the TLM link lines. If the time step in both schemes is set to $\Delta t = \Delta l/(c\sqrt{2})$, the current impulses cross the cell boundaries at exactly half-time between the updates of E_y (when the magnetic

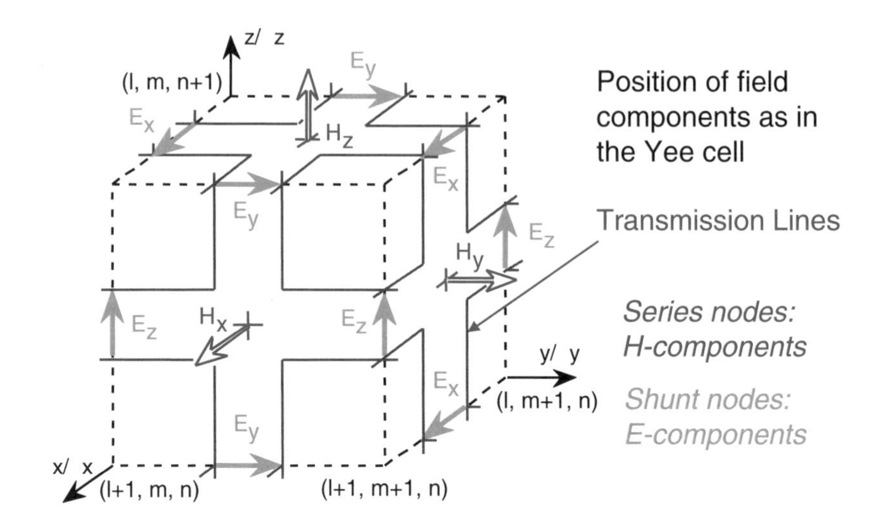

Figure 3.25 Three-dimensional expanded node TLM cell consisting of alternate 2D shunt and series nodes. The staggered positions of the field component samples are identical to those of the Yee cell (Figure 3.21).

fields are updated), and both schemes are equivalent, yielding identical results. Note that the velocity of impulses on the TLM link lines is $c\sqrt{2}$ in 2D TLM meshes, and $2c$ in 3D TLM meshes. By virtue of duality, we can also use a series-connected TLM mesh to model the TE case. The properties of the series-connected TLM node are described in [17].

3.12.2 The Three-Dimensional Expanded TLM Mesh

The expanded node TLM network, presented by Akhtarzad and Johns [18] in 1974, is an intricate 3D lattice of shunt- and series-connected transmission lines (see Figure 3.25). The unit cell has the same topology as the Yee cell shown in Figure 3.21, and it yields identical solutions for the six field components when the time step in the Yee algorithm is set to $\Delta t = \Delta l/(2c)$ (free space, cubic cell). However, in contrast to the strictly mathematical formulation of FDTD, the TLM model is a "hardwired" network (albeit conceptual rather than material) to which all known techniques of circuit and transmission line analysis can be applied, both in frequency and time domains. However, like the FDTD method, TLM is predominantly used in the time domain.

It is a disadvantage of both Yee's FDTD algorithm and of the expanded node TLM formulation that the electric and magnetic field components are staggered in space and time. This makes the modeling of complex boundary conditions and interfaces more difficult and can cause errors.

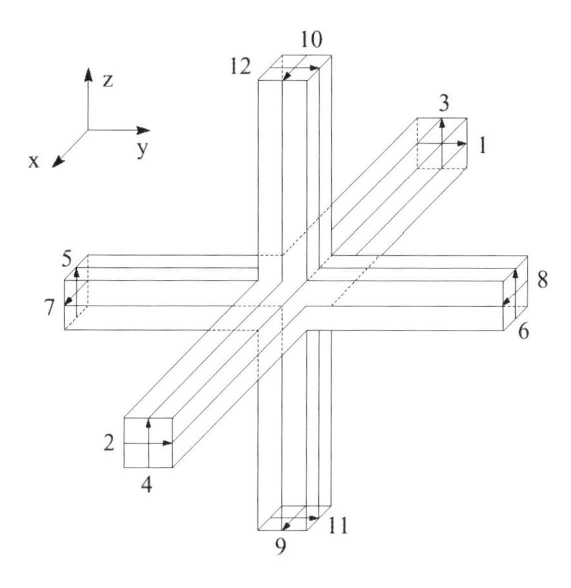

Figure 3.26 Three-dimensional symmetrical condensed TLM node proposed by P. B. Johns in 1986. This node and its variants are used in modern TLM time domain simulators.

3.12.3 The Symmetrical Condensed Node TLM Mesh

The *expanded node* has been superseded by the *symmetrical condensed node* [19] in 1986, and several new TLM formulations, from the *hybrid* and *supercondensed* nodes [20] to the *alternating* [21] and *alternating rotated* [22] TLM models have subsequently been developed. These condensed nodes have the advantage that all six field components are available simultaneously at the center of the TLM cell as well as at the cell boundaries, providing maximum flexibility for embedding devices and complex boundaries in the TLM field model. The unit cell of the symmetrical condensed TLM scheme is shown in Figure 3.26. It contains a hybrid junction of 12 transmission lines (the node) which is characterized by a 12×12 scattering matrix. For a homogeneous, lossless and isotropic medium, all transmission lines of a cubic cell have the same characteristic impedance. The 12×12 voltage impulse scattering matrix $\boldsymbol{S}$ is then

$$\boldsymbol{S} = \begin{bmatrix} 0 & \boldsymbol{S}_0 & \boldsymbol{S}_0^T \\ \boldsymbol{S}_0^T & 0 & \boldsymbol{S}_0 \\ \boldsymbol{S}_0 & \boldsymbol{S}_0^T & 0 \end{bmatrix} \tag{3.60}$$

where the submatrix $\boldsymbol{S}_0$ is given by

$$\boldsymbol{S}_0 = \begin{bmatrix} 0 & 0 & 0.5 & -0.5 \\ 0 & 0 & -0.5 & 0.5 \\ 0.5 & 0.5 & 0 & 0 \\ 0.5 & 0.5 & 0 & 0 \end{bmatrix} \tag{3.61}$$

and $\boldsymbol{S}_0{}^T$ is the transpose of $\boldsymbol{S}_0$. The numbering of the rows and columns of the matrices correspond to the port numbering scheme in Figure 3.26. Since the transit time Δt of the pulses is linked to the space step Δl by the pulse propagation velocity along the transmission lines, the TLM process is unconditionally stable.

3.12.4 Inhomogeneous Materials and Losses

Dielectric or magnetic materials can be modeled by loading the nodes situated inside these materials with reactive shunt stubs of appropriate normalized characteristic admittance and a length $\Delta l/2$ [17]. An open-circuited shunt stub will produce the effect of additional capacitance at the node, while a short-circuited series stub creates additional inductance. The resulting storage of reactive energy reduces the phase velocity and alters the intrinsic impedance in the structure. The interface conditions at the boundary between different materials are automatically fulfilled. Each cell can have a different set of stubs (three permittivity and three permeability stubs), thus allowing the modeling of inhomogeneous anisotropic materials with diagonal permittivity and permeability tensors. The six stubs add six more ports to the node, and as a result, $\boldsymbol{S}$ becomes an 18×18 matrix. Losses can be modeled by connecting so-called loss stubs to the nodes. The loss stubs are matched transmission line sections that extract a fraction of the energy scattered at the node at each time step. Since no pulses travel back into the nodes on these stubs, they only modify the elements of S without increasing its size. More sophisticated approaches are required for the modeling and embedding of dispersive and nonlinear materials [24] and devices [25].

In the hybrid and supercondensed nodes, stubs are substituted by equivalent modifications of the link line properties which modify the node impulse scattering matrix [20]. The electrical dimensions and constitutive parameters of each individual unit cell can also be modified by these measures. These modeling features are transparent to the user of a TLM simulator, but they nevertheless have an impact on the dispersion error.

3.12.5 Initial and Boundary Conditions

At the start of a computation the magnitude of all pulses incident on all link lines must be initialized. The field components are uniquely determined in the center of the nodes by a linear combination of these pulses at the moment of scattering [17]. When the pulses transit from one cell to the next ($t/\Delta t = k + 1/2$) the tangential components of the fields are obtained in the cell boundaries as well. By imposing the pulse values (and hence the corresponding electric and magnetic field values) in certain regions at each time step, or by adding a predetermined amount to the existing impulses in a subregion, hard and soft sources with arbitrary time and space dependence can be modeled.

Boundary conditions can be imposed either in the center of the nodes or in the cell boundaries. In the latter case, boundaries are represented by means of impulse reflection coefficients. Electric walls reflect impulses with a reflection coefficient of -1, while magnetic walls have a reflection coefficient of $+1$. Lossy resistive boundaries have impulse reflection coefficients less than unity in magnitude. More complex boundary conditions such as wideband absorbing walls or frequency dispersive boundaries are treated in the same way as FDTD boundaries with the difference that the boundary operators are applied to the incident pulses rather than to the field quantities at the boundaries. It is straightforward to implement nonrecursive and recursive convolution techniques for the modeling of frequency dispersive boundaries and for partitioning large computational domains using time domain diakoptics [23].

3.12.6 Stability

Since TLM is a "hardwired" passive network model of Maxwell's equations, the TLM algorithms always operate under stable conditions. In fact, due to the well-defined propagation velocity of the impulses on the link lines, the time step is always shorter or equal to the value defined in (3.57).

3.13 OUTPUT FROM ELECTROMAGNETIC SIMULATORS

The primary result of an electromagnetic simulation is always a field, potential or current distribution in space (and time, in the case of a time domain solver). More specifically, we obtain numerical values for the coefficients of the known expansion functions, which allow us to construct the approximate field or current distribution.

If the simulator uses full-domain expansion functions, these must be added together with the appropriate weight (coefficient) to yield the total field, potential or current distribution (see Figure 3.5). The reconstruction of the solution is much easier when the simulator uses subsectional expansion functions (rooftop functions in the MoM, piecewise linear functions in FEM, or pulse functions in FE, FDTD, or TLM). In this case, the total solution is simply sampled by the coefficients at the center of the subsections (see Figure 3.6). It is then straightforward to reconstruct the solution between the samples by linear or higher-order interpolation. This field or current distribution can be visualized directly in false color, wiremesh, or vector display form. Chapter 5 contains many examples of current distributions computed with MoM solvers, while Chapter 6 features some typical displays generated with FEM and FDTD/TLM solvers.

Time-harmonic field distributions can be animated by generating a sequence of distributions for discrete phase angles. Transient simulations naturally yield a time sequence of distributions that can be made into a movie, or displayed as they are generated by the solver. Depending on the size of the problem and the speed of the processor, this type of solution-generated animation can be too fast for human observation, so that the computation must actually be slowed down.

In many cases one is only interested in the transfer functions (such as *S*-parameters) between specific locations or ports of a structure. Such functions usually involve definitions of characteristic impedance, modal voltages, and currents. It is important to remember that electromagnetic solvers do *not* compute voltages, currents, impedances, capacitances, or *S*-parameters. These secondary quantities must be computed from the primary field data by additional processing, usually some type of numerical integration in the port cross-section. This is a very important distinction that is relevant to the discussion of errors. For example, the field data can be very accurate, but if the path of integration chosen for computing a voltage, or the definition of a characteristic impedance, is different from that used in the test environment, the field-solver is not responsible for the disagreement of results, but rather the post-processor. On the other hand, an integral of a field solution can be much more accurate than the field solution itself, which means that the secondary parameter can be determined satisfactorily from a rather coarse and quick field solution. Issues involving ports and de-embedding are discussed in Chapter 8.

Finally, the extraction of frequency-domain parameters from transient field solutions is an important feature in FDTD and TLM solvers. A typical time response is simply a sequence of numbers that sample the port quantity at each time step. The complex spectral response is found by performing a discrete Fourier transform of this sequence, yielding its real and imaginary part as follows:

$$Re[F(\Delta l/\lambda)] = \sum_{k=1}^{N} {}_kA\,cos(2\pi k(\Delta l/\lambda)) \tag{3.62}$$

$$Im[F(\Delta l/\lambda)] = -\sum_{k=1}^{N} {}_kA\,sin(2\pi k(\Delta l/\lambda)) \tag{3.63}$$

where $F(\Delta l/\lambda)$ is the frequency response, ${}_kA$ is the value of the impulse response sample at time $t = k\Delta l/\lambda$, and N is the total number of time steps. Figure 3.27 shows a typical transient response of a two-resonator waveguide bandpass filter to a Gaussian-modulated sine (band-limited) excitation; and Figure 3.28 shows its *S*-

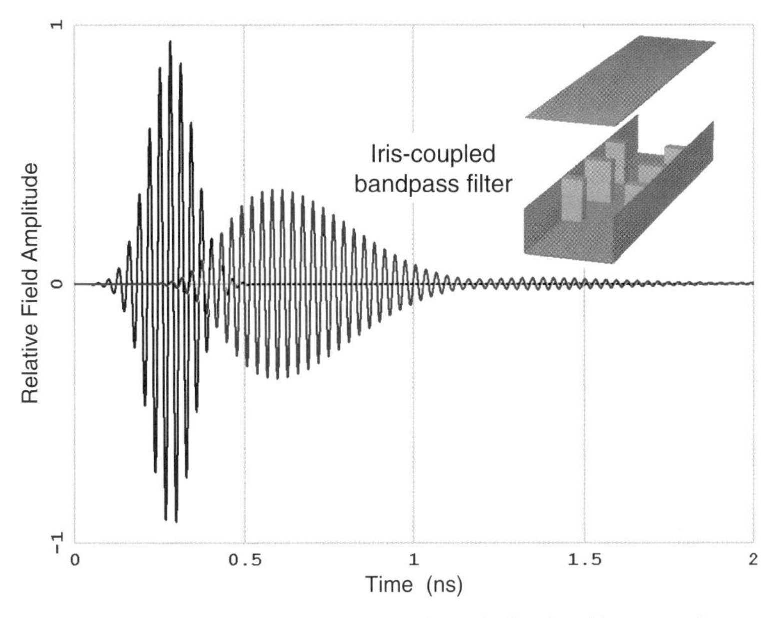

Figure 3.27 Time domain excitation (black trace) and transfer function (blue trace) of a two-resonator waveguide bandpass filter obtained with a TLM simulator (MEFiSTo-3D Pro). The excitation is a band-limited Gaussian-modulated sine wave.

parameters obtained after performing the above discrete Fourier transform. Fourier-transformed data can also be displayed in a Smith Chart format or, after near-to-farfield transformation, as a radiation pattern.

When equipped with these processing and visualization features, a modern electromagnetic field simulator can emulate all major instrumentation available in a microwave laboratory, such as an oscilloscope, time domain reflectometer, spectrum analyzer, network analyzer, and antenna test range.

3.14 DISCUSSION AND CONCLUSION

None of the numerical methods discussed in this chapter is capable of solving all electromagnetic modeling problems. The methods are either limited by the available computer memory and/or by computer run time, or the numerical model can simply not be applied to the structure at hand. For example, the method of moments is not applicable to structures with inhomogeneous or nonlinear dielectrics and enclosures of complex shape. The FDTD or TLM method is difficult to implement when fine geometrical detail must be resolved within a structure of large dimensions. The discretization size chosen for the smallest feature determines the time step and the total number of cells. Similarly, the finite element method cannot effi-

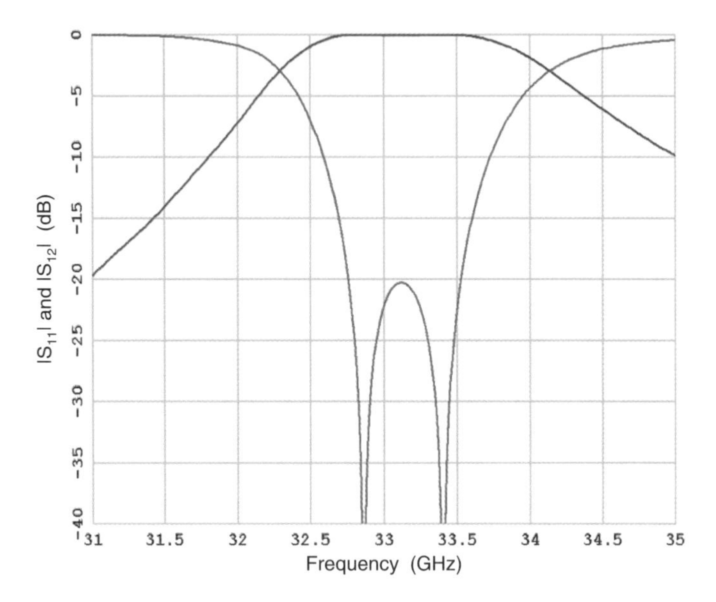

Figure 3.28 Magnitude of S-parameters extracted from the time response via discrete Fourier transform. The phase information is also available (not shown).

ciently model large open radiation and scattering problems because of the large computational space that would have to be discretized.

It is thus impossible to make a general statement as to the superiority of one method over another. Rather, one should look at the various methods as a diversified tool set. The hallmark of a skilled engineer is not to perform every task with the same tool, but to select the most appropriate tool from the available set, and to combine the strengths of several tools if necessary. The different numerical techniques and the simulators based on them cover indeed a wide variety of electromagnetic problem scenarios with sufficient overlap to provide meaningful corroboration of results.

To use electromagnetic simulators with confidence and success, the user must:

(a) Develop a good understanding of the type of problems best handled by the various available simulators;

(b) Understand the sources of error, know the order of magnitude of the error, and learn how to minimize it;

(c) Learn how to achieve a good trade-off between accuracy and speed.

The remaining chapters will help the reader to develop this expertise, and we hope that this rather general and succinct discussion of the major numerical methods provides a useful launching pad for further study and deeper understanding of the principles that govern modern electromagnetic simulators.

3.15 FURTHER READING

The literature on discrete mathematics, computational electromagnetics and field modeling is enormous, and only a few key references could be cited within the text of this chapter. A good way to access the extensive knowledge base is to consult the many excellent books and collections of reprinted key papers. Very accessible introductory texts on numerical methods are the books by Booton [8] and by Sadiku [26]. More advanced texts on numerical techniques are the books edited by Itoh [27] and by Yamashita [28]. Sorrentino [29], Miller, *et al.,* [30] and Itoh and Houshmand [31] have assembled excellent reprint collections of key papers on numerical techniques. Furthermore, various journals and conferences are devoted to the topic of numerical modeling of electromagnetic fields.

The monograph *Field Computation by Moment Methods* by Harrington [1] is a classic. Among the many excellent books on the finite element method, the texts by Silvester and Ferrari [4] and by Salazar-Palma, et al., [32] contain extensive information and references.

FDTD is the subject of books by Kunz and Luebbers [33], Taflove and Hagness [11] and Sullivan [34] are excellent sources of information on all aspects of FDTD modeling and contain extensive bibliographies on the theory, implementation, and application of the FDTD method. Together with Yee's seminal paper [7] they are good starting points for exploring the extensive literature on FDTD theory and applications. A complete FDTD bibliography with search capability is available on the Internet at http://www.fdtd.org/.

Johns' seminal papers [15, 18, 19] are good starting points for exploring the world of TLM modeling, as are an introductory chapter on TLM by Hoefer [17] and a book by Christopoulos [35]. They contain many references and describe the implementation and applications of TLM in detail.

There are many more excellent books and articles on the subject, which the reader will find in the extensive bibliographies of the works cited above.

References

[1] Harrington, R. F., *Field Computation by Moment Methods*, New York, NY: Macmillan, 1968.

[2] Burke, G. J., and A. J. Poggio, "Numerical Electromagnetics Code (NEC-2)," Lawrence Livermore Laboratory, January 1981.

[3] Mosig, J. R., "*Integral Equation Technique,*" Chapter 3 of *Numerical Techniques for Microwave and Millimeter-Wave Passive Structures,* T. Itoh (ed.) New York: J. Wiley & Sons, 1989.

[4] Silvester, P. P., and R. L. Ferrari, *Finite Elements for Electrical Engineers,* 3rd Edition, New York: Cambridge University Press, 1996.

[5] HFSS, Hewlett-Packard, Santa Rosa, CA and Ansoft, Pittsburgh, PA.

[6] Krumpholz, M., C. Huber, and P. Russer, "A Field Theoretical Comparison of FDTD and TLM," *IEEE Trans. Microwave Theory Tech.*, Vol. MTT-43, No. 8, 1995, pp. 1934–1950.

[7] Yee, K. S., "Numerical Solution of Initial Boundary-Value Problems Involving Maxwell's Equations in Isotropic Media," *IEEE Trans. Ant. Prop.*, Vol. AP-14, No. 5, 1996, pp. 302–207.

[8] Booton, Jr., R. C., *Computational Methods for Electromagnetics and Microwaves,* New York: John Wiley & Sons, 1992.

[9] Taflove, A., and M. E. Brodwin, "Numerical Solution of Steady-State Electromagnetic Scattering Problems Using the Time-Dependent Maxwell's Equations," *IEEE Trans. Microwave Theory Tech.,* Vol. MTT-23, No. 8, 1975, pp. 623–630.

[10] Weiland, T., "A Discretization Method for the Solution of Maxwell's Equations for Six-Component Fields," *Electronics and Communication (AEU),* Vol. 31, 1977, p. 116.

[11] Taflove, A., and S. Hagness, *Computational Electromagnetics: The Finite-Difference Time-Domain Method*, Norwood, MA: Artech House, 2000.

[12] Mur, G., "Absorbing Boundary Conditions for the Finite-Difference Approximation of the Time Domain Electromagnetic-Field Equations," *IEEE Trans. Electromagnetic Compatibility.*, Vol. EMC-23, No. 4, 1981, pp. 377–382.

[13] Higdon, R. L., "Numerical Absorbing Boundary Conditions for the Wave Equation," *Math. Comput.*, Vol. 49, No. 179, 1987, pp. 65–90.

[14] Berenger, J.-P., "Perfectly Matched Layer for the FDTD Solution of Wave-Structure Interaction Problems," *IEEE Trans. Ant. Prop.*, Vol. AP-44, No. 1, 1996, pp. 110–117.

[15] Johns, P. B., and R. L. Beurle, "Numerical Solution of 2-Dimensional Scattering Problems Using a Transmission-Line Matrix," *Proc. Inst. Electr. Eng.*, Vol. 118, No. 9, 1971, pp. 1203–1208.

[16] Whinnery, J. R., and S. Ramo, "A New Approach to the Solution of High Frequency Field Problems," *Proc. I.R.E.*, Vol. 32, 1944, pp. 284–288.

[17] Hoefer, W. J. R., "*The Transmission Line Matrix (TLM) Method,*" Chapter 8 of *Numerical Techniques for Microwave and Millimeter-Wave Passive Structures,* T. Itoh (ed.) New York: John Wiley & Sons, 1989.

[18] Akhtarzad, S., and P. B. Johns, "Solution of 6 Component Electromagnetic Fields in 3 Space Dimensions and Time by the TLM Method," *Electron. Lett.*, Vol. 10, 1974, pp. 535–537.

[19] Johns, P. B., "A Symmetrical Condensed Node for the TLM Method," *IEEE Trans. Microwave Theory Tech.*, Vol. MTT-35, No. 4, 1987, pp. 370–377.

[20] Trencic, V., C. Christopoulos, and T. M. Benson, "New Symmetrical Super-Condensed Node for the TLM Method," *Electron. Lett.,* Vol. 30, No. 4, 1995, pp. 329–330.

[21] Russer, P., and B. Bader, "The Alternating Transmission Line Matrix (ATLM) Scheme," *IEEE MTT-S Int. Microwave Symposium Digest*, Orlando, FL, May 16–18, 1995, pp. 19–22.

[22] Russer, P., "The Alternating Rotated Transmission Line Matrix (ARTLM) Scheme", *Electromagnetics*, Vol. 16, No. 5, 1996, pp. 537–551.

[23] Hoefer, W. J. R., "The Discrete Time Domain Green's Function or Johns Matrix - A New Powerful Concept in TLM," *Int. Journal of Numerical Modelling,* Vol. 2, No. 4, 1989, pp. 215–225.

[24] De Menezes, L. R. A. X., and W. J. R. Hoefer, "Modeling of General Constitutive Relationships in SCN TLM," *IEEE Trans. Microwave Theory Tech.*, Vol. MTT-44, No. 6, 1996, pp. 854–861.

[25] So, P. P. M., and W. J. R. Hoefer, "A TLM-SPICE Interconnection Framework for Coupled Field and Circuit Analysis in the TIme Domain," *IEEE Trans. Microwave Theory Tech.*, Vol. MTT-50, No. 12, 2002, pp. 2728–2733.

[26] Sadiku, M., *Numerical Techniques in Electromagnetics*, Boca Raton, FL: CRC Press, 1992.

[27] Itoh, T., (ed.), *Numerical Techniques for Microwave and Millimeter-Wave Passive Structures,* New York: John Wiley & Sons, 1989.

[28] Yamashita, E., (ed.), *Analysis Methods for Electromagnetic Wave Problems*, Norwood, MA: Artech House, Two volumes: Vol. 1: 1990; Vol. 2: 1995.

[29] Sorrentino, R., (ed.), *Numerical Methods for Passive Microwave and Millimeter Wave Structures*, New York: IEEE Press, 1989.

[30] Miller, E. K., L. Medgyesi-Mitschang, and E. H. Newman, *Computational Electromagnetics,* New York: IEEE Press, 1992.

[31] Itoh, T., and B. Houshmand, (eds.), *Time Domain Methods for Microwave Structures,* New York: IEEE Press, 1998.

[32] Salazar-Palma, M., et al., *Iterative and Self-Adaptive Finite Elements in Electromagnetic Modeling,* Norwood, MA: Artech House, 1998.

[33] Kunz, K. S., and R. L. Luebbers, *The Finite Difference Time Domain Method for Electromagnetics*, Boca Raton, FL: CRC Press, 1993.

[34] Sullivan, D., *Electromagnetic Simulation Using the FDTD Method,* New York: IEEE Press, 2000.

[35] Christopoulos, C., *The Transmission-Line Modeling Method TLM,* New York: IEEE Press, 1995.

Chapter 4

Alternative Classifications

In the previous chapter we focused on the fundamentals of the most widely used numerical methods. We could choose the solver we would like to use strictly on the basis of the numerical method employed. The classical, mathematical classification system would focus on the solution region, the type of equation that describes the problem and the boundary conditions [1]. But for the working engineer, it is more useful to focus on the type of geometry we are trying to solve, rather than the specific numerical method or mathematical description of the problem. We will discover that the "solution" to a design project is actually a series of solutions that may use different geometrical approximations. We can also choose a solver based on the solution domain. The choices today are typically frequency domain, time domain, and an eigenmode-solver. All have advantages and disadvantages depending on the type of problem we are trying to solve.

4.1 CLASSIFICATION BY GEOMETRY

The types of problems we are trying to solve and the numerical tools used to solve them can be divided into three broad classes. We characterize each class not by the numerical method used but rather by the order of the geometry they can analyze (Figure 4.1). Within each class, any number of different numerical methods may be used. Model building time, numerical effort, and total solution time all increase dramatically as the geometry gets more complex. So the challenge to the design engineer is to do as much work as possible with simpler, lower order models rather than approaching every problem as an arbitrary 3D geometry.

4.1.1 2D Cross-Section-Solvers

The lowest-order geometry we typically solve is a 2D cross-section (Figure 4.1(a)). This solver is suitable for waveguides, strips, or slots with uniform cross-section in the longitudinal direction. Sets of uniform cross-section lines can be found in

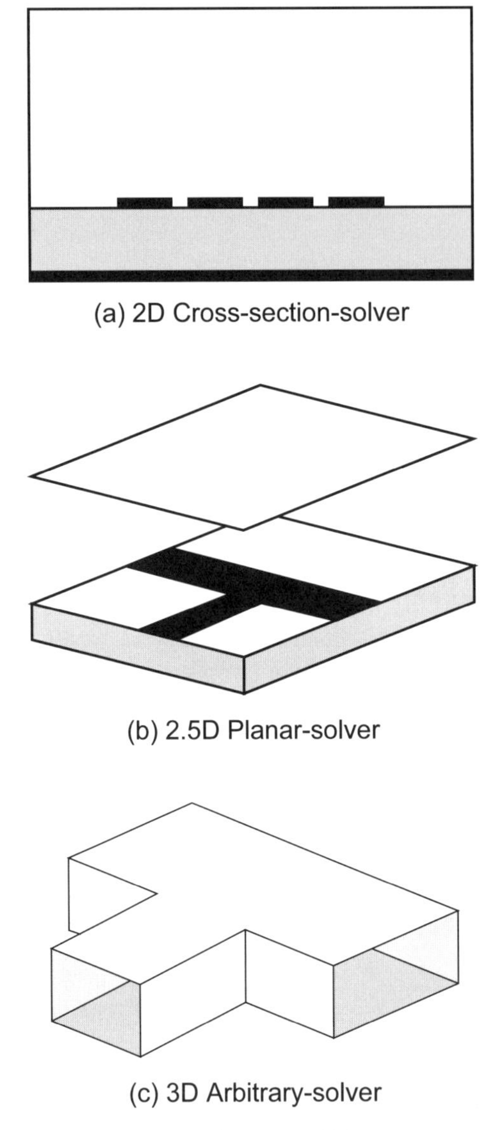

Figure 4.1 Field-solvers classified by order of geometry: (a) 2D cross-section, (b) 2.5D planar-solver, and (c) 3D arbitrary-solver.

Lange couplers, spiral inductors, interdigital capacitors, and many distributed filters. Sets of uniform cross-section lines also form the basis of digital buses. For problems with one or two signal conductors, it is quite easy to compute the impedances and phase velocities of each mode. Some example cross-sections are shown in Figure 4.2.

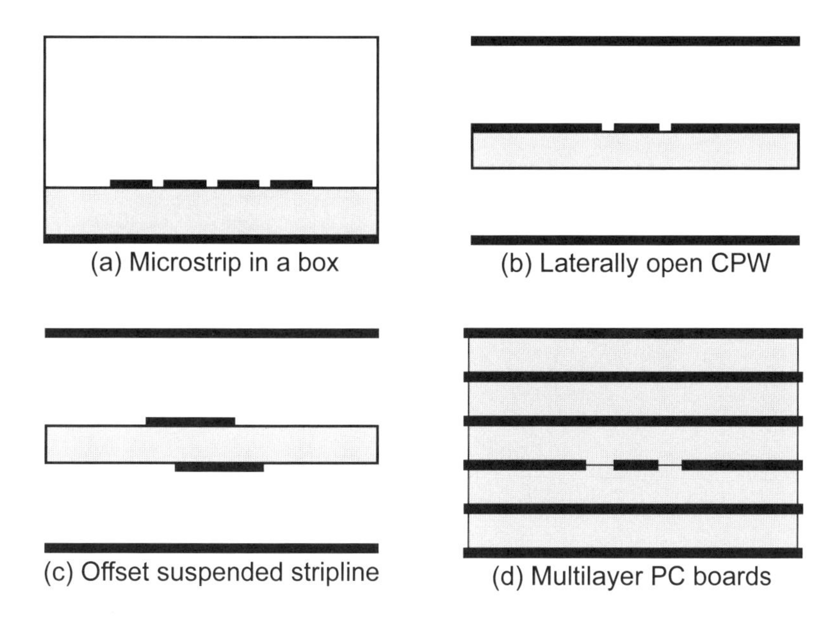

Figure 4.2 Examples of 2D cross-section problems: (a) microstrip lines in a closed box, (b) a laterally open CPW geometry, (c) offset, broadside coupled suspended stripline, and (d) a multilayer printed circuit board.

Numerically, we only have to consider a small, bounded 2D region, so solution time generally will not be an issue. We can solve for currents on the strips or the dual problem which is the voltage in the slots (sometimes referred to as magnetic currents). There are two general subclasses of problems, fully enclosed with a perfect electric conductor (PEC) boundary or laterally open. There are any number of numerical methods that can be used for this type of problem: method of moments, finite element method, finite differences, method of lines and boundary elements, among others. Many of these field-solvers are stand-alone tools while some are integrated within a linear/nonlinear simulator.

4.1.2 2.5D Planar-Solvers

If we want to solve more general planar circuits, we generally move to a 2.5D planar-solver (Figure 4.1(b)). These tools are also called 3D mostly planar-solvers by some software vendors. With these tools, an arbitrary number of homogeneous dielectric layers are allowed. An arbitrary planar metal pattern can then be placed at the interface between any pair of dielectric layers. Via metal can also be used to connect metal layers. This is where the half dimension comes from in the 2.5D description; we are somewhere in between a strictly planar structure and an arbitrary 3D structure. There are two fundamental numerical formulations for this type

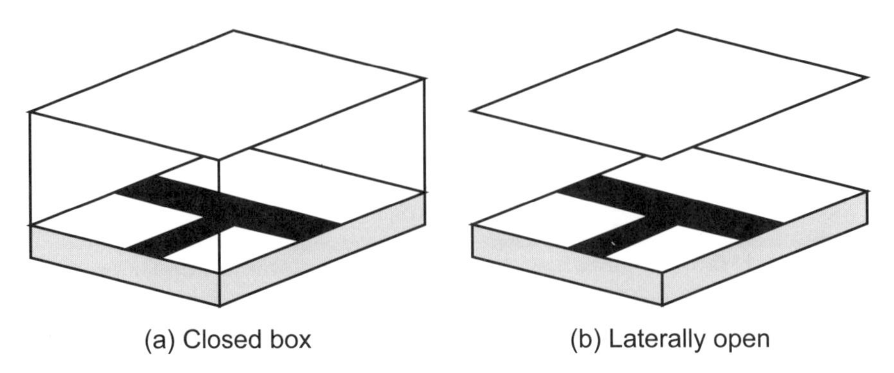

Figure 4.3 2.5D planar solvers: (a) closed box formulation, and (b) laterally open formulation.

of problem, one that assumes the circuit is in a closed, metallic box (Figure 4.3(a)) and one where no box walls are present (Figure 4.3(b)). Numerically, the software is looking for the unique current distribution on the strips that forces the tangential component of the E-field on the strips to be zero. Compared to the 2D cross-section-solvers, the numerical effort has increased dramatically and solution time becomes an issue. The numerical method used is typically the method of moments.

4.1.3 3D Arbitrary Solvers

Finally, 3D field-solvers allow us to analyze a truly arbitrary 3D structure (Figure 4.1(c)). The basic formulation for these solvers assumes a closed, metallic boundary around the solution region (Figure 4.4). However, an open environment can be approximated using various types of absorbing boundaries. While these tools offer great flexibility, the penalty is longer modeling time and solution time. Building a

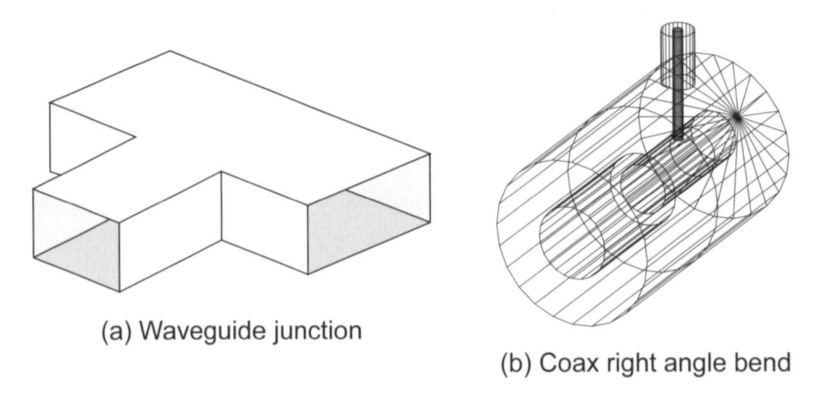

Figure 4.4 3D solver examples: (a) waveguide junction, and (b) coaxial right angle bend.

Table 4.1

Field-Solvers Classified by Geometry

High Numerical Expense

2.5D Planar–Laterally Open	**3D Arbitrary Geometry**
No fixed grid	**Arbitrary geometry**
Rectangular and triangular elements	**Basic formulation is closed box**
Numerical Green's function	**Absorbing boundaries possible**
Symmetry or walls require image theory	**Model building time can be significant**
Arbitrary spatial resolution	**Must discretize the entire volume**
2D Cross-Section-Solvers	**2.5D Planar–Closed Box**
Strips or slots with uniform cross-section	**Fixed grid**
Easy to find Z_0 and ε_{eff} for single strips	**Rectangular elements**
Two subclasses–closed and laterally open	**Analytical Green's function**
Model building time very low	**One plane of symmetry is easy**
Discretize only the 2D cross-section	**Small features (resolution) can be a problem**

Low Numerical Expense

model in 3D is considerably more difficult than 2D or 2.5D modeling. Numerically, we are forced to solve for the fields in the entire 3D volume, which leads to a dramatic increase in solution time compared to the other two classes of solvers. The numerical method used is typically the finite element, transmission line matrix, or finite difference time domain method.

4.1.4 Summary

For any given problem, one of the three general solver types will offer the most efficient solution. In the course of a design project, we might use all three types of solver at some point. If we arrange our field-solver classes in a grid, we can make some general observations (Table 4.1). At the lower left are the 2D solvers that handle the simplest geometries and are the fastest. As we move toward the upper right we can handle more complex geometries at the expense of longer solution times. Our first inclination is to pick the most general tool that will "do everything." Instead, we should constantly try to reduce the problem to the lowest-order geometry possible.

4.2 CLASSIFICATION BY SOLUTION DOMAIN

A second way to classify projects is by the solution domain. The choices are typically frequency domain, time domain, and eigenmode solution. Most of the major methods have now been formulated in both the frequency and time domains, but the commercial implementations typically only use one of the two [2].

4.2.1 Frequency Domain Solvers

Frequency domain solvers typically discretize the solution domain, build a matrix, and invert the matrix to find the solution. The matrix building and inversion steps must be repeated at each frequency of interest. The numerical method is typically FEM or MoM. If the desired solution parameters vary slowly as a function of frequency, only a few frequency points may be needed. If the solution parameters vary rapidly as a function of frequency or if broadband data is needed, then many frequency points must be computed. To overcome this drawback, most of the frequency domain codes offer some type of "fast sweep" option. This option attempts to find a rational polynomial that describes the solution behavior using a minimum set of computed frequency points. These fast sweep methods are based on asymptotic waveform evaluation (AWE) [3–5], the Padé via Lanczos method (PVL) [6–8], or an adaptive Lanczos-Padé sweep (ALPS) [9–11]. Although the robustness of these techniques continues to improve, the careful user should always check these solutions, particularly at the extremes of the frequency range. If enough frequency domain points are available, a time domain response can be obtained using an inverse Fourier transform process. However, the inverse solution must be carefully checked for causality problems.

4.2.2 Time Domain Solvers

Time domain solvers typically discretize the solution domain, then excite the problem space with an impulse of energy. Algebraic equations are used to update the field quantities as a function of space and time until convergence is reached. A stored record of the time response at a port or other point of interest can be converted into the frequency domain using a discreet Fourier transform (DFT) or fast Fourier transform (FFT) process. This solution process can be quite useful when broadband frequency data is needed or when the exact location of responses in the frequency domain are not known. RCS problems are a good example. When a ship or aircraft is hit with a radar pulse, we would like to know the "signature" of the object in the frequency domain, but it is difficult to predict exactly where responses will occur. The same solution process has drawbacks for high Q structures with closely spaced resonances, like microwave filters. The high Q condition forces a long run time to convergence and closely spaced resonances require many time samples for the Fourier transform process to converge. The time domain also has advantages when we would like to see the evolution of fields or derived quantities as a function of time. A classic example is a time domain reflectometry (TDR) measurement where the response in the time domain conveys qualitative and quantitative information to the observer. Finally, many nonlinear problems are handled easier in the time domain than in the frequency domain. Since field quantities are known everywhere in space and time, nonlinear properties can be updated at each time step according to the local field strength and direction.

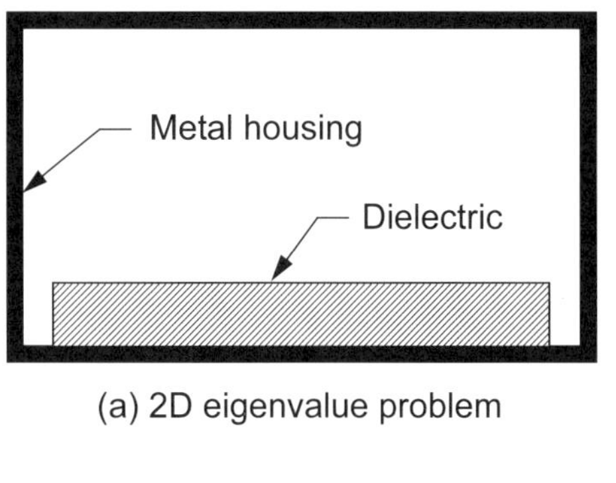

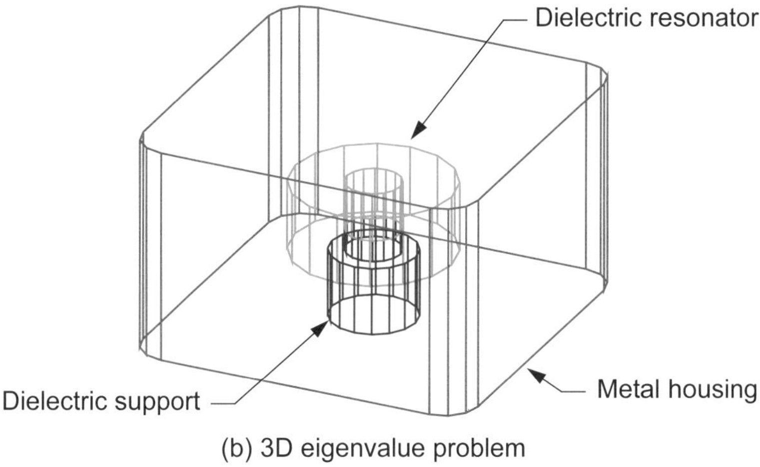

Figure 4.5 Examples of eigenvalue problems: (a) finding the cutoff frequency of a metal housing partially filled with dielectric (microstrip), and (b) finding the resonant frequency of a dielectric resonator (green) sitting on a dielectric support (blue) in a metal housing (red).

4.2.3 Eigenmode-solvers

Our discussion of frequency domain and time domain solvers assumed we were looking at a "driven" problem with an external source of energy, typically at a port. With both types of solvers we can also formulate sourceless, variational type problems called eigenvalue problems. The solutions are typically some stationary field configuration and the derived quantities might be a cutoff frequency of a waveguide type structure, a propagation constant for a waveguide type structure, or the resonant frequency of a resonator. Here "waveguide type structures" include classic waveguides, higher-order modes for microstrip cross-sections in a closed box, and any cross-section of a package that is enclosed by metal. Eigenvalue problems can be formulated in 2D and 3D. Figure 4.5(a) shows a dielectric substrate that partially fills the housing cross-section. A 2D eigenmode solution can tell us the cutoff frequency of this geometry. Figure 4.5(b) shows a cylindrical dielectric resonator sitting on a dielectric support. A 3D eigenmode solution can tell us the resonant frequencies of this geometry.

References

[1] Sadiku, M., *Numerical Techniques in Electromagnetics,* Second Edition, Boca Raton, FL: CRC Press, 2001, pp. 14–22.

[2] Hafner, C., *Post-modern Electromagnetics*, Chichester, UK: John Wiley & Sons, 1999, pp. 171–199.

[3] Yuan, X., and Z. Cendes, "A Fast Method for Computing the Spectral Response of Microwave Devices Over a Broad Bandwidth," *Proc. IEEE AP-S/URSI Int. Symp. Dig.*, Ann Arbor, MI, June 1993, p. 196.

[4] Pillage, L. T., and R. A. Rohrer, "Asymptotic Waveform Evaluation for Timing Analysis," *IEEE Trans. Computer-Aided Design*, Vol. 9, No. 4, 1990, pp. 352–366.

[5] Bracken, J. E., D. Sun, and Z. J. Cendes, "*S*-domain Methods for Simultaneous Time and Frequency Characterization of Electromagnetic Devices," *IEEE Trans. Microwave Theory Tech.*, Vol. 46, No. 9, 1998, pp. 1277–1290.

[6] Zhang, X., and J. Lee, "Application of the AWE Method with the 3-D TVFEM to Model Spectral Responses of Passive Microwave Components," *IEEE Trans. Microwave Theory Tech.*, Vol. 46, No. 11, 1998, pp. 1735–1741.

[7] Feldmann, P., and R. W. Freund, "Efficient Linear Circuit Analysis by Padé Approximation Via the Lanczos Process," *IEEE Trans. Computer-Aided Design*, Vol. 14, No. 5, 1995, pp. 639–649.

[8] Feldmann, P., and R. W. Freund, "Reduced-order Modeling of Large Linear Subcircuits Via a Block Lanczos Algorithm," *Proc. 32nd ACM/IEEE Design Automation Conf.*, 1995, pp. 474–479.

[9] Sun, D.-K., "ALPS–An Adaptive Lanczos-Padé Spectral Solution of Mixed-Potential Integral Equation," in *USNC/URSI Radio Sci. Meeting Dig.*, July 1996, p. 30.

[10] Sun, D.-K., "ALPS–An Adaptive Lanczos-Padé Spectral Solution of Mixed-Potential Integral Equation," *Comput. Methods Appl. Mech. Eng.*, Vol. 169, 1999, pp. 425–432.

[11] Sun, D.-K., J.-F. Lee, and Z. Cendes, "ALPS–A New Fast Frequency-Sweep Procedure for Microwave Devices," *IEEE Trans. Microwave Theory Tech.*, Vol. 49, No. 2, 2001, pp. 398–402.

Chapter 5

Moment Method Simulators

Most multilayer board design problems that RF and digital design engineers face today can be described as "mostly planar" problems. That is, most of the metal lies in several horizontal planes, homogeneous dielectric layers separate those planes, and vias connect the metal layers together at various points. Many types of packaging problems also fall into this category. The 2.5D moment method codes are ideally suited for this type of problem. They concentrate their numerical energies only on the metal conductors. We have already discussed some of the theoretical aspects of the closed box and laterally open method of moments formulations in an earlier chapter.

5.1 CLOSED BOX MOMENT METHOD—STRENGTHS

The closed box moment codes are characterized by a fixed resolution grid and an analytic Green's function. The finite resolution of the grid is the price to be paid for the fast computation of the Green's function, a sum of cosine terms [1]. Symmetry can be easily implemented by ignoring the summation terms that are zero on the symmetry plane. Box walls are included in the basic formulation; the interaction of a circuit with its package is obvious. The box walls also provide an unambiguous ground reference for port calibration. The source is connected across an infinitesimal gap between the box wall and the input strip. The de-embedding scheme is very self-consistent, the solution mesh and the de-embedding mesh are very similar.

5.2 CLOSED BOX MOMENT METHOD—WEAKNESSES

Resolution can be a problem with the fixed grid. The grid limitations can be overcome somewhat with interpolation or extrapolation techniques. The fixed grid also makes it difficult to import old designs. If the old design was not laid out with this particular method in mind, it may not be possible to find a grid that fits it well. Only

Table 5.1

Closed Box Method of Moments

Strengths	*Weaknesses*
Analytic Green's function - fast computation	Resolution can be a problem with fixed grid:
Box walls included in basic formulation:	• Can be overcome somewhat with interpolation or extrapolation
• Interaction of circuit with package is obvious	Difficult to import old designs with fixed grid
• Box walls provide unambiguous ground reference for port calibration	Only rectangles and special 45-degree elements available
Numerical implementation of symmetry is easy	Open structures must be approximated:
Very self-consistent de-embedding scheme	• Move box walls away
Multistrip de-embedding is easy	• Set cover to 377 ohms
	Box resonances can appear

rectangles and special 45-degree elements are available to define the geometry. Open structures must be approximated by moving the box walls away and setting the cover to 377 ohms (the free space impedance). In millimeter-wave structures, the closed box may support resonant modes that interfere with the analysis. The strengths and weaknesses of the closed box MoM formulation are summarized in Table 5.1.

5.3 LATERALLY OPEN MOMENT METHOD—STRENGTHS

In the laterally open moment method codes there is no fixed grid; the user basically has arbitrary resolution. Rectangles and triangles are available to define the geometry. The addition of triangles makes it possible to approximate smooth arcs. This formulation is well suited to patch antenna and other fundamentally open problems. Most tools can do multistrip de-embedding.

5.4 LATERALLY OPEN MOMENT METHOD—WEAKNESSES

The penalty to be paid for infinite resolution is a Green's function that must be numerically integrated. In general, this means the Green's function computation is slower than the closed box case. However, it is possible to precompute some coefficients in the laterally open case. Image theory must be invoked to implement symmetry, this implies another infinite summation that must be truncated at some point. Image theory is also required to implement box walls. De-embedding is more difficult compared to the closed box formulation; there is no box wall to provide a ground reference. The actual ground may be quite close to the port or quite far away, the developer and the software have no way of knowing what the user might

Table 5.2

Laterally Open Method of Moments

Strengths	*Weaknesses*
No fixed grid - infinite resolution	Green's function requires numerical integration:
Rectangles and triangles available to define geometry	• Slower to compute
	• Some coefficients can be precomputed
Well suited to patch antenna and other fundamentally open problems	Symmetry requires image theory
	Box walls require image theory
	De-embedding is more difficult:
	• No box walls for ground reference
	• Need separate 2D solution to find impedance

specify. In most cases the port defined by the user is extended by three to five cells of metal and a current excitation is applied at the end of the extension. A separate 2D solution is used to find impedance and phase velocities for de-embedding; this can introduce inconsistencies into the solution process. The 2D impedance solution must match the impedance computed by the 2.5D engine to maintain an accurate solution. The strengths and weaknesses of the laterally open MoM formulation are summarized in Table 5.2.

5.5 ISSUES COMMON TO BOTH MOM FORMULATIONS

There are some common characteristics that both the closed box and laterally open formulations share. In all MoM codes, there is a large, dense matrix to invert. In general, the matrix is smaller than the equivalent FEM problem matrix. But because it is a dense matrix, there are no easy ways to speed up the matrix inversion process. The matrix fill time is roughly proportional to N^2, and the matrix inversion time is roughly proportional to N^3. Most MoM codes do not use the full 3D Green's function. Instead, they solve for the X-Y currents in each plane using one set of basis functions. For the via metal that connects planes vertically, only Z-directed currents are allowed and a different basis function is used. MoM codes are in general limited to homogeneous, layered dielectrics. It is possible to formulate dielectric "bricks," but they are numerically expensive to compute. The basic MoM formulation assumes infinitely thin strips. Thick conductors can be approximated using double layers of metal separated by a thin layer of air or other dielectric. However, there will be a large impact on solution time since the number of cells dedicated to that conductor has been doubled. Some codes support a dual formulation that solves for "magnetic currents" in a slot. While magnetic currents do not actually exist, this is equivalent to solving for the voltage across the slot. Slot and strip formulations may not be mixed in the same layer. In most MoM codes mesh generation is automatic, but there is no automatic mesh refinement. There is no fundamental aspect of the

Table 5.3

Issues Common to Both MoM Formulations

Large, dense matrix to invert: • Matrix fill proportional to N^2 • Matrix inversion proportional to N^3 Most codes use one set of basis functions to solve for X-Y currents Different basis function used for Z directed currents Limited to homogeneous, layered dielectrics Dielectric bricks are possible, but numerically expensive Dual formulation of voltage in a slot is possible Strip and slot formulations cannot be mixed in same layer No automatic mesh refinement	Basic formulation assumes infinitely thin strips: • Finite thickness can be approximated using double metal layers • Surface impedance calculation includes thickness information Thin dielectric layers can cause numerical precision problems: • Mesh must align on lower and upper plates of capacitors Solving at very low frequencies can cause numerical precision problems For best accuracy: • Small, square cells or equilateral triangles • Limit aspect ratio to 4:1

MoM formulation that indicates where to place the next cell in order to reduce the solution error. Automatic mesh refinement is currently a topic of research. Best accuracy is generally obtained with small, square cells or small, equilateral triangles. Cells or triangles with an aspect ratio greater than 4:1 should be avoided if possible. The impact of cell or subsection aspect ratio on accuracy was explored in [2]. The issues common to both MoM formulations are summarized in Table 5.3.

5.6 EXCEPTIONS TO GENERAL MOM COMMENTS

As always, there are exceptions to every rule. Both Ansoft Ensemble and Zeland IE3D use a mixed potential integral equation (MPIE) formulation. Both codes retain all the vector components and the scalar component of the Green's function in all calculations, which may impact solution time. But this allows placement of metal patches at arbitrary angles, rather than strictly horizontal and vertical. However, they are still limited to layered, homogeneous dielectrics. Other exceptions to the general comments can be found in Ansoft Ensemble. In Ensemble, only triangles are used for meshing and some form of automatic mesh refinement has been implemented. The features and details of all of these codes are in a constant state of flux; the user should check with the vendors on a regular basis.

5.7 50-OHM MICROSTRIP LINE

Now that we have summarized the general characteristics of MoM codes, let us look at them more detail. We will begin our exploration with a simple 50-ohm

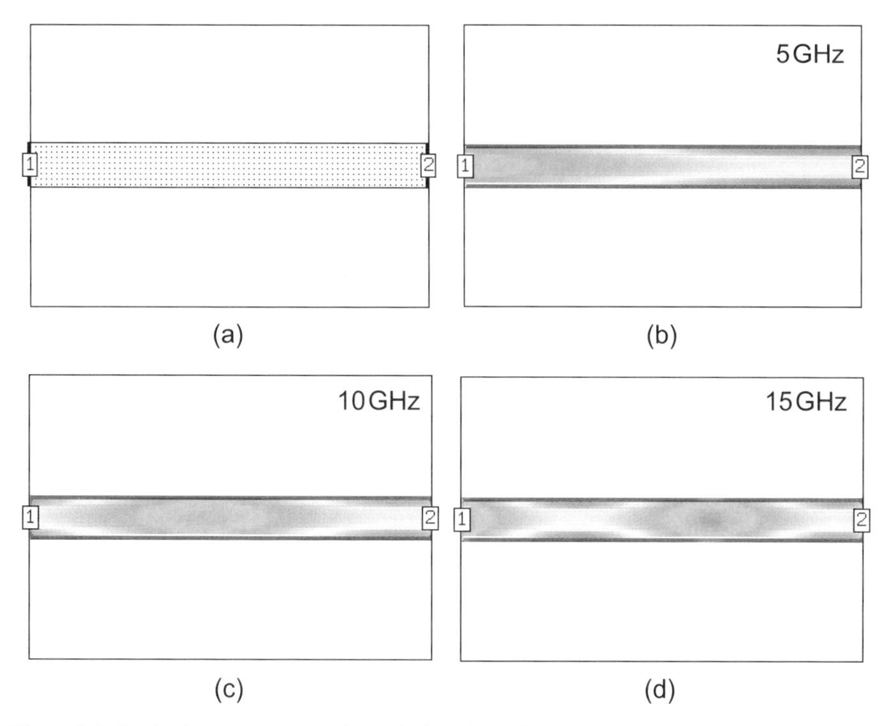

Figure 5.1 Conduction current on a mismatched 50-ohm microstrip line: (a) microstrip geometry, (b) current at 5 GHz, (c) current at 10 GHz, and (d) current at 15 GHz. In all cases we are viewing the time averaged, vector magnitude of the X- and Y-directed conduction currents.

microstrip line [3]. In this case, the line is 24-mil wide and 225-mil long on a 25-mil thick alumina substrate, although any substrate material and thickness could be used. Figure 5.1(a) shows a top view of the microstrip line. There is a matched source at port one and a 25-ohm load at port two, so we expect to see some kind of standing wave behavior. The additional plots in Figure 5.1 show the time averaged, vector magnitude of the X- and Y-directed conduction currents on the microstrip line; high values are red and low values are blue. We used a uniform grid of 2-mil square cells to obtain a high resolution plot.

At 5 GHz, the line is roughly 90 degrees long (Figure 5.1(b)). There is a clear maximum on the right and a minimum on the left. At 10 GHz the line is nearly 180 degrees long (Figure 5.1(c)). Starting from the load at the right we see a maximum, a minimum, and a maximum again. In Figure 5.1(d) we analyze the line at 15 GHz, where the line is roughly 270 degrees long and the current pattern due to guide wavelength is quite evident. With these three plots we are confident we have set up the problem correctly because we can observe a pattern in the current that we can relate to guide wavelength. The skeptical reader could even measure the dis-

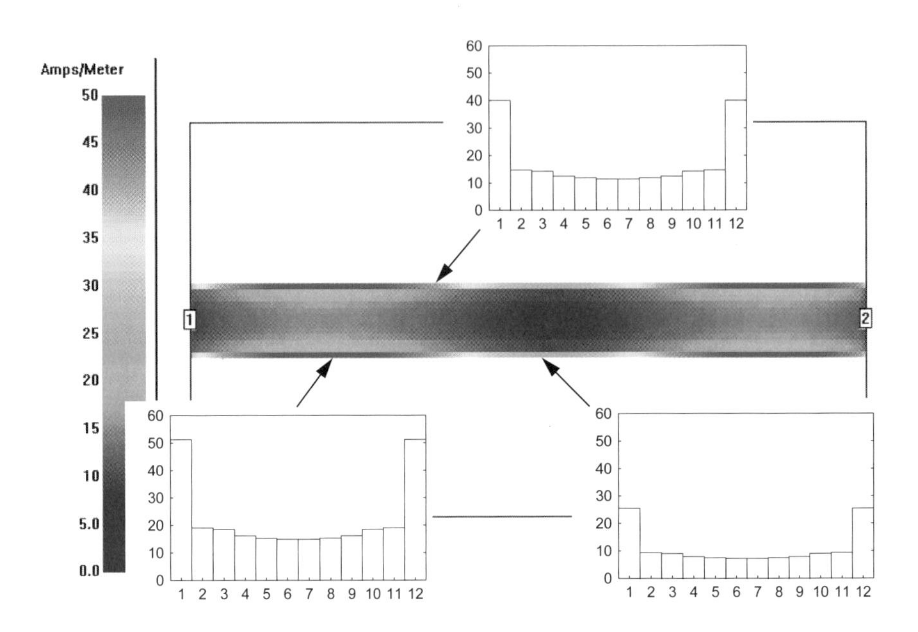

Figure 5.2 Conduction current on 50-ohm microstrip line. Cell currents across the width of strip are sampled at three positions.

tance between peaks and nulls in the plot and perform a wavelength calculation. In the current minimum regions, the currents on the edges of the strip do not show a minimum due to the time average nature of the plot. In an animation, we would see a full null across the width of strip due to the standing wave pattern.

In these three current plots we can also observe a very strong variation in the current distribution across the width of the strip. The current is forced to the surface of the conductor and then to the edges of the strip by skin effect, on a cylindrical conductor we would get a uniform distribution of current on the surface. We will call this variation across the width of the strip *spatial wavelength.* This variation is perpendicular to the direction of propagation and is typically not a function of frequency. It also requires a much finer discretization or meshing than the longitudinal guide wavelength. The concept of spatial wavelength on our microstrip line is probably new to many readers. We have not considered this variation across the width in the past because, unlike guide wavelength, the concept is not needed in circuit-theory-based CAD. In Figure 5.2, there are again 12 cells across the width of the line. We have sampled the current across the width of the line at three points and plotted the magnitude of the current on each cell. Although the absolute values vary, we see the same, nonuniform distribution in all three sample regions. The current is highest on the edge cells and nearly constant across the remaining cells.

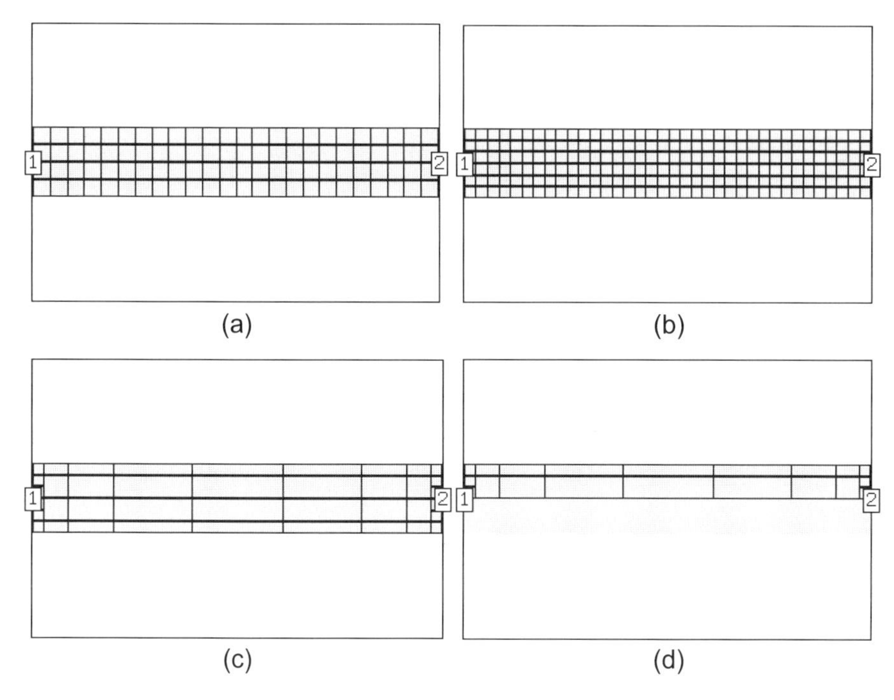

Figure 5.3 Meshing options for 50-ohm microstrip line using closed box MoM formulation: (a) cell size of 6 mil, (b) cell size of 4 mil, (c) some cells recombined into subsections, and (d) application of symmetry to the problem.

5.8 MOM—CELLS AND SUBSECTIONS

All the 2.5D moment method codes solve this type of problem by subdividing the metal patterns into smaller units. The closed box or fixed grid codes call these basic units "cells." If we can solve for the current on the conductors, we can derive more familiar units like *S*-parameters. In Figure 5.3(a) the 50-ohm line is divided into cells that are 6 mil on a side. The software solves for the X- and Y-directed currents on each cell. To get more accuracy we can use smaller cells (Figure 5.3(b)) to resolve the actual current distribution with more fidelity. Unfortunately, the N^2/N^3 effect will cause the solution time to increase very rapidly as we increase the number of cells. One rule of thumb is that we want roughly 20 cells per wavelength at the highest frequency of interest. One way to speed up the solution is to combine cells into "subsections," as shown in Figure 5.3(c). The subsection dimensions must be an integer multiple of the cell size. This results in a much smaller matrix to invert. If the subsections are no larger than $\lambda/20$ at the highest frequency of interest, then we usually have enough accuracy. In Figure 5.3(c) the 50-ohm line has been subsectioned using the $\lambda/20$ rule at 15 GHz. Another way to reduce the solu-

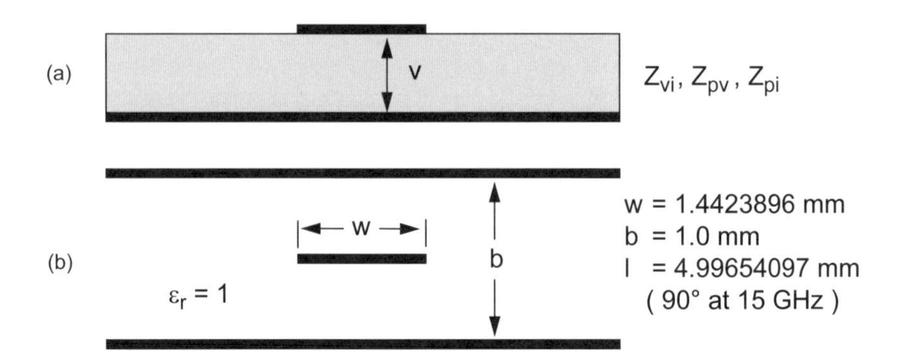

Figure 5.4 Potential validation standards: (a) microstrip with integration path to find voltage, and (b) stripline standard with dimensions for 50 ohm.

tion time is to use symmetry whenever possible. This can literally cut the problem size by two and may decrease the solution time by a factor of four to eight. However, this only works if all the ports are on the symmetry plane, as in Figure 5.3(d). If taking a symmetry plane deletes any ports, then we must do multiple solutions of the reduced port project to get a full set of *S*-parameters (see Section 8.8). Symmetry also suppresses all field solutions that do not satisfy the boundary conditions imposed by the symmetry plane. For example, placing an open-circuit plane (magnetic wall) in the center of the cross-sections in Figure 5.4 suppresses all modes with an odd symmetry of the electric field.

5.9 MOM—VALIDATION STRUCTURES

At some point, a critical user will question the ultimate accuracy of these tools. To measure absolute accuracy we need some kind of validation structure. Most of our RF and microwave circuits use transmission line structures. To use a transmission line we need to know impedance and phase velocity. Is microstrip a good candidate for a validation structure? Unfortunately, the answer is probably no. There are three possible definitions for microstrip impedance based on different combinations of computed voltage, current, and power. To find the voltage between strip and ground we must compute a line integral (Figure 5.4(a)). Unfortunately, the solution of this integral is not unique; it depends on the integration path (see Section 8.1).

Air-filled stripline (Figure 5.4(b)) is probably a better choice for a validation structure. This approach was proposed by Rautio [4]. In an air-filled, homogeneous TEM line, we know the phase velocity is the speed of light, *c*. There is also a very accurate analytical equation for stripline impedance, due to the homogeneous

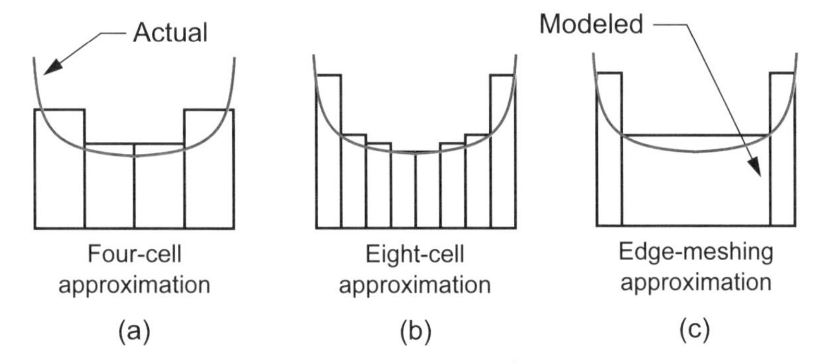

Figure 5.5 The typical microstrip current distribution (red) is approximated in three ways: (a) a rather coarse four-cell approximation, (b) a finer eight-cell approximation, and (c) the edge-meshing approximation.

dielectric and the symmetry of the structure. To first order, the total error is a combination of impedance error and phase error:

$$\text{Error}(\%) = (100 \times |S_{11}|) + \frac{|90 + Ang(S_{21})|}{0.9} \tag{5.1}$$

For any reader uncomfortable with the previous definition of error there is an alternative approach. First we convert the *S*-parameters for the validation line to *ABCD*-parameters. Then the computed phase length and impedance can be extracted from the transmission line equations in *ABCD* form.

$$\begin{bmatrix} S_{11} & S_{12} \\ S_{21} & S_{22} \end{bmatrix} \Rightarrow \begin{bmatrix} A & B \\ C & D \end{bmatrix} \tag{5.2}$$

$$\begin{bmatrix} A & B \\ C & D \end{bmatrix} = \begin{bmatrix} \cos(\beta L) & jZ_0 \sin(\beta L) \\ \dfrac{j\sin(\beta L)}{Z_0} & \cos(\beta L) \end{bmatrix} \tag{5.3}$$

First we solve for βL using the A or D term, then we can solve for Z_0 using the B or C term. With this approach we can plot impedance error and phase error independently.

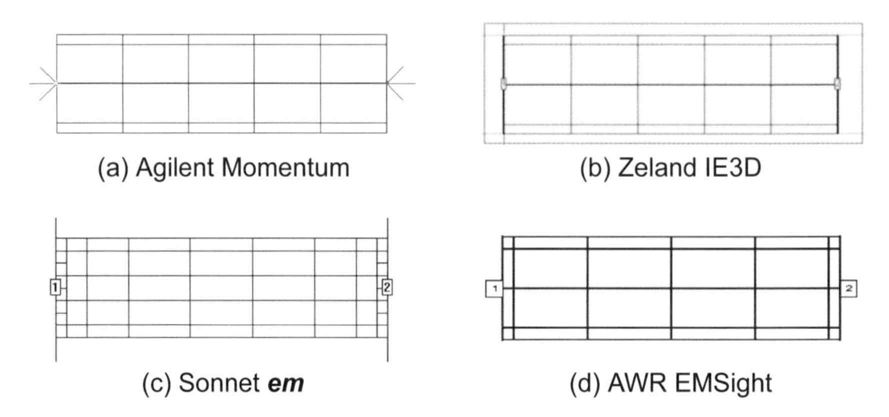

Figure 5.6 Default meshing for the stripline standard using several different MoM simulators. Although the details vary slightly, each simulator uses the edge-meshing concept.

5.10 MOM MESHING AND CONVERGENCE

We noted earlier that the currents on a microstrip line maximize on the edges of the strip. Again, for a low-loss line, the charges that make up the current repel and we get the charge/current distribution shown in Figure 5.5. How the field-solver approximates this current distribution can have a large impact on the final solution.

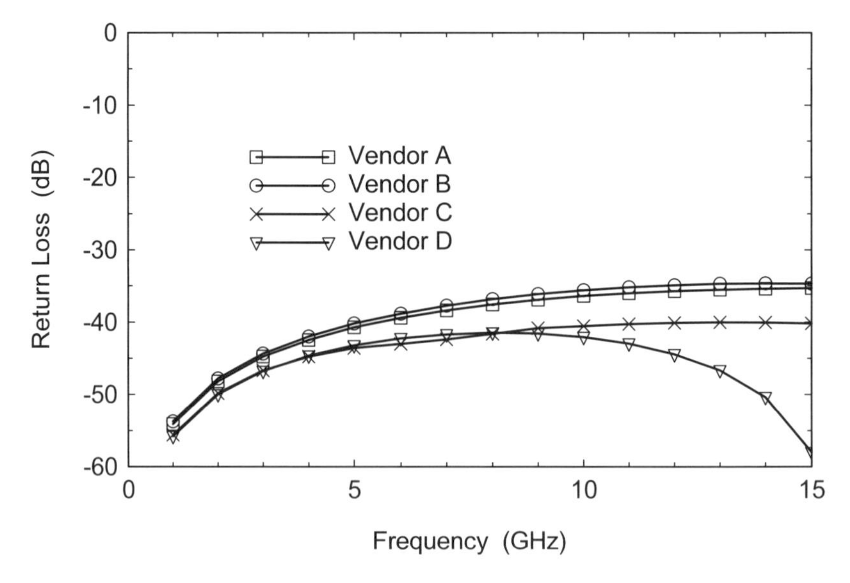

Figure 5.7 The stripline standard is analyzed for the four MoM simulators shown in Figure 5.6. The order of the results has been randomized.

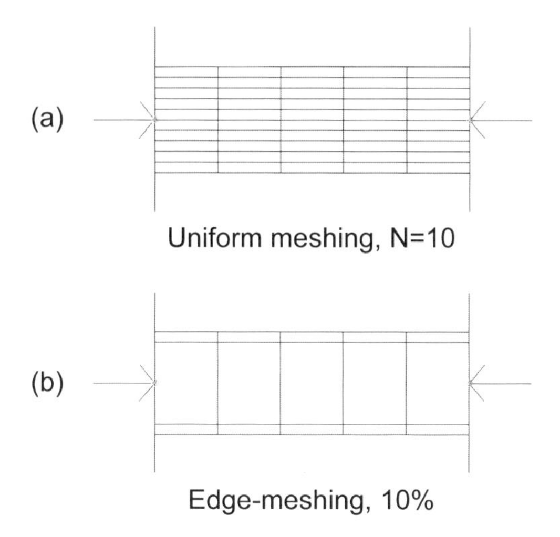

Figure 5.8 A comparison between uniform meshing and edge-meshing: (a) uniform meshing, $N = 10$, and (b) equivalent edge-mesh, 10% of strip width (Agilent Momentum, ADS Ver. 1.3).

If we think in terms of equal width cells or subsections, we clearly get a better approximation with more cells across the width of the strip. Notice that in the eight subsection case, all the center subsections have nearly the same value. This leads us to a concept most software vendors now call *edge-meshing*. We retain narrow cells on the edges of the strip to capture the singularity and use a single cell in the middle of the strip to model the remaining current. The edge-meshing scheme is particularly effective for tightly coupled structures, which tend to magnify the edge singularity. In Figure 5.6 we see the default mesh produced by several MoM based simulators. Although the details vary, each one is clearly applying the edge-meshing concept. With a little user intervention, we could force each simulator to produce exactly the same mesh.

The stripline standard has been analyzed using the four simulators shown in Figure 5.6. In Figure 5.7 we see the return loss results for these simulations, using the default mesh in each case. All the simulators agree that our standard has a minimum return loss of –35 dB up to 15 GHz. The most significant deviation is for Vendor D, which shows a null near 15 GHz. The null could be due to residual errors in the de-embedding scheme which cancel due to the $\lambda/4$ spacing. The null could also be cancellation due to the particular aspect ratio of the cells that were analyzed [2].

5.10.1 Uniform Versus Edge-Meshing

If we apply uniform meshing to a transmission line, we expect the impedance error to get smaller as we increase the number of cells across the width. But how does the edge-meshing approach behave under similar conditions? We can compare the two

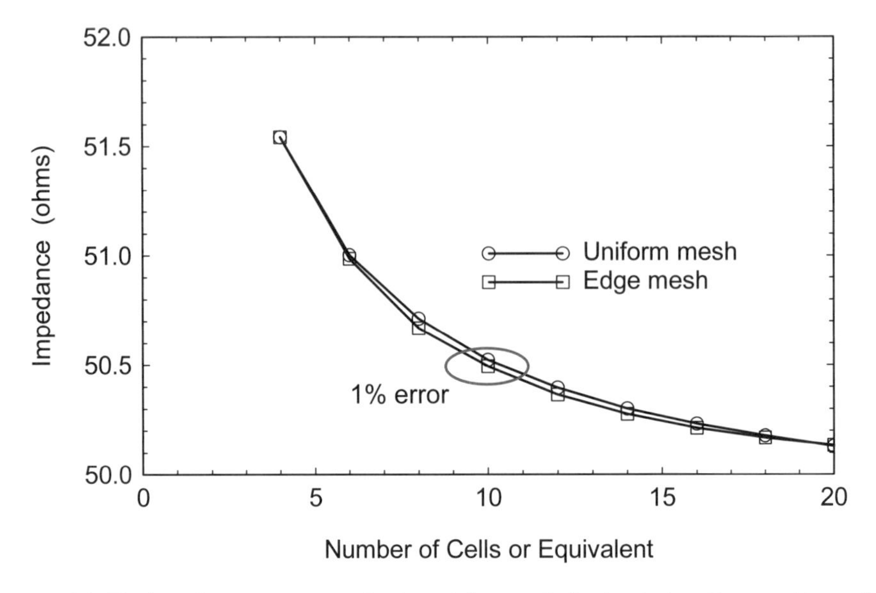

Figure 5.9 The impedance convergence for the stripline standard using both uniform meshing and edge-meshing (Agilent Momentum).

by computing the line impedances for both cases (Figure 5.8). The equivalent edge-mesh will have the same width cell on the strip edge as the uniform mesh. In Figure 5.9 we have plotted the impedance convergence for both uniform meshing and edge-meshing of the stripline standard. The impedance error falls to about 1% at 10 uniform cells or 10% edge-meshing. The recommendation from Agilent EEsof EDA is 10% to 15% of the strip width for edge-meshing. If the edge cell becomes too small, it is possible to observe a sudden divergence in the computed impedance. Of course, another advantage of edge-meshing is solution time. With edge-meshing, the number of unknowns stays constant in this exercise. For uniform meshing, the number of cells grows as we increase the resolution. For the range of uniform meshing shown in Figure 5.9, the solution time can increase by two orders of magnitude.

5.10.2 Microstrip Convergence

If our chosen medium is microstrip, we should probably do a convergence study in that medium. Assuming uniform meshing, we can repeat our experiment on computed impedance as a function of the number of subsections across the width of the strip. With only one or two subsections across the width the error is roughly 5%. With four to six subsections across the width the error drops to about 1%; this may be "good enough" for many engineering applications. To get to 0.1% error we may

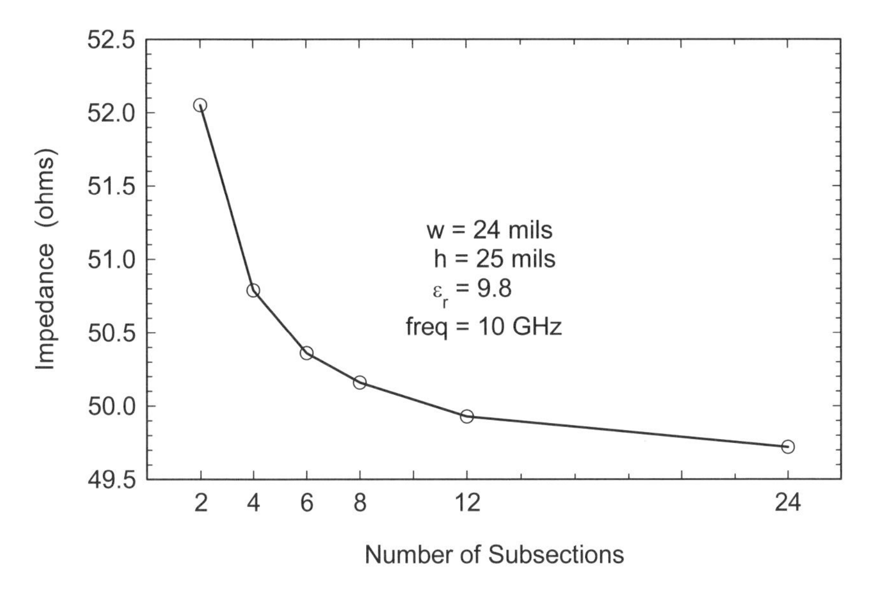

Figure 5.10 Convergence of a microstrip line on a ceramic substrate using uniform meshing (Sonnet *em* Ver. 8).

need 12 or more subsections across the width. As sophisticated users of field-solver software, we must constantly be conscious of convergence issues [5]. The results of this convergence experiment are shown in Figure 5.10.

5.10.3 Summary for Meshing and Impedance Convergence

Impedance and phase are the two most basic quantities for distributed RF and microwave circuits. We need to understand the behavior of our CAD tools regarding these two basic quantities. We can explore the absolute error for any field-solver using the stripline standard. However, it is probably more important to study the medium we are actually working in, whether it be microstrip, coplanar waveguide (CPW), or some other configuration. The user needs to develop some intuition for the accuracy of the mesh generated by the software. Error levels of 5% to 10% are probably not good enough; we should be able to do that well with circuit theory alone. An error level of 1% may be a good compromise between accuracy and solution time. Typical tolerances and errors in manufacturing are probably 1% or greater as well. Achieving error levels of 0.1% may be of academic interest, but will be a waste of resources in most engineering environments. The convergence behavior of all MoM simulators should be similar. In general, edge-meshing will be more efficient than uniform meshing. If we are going to import our field-solver results to a linear simulator, we need agreement between the two tools on imped-

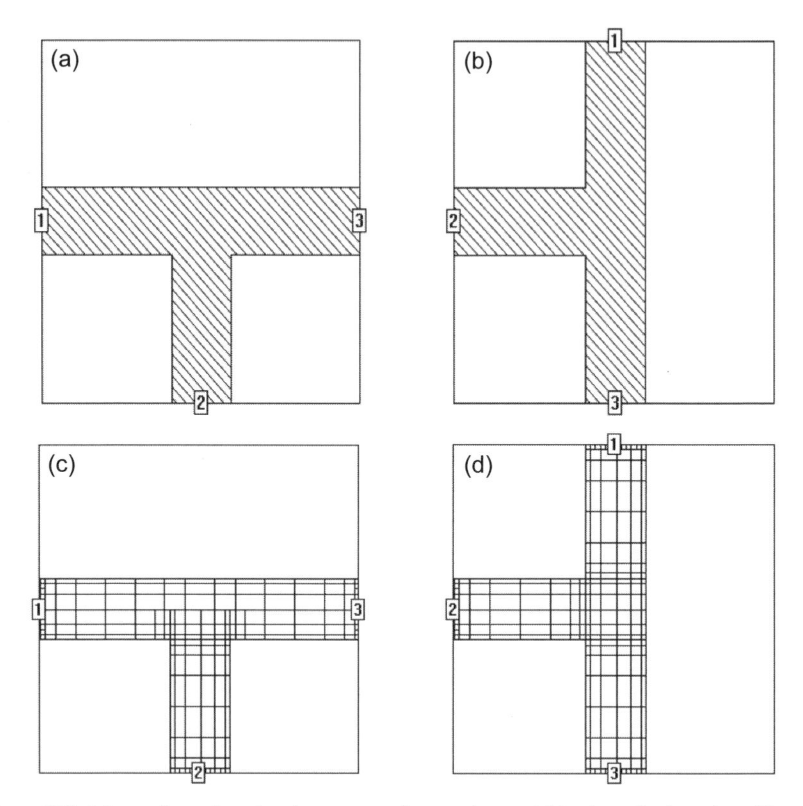

Figure 5.11 Microstrip tee-junction drawn as one large polygon: (a) horizontal orientation, (b) vertical orientation, (c) resulting mesh for horizontal orientation, and (d) resulting mesh for vertical orientation. The line widths are equal at all three ports (Sonnet ***em*** Ver. 8.0).

ance and phase velocity. For de-embedding to be accurate and useful, the two tools must agree on the characteristics of a given structure. There are other test structures that have been proposed for evaluation of electromagnetic field-solvers [6]. The interested reader is encouraged to pursue some of these additional test circuits.

5.11 CONTROLLING MESHING

How we draw our projects can have a big impact on the mesh generated by the field-solver. We should always keep in mind the nonuniform current distribution across strip type transmission lines. Corners, steps, and other features that cause abrupt change in the direction of current flow should also be noted. In general, we want to provide small, square cells or small, equilateral triangles in regions where the current changes direction abruptly. If our structure is symmetrical and we expect symmetrical *S*-parameters, then our mesh has to be symmetrical as well. As

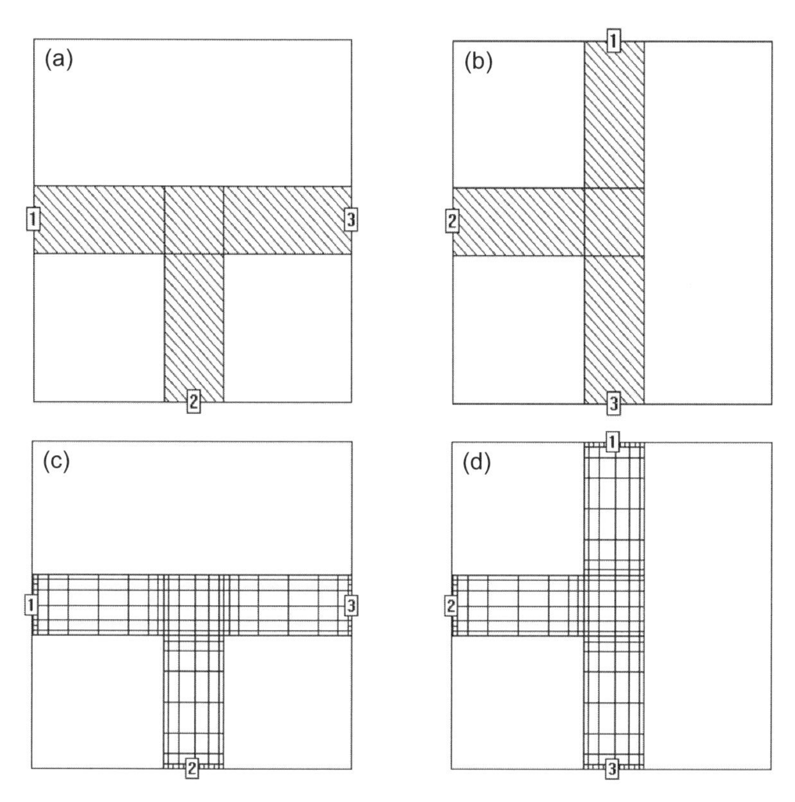

Figure 5.12 Microstrip tee-junction drawn as four polygons: (a) horizontal orientation, (b) vertical orientation, (c) resulting mesh for horizontal orientation, and (d) resulting mesh for vertical orientation (Sonnet ***em*** Ver. 8.0).

a rule, we want to avoid drawing our metal patterns as single, large polygons. Breaking the description into smaller rectangles and triangles gives us better control over the meshing process.

5.11.1 Meshing a Microstrip Tee-Junction

In Figure 5.11 we have drawn a simple microstrip tee-junction in two different orientations. The line widths are equal at all three ports. In both cases, we draw the discontinuity as one big polygon (Figure 5.11(a, b)). The resulting mesh in Figure 5.11(c) is slightly asymmetric, while the mesh in Figure 5.11(d) is perfectly symmetric. The final mesh is clearly dependent on the orientation of our object. In this simple example, the *S*-parameters for the two meshes are probably not radically different. But, in a larger project, changes in the mesh due to orientation could modify the computed results in an unpredictable way.

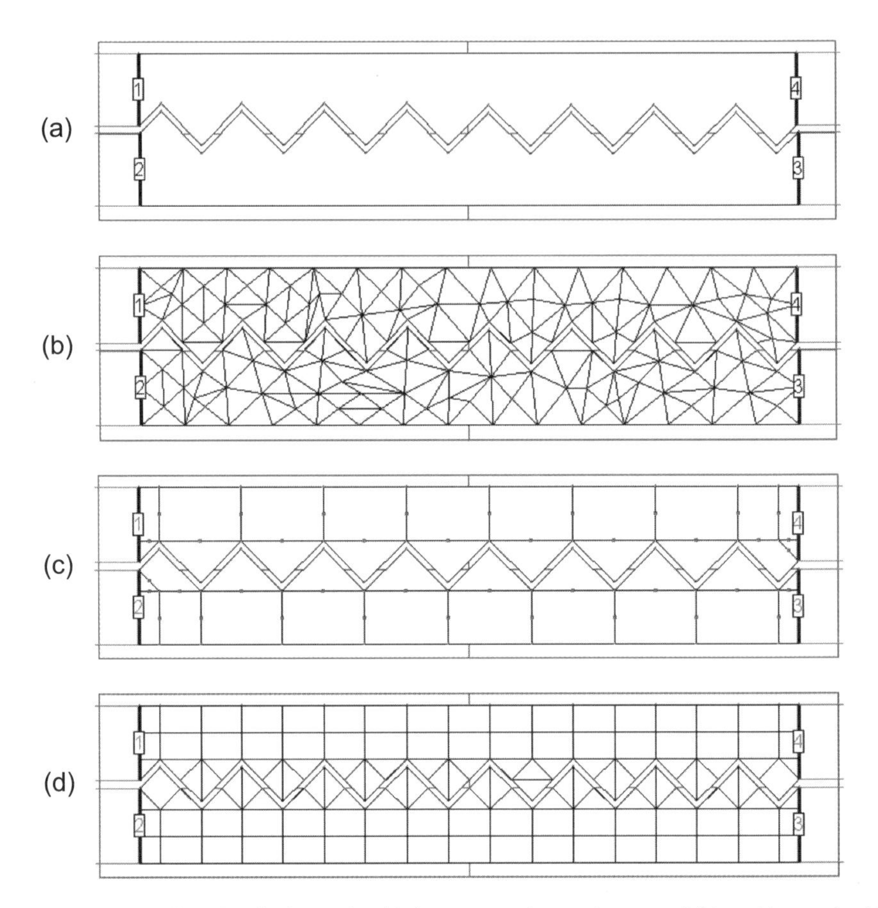

Figure 5.13 Meshing the wiggly coupler: (a) drawn as two large polygons, and (b) resulting mesh with large polygons, 290 cells, 392 unknowns, 55 sec/freq; (c) drawn as smaller rectangles and triangles, and (d) resulting mesh with smaller polygons, 204 cells, 310 unknowns, 31 sec/freq (Zeland IE3D Ver. 7.0).

It is just as easy to draw the tee-junction as four smaller polygons, rather than one big one. Figure 5.12 shows the same project drawn as four polygons. If we examine the meshes in Figure 5.12(c, d), we observe a perfectly symmetrical mesh regardless of the orientation. We get more consistent results with the smaller polygons because the meshing algorithm "starts again" at the boundary of each new polygon. By controlling the size and shape of the individual polygons that we draw, we can give the meshing algorithm strong hints as to what we want the final mesh to look like. Although the changes in the mesh for this example are rather minor, you should subdivide your geometries into smaller units as you draw. You should also examine the mesh carefully for areas of possible improvement before starting a simulation. We would also like to emphasize that we do not consider this type of behavior in the meshing algorithm to be a bug. Any routine designed to mesh a project must start at some boundary and scan across the existing geometry.

5.11.2 Meshing a Wiggly Coupler

The next example of mesh control comes from a laterally open MoM code. The geometry input into Zeland IE3D is part of a "wiggly" coupler designed to improve directivity in a microstrip environment. The pitch and depth of the coupler "teeth" are the variables we need for directivity improvement. This is clearly a case where the fine geometrical resolution we can achieve with the laterally open MoM codes is an asset. If we draw the coupler as two large polygons (Figure 5.13(a)) we get a mesh filled with many triangles (Figure 5.13(b)). This is valid mesh and we may get accurate S-parameters, but it is not the most efficient mesh for this geometry. The solution time is roughly 55 seconds per frequency point. In general, triangular cells take longer to compute than rectangular cells. We can get equally good results with far fewer cells.

After some experimenting, a new drawing was made for the coupler using several smaller polygons and triangles (Figure 5.13(c)). The new mesh that results is shown in Figure 5.13(d). The meshing parameters, upper frequency, and cells per wavelength were the same for both examples. The solution time is now 31 seconds per frequency point. This new mesh is much more efficient than the previous one; the solution time has been nearly cut in half. If we look at the predicted directivity for both meshes, there are small differences at the –40 dB level.

5.11.3 Meshing a Printed Spiral Inductor

A more complex example of meshing and convergence can be found in the pseudo-lumped filter [7] shown in Figure 5.14(a). The filter is based on the "dumbbell" or "tubular" topology that is often built in coaxial form. The large metal patches realize shunt capacitances with a smaller series capacitance across the gap. A chip capacitor is added across the outermost gaps (Figure 5.14(b)) because we cannot realize enough series capacitance using the gap alone. Series inductors, in this case printed spiral inductors, connect the capacitive regions. Different meshings were used for the spiral inductor and the error between the measured and computed center frequencies was measured. Measuring the center frequency of the filter is much easier than doing a careful de-embedded measurement of a single spiral inductor; however, there may be other confounding factors that spoil the experiment.

We can mesh the spiral inductors in any number of ways. In Figure 5.14(c) we have a spiral with 4-mil wide traces and 2-mil gaps between traces. The grid was set at 1 mil and we forced uniform meshing across the width of the trace. The bond-wire is approximated with a 4-mil wide metal trace (not shown) on a 1-mil thick layer of air just above the substrate layer. A thicker air layer provides separation from the cover. Via metal connects the upper trace to the lower traces. Note that we

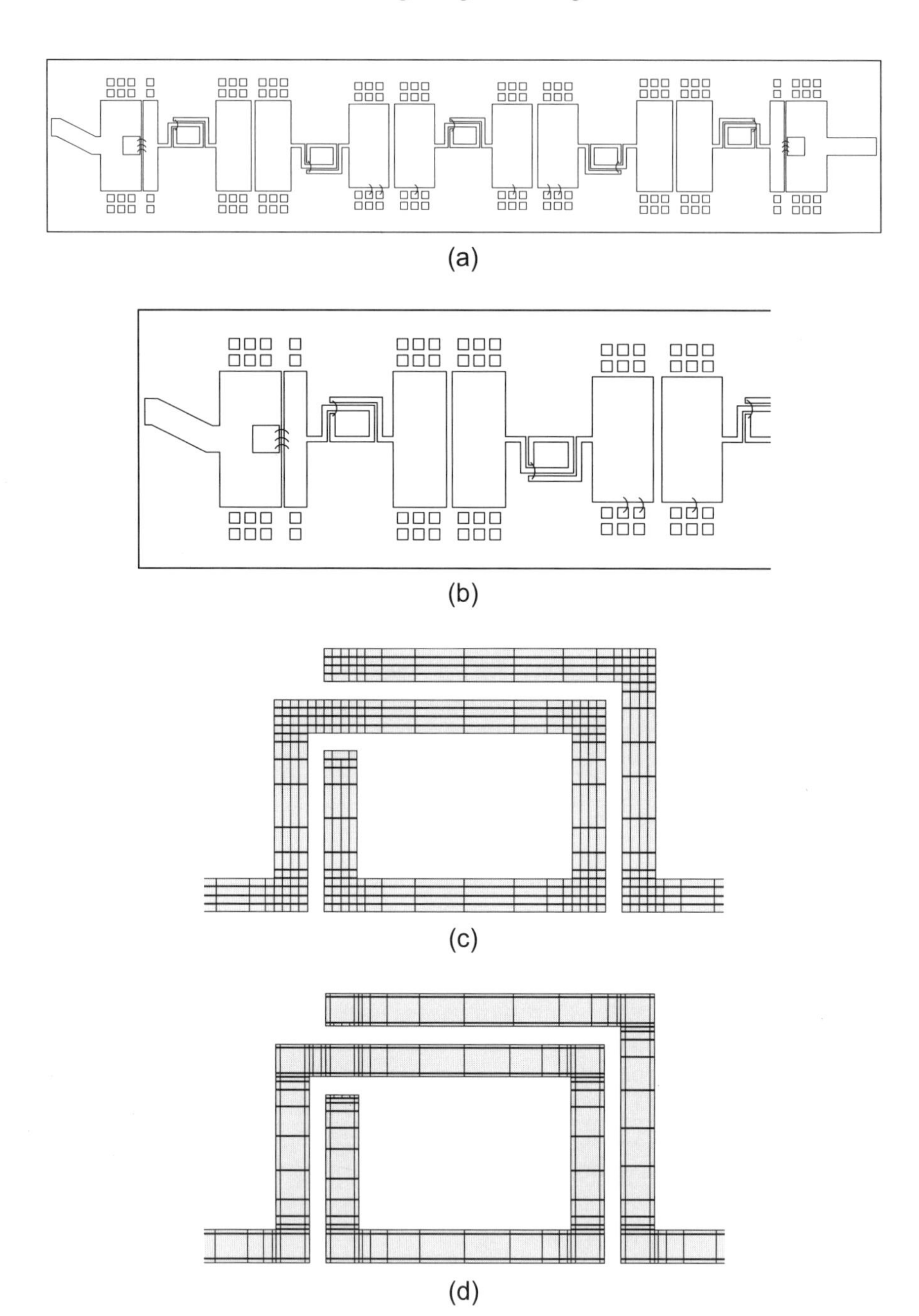

Figure 5.14 Printed bandpass filter: (a) top view of full substrate, dimensions are 94-mil long by 190-mil wide; (b) zoomed in view near the input; (c) a spiral inductor with 4-mil lines and 2-mil gaps, uniform meshing on 1-mil grid; and (d) a spiral inductor with 4-mil lines and 2-mil gaps, edge-meshing using a 0.5-mil grid. © 1995 John Wiley & Sons, Inc. [7].

Table 5.4

Printed Spiral Inductor Meshing Experiments

Trace width (mil)	*Pattern of subsection widths* (mil)	*Grid size* (mil)	*Number of subsections*	*Solution time** (min:sec)	*Filter f_0 error*
2	1-1	1.0	298	1:25	2.7%
3	1-1-1	1.0	556	1:40	2.0%
4	1-1-1-1	1.0	844	2:38	1.3%
4	1-2-1	1.0	599	7:34	1.3%
4	0.5-1-1-1-0.5	0.5	1,416	10:53	0.8%
4	0.5-3-0.5	0.5	705	4:18	0.8%

*50-MHz Sparc-10 with 64 MB RAM, circa 1994

have forced small, square cells in the corners where we know the current must change direction rapidly. A second possible meshing of the same spiral inductor geometry can be found in Figure 5.14(d). In this case we have set a 0.5-mil grid and forced edge-meshing across the width of the traces.

In both meshing examples we have also forced a particular mesh alignment where the "bondwire" metal crosses over the lower trace. For maximum accuracy, we want the meshes on both layers to be identical and in perfect registration in the region where the metals overlap. This type of mesh alignment is even more important for thinner dielectric layers, like those found in metal-insulator-metal (MIM) capacitors and in monolithic integrated circuits (MICs) in general.

Table 5.4 is a summary of several meshing experiments performed on the printed spiral inductors. In each case, a filter was built and the measured center frequency compared to the predicted center frequency. In the first three rows we are increasing the width of the spiral trace with uniform meshing applied in each case. The center frequency improves dramatically because the resolution of our mesh across the width of the traces is improving. Rows 3 and 4 of the table indicate that uniform meshing and edge-meshing give us the same accuracy for the same grid resolution. Rows 5 and 6 of the table demonstrate that increasing the resolution of the mesh further improves the accuracy of the center frequency prediction. An edge cell width of 0.5-mil is close to the 10% of trace width rule of thumb. We would expect the solution times to improve dramatically on a modern computer.

5.11.4 Meshing Printed Capacitors

Next we focus on the large metal patches in Figure 5.14(a) that realize a pi-network of capacitors. These are essentially large, coupled line structures, so we would expect the meshing techniques used on the spiral inductors to also work for the

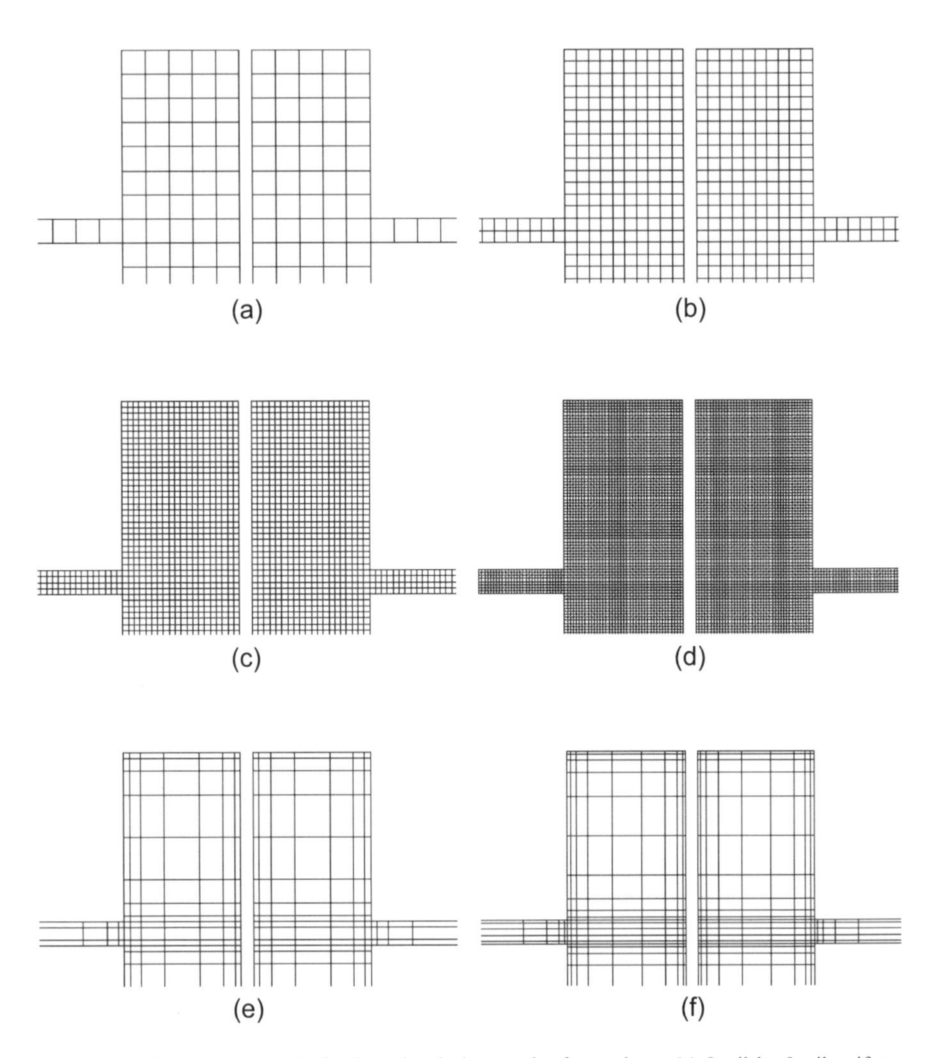

Figure 5.15 Convergence study for the printed pi-network of capacitors: (a) 8 mil by 8 mil uniform meshing; (b) 4 mil by 4 mil uniform meshing; (c) 2 mil by 2 mil uniform meshing; (d) 1 mil by 1 mil uniform meshing; (e) edge-meshing with 2-mil edge cell; and (f) edge-meshing with 1-mil edge cell (Sonnet ***em*** Ver. 7).

printed capacitors. Just to be sure, we can perform a small convergence study to verify the correct meshing procedure. In Figure 5.15(a) we start with a rather coarse, uniform mesh of 8-mil by 8-mil cells. The mesh size is then cut in half three times until we reach a uniform mesh of 1-mil by 1-mil cells (Figure 5.15(d)). Next, we tried two edge-meshing experiments. One analysis used an edge cell width of

Table 5.5

Printed Capacitor Convergence Study

Meshing pattern	C_1 (pF)	C_{12} (pF)	*Number of subsections*	*Solution time** (sec)
Uniform 8 × 8mil	0.7991	0.1788	172	1
Uniform 4 × 4mil	0.8060	0.1948	610	2
Uniform 2 × 2mil	0.8092	0.2070	2,576	16
Uniform 1 × 1mil	0.8111	0.2137	10,612	721
2-mil edge-mesh	0.8095	0.2071	304	1
1-mil edge-mesh	0.8111	0.2135	544	3

*1.13-GHz Pentium III notebook with 1 GB RAM, circa 2002

2 mil (Figure 5.15(e)) and the final run used an edge cell width of 1 mil (Figure 5.15(f)). The results of this convergence study are summarized in Table 5.5.

In each case we de-embedded (electrically removed) the series feed lines down to the edges of the large metal patches. We assume that the equivalent circuit for the structure is then two shunt capacitors with a smaller series capacitor between them. We can write the *Y*-matrix for this simple network by inspection,

$$\mathbf{Y} = \begin{bmatrix} j\omega C_1 + j\omega C_{12} & -j\omega C_{12} \\ -j\omega C_{12} & j\omega C_2 + j\omega C_{12} \end{bmatrix} \tag{5.4}$$

where C_1 and C_2 are the shunt capacitors and C_{12} is the series capacitor. If we ask for *Y*-parameters from the field-solver, rather than *S*-parameters, we can extract the capacitor values directly with a pocket calculator.

$$C_1 = \frac{y_{11} + y_{12}}{2\pi f} \tag{5.5}$$

$$C_{12} = \frac{-y_{12}}{2\pi f} \tag{5.6}$$

In the previous section we looked at a printed spiral inductor. If we assume the first order model is a series inductance with a shunt, parasitic capacitor at each end, we can use this same *Y*-matrix technique to extract the element values. This direct approach is often more reliable than trying to match a model to the computed data using optimization; the optimization results often depend on the starting point.

Table 5.6

Richardson Extrapolation of C_1 and C_{12}

			Extrapolations	
Cell size (mil)	C_1 (pF)	E_1	E_2	E_3
8	0.7991			
4	0.8060	0.8129		
2	0.8092	0.8124	0.8122	
1	0.8111	0.8130	0.8132	0.8133

			Extrapolations	
Cell size (mil)	C_{12} (pF)	E_1	E_2	E_3
8	0.1738			
4	0.1948	0.2158		
2	0.2070	0.2192	0.2203	
1	0.2137	0.2204	0.2208	0.2209

Returning to Table 5.5, note that for the uniform meshing examples both C_1 and C_{12} are converging as the cell size decreases. The results for 2-mil edge-meshing are virtually identical to the uniform meshing results for the same cell size. The same conclusion can be drawn for 1-mil uniform and edge-meshing. Of course, the number of subsections and the solution times are much lower for the edge-meshing experiments.

Although the finely meshed solutions in Table 5.5 appear to be well converged, how do we know for sure? We could run several more meshing experiments using a finer and finer mesh. But at some point, numerical precision or solution time would make this impractical, particularly for larger problems. Or in the case of edge-meshing, the edge cell may become too small and the solution could suddenly diverge. Another option is to apply Richardson extrapolation [8] to the data already available in Table 5.5. The basic idea is to successively halve the distance between data points, fit a polynomial curve to the data and compute the solution at $x = 0$. In this case x is the cell size. Booton [9] is one of the few texts on numerical electromagnetics that discuss this technique. After fully developing the method, Booton presents a simple recursion formula (equation (1-9), p. 10):

$$E_n(x) = \frac{2^n E_{n-1}(x) - E_{n-1}(2x)}{2^n - 1} \tag{5.7}$$

Richardson extrapolation was applied to the uniform meshing data for C_1 and C_{12} in Table 5.5; the results can be found in Table 5.6. The extrapolated value of C_1 is

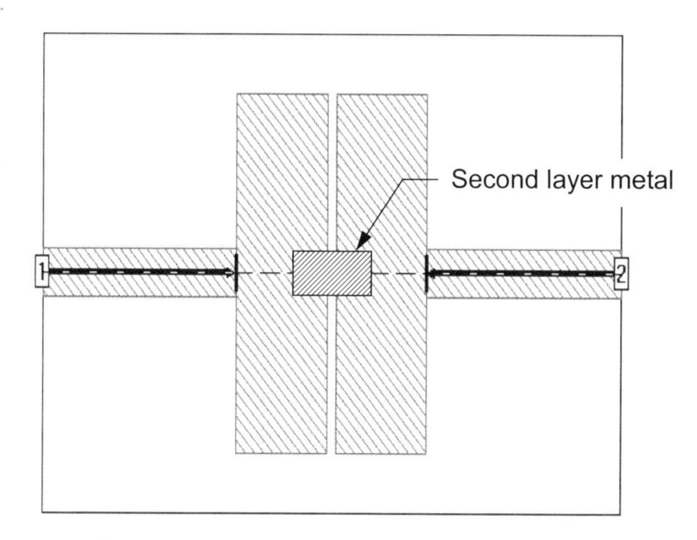

Figure 5.16 Printed pi-network of capacitors with overlay or gap capacitor.

less than 0.3% higher than the value computed with a 1-mil uniform mesh. The extrapolated value of C_{12} is about 3% higher than the value computed with a 1-mil uniform mesh. Additional information on analysis of planar capacitors can be found in [10, 11].

5.11.5 Meshing Overlay and MIM Capacitors

The final topic on MoM meshing relates to overlay and MIM capacitors. The planar filter in Figure 5.14(a) has chip capacitors across the gaps of the outer pi-networks because the required capacitance cannot be realized using only the gap between the larger patches. How do we approximate the chip capacitor in the EM simulation? The actual chip capacitor is a small, thin square of high dielectric constant material with metal plates on the top and bottom. Any dielectric layer we place in the MoM simulator must be homogenous across the entire simulation region. So any new dielectric layer above the first metal layer will modify our results, unless that new layer is air. One way to approximate the real physical size and capacitance density of the chip capacitor is to make the new air layer very thin, in this case 0.020 mil. The bondwires can be approximated using metal in the new layer and via metal. Or, a "gap capacitor" can be realized, which bridges the gap with the upper metal plate.

Figure 5.16 shows how a gap capacitor in the EM simulation was used to approximate the chip capacitor. The first time a structure like this was simulated and built, in the early days of the planar solvers, the results were very poor indeed. To quantify that statement somewhat, the error in the realized versus computed

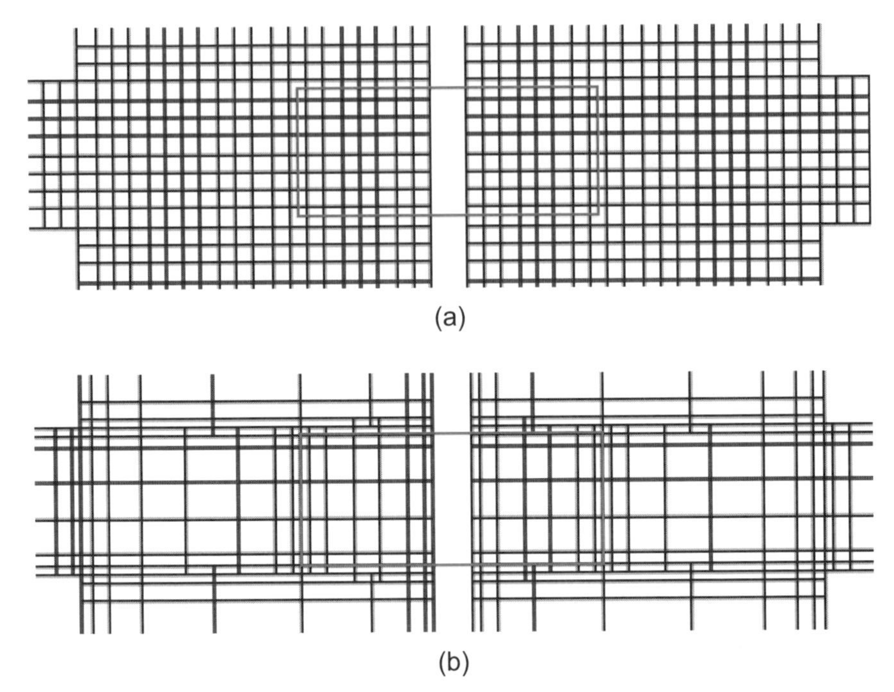

Figure 5.17 Two meshings for the overlay capacitor, the upper plate is outlined in red: (a) a uniform grid of 2 by 2-mil cells is applied to the upper and lower plates, but the upper plate is offset by half a cell in both X and Y; and (b) edge-meshing is applied to the upper and lower plates, with forced mesh alignment between the layers.

series capacitance was typically 25% to 50%. Several early users of the these codes noted this problem and came up with a solution at about the same time.

When an MoM simulator encounters metal plates separated by a very thin dielectric layer, it is important that the mesh on both metal plates be in perfect alignment. Figure 5.17 shows two numerical experiments that were designed to demonstrate this problem. In both cases the upper plate is outlined in red, but the mesh on the upper plate is not shown. In Figure 5.17(a) a uniform 2mil by 2mil mesh is applied to the upper and lower plates. We know from our earlier meshing experiments that this should give us a fairly accurate solution. However, the mesh on the upper plate is offset by half a cell in both X and Y. In Figure 5.17(b) an edge-meshing algorithm was applied to both layers and the mesher also forced mesh alignment between the two layers.

The results for these two experiments can be found in Table 5.7; a symmetry plane was applied in both cases. If we assume that the aligned mesh result is correct, the offset mesh result is too low by almost 24%. The parallel plate capacitance is virtually the same in both cases; either mesh is adequate to capture the parallel plate component and the fringing around the outer edges.

Table 5.7

Results of Mesh Alignment Experiments for Overlay Capacitors

Mesh type	C_1 (pF)	C_{12} (pF)	*Number of subsections*	*Solution time** (sec)
Aligned, 1 mil edge-mesh	0.7958	3.046	812	4
Offset, 2 mil uniform mesh	0.7937	2.326	3,055	29

*1.13-GHz Pentium III notebook with 1 GB RAM, circa 2002

The same mesh alignment is required in MIM capacitors in radio frequency integrated circuits (RFICs). Baluns, multilayer spiral inductors, multilayer transformers, and other tightly coupled structures would also require this type of mesh alignment.

After this effect was discovered and quantified by several users, the results were reported back to the software vendors. The vendors modified their meshing algorithms to force this type of mesh alignment between layers. This alignment between layers is now the default mode for most MoM meshing algorithms. However, this strict alignment is not needed for normal microstrip and stripline dimensions. In a multilayer printed circuit board, forcing this alignment between all layers can dramatically increase the problem size with no apparent increase in accuracy. The sophisticated user must be prepared to evaluate his or her particular problem and decide which mode of meshing is more appropriate.

5.11.6 Exceptions to Mesh Control Discussion

One notable exception to our mesh control discussion is Agilent Momentum. The meshing algorithm in Momentum recombines smaller polygons into larger polygons. It then searches these larger polygons for standard geometries, like bends and tee-junctions. Finally, the software applies precomputed, internally coded meshing rules to the geometries it has identified. The goal, of course, is to make the software more transparent to the average user, but it does prevent the advanced user from controlling the process.

5.11.7 Summary for Mesh Control

The fundamental goal of surface meshing is to capture the current distribution with enough resolution to provide an accurate solution. A mesh of small, equilateral square or triangular cells would be a safe bet, but not the most efficient mesh for large problems. Edge-meshing captures the edge singularity in strip type geometries and leads to a more efficient mesh. In most simulators, subdividing the geometry into smaller polygons forces the meshing algorithm to reset and gives the user

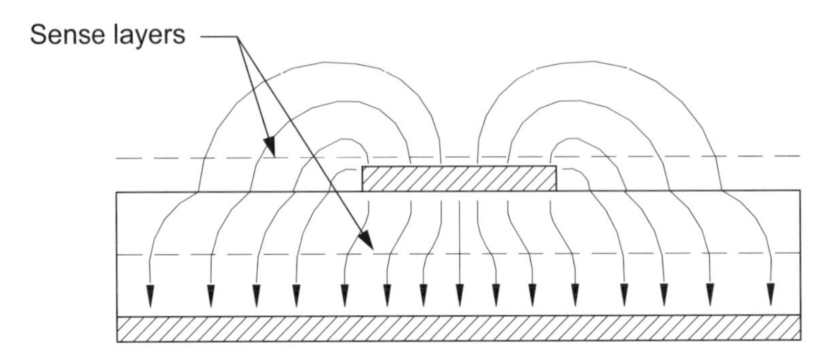

Figure 5.18 End view of a microstrip line with typical E-field distribution. Voltage sense layers have been placed just above the microstrip line and between the line and the ground plane.

control over the final mesh. The solution process almost always includes some experimenting with the drawn polygons to optimize the mesh. A quick convergence study, or a Richardson extrapolation gives the user confidence that some level of accuracy has been achieved.

Thin dielectric layers are a particular challenge for MoM simulators. When two metal plates are separated by a thin dielectric layer, the mesh on both plates must be in perfect alignment. Thin dielectric layers can also cause numerical problems in the Green's function computation. Projects with multiple thin layers, like a passivation layer over an MIM structure, should be examined carefully.

5.12 MOM—DISPLAYING VOLTAGE

The fundamental quantity that all MoM codes compute is the conduction current on the planar metals. It is sometimes helpful to also view voltages in the structure. We can get an indirect view of the voltage in certain planes by using a very old trick. A one megohm per square sheet resistor will sense the voltage in an X-Y plane which is proportional to the tangential E-field in that plane. This is very similar to the "resistance paper" you may have used in your first year physics lab. In Figure 5.18 we have placed sheet resistors in the X-Y plane at two different heights.

Figure 5.19 contains the results of this experiment. In Figure 5.19(a) we show the current on the microstrip line for reference. In Figure 5.19(b) the sense layer is 1 mil above the conductor. Note the voltage is 90 degrees out of phase with the current, as we would expect. The voltage scale is 100 to 1,000 volts/meter. In Figure 5.19(c) the sheet resistor is halfway between the strip and the ground plane. Note how far the voltage (E-field) components extend beyond the edges of the strip. The

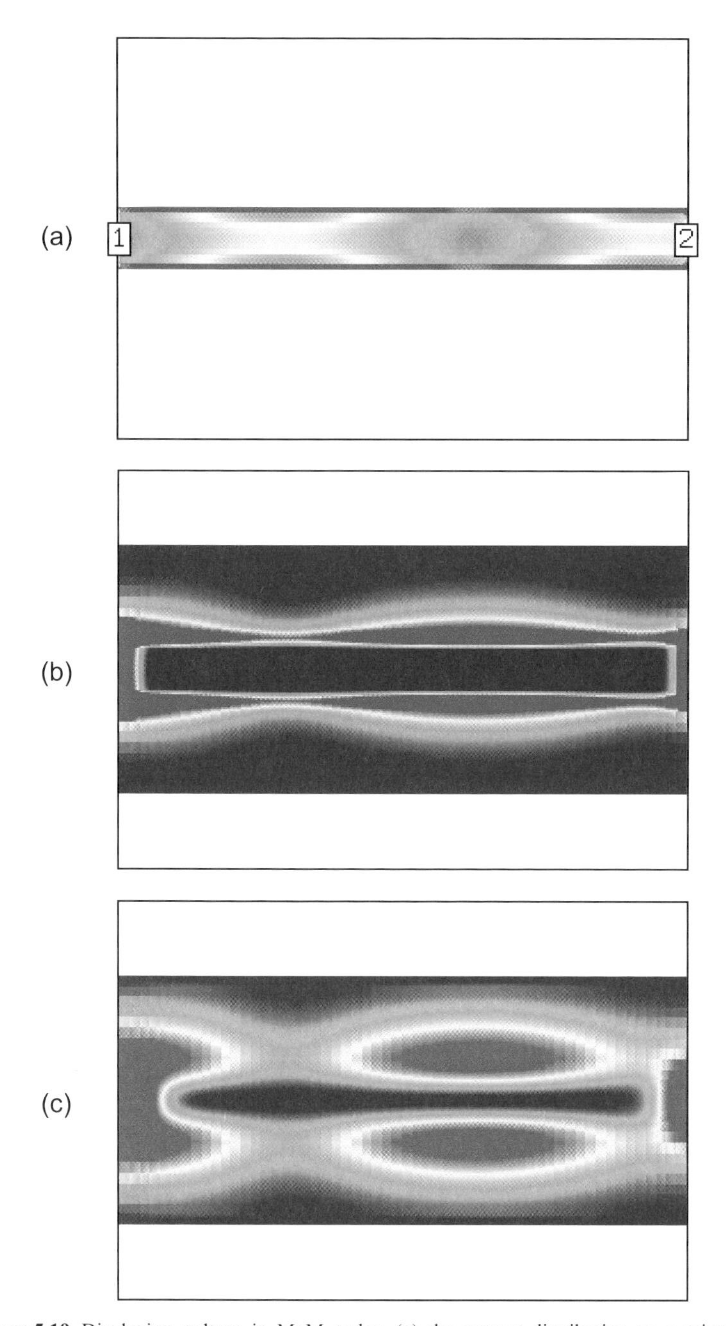

Figure 5.19 Displaying voltage in MoM codes: (a) the current distribution on a microstrip line at 15 GHz; (b) the tangential component of the voltage 1 mil above the strip; and (c) the tangential component of the voltage halfway between the strip and the ground plane.

voltage scale is 30 to 300 volts/meter. Of course, the disadvantage of this display trick is that we get a null value where the E-field is normal to the conductor.

5.13 MOM—CALIBRATION STRUCTURES

Whenever we acquire a new CAD tool it is very tempting to immediately begin working on a fairly difficult problem. If the CAD tool is a field-solver, the analysis of a difficult problem might take hours or even days. It is more helpful to start with a few simple, well-understood projects that will allow the user time to experiment with the various features of the software. The goal is to "calibrate" or train the user, not to measure or improve the accuracy of the software. While solving these simple projects, the user should experiment with:

- Meshing;
- Convergence;
- De-embedding;
- Visualization.

For the moment method solvers, a set of planar circuits is the most obvious choice. With the goal of "calibration" in mind, perhaps there is a strong analogy to the collection of standards used to calibrate a vector network analyzer out in the lab. With this similarity in mind, we have chosen:

- Microstrip through line;
- Microstrip ideal short circuit;
- Microstrip open circuit
- Microstrip 50-ohm termination

Again, these structures should be so simple that they compute in minutes, if not tens of seconds. They should also be so simple that we think we know what the result will be before the computation begins. These very basic structures also train the user to recognize "normal" current distributions or field patterns. Then, when something goes wrong, the sophisticated user immediately recognizes an abnormal current distribution or field pattern.

5.13.1 Microstrip Ideal Short Circuit

We have already looked at a microstrip through line in some detail, so we will move on to the next example. The second "calibration" structure in our orientation is an

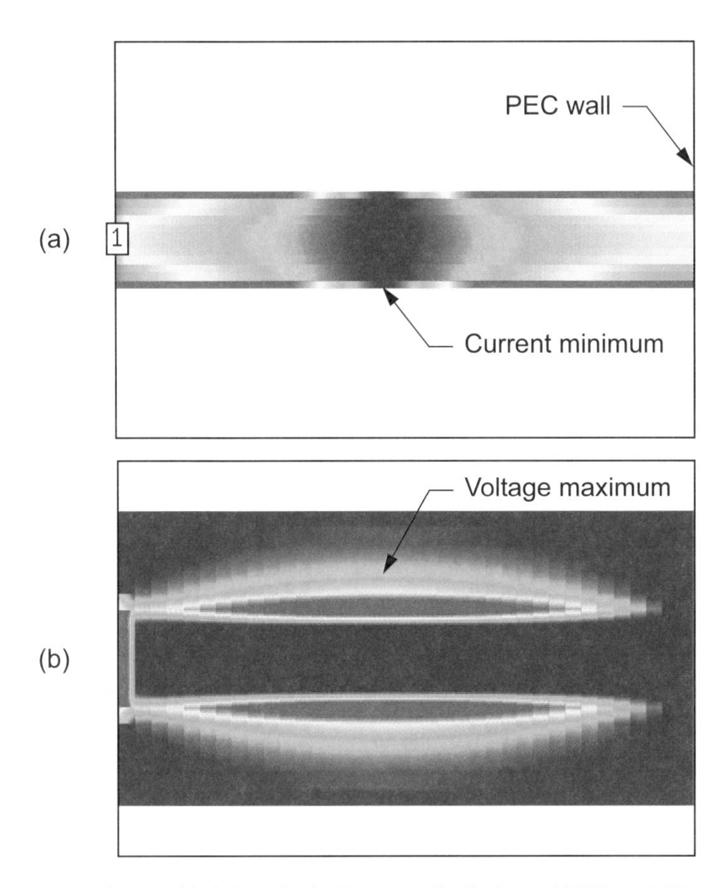

Figure 5.20 Microstrip ideal short circuit: (a) current distribution at 15 GHz; and (b) tangential voltage 1 mil above the strip.

ideal short circuit (Figure 5.20(a)). The generator is at port one and the line terminates in an ideal conductive wall on the right. We expect to see a current maximum at the short circuit. Before we generated the current plot we assumed that the entire right hand end of the strip would be bright red. Why doesn't the current maximum spread across the full width of the strip at the wall? The current follows the shortest path to ground and stays on the edges of the strip. Or in other words, if the potential is the same on both edges of the strip, there is no potential difference to drive the current sideways. Transverse current flow also implies energy storage or non-TEM behavior.

If the current maximizes at the wall, we expect a null due to the standing wave pattern 90 degrees to the left of the maximum. Figure 5.20(b) shows the voltage 1 mil above the line. As we expect, it is 90 degrees out of phase with the current. The voltage scale is 200 to 2,000 volts/meter.

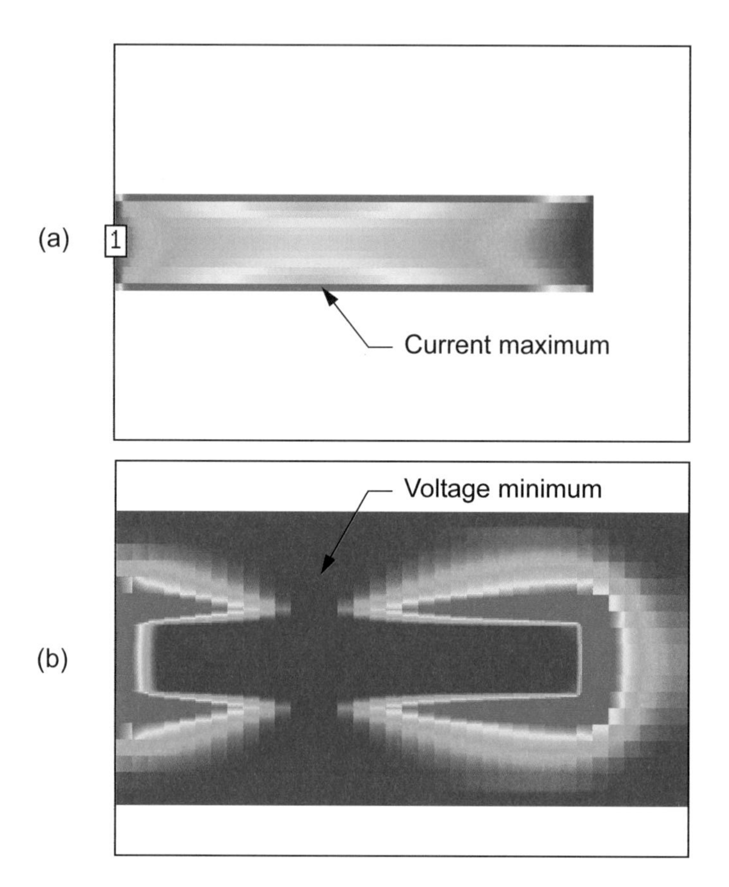

Figure 5.21 Microstrip open circuit: (a) current distribution at 15 GHz; and (b) tangential voltage 1 mil above the strip.

5.13.2 Microstrip Open Circuit

The next calibration element is a microstrip open circuit (Figure 5.21(a)). Transmission line theory tells us that the conduction current should be zero and voltage maximum at the open-end. The current maximizes 90 degrees to the left of the open end. The voltage one mil above the line, Figure 5.21(b) maximizes at the open end and is 90 degrees out of phase with the current. The voltage scale is 200 to 2,000 volts/meter.

5.13.3 Microstrip Thin-Film Resistor

Our final calibration element is a 50-ohm thin-film resistor (Figure 5.22). A generator is connected to port one and the resistor is terminated by an ideal conductive

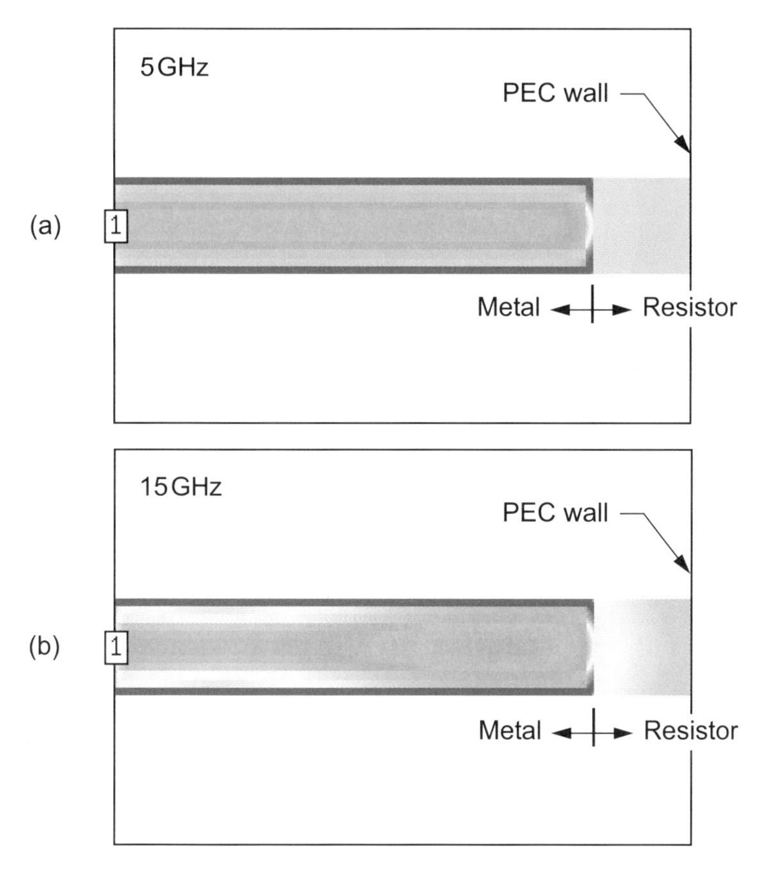

Figure 5.22 Microstrip thin-film resistor: (a) current distribution at 5 GHz; and (b) current distribution at 15 GHz.

wall on the right. Figure 5.22(a) shows the time averaged conduction current at 5 GHz; note the uniform current distribution (yellow) on the resistor. All the previous microstrip examples might lead us to expect a nonuniform current distribution across the width of the resistor. However, at dc we know the current distribution must be uniform; a resistor with uniform ohms/square implies a uniform voltage/square and a uniform current distribution. Even at 5 GHz it looks like we are in this "Ohm's law" type region. The demand for uniform current on the resistor also forces a large transverse current on the conductor where it joins the resistor. We can think of this transverse current as an inductive discontinuity in series with the resistor. A transverse notch in a microstrip line [12] forces this same current pattern and is modeled as a series inductance. Again, any time current flows perpendicular to the direction of propagation, it implies energy storage in a discontinuity or some other non-TEM behavior.

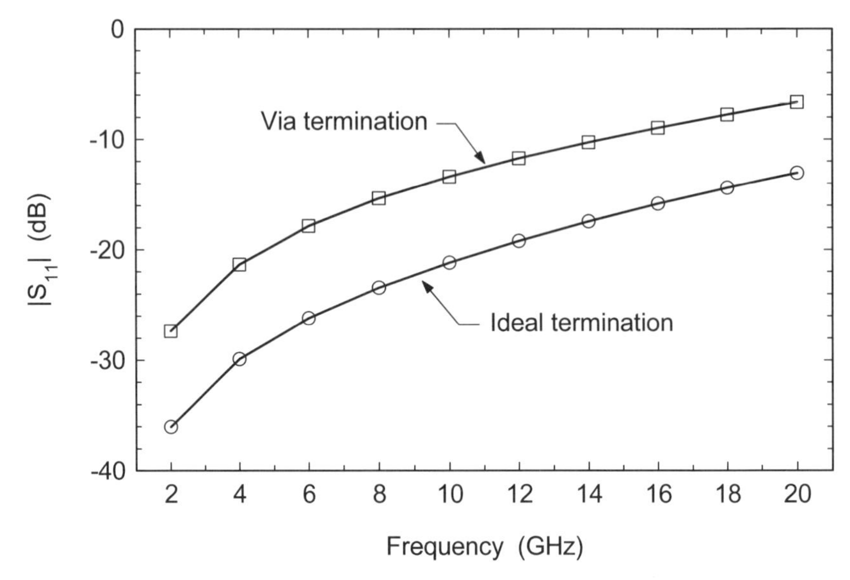

Figure 5.23 Return loss of thin-film resistor for ideal termination (box wall) and via hole termination.

Figure 5.22(b) shows the current distribution at 15 GHz. As we go higher in frequency, transmission line theory and guide wavelength come into play. The microstrip line is now long enough to see some longitudinal variation in the current distribution. The resistor is also getting longer in terms of wavelengths and we do see a finite current distribution along the length of the resistor. If we compare the two resistor current distributions in Figure 5.22, we are seeing the transition from lumped to distributed behavior for a printed component. The scale in both plots is 5 to 20 amps/meter.

It is fascinating how complicated the current distribution patterns on microstrip components actually are. On the microstrip conductor the skin effect forces the current to the surface and then to the edges of the strip. At low frequencies the current distribution on the resistor is quite uniform, but then it begins to change at higher frequencies where the resistor starts to behave more like a distributed component. Given a blank sheet of paper, how many engineers do you know that could draw these current distributions before they saw the plots?

This resistor is fairly small, 24 by 24 mil, and its return loss is better than 20 dB up to 10 GHz (Figure 5.23). But in most microwave circuits our ideal wall is more likely a plated via hole to the ground plane. The additional inductance of the via seriously degrades the return loss. An interesting exercise would be to compensate the resistor/via combination for better return loss. Earlier we claimed that the discontinuities in this structure were largely inductive. Looking at the Smith Chart (Figure 5.24) we see a large series inductance component and a smaller shunt capacitance component.

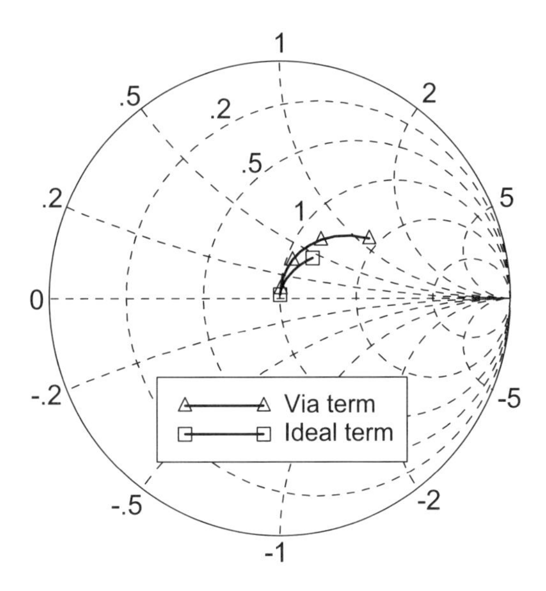

Figure 5.24 Reflection coefficient of the thin-film resistor for an ideal termination (box wall) and via hole termination.

This thin-film resistor is also an interesting example of how looking at current plots sometimes leads to a new design. A colleague looked at these plots, the light bulb came on, and he came up with a new, high performance geometry for a planar termination [13]. The key of course is the junction region between the microstrip line and the resistor. The current on the line wants to stay on the edges, but is forced to move sideways to feed the current on the interior of the resistor. The now obvious solution is to split the resistor into two, longer and thinner 100-ohm resistors whose physical width more closely matches the current distribution on the edges of the microstrip.

5.13.4 Summary for Microstrip Calibration Structures

The microstrip calibration structures are simple enough to draw fairly rapidly in any MoM simulator. In the through line, we saw, perhaps for the first time, the non-uniform current distribution due to the edge singularity. The ideal short circuit forced us to think a little about the path current will take to ground. The ideal open circuit perhaps behaved exactly as we first expected. Finally, in the case of the thin-film resistor, we found that the current distribution on the resistor and at the metal/resistor junction is more complex than we originally imagined. These very simple structures give us a chance to exercise the software, as well as our understanding. These test cases give the user an opportunity to learn what "normal" behavior looks like, so when something out of the ordinary appears, the trained eye immediately spots it.

5.14 VISUALIZATION

All the method of moments codes are solving for the currents on the conductors. Therefore, the conduction plots generated by these codes are generally of very high quality. For publication quality plots we generally use a fairly fine, uniform mesh to achieve the most pleasing results. We can display the vector magnitude of the current with a color scale applied to the magnitude. Some codes add a small vector to indicate the direction of current. When we animate the plot by rotating the phase of the input signal, the current vector generally flips back and forth in 180-degree jumps as the phase changes. If the current vector appears to rotate, this is sometimes an indication that the circuit is radiating, or would radiate in an open environment.

It is also possible to derive the E-field from the vector potential function, **A**. Some of the MoM codes offer an E-field display based on this computation. Using the trick presented in Section 5.12, any of the MoM codes can also display the component of the E-field that is tangential to a given horizontal plane. For antenna problems, these codes also generate various far field quantities and ratio type measurements. The details of exactly what types of antenna measurements are possible vary from code to code.

References

[1] Rautio, J. C., and R. F. Harrington, "An Electromagnetic Time-Harmonic Analysis of Shielded Microstrip Circuits," *IEEE Trans. Microwave Theory Tech.*, Vol. 35, No. 8, 1987, pp. 726–729.

[2] Rautio, J. C., "An Investigation of an Error Cancellation Mechanism with Respect to Subsectional Electromagnetic Analysis Validation," *Int. J. Microwave Millimeter-Wave CAE*, Vol. 6, No. 6, 1996, pp. 430–435.

[3] Rautio, J. C., "Educational Use of a Microwave Electromagnetic Analysis of 3-D Planar Structures," *Computer Applications in Engineering Education*, Vol. 1, No. 3, 1993, pp. 243–253.

[4] Rautio, J. C., "An Ultra-High Precision Benchmark for Validation of Planar Electromagnetic Analysis," *IEEE Trans. Microwave Theory and Tech.*, Vol. 42, No. 11, 1994, pp. 2046–2050.

[5] Swanson, Jr., D. G., "Experimental Validation: Measuring a Simple Circuit," *1993 IEEE MTT-S Int. Microwave Symposium Workshop WSMK*, June 1993.

[6] "Evaluation of Electromagnetic Microwave Software," Publication EVAL98-01, Sonnet Software, June 1, 1998.

[7] Swanson, Jr., D. G., "Optimizing a Microstrip Bandpass Filter Using Electromagnetics," *Int. J. MIMCAE*, Vol. 5, No. 5, 1995, pp. 344–351.

[8] Richardon, L. F., *Phil. Trans. Royal Soc.*, London, pt. A, 1911, p. 307.

[9] Booton, R. C., *Computational Methods for Electromagnetics and Microwaves*, New York: John Wiley & Sons, 1992, pp. 7–12.

[10] Matthaei, G. L., and Forse, R. J., "A Note Concerning the Use of Field-Solvers for the Design of Microstrip Shunt Capacitors in Low-Pass Structures," *Int. J. Microwave Millimeter-Wave CAE*, Vol. 5, No. 5, 1995, pp. 352–358.

[11] Lenzing, E. H., and Rautio, J. C., "A Model for Discretization Error in Electromagnetic Analysis of Capacitors," *IEEE Trans. Microwave Theory and Tech.*, Vol. 46, No. 2, 1998, pp. 162–166.

[12] Hoefer, W. J. R., "Equivalent Series Inductivity of a Narrow Transverse Slit in Microstrip," *IEEE Trans. Microwave Theory and Tech.*, Vol. 25, No. 10, 1977, pp. 822–824.

[13] Jain, N., and Wells, D., "Design of a DC-to-90 GHz Resistive Load," *IEEE Microwave and Guided Wave Letters*, Vol. 9, No. 2, 1999, pp. 69–70.

Chapter 6

Finite Element Method Simulators

Some problems we are interested in are truly 3D; there is no simple way to describe them with a lower order geometry. Many waveguide components fall into this category. Components that include both waveguide and coaxial elements or waveguide and microstrip elements are also good candidates. The key issues here are very similar to our previous moment method code discussion. How can the software help us to visualize the details of our problem? How does the software converge to the correct solution? Can I de-embed down to a meaningful reference plane for my problem?

Let us take a few moments to compare FEM and MoM using our familiar microstrip through line (Figure 6.1(a)). In FEM, we have "wave ports" at the front and rear faces of the cube. The software will perform a 2D eigenmode solution on the complete face and find the lowest order mode and higher order modes if desired. The MoM codes compute total voltage and current at the ends of the microstrip; there is no modal information. The finite element method discretizes the problem using tetrahedra of various sizes. In Figure 6.1(b) we see the result of the discretization process. Unfortunately we can only see the faces of tetrahedra that are on the surface of the problem. But we clearly have a finer subdivision of the problem space around the microstrip line and a coarser subdivision near the box walls. The MoM codes would only subdivide the conductors inside the box. We can temporally remove the tetrahedra in the air region (Figure 6.1(c)) so we can see the surface of the substrate. The fine discretization on the strip and near the edges of the strip can be seen. Again, the mesh becomes coarser as we move away from the microstrip line, which is exactly the behavior we would like to see.

6.1 FINITE ELEMENT METHOD—STRENGTHS

Probably the greatest strength of the finite element method is its generality. It is very easy to define arbitrary geometries with various levels of resolution in the problem space. We can accurately capture some small feature in a much larger

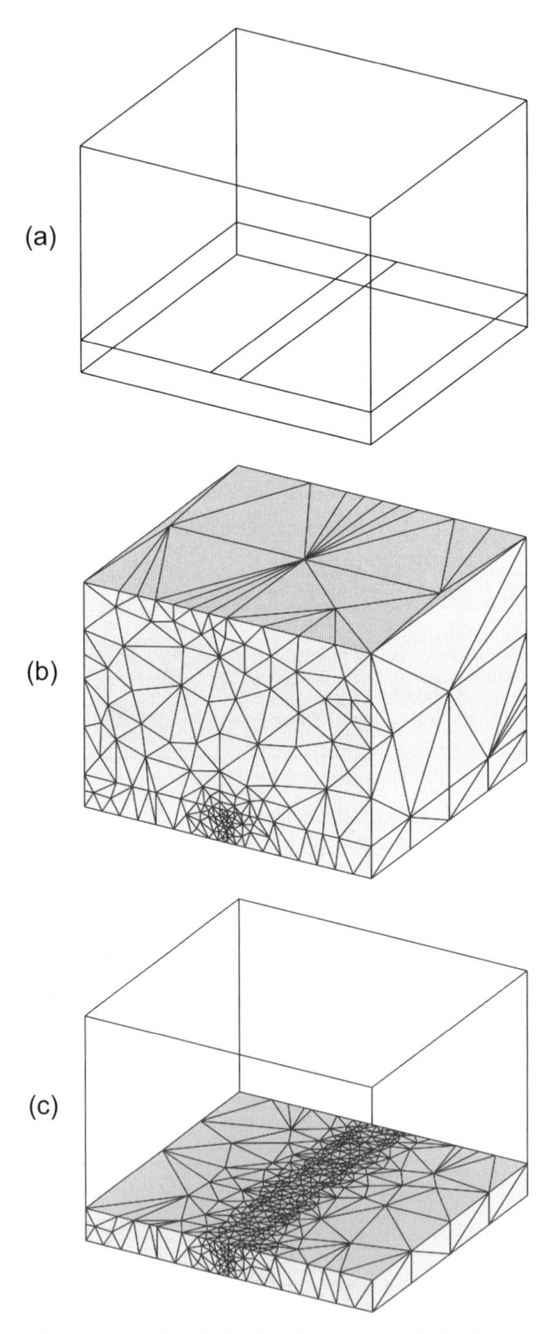

Figure 6.1 Finite element method analysis of a microstrip line: (a) basic geometry, (b) surface view of the complete mesh, and (c) air region removed, surface view of mesh on substrate and microstrip line (Ansoft HFSS Ver. 5.0).

Table 6.1

Finite Element Method

Strengths	*Weaknesses*
Easy to draw arbitrary objects	Must discretize entire volume
Resolve small details in larger problem space	Large, sparse matrix to invert
Automatic mesh refinement is unique to this method	Wave ports occupy complete "face:"
Basic formulation is closed box:	• More than one port per "face" is difficult
• Package effects are visible	• Separated 2D impedance calculation at port
• Box resonances are detected	• Multistrip de-embedding not possible
Multimode *S*-parameters available:	Approximate free space with ABCs or PMLs:
• Mode conversion can be observed	• Increased computation time
• Mode conversion can be enhanced or suppressed	Structures with high reflection at ports:
Visualization:	• Mesher may not make connection from input to output
• Large number of plot types	• Mesh seeding may be required
	Spurious modes may reappear at low frequencies

problem space and still solve the problem rather efficiently. Automatic mesh refinement is unique to this method. It is true error-based refinement made possible by the basic formulation of the method. The basic FEM formulation is closed box; we can include package effects. And we can compute multimode *S*-parameters. We can actually see mode conversion in various structures, which allows us to either optimize it or try to suppress it [1, 2].

6.2 FINITE ELEMENT METHOD—WEAKNESSES

The downside of the general nature of FEM codes is problem size. We must discretize a complete volume rather than a 2D cross-section or a planar layer of metal. The result is a large, sparse matrix to invert. In general this matrix will be larger than an equivalent MoM matrix, but because it is sparse, there are many specialized numerical methods that can be applied to the matrix inversion problem. In FEM codes the wave ports occupy a complete "face" of the problem space. Placing more than one wave port on a face is difficult. A separate 2D calculation of impedance is required at the port, which may be slightly different than the 3D computation in the interior. This can lead to errors when trying to extract small discontinuities. Multistrip de-embedding is not generally available. Free space problems must be approximated using ABCs or PMLs which leads to increased computation time. Resonant structures or other problems with high reflection at the ports are particularly difficult for the automatic meshing algorithm. In these cases, the mesher may not make a connection from input to output and seeding the mesh may be required. And at

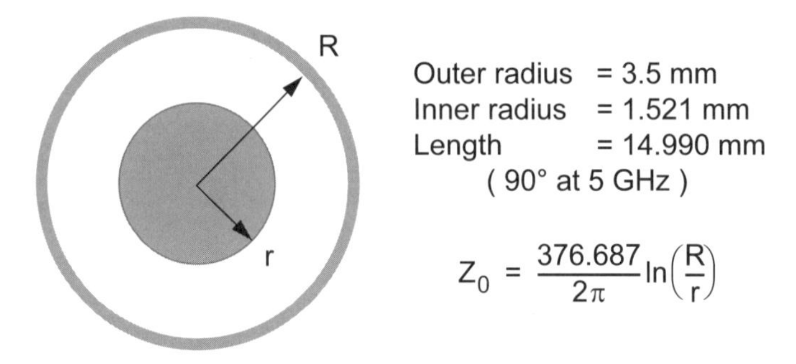

Figure 6.2 Air-filled, 7-mm coaxial transmission line. This is a potential validation structure for 3D FEM simulators.

very low frequencies, spurious modes may reappear in the solution. The strengths and weaknesses of the finite element method are summarized in Table 6.1.

6.3 FEM SIMULATORS—VALIDATION STRUCTURES

If we are interested in absolute accuracy, the stripline standard we studied in a previous chapter would also be a valid test case for the FEM simulators. Another potential validation structure is the air-filled coaxial transmission line shown in Figure 6.2. Like the stripline standard, it is pure TEM; there is an exact analytical equation for impedance and we know the phase velocity exactly for the air-filled case. We will use the coaxial standard in the following sections to explore various aspects of FEM meshing.

6.4 CONTROLLING MESHING

In a 3D model, mesh generation is much more difficult than the 2D and 2.5D cases we have already examined. The major tasks for the mesh are to capture the geometry of the problem at some finite resolution, capture changes in the fields due to guide wavelength, and capture finer grain variations in the fields that we have labeled spatial wavelength. We would also like to control the aspect ratio of the tetrahedra in the mesh. Small, equilateral tetrahedra will provide the most accurate approximation to the fields. If we allow large aspect ratios in the mesh, the accuracy of the field approximation varies as a function of position and direction. In most cases, the software will find a suitable mesh on its own, but an intelligent user can always steer the process and find a more efficient mesh for a given problem.

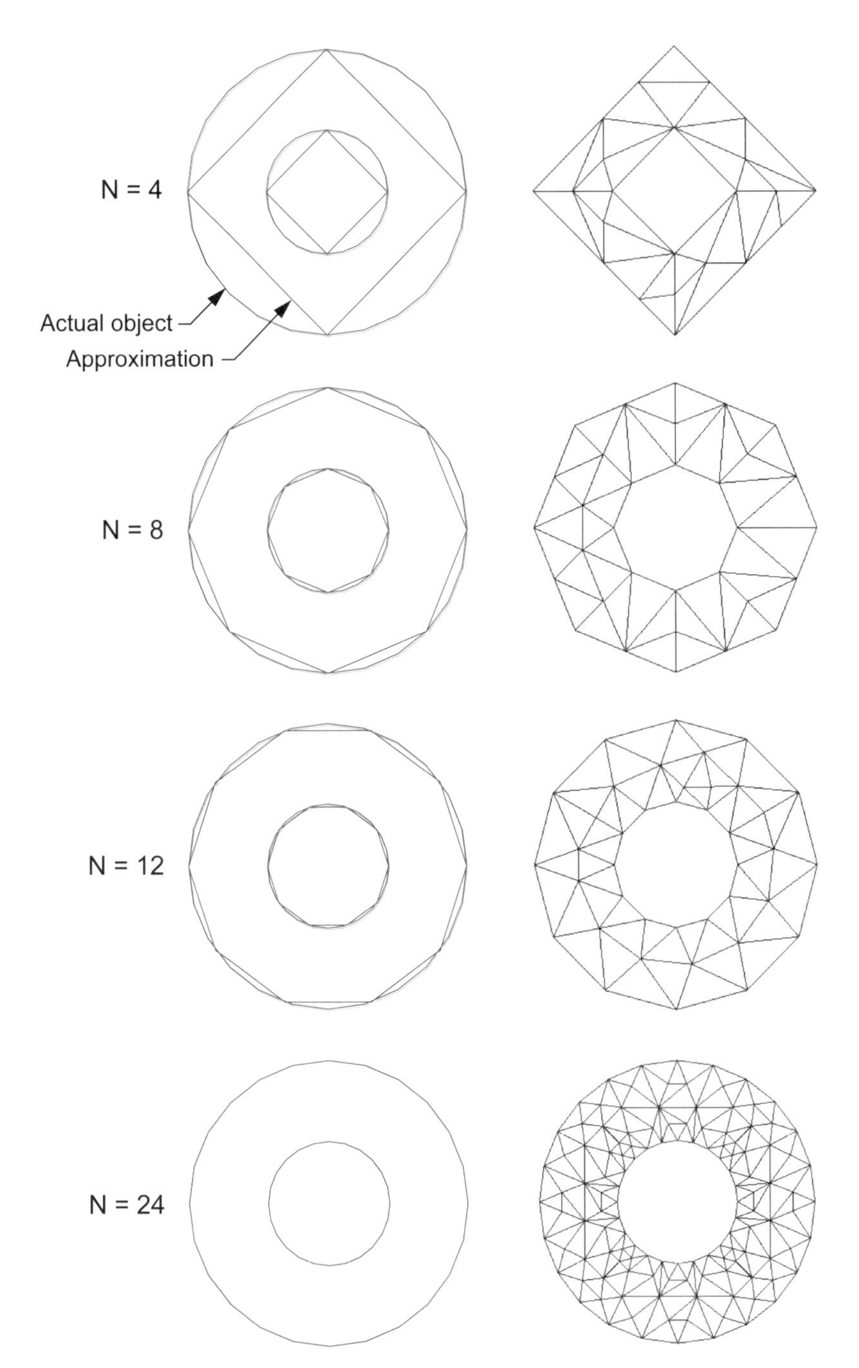

Figure 6.3 On the left we approximate the coaxial line cylinders using 4 through 24 segments. On the right is the resulting mesh at the ports after three adaptive passes (Agilent HFSS Ver. 5.6).

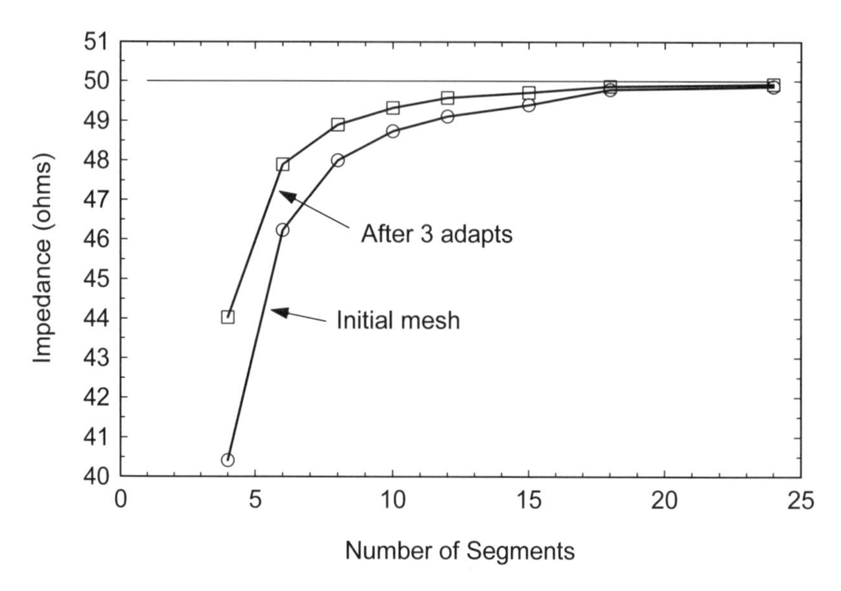

Figure 6.4 Impedance convergence of 7-mm coax validation structure as function of number of segments used to define the cylinders.

6.4.1 Meshing The Coaxial Standard—Geometrical Resolution

One reason we often move to 3D codes is to accommodate objects with curved surfaces or boundaries. Any kind of coaxial structure would be one example of this type of problem. When we build these objects they have smooth surfaces with no obvious approximations to the arcs. But in our CAD tool, we have to approximate these smooth arcs with a number of straight line segments. The number of segments we choose will affect the size of the mesh and therefore the solution time. Even though we have automatic mesh refinement, the decisions we make when drawing circular objects will have a great impact on mesh size.

One simple test case we can use to explore convergence as a function of drawing or geometrical resolution is the 7-mm coaxial transmission line shown in Figure 6.2. When we draw the outer and inner conductors of the coaxial line, we have to choose the number of segments we want to use to define each cylinder. The cross-section of the 7-mm coax is shown in Figure 6.3 with low resolution and high resolution approximations to both cylinders. The $N = 4$ approximation is obviously very coarse and we would probably never use it. The 8 and 12-segment approximations look fairly good. The 24-segment approximation is obviously very good. Figure 6.3 also shows the mesh generated at one port plane for each approximation. The mesh is the result of three adaptive passes and the data were generated by Agilent HFSS Ver. 5.6. With 12 segments the mesh is starting to look fairly uniform, with a double layer of cells between the conductors. At $N = 24$, the mesh is quite refined and we

Table 6.2

Impedance Convergence for 7-mm Coax Validation Structure

Number of segments	*Segment angle (deg)*	*Impedance (ohms)*	*Error (%)*	*Number of unknowns*	*Solution time* (seconds)*
4	90	44.03	11.9	1,884	44
6	60	47.90	4.2	2,298	49
8	45	48.90	2.2	2,530	54
10	36	49.33	1.3	3,190	61
12	30	49.58	0.8	3,832	68
15	24	49.71	0.6	4,406	80
18	20	49.87	0.3	12,390	205
24	15	49.92	0.2	16,342	317

*166-MHz Pentium II notebook with 256 MB RAM, circa 1997

might expect a very accurate solution. With just this visual examination at the port face we might be tempted to use fairly high resolution to define the geometry of all our cylinders. But we must mesh the complete object, and the mesh density we see at the port face will generally be repeated throughout the object on a per-wavelength basis. So the geometrical resolution we choose for our model objects will influence the trade-off between solution time and accuracy.

Visually we have started to get some sense of the effect of our approximations on the mesh. But what we really want to know is how the impedance converges as a function of our starting point. Figure 6.4 shows the convergence of the initial mesh and the convergence after three adaptive passes as a function of the number of segments for the coaxial validation standard. Both curves asymptotically approach 50 ohms as we would expect. With only four or six segments, the error is high, in the region of 10%. With 10 to 12 segments, the error is closer to 1%, probably good enough for most engineering applications.

Table 6.2 presents the same convergence data in tabular form. As our approximation to the true cylindrical objects gets better, the impedance error clearly goes down. We express the approximation as number of segments or segment angle, which is the way it is sometimes described in the software. If the error in impedance for $N = 24$ is very low, why not use this value all the time? The answer is in the last two columns of Table 6.2. As we increase N, at some point the number of unknowns and the solution time rise dramatically. The difference in solution time between 1% error and 0.2% error is nearly five times. Again, this is due to the time needed to fill and invert an ever larger matrix. So we often need to make a trade-off between speed and accuracy.

Although our simple coax transmission line example may seem trivial, it is an important exercise that develops our intuition for how to draw projects in the simulator. Without doing this exercise, we might be tempted to draw the project in Fig-

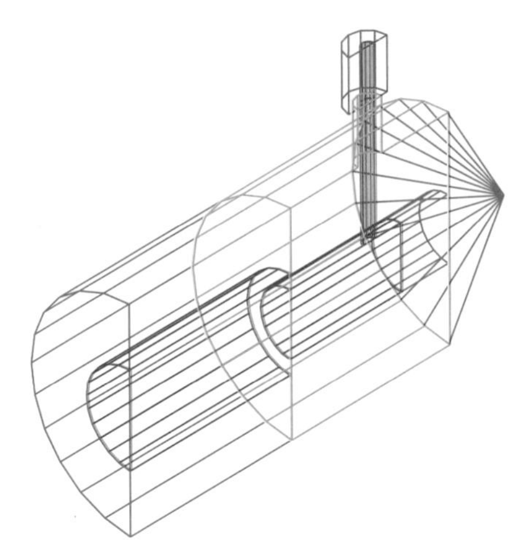

Figure 6.5 Right angle transition from a large-diameter, air-filled coax to a smaller diameter, Teflon-filled coax. The structure has been split vertically along a symmetry plane.

ure 6.5 at a very high resolution, with a resulting steep penalty in solution time. We could perform a convergence study on this project as well, and for a very critical or sensitive structure it may be worth doing. But in most cases it will be "good enough" to pick a segment angle of 30 to 45 degrees as a starting point for coaxial structures. In some cases, a starting point that is too coarse may cause "oscillations" in the convergence plot. Increasing the resolution slightly or some other use intervention may be called for in these cases.

6.4.2 Meshing a Coaxial Resonator—Dummies and Seeding

There are other strategies we can use to control the finite element mesh besides the fundamental geometrical resolution of objects. Two additional techniques for mesh control are placement of dummy objects and seeding of objects. A dummy object can be air or some other material that is part of the model. Basically, we subdivide an existing object or objects to provide a new boundary that the meshing algorithm will detect without modifying the electrical properties of the model. The concept is not unlike the subdivision of larger polygons that we advocated for the 2.5D simulators. We can also seed objects, which forces an initial mesh with a maximum cell size on the object. Recently we did a convergence study on some simple combline type resonators using the eigenmode-solver in Ansoft HFSS. The geometry is a metal post in an air-filled cylindrical cavity (Figure 6.6). The circuit theory model would be a short-circuited transmission line with capacitive loading at the open end.

The resonant frequency and unloaded quality factor (Q_u) for this geometry were published in a paper by Wang, et al., [3]. We also checked it using an 2D axi-symmetric model in FlexPDE [4]. The results from Wang were probably obtained

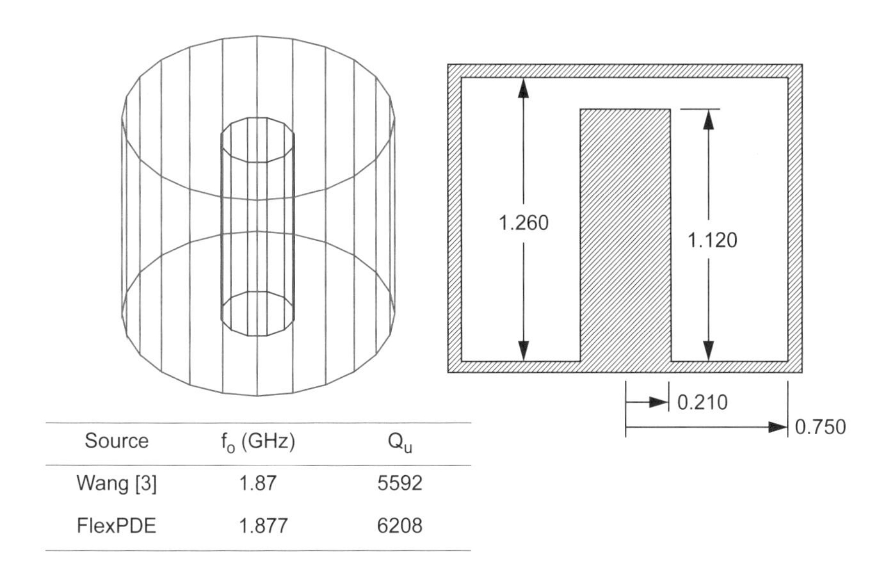

Source	f_o (GHz)	Q_u
Wang [3]	1.87	5592
FlexPDE	1.877	6208

Figure 6.6 Coaxial resonator meshing experiment with resonant frequency and Q_u from two different sources. Dimensions are in inches.

with a proprietary mode matching program. For the Q_u calculations, we assumed $\sigma = 6.17 \times 10^7$ mho/m (ideal silver plating).

We used the eigenmode-solver in Ansoft HFSS Ver. 8.0 to find the resonant frequency and unloaded Q of this geometry (Figure 6.7). Ports are normally not needed or desirable in an eigenmode solution. In all cases, the initial seeding based on wavelength is set to the default value of $\lambda/3$, which is rather coarse. The tetrahedra refinement parameter is increased from the default value of 20% to 40%. In Case 1 we left all the other control parameters set to their default values. The number of segments for the post and cavity were chosen based on past experience and the convergence data from the coaxial standard.

Case 2 places a "dummy" object made of air around the post. This will have no impact electrically, but the mesher "sees" the boundaries of this object and fits the mesh to it. Placing this dummy should improve the aspect ratio of the tetrahedra in the radial direction. We can also force a certain size mesh in an object by seeding. If this is a resonator, we know what the voltage and current look like along the post. There is a current maximum and a voltage minimum at the base of the post. At the open end of the resonator we have a voltage maximum and a current minimum. If we know the fields vary rapidly across some region, we may want to force a finer mesh than would be produced by the default meshing rules. By seeding at 0.1 inch in the dummy, we are forcing a minimum of 10 to 12 tetrahedra along the length of the resonator. We can also place an upper limit on the number of tetrahedra that are added during the seeding operation. In the absence of seeding, we could subdivide

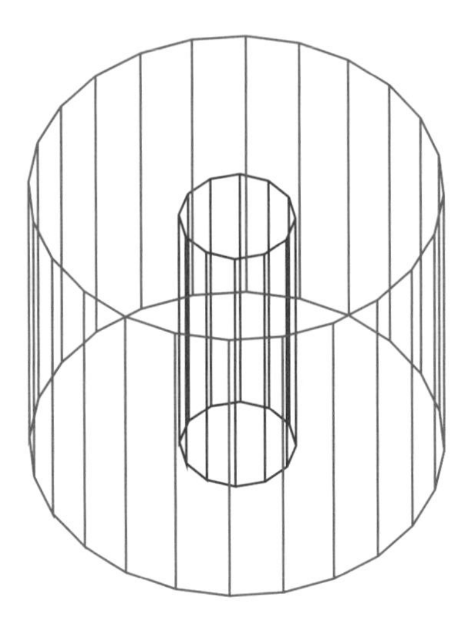

Case 1:
12 segments on post
24 segments on cavity
No seeding or dummy objects

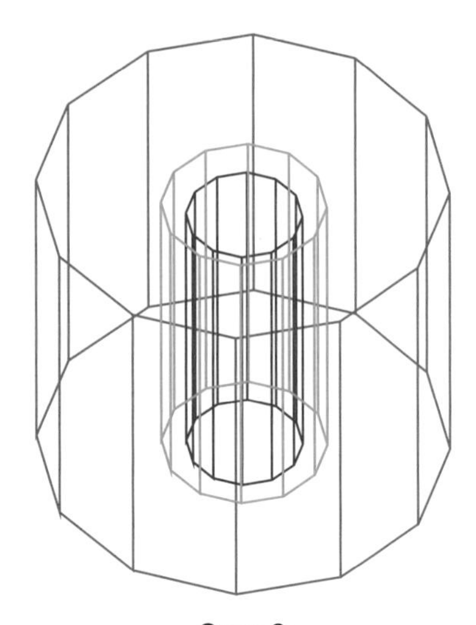

Case 2:
12 segments on all objects
Dummy object around post
Seed dummy at 0.1, max of 2,000

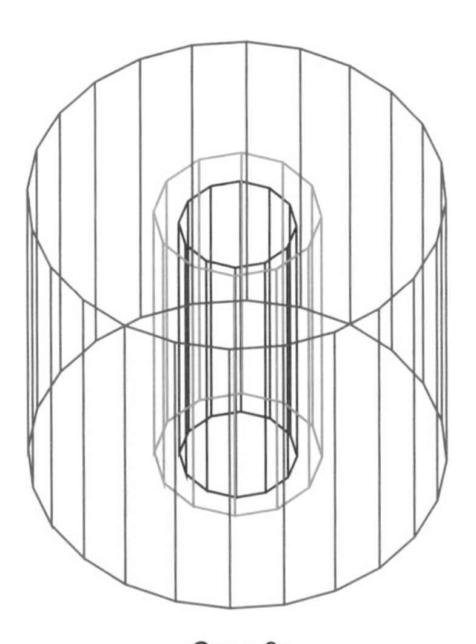

Case 3:
12 segments on post and dummy
24 segments on cavity
Seed dummy at 0.1, max of 2,000

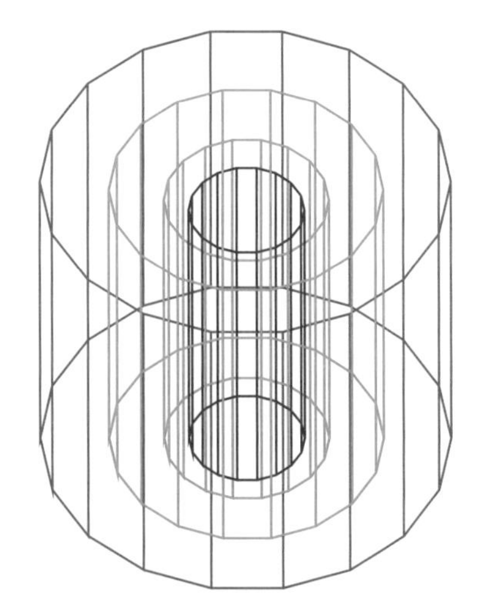

Case 4:
18 segments on all objects
Two dummy objects
Seed inner at 0.1, max of 4,000

Figure 6.7 Coaxial resonator meshing experiments. In all cases the initial seeding based on wavelength is set to the default value of $\lambda/3$, which is rather coarse. The tetrahedra refinement parameter is increased from the default value of 20% to 40%.

Table 6.3

Results of Coaxial Resonator Meshing Experiments

	Pass No.	f_o (GHz)	Q_u	No. of Tets	Δf_o (%)
Case 1	1	1.6787	4600	192	N/A
	2	1.7304	4588	256	3.02
	3	1.7621	1584	377	1.83
	4	1.8183	4309	548	3.19
	5	1.8497	4354	789	1.73
	6	1.8646	4513	1094	0.80
	Pass No.	f_o (GHz)	Q_u	No. of Tets	Δf_o (%)
Case 2	1	1.8647	5265	2390	N/A
	2	1.8822	5325	3161	0.94
	3	1.8863	5346	4240	0.22
	4	1.8887	5345	5738	0.14
	5	1.8899	5372	7793	0.44
	6	1.8904	5378	10644	0.09
	Pass No.	f_o (GHz)	Q_u	No. of Tets	Δf_o (%)
Case 3	1	1.8604	5398	2480	N/A
	2	1.8716	4762	3471	11.92
	3	1.8759	5009	4841	4.71
	4	1.8784	5304	6699	5.43
	5	1.8796	5184	9234	2.35
	6	1.8799	5430	12721	4.49
	Pass No.	f_o (GHz)	Q_u	No. of Tets	Δf_o (%)
Case 4	1	1.8656	5452	4929	N/A
	2	1.8742	5449	6631	0.51
	3	1.8777	5447	8899	0.22
	4	1.8791	5437	12282	0.27
	5	1.8800	5447	16832	0.13
	6	1.8804	5452	21837	0.08

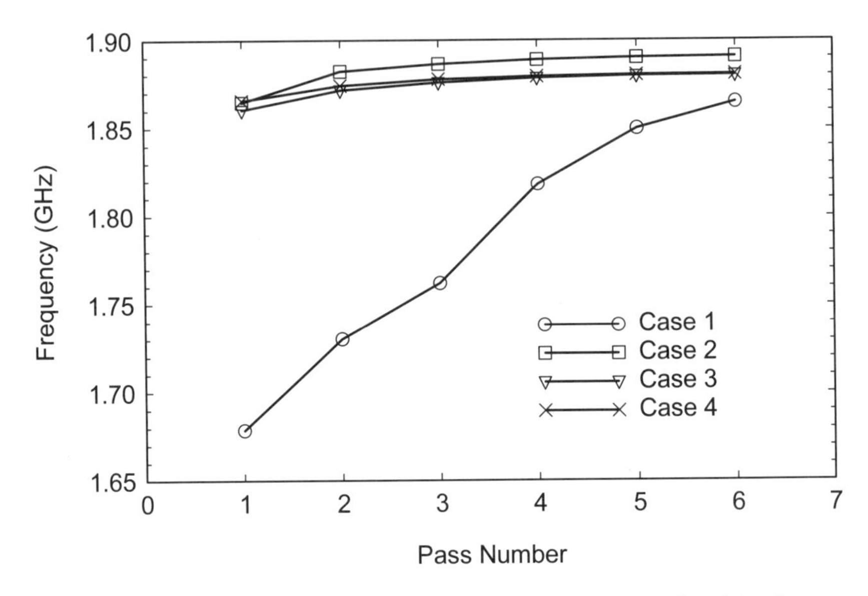

Figure 6.8 Absolute frequency convergence for the four resonator test cases. Case 1 has the coarsest starting mesh. Case 2 is converging to a slightly different solution. Cases 3 and 4 are converging to similar solutions.

the post along its length into 10 or 12 unique objects. This would also impact the density of the mesh along the post.

Case 3 is the same as Case 2 with finer resolution on the cavity (outer cylinder). Case 4 uses an intermediate resolution on all cylinders and introduces a second dummy object. The two dummy objects now force at least three distinct layers of tetrahedra between the post and the outer wall. The inner dummy object is seeded at 0.1 inch but the upper limit on tetrahedra is increased to 4,000. Case 4 is probably a good example of changing too many things at once in our numerical experiment. This might also be a good place to apply design of experiments theory.

The results for all four experiments are shown in Table 6.3. With only a coarse, lambda based seed, Case 1 is basically starved for tetrahedra. The automatic mesher is driving it towards convergence, but not very fast. Case 2 is converging but the final frequency is a little high. Cases 3 and 4 are converging to the same solution, but the number of tetrahedra for Case 4 is twice as big as Case 3.

The convergence of absolute frequency is plotted for all four cases in Figure 6.8. Case 1 has a very sparse initial mesh; its initial solution is quite far away from the final solution. Case 2 is converging to a solution that is slightly too high in frequency. We set the geometrical resolution of the outer cylinder too coarse, so the impedance of the resonator is wrong and therefore the resonant frequency will be wrong. Cases 3 and 4 are converging to a frequency that agrees with the Wang data and the FlexPDE solution.

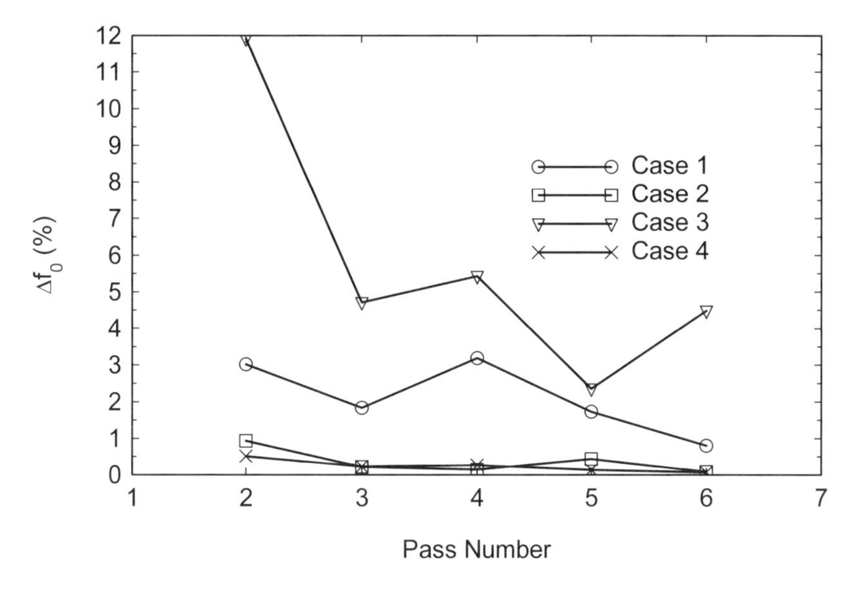

Figure 6.9 Percentage change in resonant frequency as a function of the pass number. Rapid changes in the imaginary part of the frequency, which is related to loss and Q_u, are causing the Case 3 curve to vary rapidly.

In Figure 6.9 we have plotted the percentage change in resonant frequency as a function of the pass number. The behavior of the Case 3 curve appears to be somewhat anomalous. The software is actually tracking the changes in both the real and imaginary parts of the eigenmode solution. The imaginary part is related to the loss and the Q_u of the structure. In Case 3, the real part is converging nicely but rapid changes in the imaginary part are causing the reported Δf_0 to shift radically. If this experiment was to be repeated, we could track only the real part of the resonant frequency.

These resonator examples demonstrate how the user's knowledge of the problem can have a large impact on the results obtained. The software is a very powerful general purpose tool, but it has no knowledge of the specific case we are trying to solve. A combination of dummy objects and seeding of the mesh helps the software converge to the correct solution more rapidly. Measured data or another trusted solution are always very valuable in determining if we are approaching convergence.

6.4.3 Meshing a Coaxial Step Discontinuity—Dummies and Seeding

When we make a step change in the inner or outer diameter of a coaxial line, we introduce a capacitive discontinuity. This type of discontinuity appears in many coaxial structures including lowpass filters, various types of connectors, and transi-

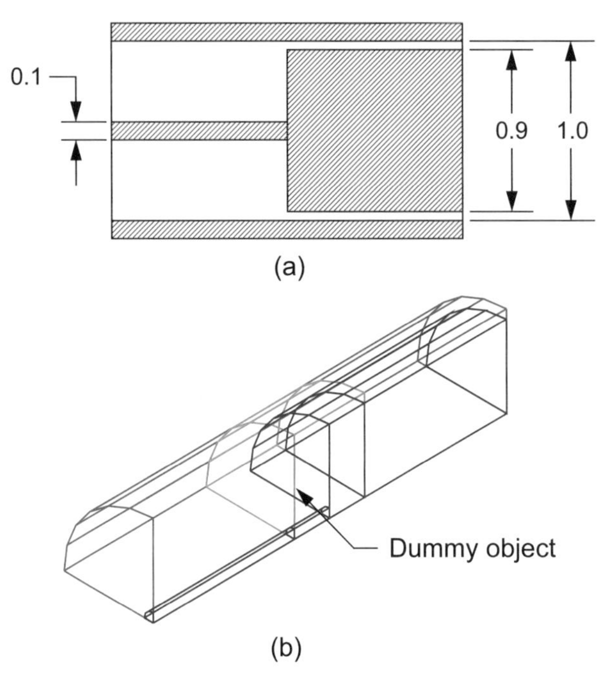

Figure 6.10 Step in inner conductor of coaxial line: (a) cross-section dimensions (inches), and (b) FEM model using quarter cut symmetry. Dummy object extends 0.2 in on either side of the step.

tions between connector types. We can also find accurate data on the coaxial step discontinuity in the literature [5–7]. All of these facts make this geometry a good candidate for a meshing experiment and it may also qualify as a 3D validation structure.

The dimensions of our test discontinuity are shown in Figure 6.10(a). In this case we will compare our computed data against those found in Somlo [7]. The Somlo data were generated using the mode matching method. For each geometry, data using the first 28, 29, ... 40 modes were computed, then least squares fit to a hyperbolic curve of the form $C = A/m + B$, where A and B are constants. The asymptotic value B of the first-order hyperbola was assumed to be the solution. The Somlo data have been used in lowpass filter design and connector compensation for many years and are considered to be reliable. However, the aspect ratio of our example is at one extreme of the Somlo data and falls outside the recommended range of the equation that was fit to those data. For this geometry, we extracted a value of 0.8698 pF from Figure 1 in the Somlo paper.

In a 3D field-solver, it is convenient and faster to analyze one quarter of the actual geometry (Figure 6.10(b)) by applying magnetic walls to the X-Y and X-Z planes. We also have the option of adding dummy objects and applying seeding as we did in the previous resonator experiment. We performed nearly 50 numerical

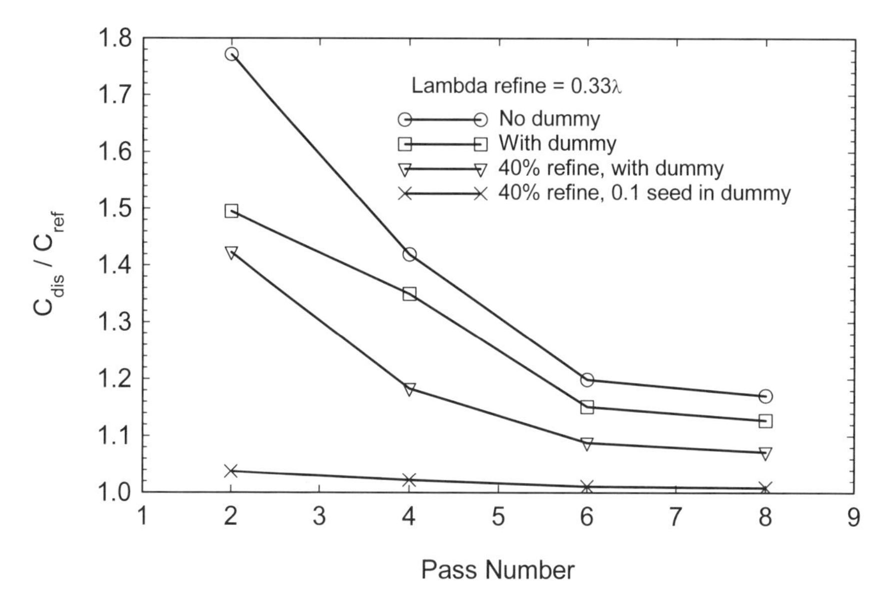

Figure 6.11 The ratio of computed capacitance to the reference value from Somlo. The base solution uses the default meshing parameters. Adding the dummy object and seeding the dummy improves the rate of convergence dramatically.

experiments on this test geometry using Agilent HFSS Ver. 5.6, Ansoft HFSS Ver. 8.0, CST Microwave Studio Ver. 3.4, and FlexPDE Ver. 3.0. The test frequency for the fullwave tools was 1 GHz.

The following results are from Ansoft HFSS Ver. 8.0. The first set of experiments starts with the default meshing parameters. The default target for wavelength based refinement is 0.33λ, which is rather coarse. In Figure 6.11 we are plotting the ratio of the computed discontinuity capacitance, C_{dis}, over the reference value from Somlo, C_{ref}. The curve with circle markers is the base solution using the default meshing parameters. Adaptive meshing reduces the error from 80% to 20% after eight passes. Next we place an air dummy object around the step region. It is a 1.0 in diameter cylinder that extends 0.2 in on either side of the step. With the dummy object in place (square markers), the error starts at 50% and is reduced to about 12% after automatic refinement. In the third experiment (triangle markers), we increased the mesh refinement parameter from the 20% default to 40%. This allows the software to add more tetrahedra at each adaptive pass. Now the error runs from about 44% to 7%.

The most dramatic improvement is obtained when we seed the dummy region at 0.1 in by length (cross markers). If we assume there is a large amount of evanescent energy near the step, we need to sample the fields with a finer resolution than the lambda-based meshing. Seeding at 0.1 in the dummy forces a minimum of five to six elements across the radius of the outer conductor in the vicinity of the step.

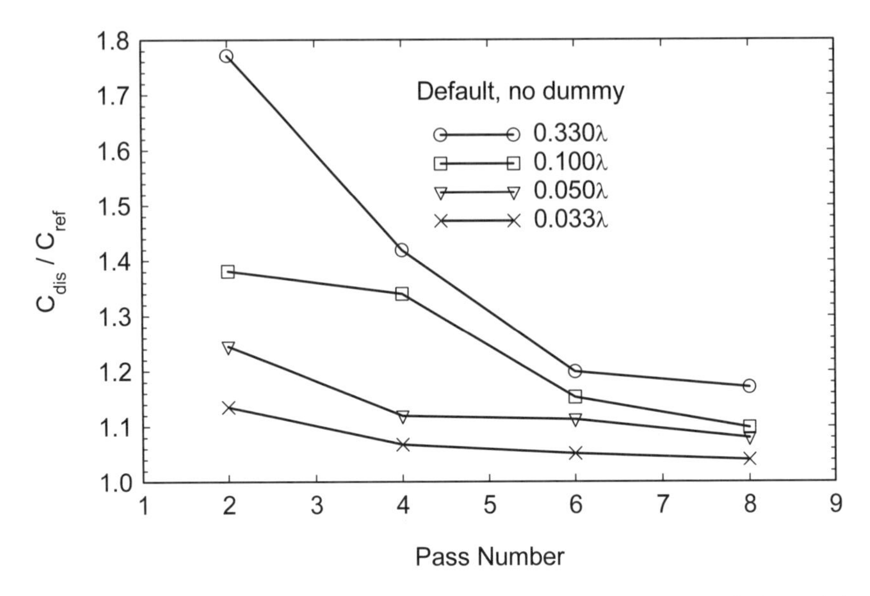

Figure 6.12 The ratio of computed capacitance to the reference value from Somlo. The base solution uses the default meshing parameters. Using a finer, wavelength-based mesh as a starting point increases the convergence rate dramatically.

For this last experiment, the error starts near 4% and drops below 1% after eight adaptive passes. Note also that each curve seems to be converging to a different final value, depending on the starting point.

Lambda-based meshing for most field-solvers is typically $\lambda/10$ to $\lambda/30$. The next set of experiments (Figure 6.12) explores the impact of meshing the step discontinuity on a per-wavelength basis only. The curve with circle markers is again the default starting point. In the next experiment (square makers), we decreased the initial mesh size to 0.10λ. The initial error is about 38% and drops to 10% after eight adaptive passes. The third experiment (triangle markers) decreases the initial mesh size to 0.05λ. The initial error of 25% is reduced to about 8% after eight passes. The final experiment (cross markers) uses an initial mesh size of 0.033λ. The initial error is about 14% and drops to about 4%. Again, note that we seem to be converging to different final values, depending on the starting point.

The next logical step is to combine a fine, wavelength-based initial mesh with a seeded dummy object around the step region. Figure 6.13 shows the results of combining a 0.033λ initial mesh with various combinations of dummy objects and seeding. The curve with circle markers is the first experiment with just the wavelength-based initial mesh. The initial error is about 13%, which then drops to about 4% after eight adaptive passes. In this case, adding the dummy (square markers) has little impact on the results. The mesh refinement parameter was increased to 40% in the third experiment (triangle markers). Finally, we seed the dummy object

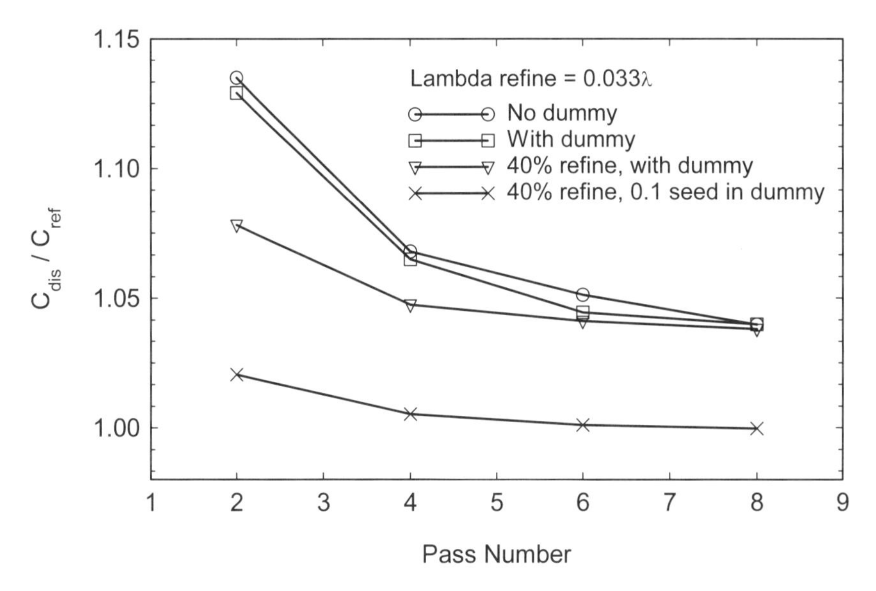

Figure 6.13 The ratio of computed capacitance to the reference value from Somlo. The base solution uses a lambda-based initial mesh of 0.033λ. Using a finer, wavelength-based initial mesh in combination with a seeded dummy object results in a very accurate solution.

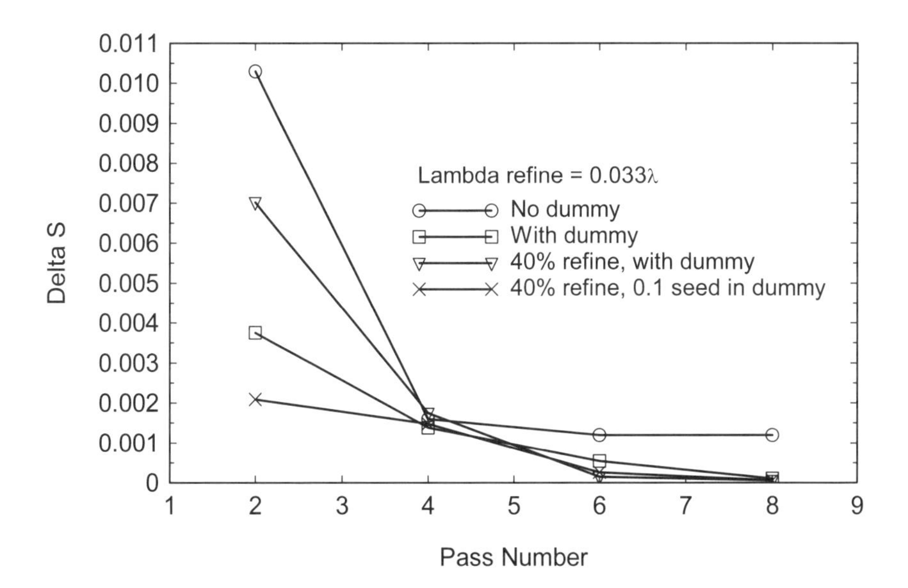

Figure 6.14 The change in *S*-parameters (Delta *S*) for each of the experiments in Figure 6.13. Each curve exhibits strong convergence behavior and the absolute Delta *S* is very low. However, we know the curve with cross markers corresponds to the more accurate solution.

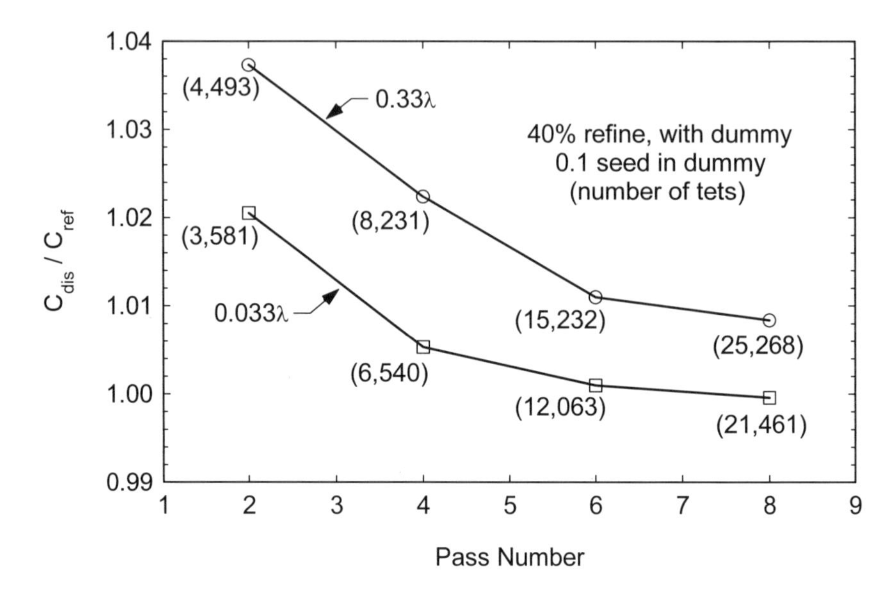

Figure 6.15 The best results from Figure 6.11 and Figure 6.13 are plotted together along with the number of tetrahedra at each pass. Starting with a finer, lambda-based mesh results in more rapid convergence with fewer tetrahedra.

at 0.1 in by length. Note that at the solution frequency of 1 GHz, this is equivalent to a wavelength-based mesh of roughly $\lambda/100$. In this last experiment (cross markers) the initial error is about 3% and falls to nearly zero after eight passes. We should point out that the first three experiments in Figure 6.13 are converging to a similar solution. The fourth solution, with seeding, is obviously converging to a different, more accurate solution. Clearly it is the seeding that captures the spatial wavelength details near the step discontinuity.

In most 3D FEM codes, the convergence process is monitored by tracking the changes in the magnitude and phase of the *S*-parameters at the ports. When the change in *S*-parameters (Delta *S*) drops below a predetermined, user defined level, the automatic meshing process is terminated. Figure 6.14 plots Delta *S* for the four experiments in Figure 6.13. All of the Delta *S* curves in Figure 6.14 exhibit strong convergence behavior and absolute values well below the default termination point. However, we know that one of the experiments, number four, has converged to a solution that is more accurate than the others. We can only conclude that Delta *S* is a good indicator of relative convergence within a given project, but it does not guarantee absolute convergence. It is also difficult to predict the absolute value of Delta *S* that will be achieved for a given project.

Finally, in Figure 6.15 we compare the results from two previous experiments. Both use the seeded dummy object and 40% mesh refinement. One project starts with a lambda-based mesh of 0.33λ, while the other starts with an initial lambda-based mesh of 0.033λ. The number of tetrahedra used is also plotted for each curve. Both experiments achieve less than 1% error after eight adaptive passes. But the experiment with the finer starting point (square markers) falls below 1% error after only four passes and the solution process could have been terminated at that point with acceptable accuracy. Using a finer lambda-based starting mesh seems to offer quicker convergence, with fewer tetrahedra when used in conjunction with intelligent seeding of the project.

6.4.4 Solving the Step Discontinuity in 2D

Another software tool we explored for the coaxial step problem was FlexPDE, a 2D FEM solver. In this case we can set up a rotationaly symmetric or axisymmetric problem in cylindrical coordinates; the software can automatically integrate around the circumference. If we take finer and finer slices of our 3D geometry, this 2D solution is the lower limit of that process. In Figure 6.16 we show the finite element mesh and the normalized E-field vectors in the air region between the conductors.

To find the step capacitance we take the capacitance of the complete structure and subtract the contributions of the high- and low-impedance lines on either side. This may be a source of error, as we are taking the difference of small numbers of similar magnitude. The E-field vectors also give us an appreciation for how large the evanescent or non-TEM region is. In a TEM region the E-field vectors must be perpendicular to both the inner and outer conductors at all times. We observe that the non-TEM region extends about one outer radius away from the step discontinuity on the high impedance side of the step.

In Figure 6.17 we plot the convergence of the computed step capacitance as a function of the number of 2D triangles or elements. The convergence is quite rapid and less than 1% error is easily achieved. Reducing the problem to 2D has obvious advantages. We are applying our computer resources to a smaller problem and the 2D axisymmetric formulation further restricts the orientation of the fields. In 3D the fields can have any arbitrary orientation. In this 2D problem, the E-field is restricted to the plane of the page while the H-fields are restricted to the phi-axis which goes into the page. The disadvantage of the simpler, 2D solution is that it is quasi-static; frequency does not appear as a variable anywhere in the solution process. The capacitance of the step does vary slightly as a function of frequency. There is a full-wave, FDTD based axisymmetric program dedicated to connector design available from QWED [8].

6.4.5 Mesh Control Summary

In the early days of numerical electromagnetics, generating a FEM mesh was a tedious, largely manual process. The addition of automatic mesh refinement to FEM codes was a welcome addition. However, the sophisticated user should not rely too heavily on adaptive refinement of the mesh. The user should always know more than the software about the fundamental behavior of his or her project. Even

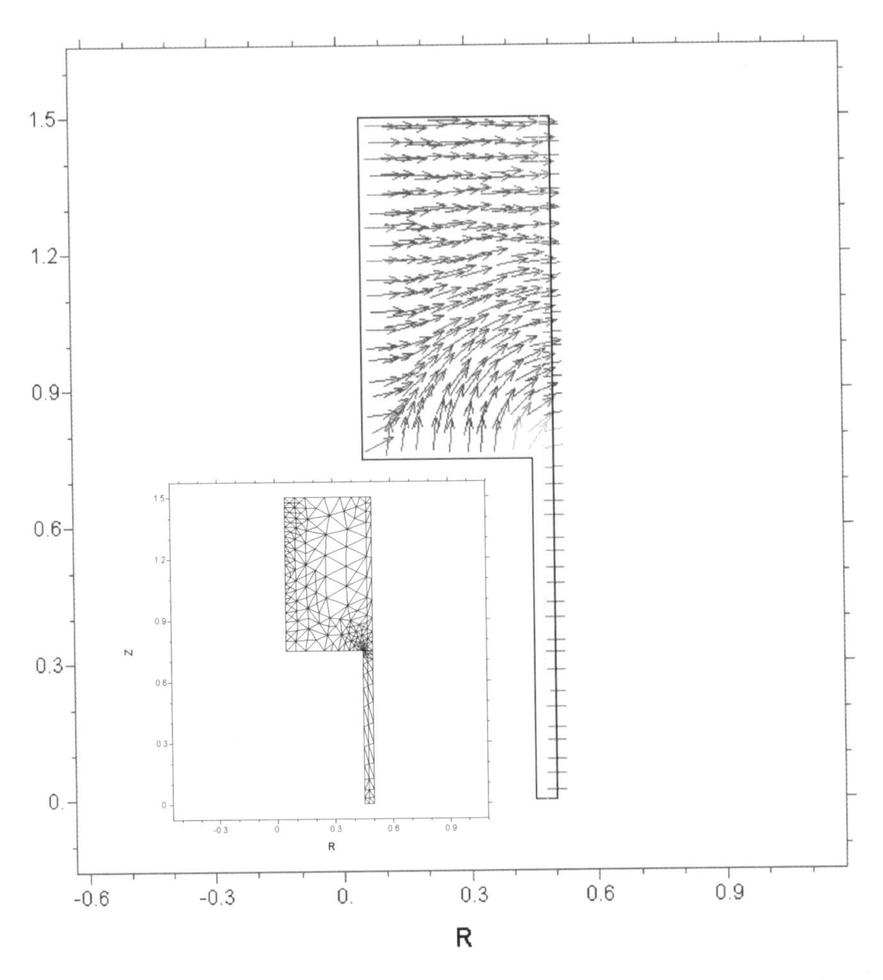

Figure 6.16 The step discontinuity solved as a 2D axisymmetric problem. The main plot shows the normalized E-field vectors in the region of the step. The insert is the 2D FEM mesh (FlexPDE Ver. 3.0).

if the key aspects of the project are not immediately obvious, some simple convergence experiments will soon lead to a deeper understanding of the problem.

As we guide the software in the meshing process we need to consider geometrical resolution, guide wavelength, and spatial wavelength. The geometrical resolution we use for various objects in the model is a trade-off between accuracy and solution time. Setting a realistic lambda-based starting mesh helps the software sample the problem based on the guide wavelength. Finally, there will be some regions in the problem where intelligent application of dummy objects and seeding capture finer grain details that we have labeled spatial wavelength.

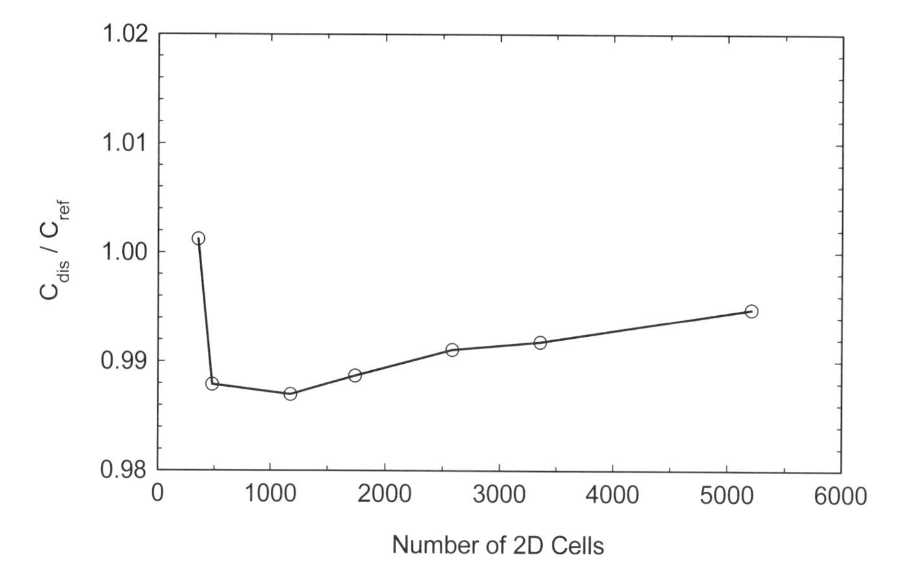

Figure 6.17 Convergence of the 2D FEM solution for the step discontinuity as a function of the number of 2D cells (FlexPDE Ver. 3.0).

6.5 FEM CALIBRATION STRUCTURES

What kind of structures should we choose for "calibration" with an FEM tool? Remember, the goal is to get the new user more familiar with the tool and encourage experimentation with the various control parameters in the software. We could repeat our series of microstrip structures from the previous chapter, which might be useful for comparing methods. A set of waveguide structures is another obvious choice. The waveguide through line and ideal short are easy to construct. But a waveguide open requires a radiation boundary, which may be too complex at this stage. Also, it may not be obvious how to build a good, broadband waveguide termination.

A coaxial version of our standard calibration kit might be a better choice. Again, the through line and ideal short standards are trivial. We can make a shielded open by creating a gap in the center conductor, so no radiation boundaries are required. The termination will be realized as a simple cylindrical resistor.

6.5.1 7-mm Coaxial Through Line

Setting up the coaxial through line is very quick and easy (Figure 6.18). The starting point for an FEM model is a solid block of metal. We remove a cylinder of air from that block to form the outer boundary of the coax. Another cylinder of metal defines the inner conductor. With this basic approach the outer boundaries of the

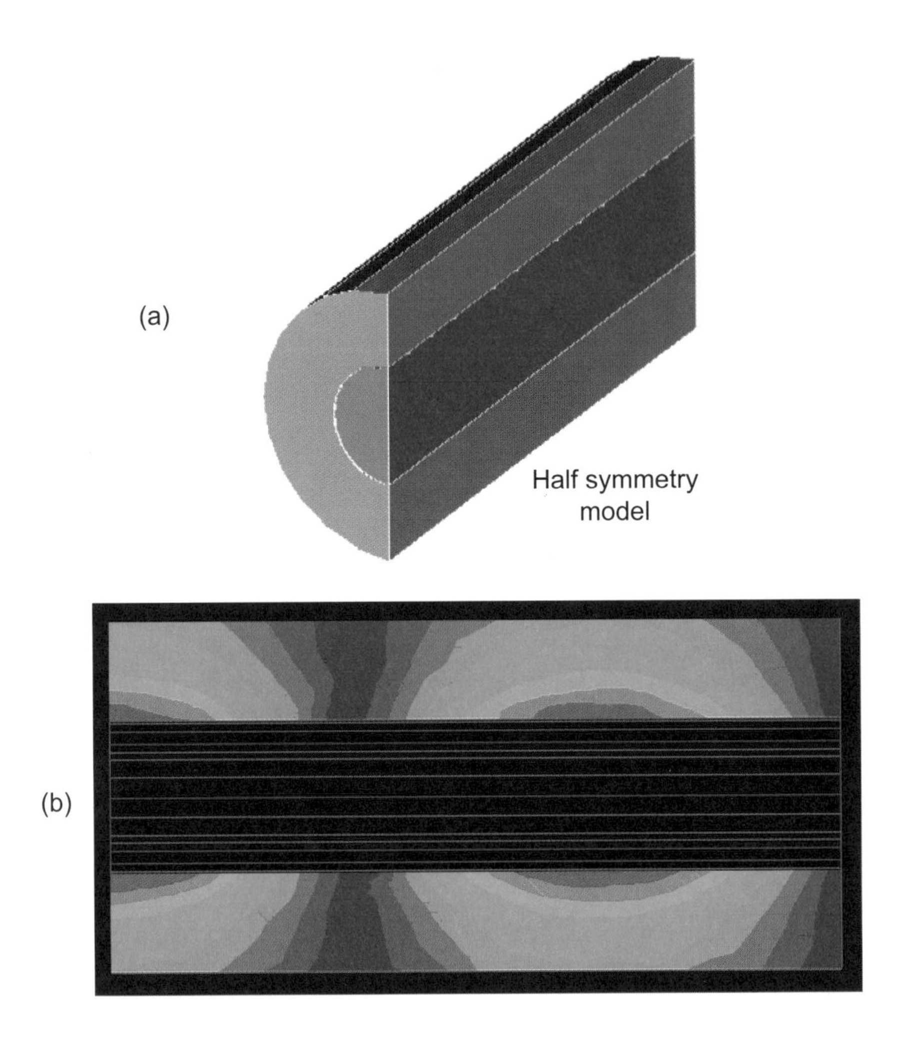

Figure 6.18 Calibration structure, 7-mm coaxial through line: (a) half symmetry model; and (b) E-field magnitude in the X-Z plane at 15 GHz (Ansoft HFSS Ver. 5.0).

problem are always a perfect electric conductor (PEC) by default. Ports are then defined at both ends of the structure. If we are only interested in *S*-parameters, we could analyze one-half, one-quarter, or even one-eighth of the structure. The software would help us correct the final results. If we want to experiment with visualization, half-symmetry might be a good choice. A magnetic wall (no such thing physically, but very useful mathematically) defines the symmetry plane. Note that only the higher order modes with even E-field symmetry about the symmetry plane will be modeled.

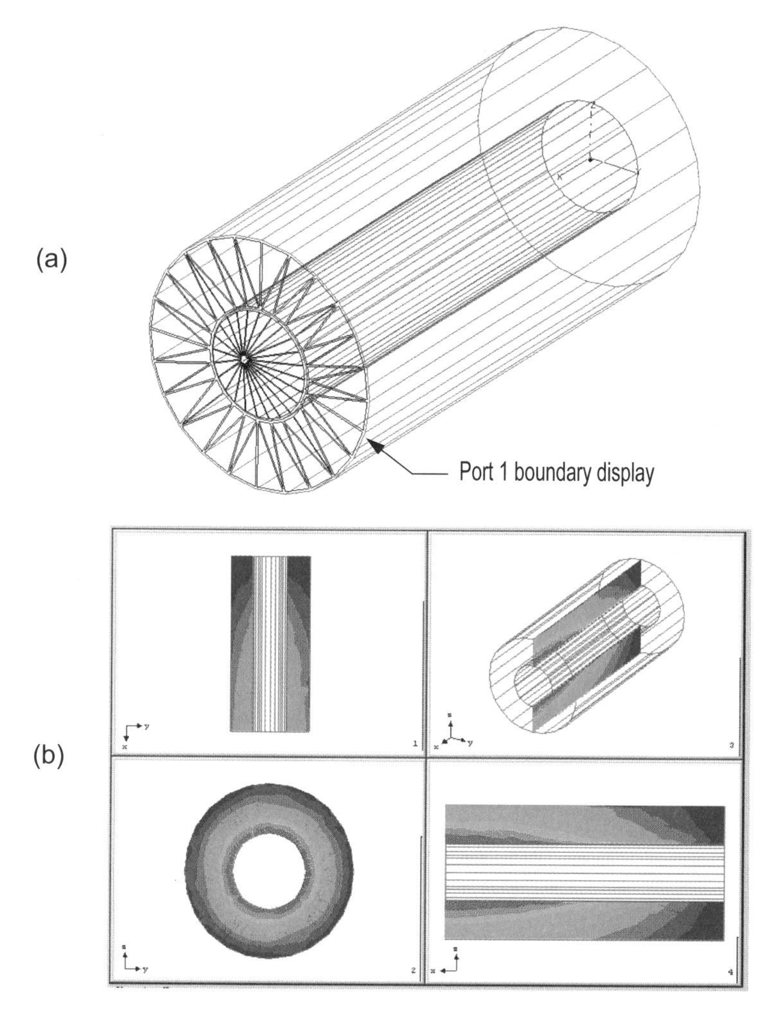

Figure 6.19 Calibration structure, 7-mm coaxial short: (a) boundary display of the port region; and (b) four different views of the E-field magnitude at 15 GHz (Ansoft HFSS Ver. 5.0).

6.5.2 7-mm Coaxial Short

The coaxial short (Figure 6.19(a)), can be created by copying and modifying the through line model. All we have to do is remove one port. One useful error checking feature of the FEM codes is boundary display. We can ask the solver to display the different boundary types, including ports, to make sure we have set up the problem correctly. Another strength of the FEM codes is their very flexible options for visualization. We can generate displays on any plane of interest in the structure.

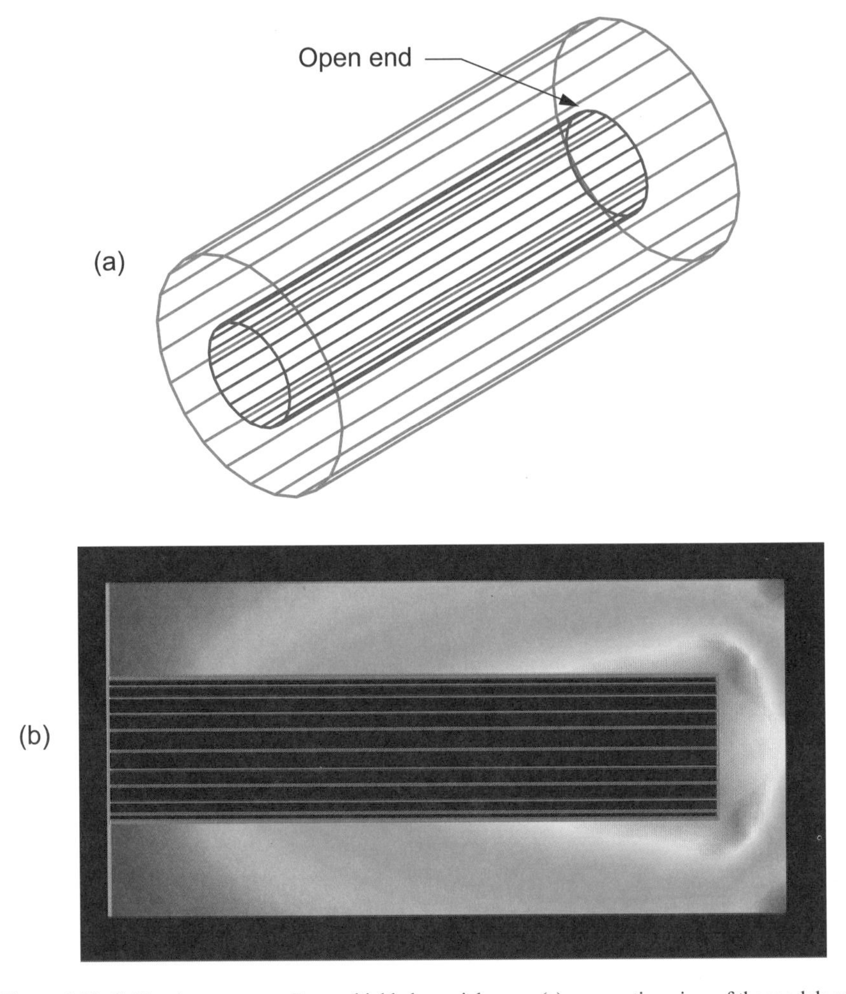

Figure 6.20 Calibration structure, 7-mm shielded coaxial open: (a) perspective view of the model; and (b) E-field magnitude in the X-Y plane at 15 GHz (Agilent HFSS Ver. 5.3).

6.5.3 7-mm Shielded Coaxial Open

The shielded coaxial open structure (Figure 6.20(a)) can be created by copying and modifying the coaxial short model. All we have to do is pull the center conductor back from the end wall. The singularity created by the edges is apparent in Figure 6.20(b). To make this an unshielded open we would have to create a cylinder of air around the original model and set the boundary of that cylinder to be absorbing. We must also be careful to place the absorbing boundary far enough away from the

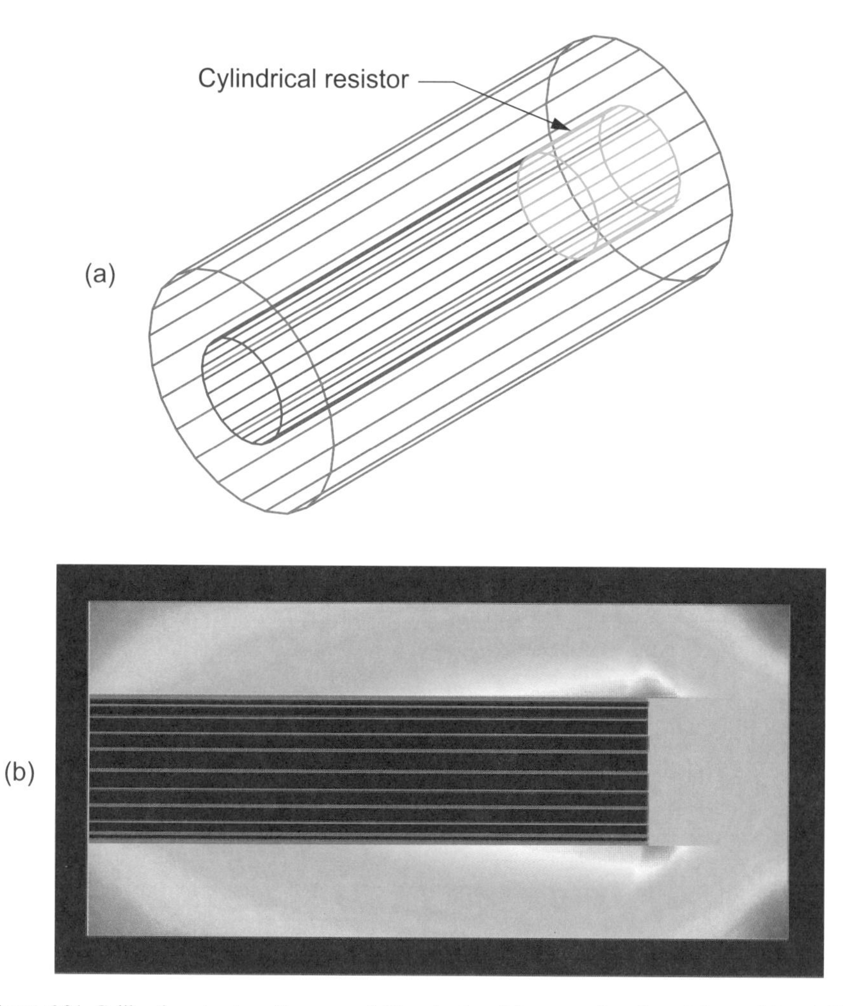

Figure 6.21 Calibration structure, 7-mm coaxial termination: (a) perspective view of the model; and (b) E-field magnitude in the X-Y plane at 15 GHz (Agilent HFSS Ver. 5.3).

open end. The rule of thumb for absorbing boundaries is to place them at least $\lambda/4$ away from the modeled structure.

6.5.4 7-mm Coaxial Termination

In the coaxial termination, the resistor is a cylinder of ceramic with a resistive material on the surface (Figure 6.21(a)). The resistivity of the material was adjusted for good match at a lower frequency. In a real broadband termination [9] the outer

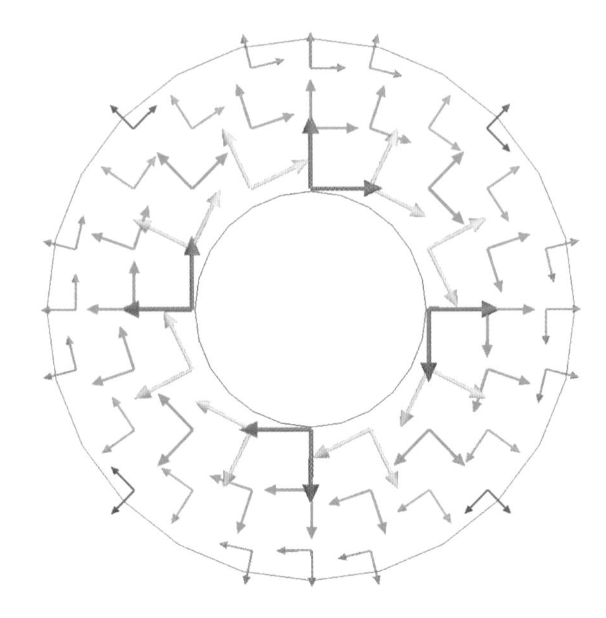

Figure 6.22 Vector E-field (radial direction) and vector H-field (tangential to conductors) in a plane perpendicular to the direction of propagation (Ansoft HFSS Ver. 8.0).

conductor would taper down to the shorted end using a fairly complicated spline curve. The resistive material is a lossy metal and allows some field to penetrate inside the ceramic body. The discontinuity at the center conductor/resistor boundary is also quite evident (Figure 6.21(b)).

6.5.5 7-mm Coax—TEM Behavior

Any one of the coaxial calibration structures also gives us the opportunity to ask the question, "What does TEM look like?" We can display the vector E-field and vector H-field in a plane perpendicular to the direction of propagation (Figure 6.22). When we animate this plot, we see that the E and H vectors are in phase and perfectly transverse to the direction of propagation. The same plot in the vicinity of a discontinuity would show some of the vectors rotating out of the plane shown here. In other words, there would be components that are no longer purely transverse.

6.6 VISUALIZATION

The most common FEM formulations solve for the vector E-field (Section 3.9). The H-field can then be derived from E-field using the Maxwell curl equation. Therefore, the H-field solution is slightly less accurate. Currents on conductors are approximated from the tangential component of the H-field near the conductor. The quality of the resulting current plot depends on the mesh density near the conductor and the smoothing algorithms that may or may not be applied to the plot.

All the of field components in any plane are generally available from an FEM solution. So additional quantities like the Poynting vector (direction and magnitude of power flow) can be computed. Starting with the earliest versions, Ansoft HFSS has offered a post processor "calculator" that understands field quantities and vector operators like cross product. The calculator can compute integrals along a user defined path, which can be quite useful for voltage breakdown estimates.

References

[1] Eisenhart, R. L., "A Novel Wideband TM01-to-TE11 Mode Converter," *IEEE MTT-S Int. Microwave Symposium Digest*, Baltimore, MD, June 7–12, 1998, pp. 249–252.

[2] Jain, N., and N. Kinayman, "A Novel Microstrip Mode to Waveguide Mode Transformer and its Applications," *IEEE MTT-S Int. Microwave Symposium Digest*, Phoenix, AZ, May 20–25, 2001, pp. 623–626.

[3] Wang, C., et al., "Dielectric Combline Resonators and Filters," *IEEE MTT-S Int. Microwave Symposium Digest*, Baltimore, MD, June 7–12, 1998, pp. 1315–1318.

[4] FlexPDE, PDE Solutions, Antioch, CA.

[5] Green, H. E., "The Numerical Solution of Some Important Transmission-Line Problems," *IEEE Trans. Microwave Theory and Tech.*, Vol. 13, No. 5, 1965, pp. 676–692.

[6] Green, H. E., "The Numerical Solution of Some Important Transmission-Line Problems (Correction)," *IEEE Trans. Microwave Theory and Tech.*, Vol. 23, No. 5, 1975, p. 455.

[7] Somlo, P. I., "The Computation of Coaxial Line Step Capacitances," *IEEE Trans. Microwave Theory and Tech.*, Vol. 15, No. 1, 1967, pp. 48–53.

[8] QWCX, QWED, Warsaw, Poland.

[9] MacKenzie, T. E., and A. E. Sanderson, "Some Fundamental Design Principles for the Development of Precision Coaxial Standards and Components," *IEEE Trans. Microwave Theory and Tech.*, Vol. 14, No. 1, 1966, pp. 29–39.

Chapter 7

FDTD and TLM Simulators

Like the FEM simulators, FDTD and TLM simulators are volume-meshing codes. Unlike FEM, the basic computation in FDTD and TLM takes place in the time domain rather than the frequency domain. In FDTD, Maxwell's time-dependent curl equations are converted to central difference equations and solved in the time domain [1–5]. Once the problem space has been gridded, the electric and magnetic fields are initialized throughout the grid at a given time. Time is then advanced one step and fields are computed again. Convergence is reached when the field quantities have reached a steady state (resonant and periodically driven problems) or have died down to negligible values (transient problems).

The TLM method is based on a transmission line analogy [6–9]. The problem space is subdivided into many small transmission line elements connected at nodes on a grid. One or several nodes of the grid are excited and the impulses "scatter" across the problem space. At each node, part of an incident impulse is transmitted to adjacent nodes and part of it is reflected back. At each time step, we compute the scattering at every node in the problem space. The field quantities are modeled by voltage and current impulses on the analogous transmission line network.

In both FDTD and TLM, if we record the time domain response at a given point in problem space, or at a port, we can obtain frequency domain data by running the time domain data through a fast Fourier (FFT) or a discrete Fourier (DFT) transform. While FDTD and TLM are derived and formulated from very different points of view [10–12], they are closely related and have many similar features.

7.1 FDTD AND TLM — STRENGTHS

Although FDTD and TLM start from very different approximations of the real world, when both methods have been turned into computer code they are strikingly similar. The biggest difference is that TLM works on a single grid and uses a scattering formalism, while FDTD requires a dual grid for E and H and uses a differential formalism. The storage requirements for FDTD and TLM are considerably lower than for MoM or FEM; there is no large matrix to invert. Therefore, solution time tends to scale more linearly as the problem gets bigger in terms of the number of unknowns. The time domain response is subject to a fast or discrete Fourier transform to generate broadband frequency domain data in a single simulation. Thus it is easy to find widely separated resonances without any prior knowledge of where they are located. These characteristics are a very good fit for RCS studies of ships and airplanes. These objects are large in terms of wavelengths and their radar returns come back at nearly random, widely separated frequencies.

The ability to inject arbitrary waveforms into a discretized structure allows us to simulate high-speed digital circuits and EMC scenarios under transient conditions. The time responses are similar to those of time domain reflectometers, are always causal, and allow us to pinpoint the location of discontinuities, scatterers, and boundaries.

Another strength of time domain simulators is their ability to model nonlinear materials, boundaries, and devices in a natural way. Since field quantities are known everywhere in space and time, nonlinear properties can be updated at each time step according to the local field strength and direction.

Since the updating of field components in FDTD, or the scattering of impulses in TLM are strictly local operations, we could imagine assigning each node in the grid to an individual processor. In other words, FDTD and TLM are easily adapted to parallel processing. The missing element in this scenario has been low-cost, standardized hardware and software to implement this approach. Several vendors of massively parallel computer hardware have appeared and quickly disappeared in the 1980s and 1990s.

7.2 FDTD AND TLM — WEAKNESSES

The convergence of FDTD depends on the size of the mesh, the number of required time steps, and the Q of the structure. The most basic mesh uses uniform cubic cells. This means we must use a staircase approximation for curved surfaces. If we need to resolve a small feature somewhere in the problem space, this may force us to use a very fine grid in the entire space. There are more sophisticated gridding techniques such as variable gridding or subgridding. While these techniques help the grid adapt to a smaller feature, smaller cells still tend to propagate into regions where they are not needed. Furthermore, due to the stability criterion, which requires that the time step must always be smaller than the time it takes for the signal to travel through the main diagonal of a cell, a graded mesh imposes the time step for the smallest cell upon the entire grid. On the other hand, multigridding or subgridding allows for separate time steps in the finer submesh and the main grid, but time and space averaging must be performed at the boundary between the subdomains, which can introduce errors due to spurious reflections and aliasing.

Table 7.1

FDTD and TLM Methods

Strengths	*Weaknesses*
Storage requirements lower than MoM or FEM	Convergence is function of mesh, run time, and Q
No large matrix to invert	Time step limited by stability criterion
Solution time scales more linearly with number of unknowns	Basic discretization is uniform cubic cells:
Easy to generate broadband data with FFT or DFT	• Staircase approximation for curves
Easy to locate widely separated resonances	• Large deltas in resolution a problem
Ability to model nonlinear materials, boundaries, and devices	• More sophisticated gridding is available
Ability to model dispersive materials and boundaries to yield wideband data with a single simulation	• Local mesh modification improves accuracy without computational penalty
Ability to model transient operation (TDR)	Multiple solution runs needed for multiport S-parameters
Bandwidth of excitation can be tailored to circuit bandwidth	Hard to resolve closely spaced and high Q resonances:
Easily adapted to parallel processing	• Not a good choice for filters
Single grid for TLM, dual grid for FDTD	• Convergence can be accelerated using the Prony's method or System Identification
	Open boundaries require ABCs or PMLs
	FDTD dual grid can cause ambiguities at interfaces

For problems with very large aspect ratios (small details in a larger problem space) it is sometimes easier to generate an FEM mesh than an FDTD or TLM mesh. In recent years, however, many of these weaknesses have been overcome by the development of special conformal boundary cells, quasi-static singularity corrections, and local cell modifications that avoid the penalty of reduced time step without sacrifice in accuracy.

The time domain codes are quite efficient for passive, reciprocal two-port networks; only one analysis run is needed to obtain all four S-parameters. For two-ports, solution times for the time domain codes and the FEM codes are comparable. However, for networks with more than two ports, if the full set of S-parameters is required, the time domain codes must do a complete solution for each port in the network. The FEM codes can obtain the full S-parameter matrix with one solution run at each frequency point.

The time domain codes are generally quite efficient at generating broadband frequency data using a Fourier transform. However, basic Fourier transform theory tells us it will be hard to resolve closely spaced resonances; so FDTD and TLM are probably not the best choice for filters (see Section 17.2). Convergence for high Q resonant structures and closely spaced resonances can be accelerated using the Prony method or System Identification.

The dual grid for E and H used by FDTD can sometimes cause ambiguity at material interfaces and requires additional processing. And similar to FEM, open boundaries require absorbing boundary conditions (ABCs) or perfectly matched layers (PMLs) [13, 14]. The latter are easier to implement in FDTD than in TLM. On the other hand, the implementation of convolution and general dispersive boundaries is straightforward in TLM due to the network nature of the mesh and the collocation of tangential electric and magnetic field components at the boundaries and in nodes. However, this requires storing twice as many numbers per TLM cell than per FDTD cell. While these differences are apparent to developers of FDTD and TLM codes, users will rather perceive their strong similarities. The strengths and weaknesses of the FDTD and TLM methods are summarized in Table 7.1.

7.3 FDTD AND TLM—VALIDATION STRUCTURES

Once again we must address the question of a proper validation structure. If we want to look at impedance and phase convergence for a strip type structure, then the stripline standard is still a good choice. However, this will require absorbing boundaries at the ports and proper meshing of a thin strip may not be obvious to the new user. In other words, it would not be possible to separate and quantify the respective contributions of the absorbing boundaries, the numerical dispersion of the grid and of the edge singularity to the total simulation error. We could reevaluate the coaxial standard used for the FEM codes. This again requires absorbing boundaries at the ports and meshing of curved boundaries. But compared to stripline, there is no edge singularity to worry about. The classic validation structure used most often in the technical literature to evaluate the dispersion error is a simple rectangular resonator. There are known, analytical solutions for comparison and we only have to deal with perfectly conducting walls. There are also no singularities in such a resonator. One potential difficulty for the new user will be the position and orientation of the excitation. The excitation must be correctly oriented to excite the desired mode. The resonator should also be rectangular, rather than cubic to avoid degenerate modes.

To calibrate the approximation of curved boundaries, a circular or elliptical cavity resonator can be used, for which exact analytical solutions exist as well. Resonant TEM strip transmission line and square coax sections can be used to evaluate the effect of sharp edge or corner singularities on the accuracy of the phase velocity, since it should ideally be identical to the wave velocity in the medium that these sections contain. Finally, the quality of dispersive wideband absorbing boundaries can be validated by computing the S-parameters of a standard rectangular waveguide section over its entire operating bandwidth using a transient analysis.

A typical rectangular resonator geometry is defined in Figure 7.1(a). There are two sets of modes to consider, the transverse electric (TE) and the transverse magnetic (TM). Each mode has an index which indicates the integer number of half-

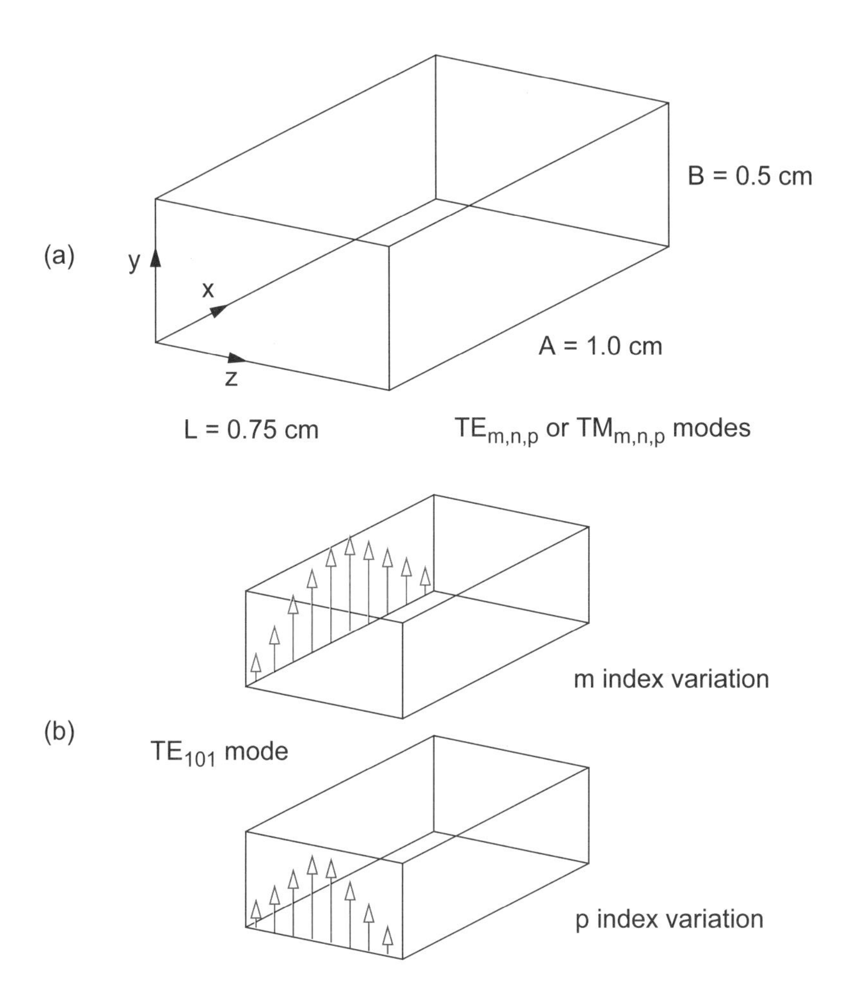

Figure 7.1 Typical rectangular resonator geometry: (a) basic geometry used to find TE or TM modes; and (b) field variation for the TE_{101} mode.

period variations in the field for each coordinate. The field variation for the TE_{101} mode is shown in Figure 7.1(b). A simple sketch like this will help us visualize where the excitation should be located and how it should be oriented. The mode frequencies for the rectangular resonator on the previous page can be calculated using a simple analytical formula

$$f = \frac{c}{\sqrt{\mu_r \cdot \varepsilon_r}} \cdot \sqrt{\left(\frac{m}{2 \cdot A}\right)^2 + \left(\frac{n}{2 \cdot B}\right)^2 + \left(\frac{p}{2 \cdot L}\right)^2} \tag{7.1}$$

Table 7.2

Resonant Frequencies and Eigenvalues for the Resonator in Figure 7.1(a)

Mode	*Frequency* (GHz)	*Eigenvalue* (k_0, 1/cm)
TE_{101}	24.983	5.236
TM_{110}	33.519	7.025
TE_{011}	36.031	7.551
TE_{201}	36.031	7.551
TM_{111}	39.025	8.179
TE_{111}	39.025	8.179
TM_{210}	42.398	8.886
TE_{102}	42.692	8.947

Sometimes in the technical literature you will find the resonant modes expressed as eigenvalues or wave numbers

$$k_{mnp} = \sqrt{\left(\frac{m \cdot \pi}{A}\right)^2 + \left(\frac{n \cdot \pi}{B}\right)^2 + \left(\frac{p \cdot \pi}{L}\right)^2} \tag{7.2}$$

Whichever nomenclature is used, we can easily compute the expected results for a rectangular resonator. Table 7.2 lists the resonant frequencies and eigenvalues for the first eight modes of the resonator shown in Figure 7.1(a). Note that even with this rectangular geometry some modes are degenerate; they share the same frequency.

7.3.1 TE_{101} Mode Convergence

If we choose the lowest order mode from Table 7.2, we can set up an FDTD problem to simulate that mode in our validation cavity. We must be careful to choose the location and the orientation of the source and sense points so that the desired mode will be excited and detected. Next, the mesh size should be adjusted to provide enough samples in all three dimensions. In addition to meshing issues, the user must decide when to truncate the time stepping. This is usually done by watching one of the output parameters. In Figure 7.2(a) we see the output of the resonator simulation after 10,000 time steps. The TE_{101} mode resonance is visible at the left,

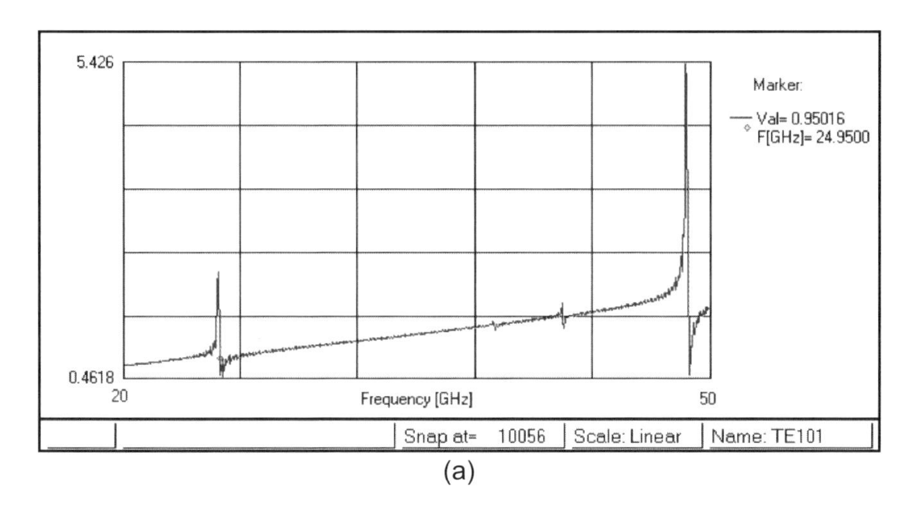

(a)

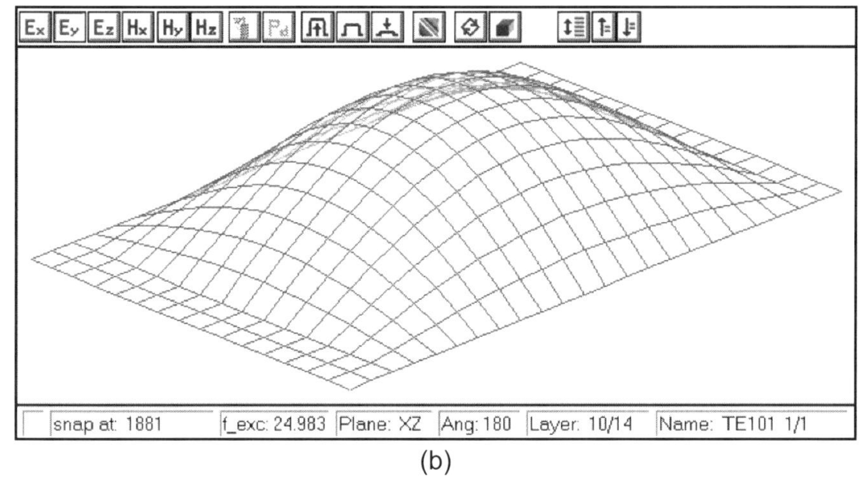

(b)

Figure 7.2 FDTD analysis of rectangular resonator in TE_{101} mode: (a) FFT analysis of sense point data after 10,000 time steps; and (b) Y-component of the E-field with excitation at 24.983 GHz (QuickWave 3D Ver. 2.1).

and a higher order mode is visible at the far right. We can also identify two weak modes between 38 GHz and 44 GHz. The responses are weak either because the excitation is not strongly coupled to them or the sense point is not strongly coupled to those particular modes.

Once the modal frequencies have been identified, we can animate the fields in the box by switching to a sinusoidal excitation at a particular modal frequency. To excite only the mode of interest, the excitation should be band-limited, such as a Gaussian-modulated sine, which injects only a finite amount of energy into the resonator. When the excitation has decayed, the mode will oscillate freely and will not be contaminated by other modes. Another technique for exciting a clean single mode in the cavity is to impress the known spatial field distribution as the initial condition at the start of the simulation. The exact period of the oscillation is then obtained in only a few time steps because it is the time between two successive maxima or zeros of the resonating field.

Figure 7.2(b) shows the Y-component of the E-field for the TE_{101} mode. We can see one half-period variation along the X-axis and one half-period variation along the Z-axis. A useful convergence study would be to plot the computed mode frequency as a function both of the number of time steps and the mesh resolution. These results could then be compared to the analytical solution from Table 7.2.

7.3.2 Wideband Rectangular Waveguide Validation

Unlike frequency domain simulators that compute the field solution one frequency at a time, time domain simulators can yield information over a wide frequency band with a single transient computation, followed by a Fourier transform. This is relatively straightforward when the properties of the structure under test do not depend on frequency. However, when they are frequency dispersive, a time domain solver must model them at a more fundamental level than a frequency domain solver where parameters can simply be changed for each frequency point. Examples are matched loads in non-TEM waveguides, or skin effect losses in metal boundaries. Such features are best validated under wideband transient simulation conditions.

A suitable and useful validation structure is a rectangular waveguide section like that shown in Figure 7.1, roughly one guided wavelength long at midband frequency of the dominant mode TE_{10}. The test section can be terminated with an ideal open-circuit, short circuit, and a wideband absorbing boundary. The validation test consists of computing the complex input reflection coefficient S_{11} of the section under the three terminating conditions, and the complex transmission coefficient S_{21}. The test is to be performed for each termination using a single transient simulation. The excitation should be a band-limited transient signal that covers the entire single mode operating bandwidth of the waveguide.

Exact analytical expressions are available for validating the scattering parameters obtained by the simulator. The errors in the phase angles of S_{11} and S_{21} are measures for the accuracy of the numerical phase velocity in the discretized waveguide section, while the magnitudes of S_{11} and S_{21} yield information on the numerical noise level and the quality of the wideband absorbing boundaries. If wall and dielectric losses are included, these magnitudes also yield information on the accuracy of the frequency-dependent loss computation. The quality of the Fourier transform capability affects both the magnitude and phase of the S-parameters.

Figure 7.3 shows a top view of the waveguide section with two reference planes separated by a distance L. Assuming that the section is lossless, the three terminations are perfect, and the phase constant is

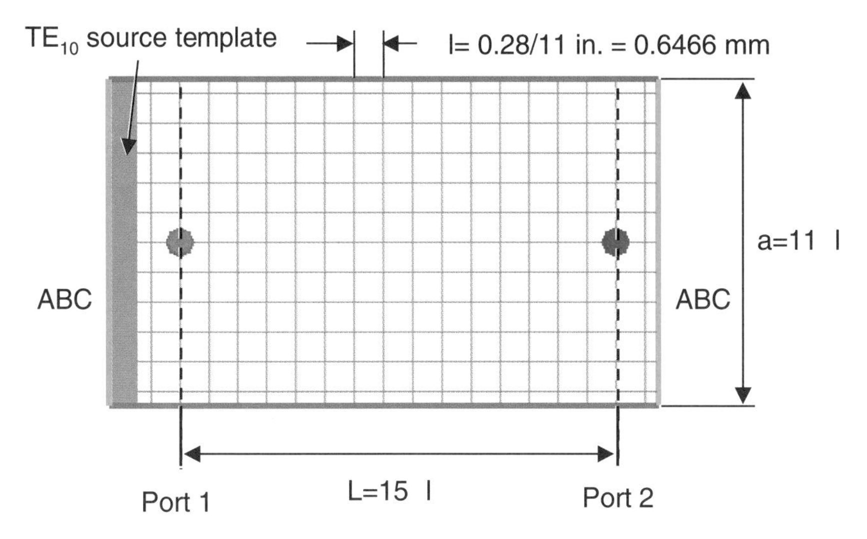

Figure 7.3 Top view of a WR-28 waveguide section terminated with wideband matched boundaries and discretized into cubic cells (MEFiSTo-3D Pro V3.0).

$$\beta = \frac{2\pi}{\lambda_g} = \frac{2\pi}{\lambda_0}\sqrt{1-(\lambda_0/2a)^2} \tag{7.3}$$

the S-parameters of the test section will be as follows:

(1) Perfectly matched source, sense point in Port 1, perfect open-circuit load placed across Port 2:

$S_{11} = 1.0\, e^{-j2\beta L}$, hence $|S_{11}| = 1.0$, $ang(S_{11}) = -2\beta L$

(2) Perfectly matched source, sense point in Port 1, perfect short-circuit load placed across Port 2:

$S_{11} = -1.0\, e^{-j2\beta L}$, hence $|S_{11}| = 1.0$, $ang(S_{11}) = \pi - 2\beta L$

(3) Perfectly matched source, perfectly matched load (position is not critical), sense points in Ports 1 and 2:

$S_{11} = 0.0$, hence $|S_{11}| = 0.0$, $ang(S_{11}) = undetermined$

$S_{21} = 1.0\, e^{-j\beta L}$, hence $|S_{21}| = 1.0$, $ang(S_{21}) = -\beta L$

The open- and short-circuited cases resemble the resonant scenario described in Section 7.3.1 and should give identical results, since the standing wave pattern in the waveguide section will be the same as those in a resonant cavity of the same cross-section. We will thus discuss only the wideband S-parameter computations for Case 3.

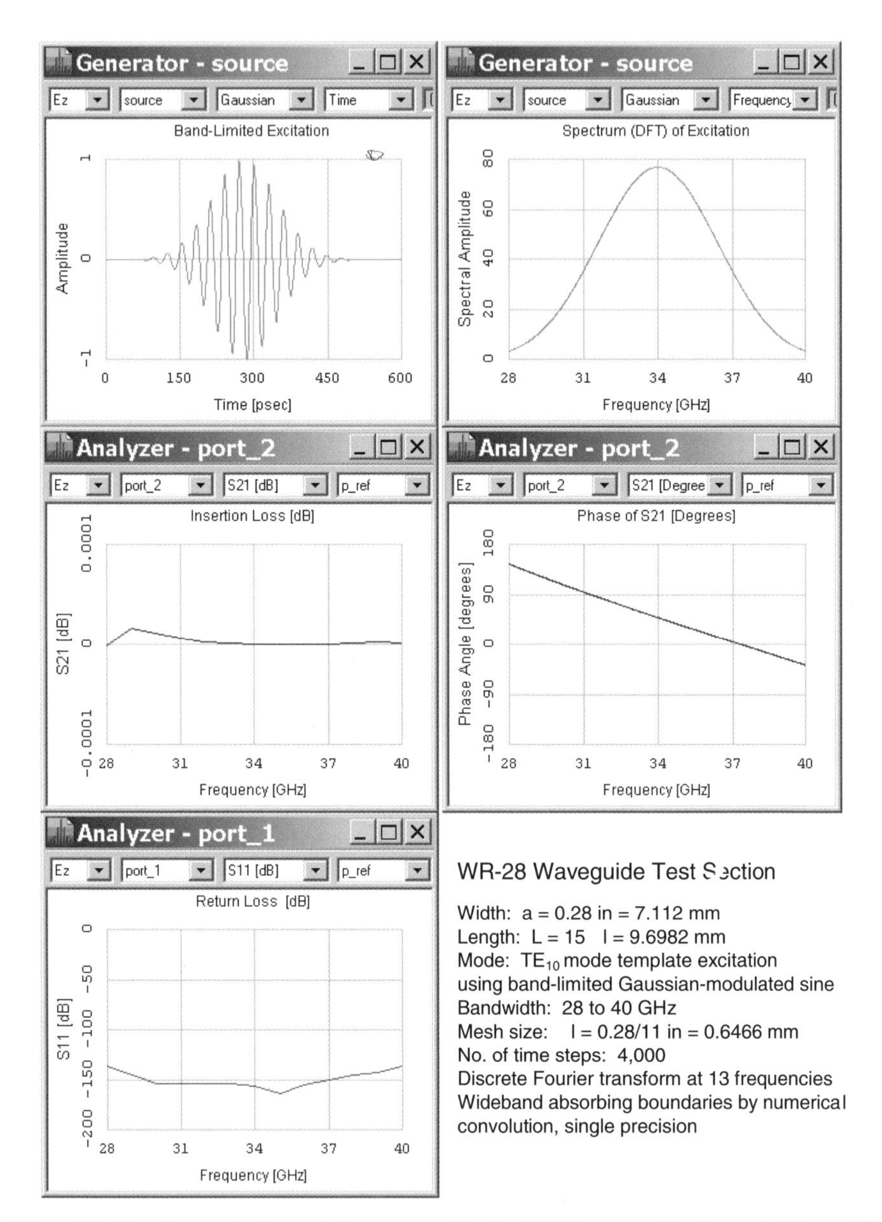

Figure 7.4 Transient excitation and *S*-parameters for the WR-28 waveguide shown in Figure 7.3 obtained with a single simulation (MEFiSTo-3D Pro, Ver. 3.0).

The WR-28 waveguide section shown in Figure 7.3 has been discretized into 11 cubic cells along the a-dimension (0.28 in = 7.112 mm). The grid size is thus

$\Delta l = 0.647$ mm. The guided wavelength at the upper frequency of the operating range (28 to 40 GHz) is 8.82 mm, ensuring a worst-case resolution of at least 13.6 cells/wavelength.

The wideband absorbing boundary conditions at both extremities of the section are modeled by convolution of the incident field with the impulse response of a semi-infinite waveguide that is generated recursively "on the fly." These ABCs are thus truly matched radiation conditions at all frequencies, even below cutoff. The TE_{10} source template acts as a current sheet that launches a modal field in the waveguide. The time responses of the waveguide section are picked up at the sense points in Ports 1 and 2 and Fourier transformed. The *S*-parameters are then extracted from these Fourier transforms.

Figure 7.4 shows the band-limited excitation waveform and its spectrum (top row) that covers the entire WR-28 operating band. The second row shows the magnitude and phase of the transmission coefficient S_{21}. The magnitude of S_{21} (insertion loss) is within 0.00002 dB of its theoretical value of 0 dB, and the phase of S_{21} is accurate within less than 0.02 degrees over the entire Ka-band from 28 to 40 GHz. Finally, the magnitude of S_{11} (return loss) is better than –130 dB.

These results demonstrate the high quality of the wideband terminations and the low directional and frequency dispersion of the numerical TLM engine. They exceed by far the accuracy margins of microwave laboratory instrumentation and inspire confidence in the accuracy of simulation results for devices tested in such a computational environment.

7.4 CONTROLLING MESHING

The most basic formulations of FDTD and TLM divide the problem space using a uniform grid of cubic cells. The smallest feature in the problem forces the cell size that will be used throughout the problem space. This is clearly inefficient and several more advanced approaches to meshing have been developed over the years. We will take a brief look at these alternative meshing schemes using a simple cylindrical object. This outline may represent the interior of a circular waveguide or a top view of a dielectric resonator.

Figure 7.5(a) shows a simple stair-case approximation using cubic cells. If we start with a fairly coarse resolution on the left, cutting the cell size in half (right hand figure) clearly gives us a better approximation, but the number of unknowns has increased dramatically. For those modes with rather uniform field distribution in the center, we probably have many more cells in the central region than we actually need. One alternative is variable or graded meshing (Figure 7.5(b)). The mesh resolution is increased near the curved boundary. The larger and smaller meshes do not have to be integer multiples of each other and some cells are no longer cubic. In three dimensions the finer mesh tends to propagate into regions where it is not needed. The smallest cell size also dictates the time step.

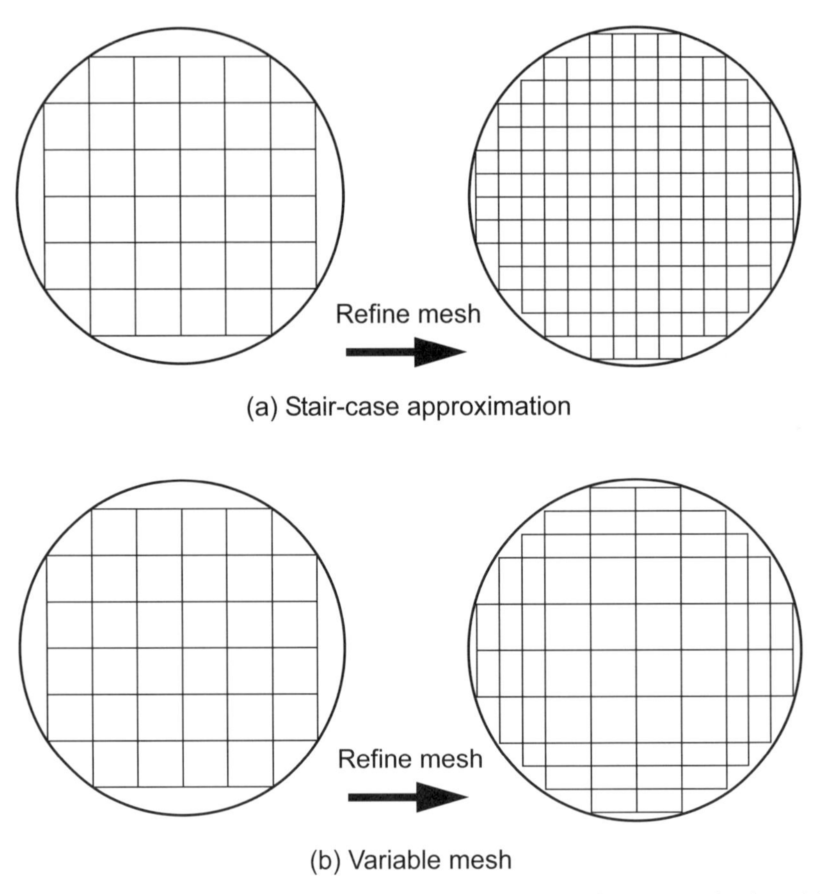

Figure 7.5 Two meshing strategies to increase resolution: (a) basic stair-case approximation, global cell size is reduced; and (b) variable mesh, cell size is reduced only in specific regions.

Another approach to meshing is called subgridding (Figure 7.6(a)). The larger mesh is divided into an integer number of smaller cells. This not only confines the dense grid better to the region where it is needed, but it also allows the different submeshes to run at different time steps. However, the interface between the submeshes requires time- and space-averaging of tangential field components, giving rise to spurious reflections, mismatch due to differences in dispersion characteristics, and sometimes instability. The coarser mesh in the center of Figure 7.6(a) may also fail to capture higher order modes which have maximum E-field or H-field in the center of the geometry.

The most sophisticated time domain meshing schemes truncate lager cells to form polygonal cells (Figure 7.6(b)). QwickWave3D can use this technique to conform to curved boundaries in one plane. CST Microwave Studio can conform to

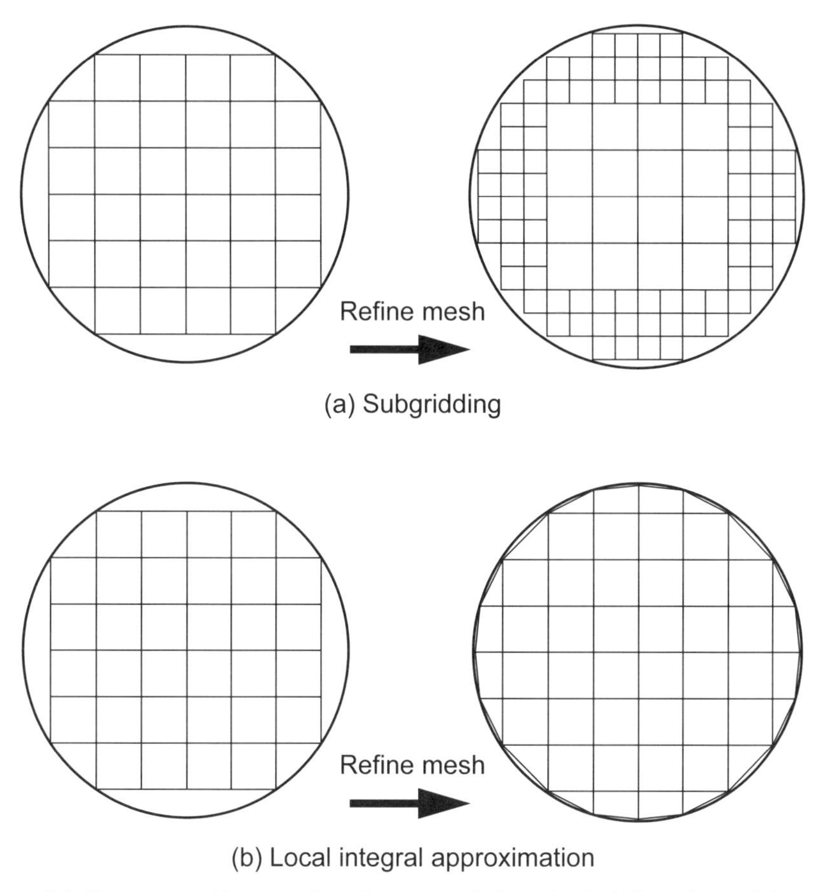

Figure 7.6 Two more meshing strategies to increase resolution: (a) subgridding, where existing cells are subdivided in integer multiples; and (b) local integral approximation, existing cells are truncated along a line forming polygonal cells.

curved boundaries in three dimensions. Advanced TLM codes, such as Mefisto-3D Pro, perform this type of approximation by adjusting the lengths and characteristic impedance of link lines between nodes. This effectively distorts the grid in the vicinity of the curved boundary.

7.4.1 Meshing the Stripline Standard

The stripline standard is a valid tool for evaluating any of the codes we are studying. In this case we will use it to demonstrate some the meshing concepts we have been discussing. In Figure 7.7, we examine uniform meshing using QuickWave3D. In Figure 7.7(a) we have uniformly meshed the problem using a 0.4-mm grid. In Figure 7.7(b) we have doubled the resolution to 0.2 mm. In both cases the grid does

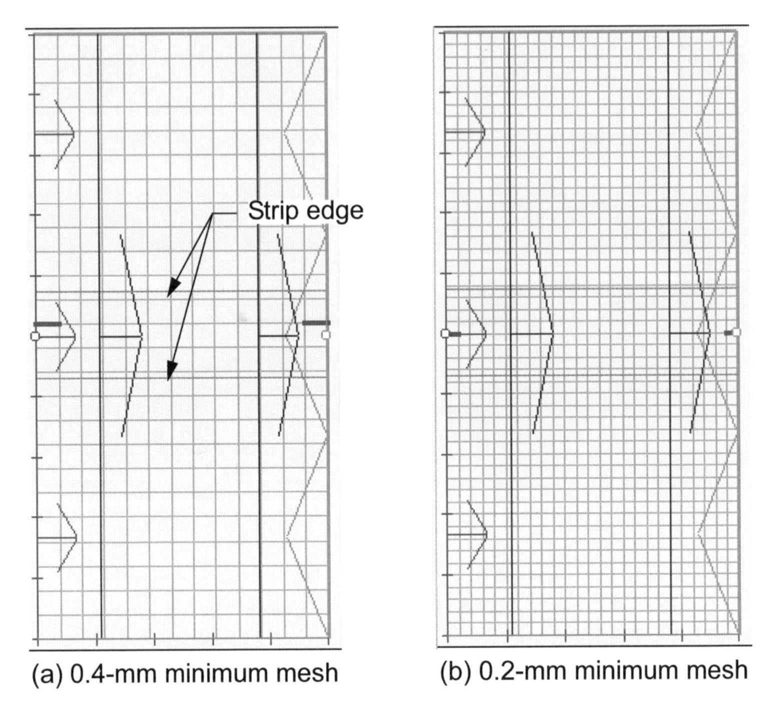

Figure 7.7 Uniform meshing of the stripline standard: (a) 0.4-mm minimum mesh size; and (b) 0.2-mm minimum mesh size. The strip runs from left to right (QuickWave3D Ver. 2.1).

not necessarily align with the edges of the strip, so we probably have not captured the position of the edge correctly. The finer grid is also placing many cells at the edges of the box where they are not needed. The red and blue arrows indicate source and sense points for *S*-parameter calculations. The green triangles on the right indicate an absorbing boundary.

We can improve on the basic mesh by adopting variable meshing. In Figure 7.8(a) we start with a uniform 0.4-mm mesh. We then apply the finer 0.2-mm mesh only to the region around the strip. Note that this finer mesh region extends from the top of the box to the bottom of the box as well. In Figure 7.8(b) we have taken the extra step of snapping the grid to edges of the stripline. This forced alignment of the grid with a physical boundary should greatly improve our approximation of the edge singularity.

7.4.2 Meshing the Coaxial Step Discontinuity

As another exercise in meshing, we can look at the same coaxial step discontinuity that we analyzed using FEM in Section 6.4.3. The concepts of dummy objects and mesh control apply equally well to the FDTD and TLM codes as they do to FEM.

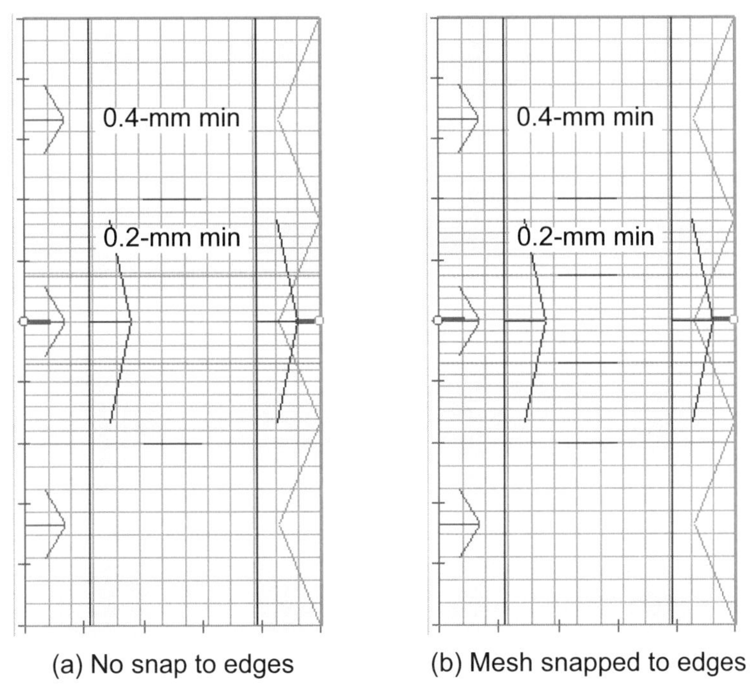

Figure 7.8 Variable meshing of stripline standard: (a) finer 0.2-mm mesh applied only to region around strip, no snap to edges of strip; and (b) mesh snapped to edges of strip (QuickWave3D).

As expert users, we assume that we can help the software converge to the correct solution by applying our knowledge of the problem.

Figure 7.9(a) shows the coaxial step geometry once again with what might be a typical starting mesh. Meshing on a guide wavelength basis is set to $\lambda/10$ at 4 GHz. We have placed dummy regions on both sides of the step just as was done for the FEM experiments. The maximum Z-axis cell size in the dummy regions is 0.050 inch, which is between $1/3$ and $1/2$ of the guide wavelength cell size. In general we would like to limit the maximum cell aspect ratio to less than 4:1. The total number of cells is 9,261.

Figure 7.9(b) shows the coaxial step with a much finer mesh. Meshing on a guide wavelength basis is now set to $\lambda/40$ at 4 GHz. The maximum Z-axis cell size is now 0.0125 inch. The total number of cells is 34,839.

Figure 7.9(c) shows the meshing in the cross-section of most of these experiments. The maximum cell size is 0.025 inch, which results in two cells across the low impedance line air gap. Cells inside the low impedance center conductor will be ignored but this meshing implies a fairly fine mesh in the air region of the high impedance line. With large changes in aspect ratio it is often difficult to keep the smallest cells confined to one particular region in the model.

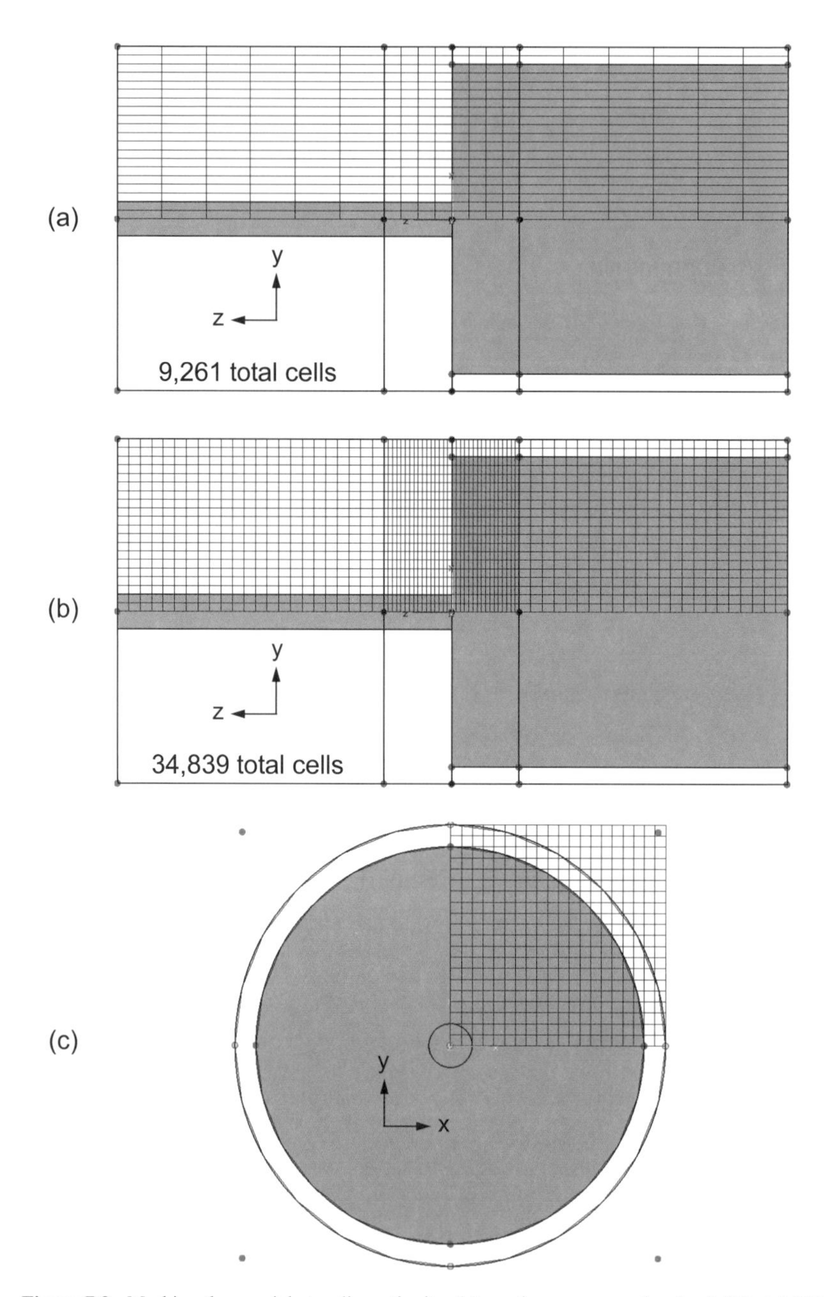

Figure 7.9 Meshing the coaxial step discontinuity: (a) a rather coarse mesh using $\lambda/10$ at 4 GHz as the default mesh size and 0.05-inch minimum in the Z-direction in the two dummy regions on either side of the step; (b) a much finer mesh using $\lambda/40$ at 4 GHz as the default mesh size and 0.0125-inch minimum in the Z-direction for the dummies; and (c) the mesh in the cross-section with 0.025-inch minimum cell size in X and Y (CST Microwave Studio).

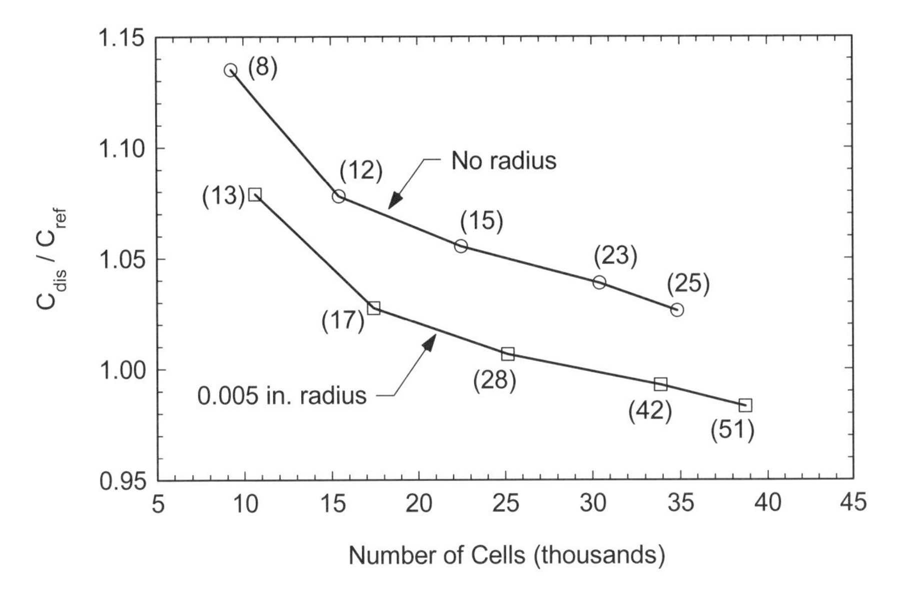

Figure 7.10 Convergence of FDTD solution for the coaxial step. Curve marked "no radius" is for the original geometry. Several meshes were also analyzed with an 0.005-in. radius on the outer step. Solution times in seconds are shown in parentheses (CST Microwave Studio Ver. 3.4).

In Figure 7.10 we compare the step capacitance computed at 1 GHz to the reference value from Somlo [15] for the meshings shown in Figure 7.9 plus several intermediate meshings. The numbers in parentheses are solution times in seconds on a 1.13-GHz Pentium III notebook computer, circa 2002. Starting from an error of just over 13% we rapidly converge to a value just over 2%. The vendor suggested that putting a small finite radius on the corner of the step might improve the convergence. This was done, but the results indicate a shift in absolute value more than an improvement in convergence. The small radius on the corner also creates some very high aspect ratio cells in the model. We also made one more run on the standard geometry with minimum X-axis and Y-axis cell sizes of 0.0125 inch. The number of cells increased by a factor of four while the computed error moved from 2.6% to 1.9%.

One factor that may be limiting the convergence of the problem is the relatively low upper frequency limit (4 GHz) that we were forced to set. The limits we set in the frequency domain are used to shape the Gaussian pulse in the time domain analysis. The cutoff frequencies for the first two modes at both ports are shown in Table 7.3. The upper frequency limit was set to one-half of the next higher order mode. If the upper frequency limit is increased, energy is transferred to higher modes and it is unclear how to extract the capacitance of the step discontinuity. But in fact this may be a problem only for this extreme geometry. In Section 17.1 we

Table 7.3

Port Modes for the Geometry in Figure 7.9

Port 1 Modes	Port 2 Modes
11.47 GHz (TE)	7.90 GHz (TE)
12.44 GHz (TM)	15.76 GHz (TE)
19.30 GHz (TM)	23.65 GHz (TE)

have results for an actual lowpass filter using this type of geometry. For the filter, the outer diameter is much smaller and the 50-ohm ports operate in the TEM mode to a very high frequency.

7.5 VISUALIZATION

In most 3D FDTD and TLM codes, all the field components are available for display in any given plane. The fairly uniform meshing that generally results from the time domain methods results in high quality field plots.

The more intriguing aspect of the time domain codes is the ability to observe the buildup of various quantities as a function of time. An inexperienced engineer may fall into trap of assuming that all effects appear instantaneously throughout the structure. A time domain code reinforces the "time and distance" nature of many effects. Another very powerful display mode is the "virtual TDR." With the time domain codes, a pulse with a very fast edge rate can be launched into a network and the resulting reflections can be observed in real time. In the "virtual TDR" the pulse is not distorted by imperfect connectors or dispersive cables like it would be in the laboratory.

References

[1] Yee, K. S., "Numerical Solution of Initial Boundary Value Problems Involving Maxwell's Equations in Isotropic Media," *IEEE Trans. Antennas and Propagation*, Vol. 14, No. 5, 1966, pp. 302–307.

[2] Taflove, A., and M. E. Brodwin, "Numerical Solution of Steady-State Electromagnetic Scattering Problems Using the Time-Dependent Maxwell's Equations," *IEEE Trans. Microwave Theory Tech.*, Vol. 23, No. 8, 1975, pp. 623–630.

[3] Gwarek, W. K., "Analysis of an Arbitrarily-Shaped Planar Circuit, a Time-Domain Approach," *IEEE Trans. Microwave Theory Tech.*, Vol. 33, No. 10, 1985, pp. 1067–1072.

[4] Gwarek, W. K., "Analysis of Arbitrarily-Shaped Two-Dimensional Microwave Circuits by Finite-Difference Time-Domain Method," *IEEE Trans. Microwave Theory Tech.*, Vol. 36, No. 4, 1988, pp. 738–744.

[5] Weiland, T., "A Discretization Method for the Solution of Maxwell's Equations for Six-Component Fields," *Electronics and Communication (AEU)*, Vol. 31, 1977, p. 116.

[6] Johns, P. B., and R. L. Beurle, "Numerical Solution of 2-Dimensional Scattering Problems Using a Transmission-Line Matrix," *Proc. Inst. Electr. Eng.*, Vol. 118, No. 9, 1971, pp. 1203–1208.

[7] Akhtarzad, S., and P. B. Johns, "Three-Dimensional Transmission-Line Matrix Computer Analysis of Microstrip Resonators," *IEEE Trans. Microwave Theory Tech.*, Vol. 23, No. 12, 1975, pp. 990–997.

[8] Hoefer, W. J. R., "The Transmission-Line Matrix Method-Theory and Applications," *IEEE Trans. Microwave Theory Tech.*, Vol. 33, No. 10, 1985, pp. 882–893.

[9] Johns, P. B., "A Symmetrical Condensed Node for the TLM Method," *IEEE Trans. Microwave Theory Tech.*, Vol. 35, No. 4, 1987, pp. 370–377.

[10] Johns, P. B., "On the Relationship Between TLM and Finite-Difference Methods for Maxwell's Equations," *IEEE Trans. Microwave Theory Tech.*, Vol. 35, No. 1, 1987, pp. 60–61.

[11] Chen, Z., M. M. Ney, and W. J. R. Hoefer, "A New Finite-Difference Time-Domain Formulation and its Equivalence with the TLM Symmetrical Condensed Node," *IEEE Trans. Microwave Theory Tech.*, Vol. 39, No. 12, 1991, pp. 2160–2169.

[12] Krumpholz, M., C. Huber, and P. Russer, "A Field Theoretical Comparison of FDTD and TLM," *IEEE Trans. Microwave Theory Tech.*, Vol. 43, No. 8, 1995, pp. 1935–1950.

[13] Berenger, J.-P., "A Perfectly Matched Layer for the Absorption of Electromagnetic Waves," *Journal Computational Physics*, Vol. 114, No. 2, 1994, pp. 185–200.

[14] Eswarappa, C., and W. J. R. Hoefer, "Absorbing Boundary Conditions for Time Domain TLM and FDTD Analysis of Electromagnetic Structures," *Electromagnetics*, Vol. 16, No. 5, 1996, pp. 489–519.

[15] Somlo, P. I., "The Computation of Coaxial Line Step Capacitances," *IEEE Trans. Microwave Theory Tech.*, Vol. 15, No. 1, 1967, pp. 48–53.

Chapter 8

Ports and De-embedding

Many problems that we would like to solve using numerical methods have ports. Ports allow us to excite a circuit or antenna and measure the results. Depending on the type of circuit analyzed, we may need several types of ports. Typical port types are single ended, differential, waveguide, microstrip, and CPW. To be really useful, ports must be calibrated. Field-solvers have numerical port discontinuities just like network analyzers and test fixtures have physical port discontinuities. The easier type of port to implement is on the boundary of the problem space. Most solvers also allow access to ports that are internal to the problem geometry. Internal ports are generally more difficult to implement and calibrate.

Sometimes when we measure an active or passive device in a fixture, we would like to remove the effects of the fixture; this process is called *de-embedding*. We also use de-embedding in field-solvers to separate numerical port and fixture effects from our device under test. De-embedding is actually easier and more flexible in a field-solver than in the laboratory. Multiport de-embedding that would be quite difficult in the lab is actually quite easy in some field-solvers.

8.1 PORTS—CONNECTING FIELDS TO CIRCUITS

In Chapter 2 we saw a simple demonstration of how to build a Y-matrix to analyze a simple network of lumped or distributed components using analytical models. To include a solution of Maxwell's equations in this type of analysis we need a consistent way to convert E- and H-fields to voltages and currents. In Figure 8.1 we show the classic low frequency definition of how voltage and current are derived from the fields at the ports of a typical component [1]:

$$V = \int_p \mathbf{E} \cdot d\mathbf{l} \tag{8.1}$$

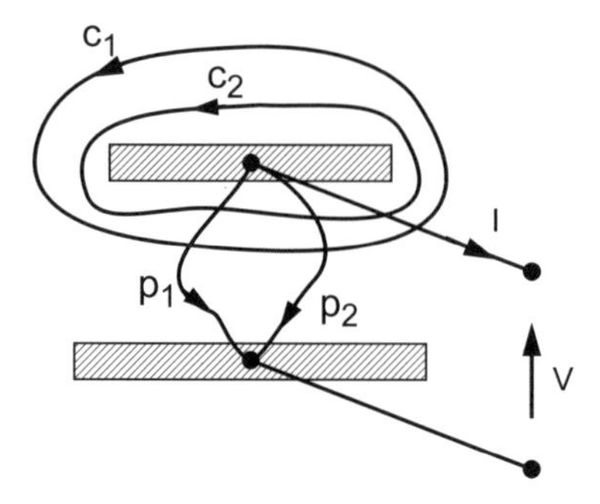

Figure 8.1 Classical circuit definition of voltage and current. Voltage is determined by integrating **E** along p_1 or p_2. Current is determined by integrating **H** along c_1 or c_2. © 1993 Oxford University Press [1].

$$I = \oint_c \mathbf{H} \cdot d\mathbf{l} \tag{8.2}$$

The voltage integral (8.1) is taken from one conductor to the other and the path (p_1 or p_2) is arbitrary. The current integral (8.2) is taken around one of the conductors and again the path (c_1 or c_2) is arbitrary. Because V and I are unique, we can define the average complex power, P_c, entering the component and impedance seen at the terminals, Z_c:

$$P_c = \frac{1}{2}VI^* \tag{8.3}$$

$$Z_c = \frac{V}{I} \tag{8.4}$$

Unfortunately, these very simple relations hold only for homogeneous geometries that are very small in terms of wavelengths. To do a more thorough analysis we need to look at the integral form of Maxwell's curl equations ($e^{j\omega t}$ time dependence is assumed):

$$\oint_{p_{tot}} \mathbf{E} \cdot d\mathbf{l} = -j\omega \iint_S \mathbf{B} \cdot d\mathbf{S} \tag{8.5}$$

$$\oint_c \mathbf{H} \cdot d\mathbf{l} = I + j\omega \iint_A \mathbf{D} \cdot d\mathbf{S} \tag{8.6}$$

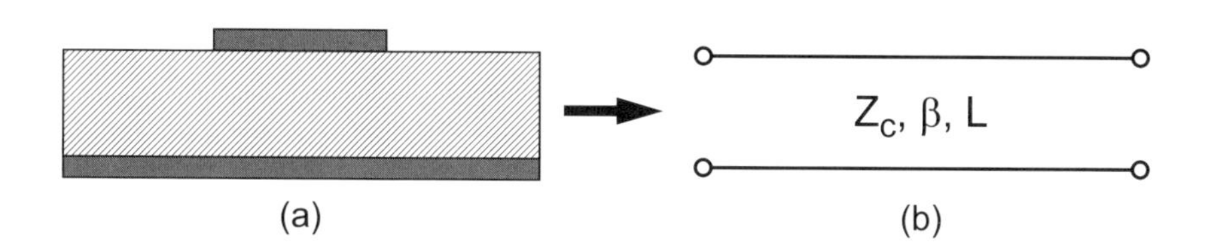

Figure 8.2 Microstrip transmission line: (a) cross-section of the physical line; and (b) circuit theory model which completely describes the line using characteristic impedance, propagation constant, and physical length.

The closed contour, p_{tot} is formed by joining p_1 and p_2. Equation (8.5) shows us that evaluating the voltage between the two conductors using p_1 or p_2 will only be equal if the flux of **B** through the surface formed by p_{tot} is negligible. In a similar fashion, (8.6) shows us that the computation of I in (8.2) is only independent of the path if the flux of **D** through surface formed by either contour is negligible.

In Figure 8.2(a) we show the cross-section of a microstrip line that we wish to make a circuit model for. In circuit theory, a transmission line (Figure 8.2(b)) is completely described by its characteristic impedance Z_c, its propagation constant β, and its physical length L. The physical microstrip line is completely described by the numerical solution of the electric field **E** and the magnetic field **H**. However, at the ports, the right-hand sides of (8.5) and (8.6) are not negligible for realistic dimensions and frequencies. Therefore, (8.1) and (8.2) cannot uniquely define the voltage and current at the ports. This ambiguity in the definition of V and I for microstrip led to a rather long running debate on the definition of Z_c [2–14]. The one quantity that can be uniquely determined at this point is the propagation constant β. We can assume that the waves in physical model and the electrical model travel with the same velocity and with the same dispersion characteristics.

The solution to the Z_c definition problem was found in the power equivalence principle [5]. We assume that the average power propagated by the physical waveguide (or the numerical model) must be the same as the average power propagated by the circuit model. With the power fixed we only need to choose a definition for V or for I (but not both) that make sense for a particular transmission line type. For microstrip, the typical choice is now the conduction current on the strip resulting in a power-current definition of the characteristic impedance (Z_{pi}). For slotline, the more logical choice is the voltage in the slot resulting in a power-voltage definition of the characteristic impedance (Z_{pv}). For a pure TEM structure with

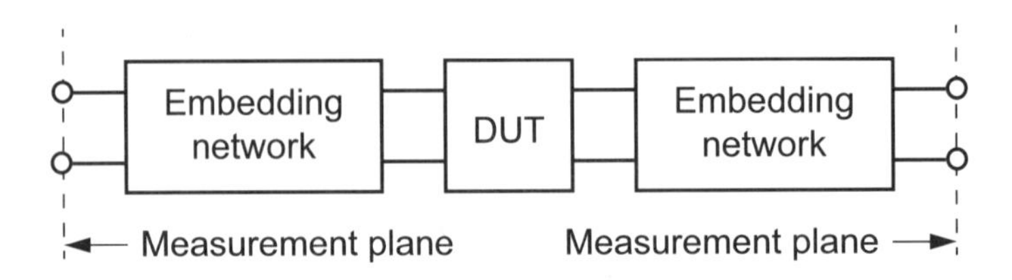

Figure 8.3 Typical measurement situation. The device under test is surrounded by embedding networks, which must be numerically removed.

small cross-section dimensions in terms of wavelengths we can still compute V and I directly; resulting in a voltage-current definition of characteristic impedance (Z_{vi}). For a pure TEM line, all three definitions should give the same result. In some field-solvers, primarily the 3D FEM solvers, the user has access to all three definitions of characteristic impedance: Z_{pi}, Z_{pv}, and Z_{vi}. Hopefully this brief discussion will help the reader understand the source and correct use of these three definitions.

8.2 DE-EMBEDDING AND UNTERMINATING

At RF and microwave frequencies it is often impractical to measure the impedance, admittance, or S-parameters of an active device directly at the device terminals. Instead the device is typically "embedded" in some form of test fixture and measurements are made at a reference plane some distance away from the actual device (Figure 8.3). In the case of two-port devices, there is an embedding network on both sides of the device under test (DUT). De-embedding is then the mathematical process of removing the embedding networks and determining the true parameters of the device under test [15, 16]. Figure 8.4(a) is a more physical picture of the problem. We have some type of packaged, three-terminal device mounted in a microstrip test fixture. We assume our automatic network analyzer (ANA) is calibrated down to the ends of the test cables, where we connect to the fixture. We would like to mathematically shift the measurement reference plane to the device terminals. We know there is some type of discontinuity and fringing fields at the transition from the connector onto the microstrip line. Then we have a uniform length of line and an additional discontinuity at the transition into the DUT.

The more difficult process is determining the properties of the embedding networks. This can be done by substituting and measuring a known set of standards for the DUT [17]. Unterminating is then the process of deducing the parameters of the embedding networks using a set of measurements of known standards. This is actually the very same process used when calibrating the ANA. A set of known standards (short, open, and 50-ohm load) is measured at the ports and the error terms for the measurement system are determined [18–20]. In fact, the error terms can be

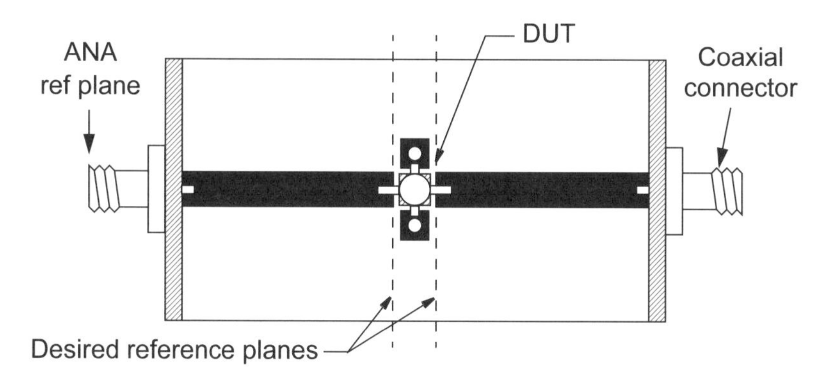

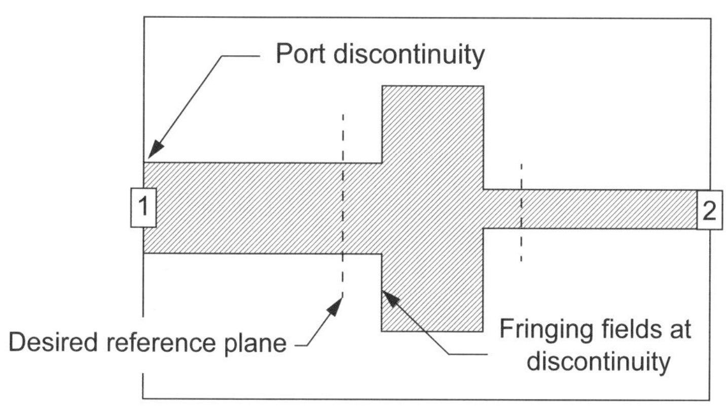

Figure 8.4 Typical de-embedding problems: (a) An active device mounted in a test fixture. The connector discontinuity and a length of uniform line must be numerically removed. (b) A similar de-embedding situation in a field-solver. The field-solver port discontinuity and a length of uniform line must be numerically removed.

thought of as a two-port network sitting in between the measurement plane and a perfect measurement system. So the ANA calibration problem and the unterminating problem for the test fixture are in fact the same.

When we compute the response of a network in a field-solver, we are also presented with a de-embedding and unterminating problem very similar to the device characterization problem we have already outlined. In Figure 8.4(b) we show a double step discontinuity that we would like to characterize. Our field-solver has discontinuities and fringing fields at the numerical ports. There are also fringing fields, and evanescent modes at the step discontinuities. If the length of line

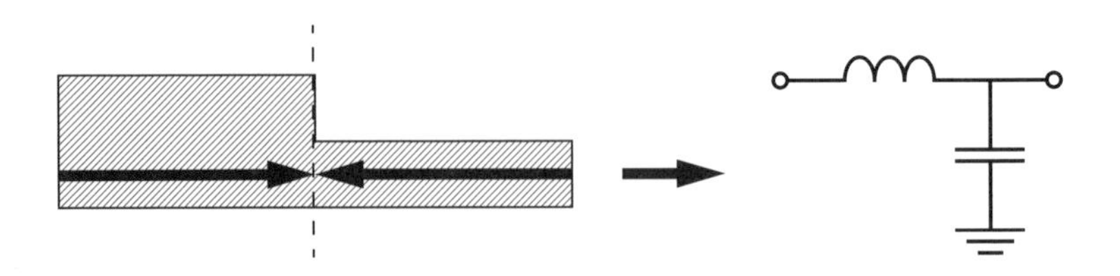

Figure 8.5 The step junction is an example of zero length de-embedding. All of the uniform transmission line is removed from both sides in order to extract a lumped element equivalent.

between the port and the first discontinuity is too short, the evanescent fields will interact and the computed results will be incorrect. Typically, we would like to remove the port discontinuity and some length of uniform line from the global field-solver solution and set a new reference plane closer to the structure of interest. The actual location of the new reference plane or planes is completely arbitrary. We tend to set them to geometrical features just because that is the easiest thing to remember. When modeling discontinuities, we tend to set the planes so the resulting models contain only lumped elements and no transmission lines.

Sometimes when we de-embed, the resulting object has zero length in one or more dimensions. For example, the step discontinuity in Figure 8.5 would typically be de-embedded to the junction from both sides in order to extract a lumped element model. If we wanted a "black box" step model based on *S*-parameters, it might be prudent to leave some uniform line on either side of the step so as not to confuse the layout tool. Other examples of "zero length" de-embedding can be found in Section 12.2 (via modeling) and Sections 16.1 and 16.3 (bandpass filter modeling).

Zero length de-embedding can also be used to evaluate the residual errors in a de-embedding algorithm. If we solve a simple 50-ohm line as a two port and zero length de-embed to the center of the line we expect $|S_{11}| = 0.0$ and $S_{21} = 1.0$ at 0.0 degrees [21].

Once we become comfortable with the concept of de-embedding, there are many experiments and measurements we can do on the field-solver that would be difficult, if not impossible, in the lab. Figure 8.6(a) is one simple example taken from a microstrip edge-coupled filter design. An edge-coupled filter can be described as a cascade of quarter wavelength coupled sections. Where these sections meet, there are two open-ends in close proximity. We typically have an analytical model for an isolated open-end in our circuit simulator. Does this model need to be modified due to the adjacent strips? Is there significant coupling between the two open-ends?

These are questions we can easily answer with the field-solver and de-embedding. We can set up a four-port problem (Figure 8.6(b)) on the field-solver and de-

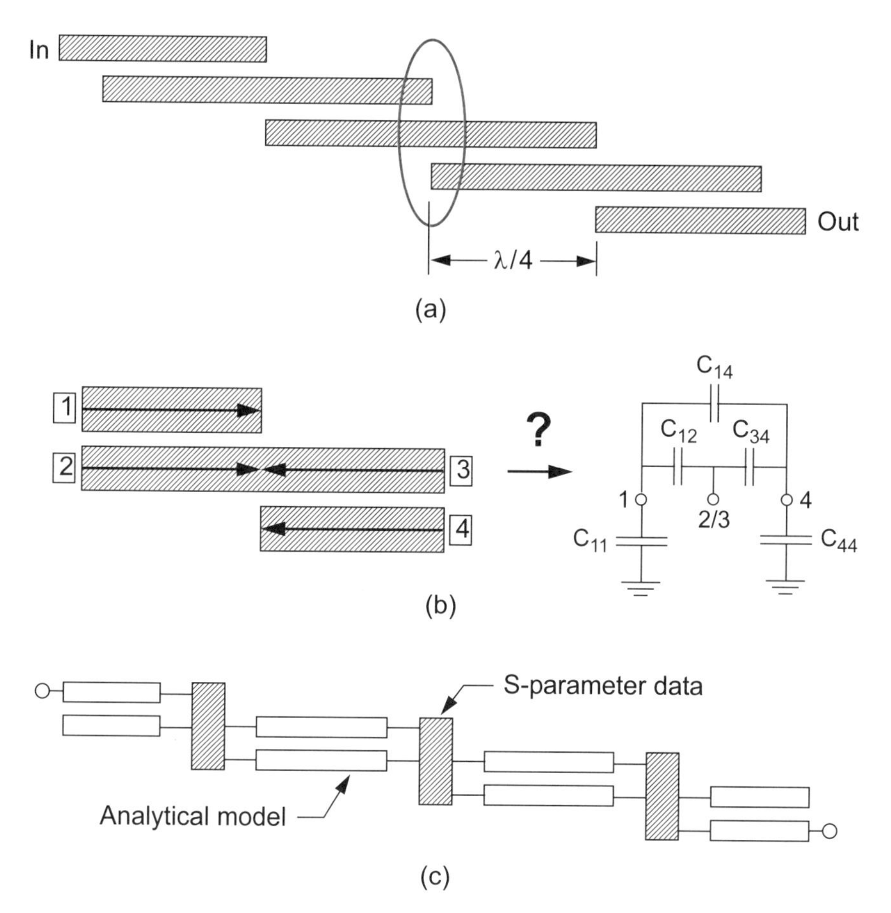

Figure 8.6 (a) Microstrip bandpass filter with open ends in close proximity. (b) Field-solver model of open ends with possible equivalent circuit. (c) The *S*-parameters for each open-end region combined with analytical coupled line models for a fast analysis of the complete filter.

embed down to the junction of the two quarter wavelength coupled sections. With our data in hand, we can begin to explore various models for this complex junction and examine the relative magnitude of the circuit elements to determine which are most important. Do C_{11} and C_{44} differ significantly from the isolated open case? Are C_{14}, C_{12}, or C_{34} significant? While it is an interesting exercise to develop a circuit model for the junction between two pairs of lines, we don't have to do this work to utilize the data from the field-solver. If our only goal is to develop a better filter design, we can use the four-port *S*-parameter data directly in our analysis and optimization of the filter.

Our filter design becomes a cascade of analytical coupled line models and our four-port "black box" data from the field-solver (Figure 8.6(c)). In this case, and in many others, we can trust the field-solver to get the details right for a fairly complex group of discontinuities while we proceed with the design task.

To summarize, de-embedding is a useful laboratory technique and an equally powerful tool when applied to field-solvers. We need de-embedding to remove the discontinuities at numerical ports and to remove lengths of line from our numerical "fixture." This is analogous to an active device measurement in the lab. But with the field-solvers we can do so much more. We can de-embed multiport structures that would be difficult, if not impossible to measure in the lab. In some software packages, lengths of multiple coupled lines can be removed as easily as single lines.

8.3 CLOSED BOX MOM PORTS AND DE-EMBEDDING

For the closed box MoM codes, the most basic port formulation places them at the box walls. Numerically, the port is an ideal voltage source connected across an infinitesimal gap between the port and the box wall. The user has control of the amplitude, phase and impedance of the source. The numerical "walls" of the simulator provide an ideal ground reference for the source. These ports are "circuit theory type" in the sense that only the total power and current are known; no information on modes is available. Simple microstrip ports are shown in Figure 8.7(a). The ground reference is the box wall and the microstrip ground plane is often the bottom of the simulation box.

In the case of CPW (Figure 8.7(b)) we have a ground strip on either side of the signal line. We can excite the desired CPW mode by driving the center line with the positive port reference and placing negative port references on both ground strips. Note that we have deliberately pulled the ground plane away from the box walls on the upper and lower edges of the simulation box. This prevents the flow of undesired transverse currents.

In high-speed digital circuits we are often interested in driving a pair of coupled lines differentially (Figure 8.7(c)). Again, this is accomplished by placing the positive port reference on one strip and the negative reference on the other. Finally, for multiple coupled lines (Figure 8.7(d)), we would like to place a unique port at both ends of each strip. Again we assume that the common ground reference is the bottom of the simulation box, or the nearest ground plane in a multilayer environment.

For the closed box MoM simulators, the de-embedding process is the same for all the port types shown in Figure 8.7. In Figure 8.8(a) we show a double step discontinuity. The location of the new reference planes on each side of the double step are indicated by the heavy black arrows. To find the port discontinuity and the electrical data for the lengths of uniform line, the simulator solves two additional problems for each unique set of ports [22]. For each port with a unique strip width, two

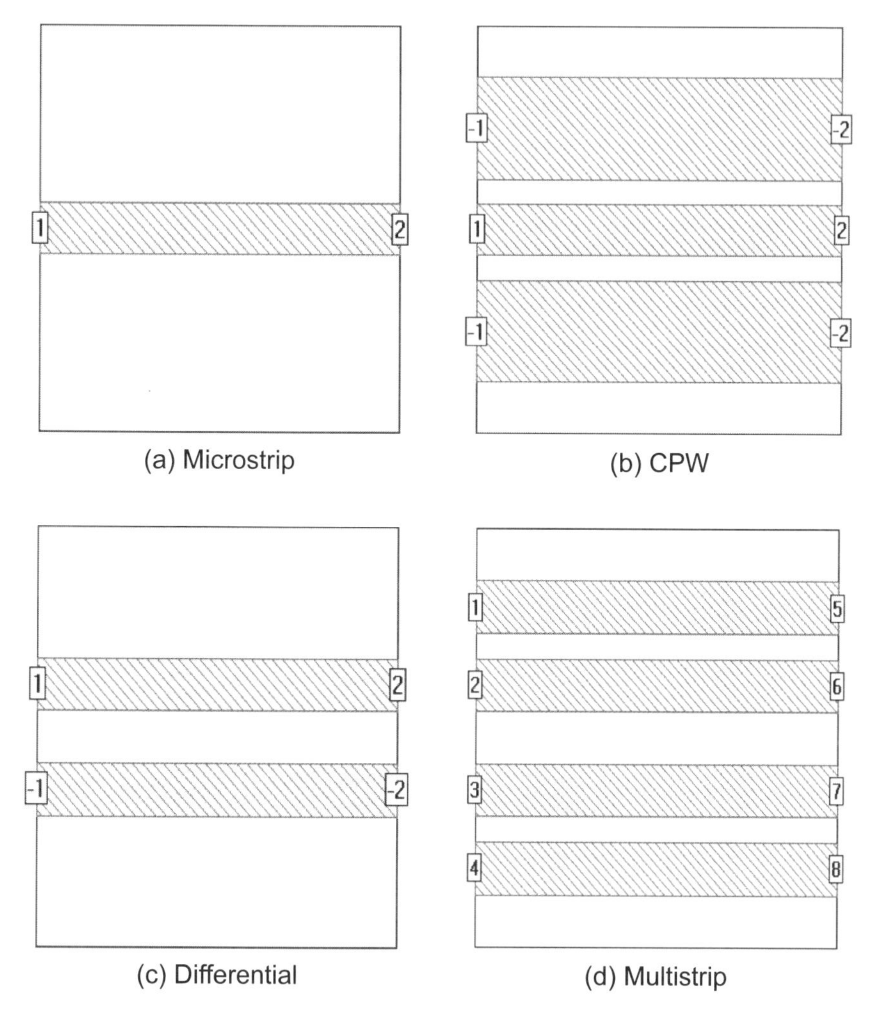

Figure 8.7 Examples of closed box MoM ports: (a) microstrip, (b) CPW, (c) a differential pair, and (d) multiple strips (Sonnet ***em*** Ver. 7.0).

through lines of different lengths are simulated. The first has length l_1 specified by the user and second is twice l_1. The necessary de-embedding data can be computed from these two numerical calibration standards. Because w_2 is different than w_1, two more problems need to be solved for Port 2 calibration and de-embedding. So the simulator first solves the complete problem, then the de-embedding problems, and finally removes the port discontinuities and the lengths of uniform line in a post processing step. Note that we are not restricted to using 50-ohm feed lines for our

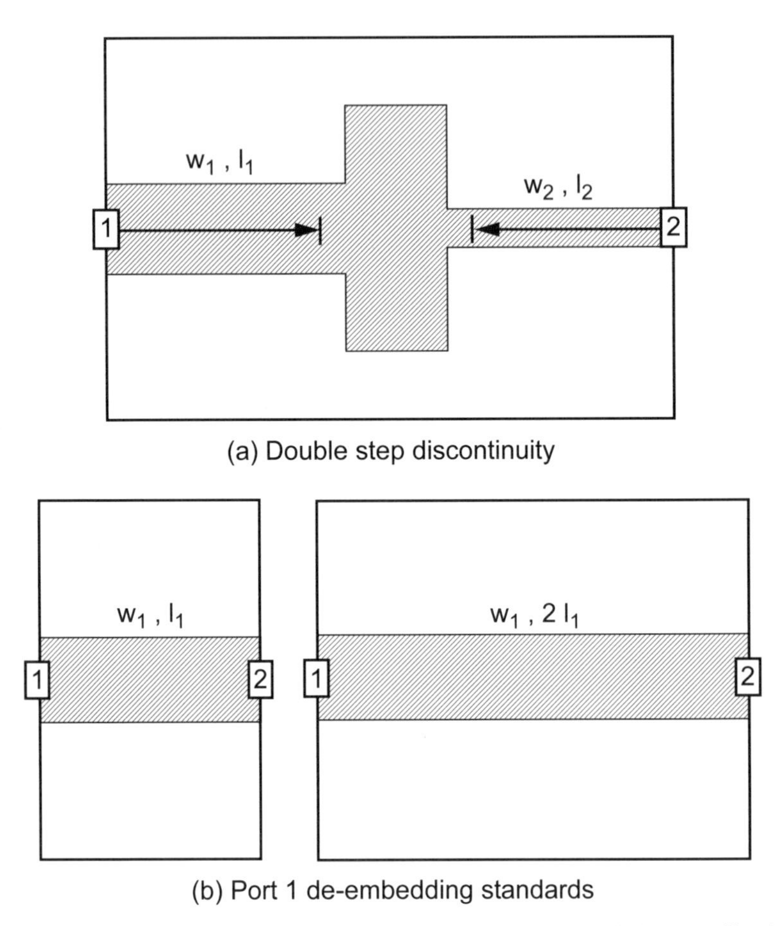

Figure 8.8 Closed box MoM de-embedding: (a) the complete problem, and (b) the two calibration/de-embedding problems that are also solved for Port 1. A similar set of standards must also be solved for Port 2.

network. The final S-parameter matrix is typically normalized to 50 ohms so that it can be imported into a circuit simulator.

The same problems that would cause this process to fail in a physical network analyzer measurement will also cause problems in the numerical simulation. If the distance between a port and a discontinuity is too small, they can interact. If the length of the first calibration standard is too short, the ports can interact. If the feed lines are too wide and are not single moded, the numerical computation will not make any sense.

One advantage of this de-embedding scheme is self-consistency. The same 2.5D computation engine is being used to find all the solutions. The meshing in the calibration problems will also be very similar to the mesh used for the full problem.

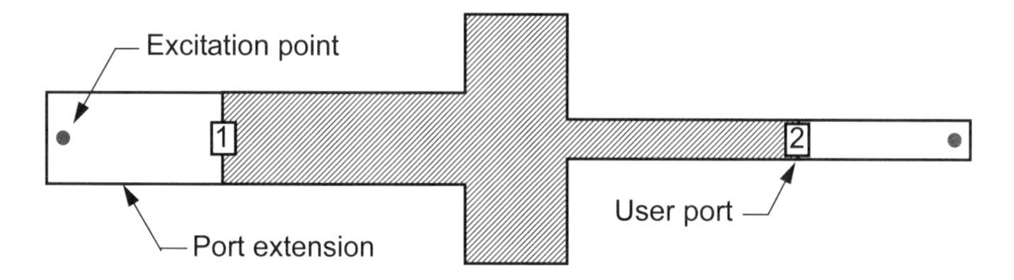

Figure 8.9 Laterally open MoM de-embedding. Extension lines are added to each user defined port. The excitation is at placed at end of the port extension. A separate 2D cross-section solution is needed to find the impedance and phase velocity of the feed lines.

Another advantage is the numerical box wall that provides an unambiguous ground reference right next to conductor or conductors we are trying to excite.

8.4 LATERALLY OPEN MOM PORTS AND DE-EMBEDDING

The laterally open MoM codes also support the port types shown in Figure 8.7. However, port calibration and de-embedding is generally more difficult for the laterally open MoM codes because we may not have a good ground reference close to the port. We can imagine connecting our numerical voltage source from the strip to the nearest ground plane. But the distance spanned by the source connection may be large or small depending on the thickness of the substrate. The software and the software developer cannot predict what the user might specify.

To overcome this uncertainty in the ground reference the laterally open MoM codes have adopted a different excitation scheme (Figure 8.9). A length of line, normally called a "port extension," is added to each user defined port location. The port extension is typically three to five cells long at the highest frequency of interest. The circuit is then excited at the far ends of the port extension with an impressed electric field or an impressed current. It is assumed that the correct current and field distribution is achieved on the strip by the time the signal reaches the original port location.

In an auxiliary process, a model is developed for the port extension lines that allows the excitation to be moved numerically back to the original ports. This is generally a 2D cross-section analysis of the feed line to determine impedance and propagation constant. This same information is used to roll the reference planes in towards the DUT during de-embedding. To compute S-parameters, some codes use the impressed current and an integrated voltage on the port extension line. In other codes, what amounts to a simple reflectometer is set up which detects forward and reverse traveling waves.

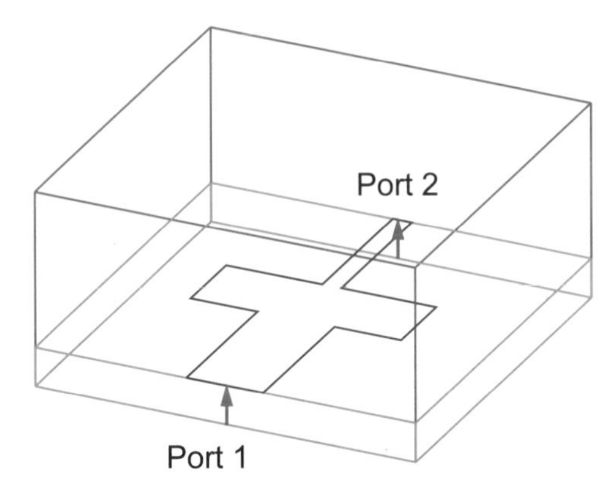

Figure 8.10 3D FEM solution of the microstrip double step. Port 1 and Port 2 each occupy an entire face of the cube. The red arrows are calibration vectors defined by the user. In this case they define the signal trace and its ground reference.

Note that we are using two different numerical engines to compute the impedance and phase velocity of the feed lines. Any differences between the 2D and 2.5D impedance and phase velocity calculations are potential sources of error. In this sense, this scheme is less self-consistent than the closed box MoM scheme. Also, the port extension lines must not interfere with any of the existing circuit. If the port extensions cross or touch other parts of the circuit, the simulation is not valid. This can be a particular problem for highly compacted circuits.

8.5 3D FEM PORTS AND DE-EMBEDDING

There are some similarities and some differences when we compare 3D FEM ports and de-embedding to the 2.5D MoM cases we have already discussed. Most of the multiconductor port types shown in Figure 8.7 are not easily handled by the basic 3D FEM port formulation. Figure 8.10 shows the same double step discontinuity we have been discussing in a 3D perspective view. The microstrip feed lines typically extend to the edges of the solution space, and Port 1 in Figure 8.10 occupies the entire face of the cube. Likewise, Port 2 occupies the entire opposite face of the cube. The red arrows indicate calibration vectors defined by the user. These vectors define the signal trace and its ground reference. They may also define the path used for voltage integration or the desired orientation of the electric field in a waveguide problem.

Given the calibration vector defined by the user, a separate 2D eigenmode solution is performed on each port face to determine what the excitation fields should look like. Although we are most often interested in the lowest order or fundamental mode, it is possible to find the correct field pattern for higher order modes at the ports as well. The impedance and phase velocity of the feed line are also determined from the 2D eigenmode solution.

Once the excitation field-pattern has been determined, a length of uniform line is added to the original problem and the computed source is applied to the end of the added line. The added length of line is to ensure that the desired mode has time to establish itself. The computed impedance and phase velocity are used to numerically move the excitation back to its original position and to move the final reference plane if de-embedding is desired.

There is some similarity between the port calibration and de-embedding process in the 3D FEM codes and the 2.5D laterally open MoM codes. Both require a separate 2D computation at the original port location to find the impedance and phase velocity of the feed line. The 3D FEM codes have the additional problem of matching the 2D excitation mesh at the port face to the 3D mesh in the volume of the full project. Because it is possible to find and separate various transmission line modes at the 3D FEM ports, we generally refer to them as "wave type." The "circuit theory type" ports in the 2.5D MoM codes always assume fundamental mode operation.

In the 3D FEM codes, single strip conductors, single coaxial conductors, and waveguide ports are easily handled by the fundamental port formulation. However, multiple signal conductors present a problem. Let's consider the coupled microstrip problem in Figure 8.11(a). If we have N strips above a ground plane, there are N unique modes. In the case of a symmetric coupled pair of conductors, we speak of the even- and odd-mode. If we tell the field-solver there are two modes at each port, it will identify the even-mode and the odd-mode. However, the result will be a four by four modal S-parameter matrix, two ports with two modes each. What the user probably wanted was a four by four nodal S-parameter matrix.

One way around this problem is to subdivide the original ports into two port regions, each with one conductor (Figure 8.11(b)). But this is clearly less than satisfactory; we are now exciting the coupled line structure with single line ports. De-embedding will not be accurate because we do not have the correct coupled line electrical data at the ports.

A more sophisticated solution to this problem was developed about a decade after the initial introduction of the 3D FEM codes. It is mathematically possible to take the modal solution from Figure 8.11(a) and convert the data to a four by four nodal S-parameter matrix. This is the so called "modes-to-nodes" problem. For homogeneous geometries or quasi-static solutions with unique $RLCG$ matrices, this conversion process is unique. For inhomogeneous geometries, like microstrip, the problem is more difficult and less well defined. The fundamental problem is the small, but finite energy in the non-TEM components; there is no definitive way to handle those components [3, 13, 23, 24].

Another challenging problem for the 3D FEM solvers is a CPW port. In Figure 8.12(a) we show a CPW port with the ground strips isolated from the side walls of the simulator. This is then a three strip problem with three possible modes. The

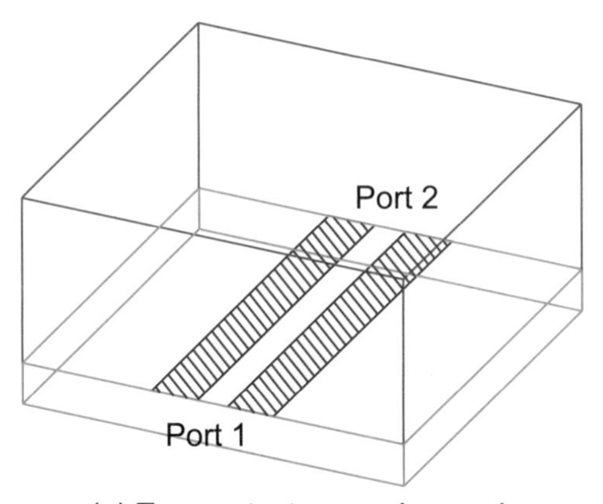

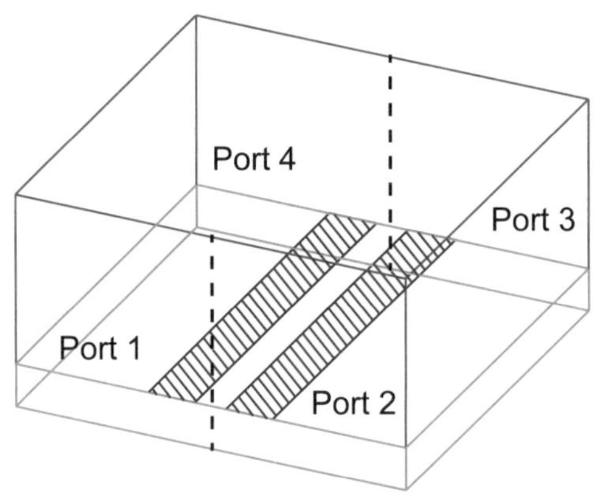

Figure 8.11 3D FEM solution of symmetric coupled microstrip lines: (a) solved as a two-port with two modes at each port, and (b) solved as a four-port with a single, uncoupled strip at each port.

CPW mode is typically denoted with a plus sign on the center strip and negative signs on the two ground strips. There is also a microstrip-like mode where all three strip are basically at the same potential above the nearest ground plane. And a slot-line type mode between the two outer strips with the center strip at some intermediate potential. If we can only define one calibration line it is difficult to ensure that we will excite the desired CPW mode. Air bridges or bondwires should be used in the physical structure and the numerical model to force the two ground strips to the same potential along their inside edges.

If we connect the two ground strips to the box walls of the simulator (Figure 8.12(b)), we have only forced the same potential at the outermost edges. We can

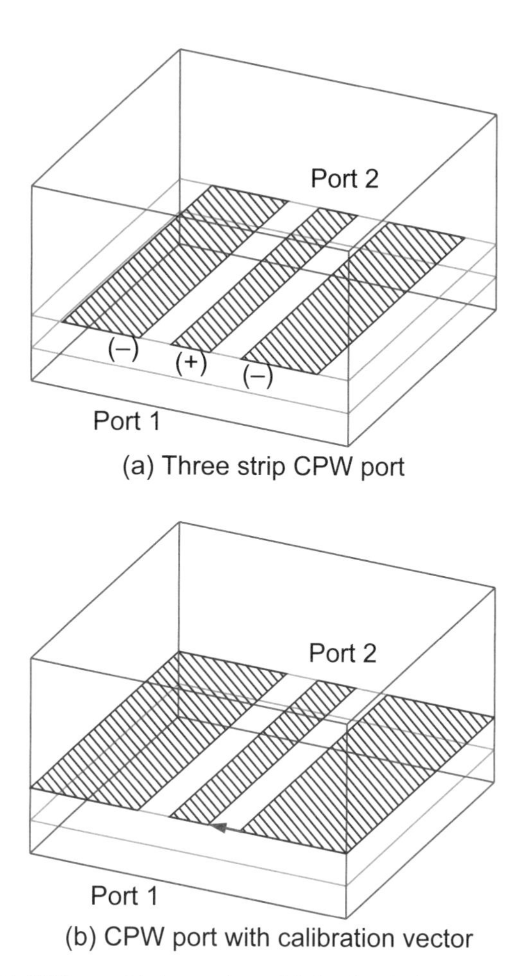

Figure 8.12 3D FEM CPW port: (a) three strip model showing desired CPW mode, and (b) ground strips connected to simulator walls. Single calibration vector implies asymmetrical, slotline type excitation.

place the calibration vector between the center strip and one of the ground strips, but this implies an asymmetrical, slotline type excitation. A potential solution to the CPW port problem will be presented in Section 8.7.

8.6 3D FDTD AND TLM PORTS AND DE-EMBEDDING

Ports in the 3D time domain simulators have evolved along the same lines as the 3D FEM solvers. For problems with strip type conductors, the simplest port to implement is an ideal voltage source connected at one of the mesh points (Figure

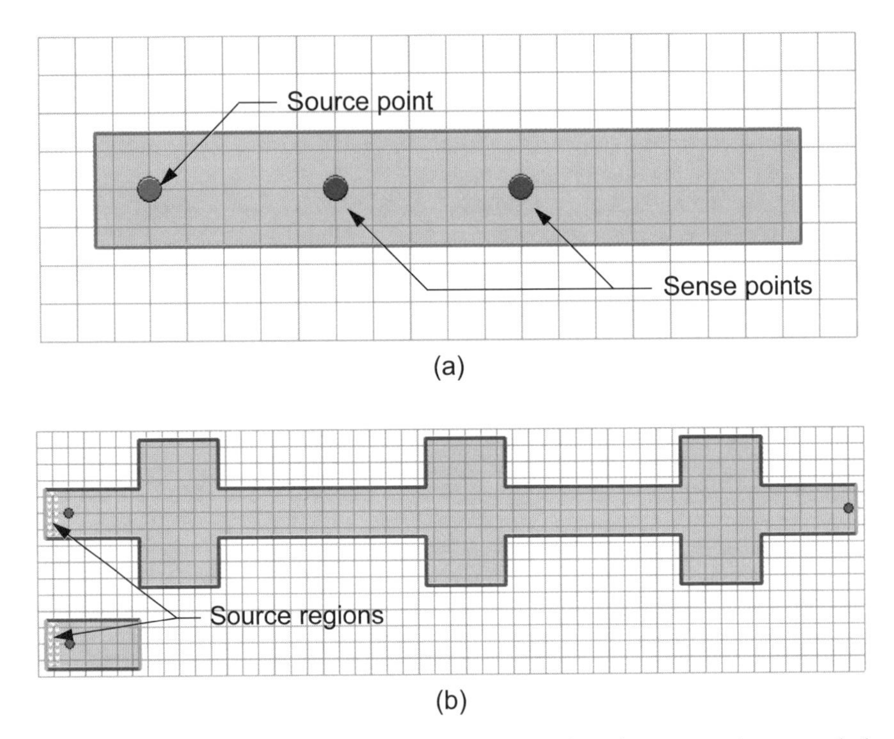

Figure 8.13 (a) Simple time domain excitation using a source point and two sense points to sample the traveling waves for *S*-parameter computations. (b) A source region can be used with defined distribution across the width of the port. The smaller structure at the lower left is a reference line for *S*-parameter calculations; it samples the incident wave.

8.13(a)). We assume that the proper mode forms five to 10 cells away from source point. We can then sample the forward and reverse travel waves at two points on the feed line to find the *S*-parameters.

In Figure 8.13(b) we have a more sophisticated excitation that puts a distributed source region backed by absorbing boundaries at the end of the strip. In this microstrip lowpass filter the planar waveguide model [25] is assumed, which puts magnetic walls at the edges of the strip. The excitation is then uniform across the width of the strip. If this were a waveguide problem, a sin(x) distribution could be specified across the width of the guide. The extra feature in the lower left of Figure 8.13(b) is a reference line used to sample the incident wave for the *S*-parameter calculations.

The most sophisticated port implementation for 3D time domain simulators is a "wave type" port, very similar to the 3D FEM simulators. Figure 8.14 shows a top view of the stripline standard from Section 5.9; the excitation is at the bottom of the figure. At the port face indicated by the red arrows, a 2D eigenmode solution is

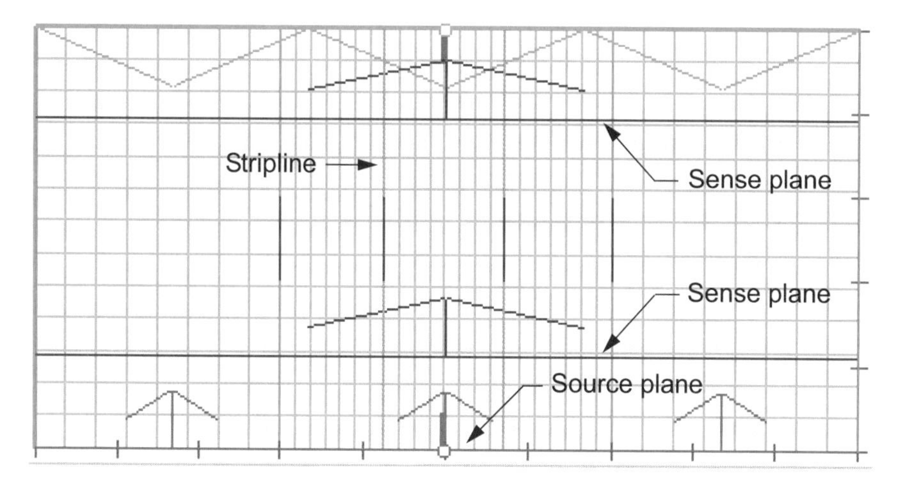

Figure 8.14 3D time domain solver using "wave type" port. An eigenmode solution at the source plane finds the field distribution for the desired mode. Sense planes three cells away from the port are used for *S*-parameter calculations.

used to find the field distribution for the desired mode. Three cells away, at the blue arrow, is a sampling plane for the *S*-parameter calculation. The same mode template used at the source can also be used at the sense plane. Another sense plane towards the top of the figure is needed for the transmission calculation. The green triangles at the top of the boundary indicate an absorbing boundary.

The "wave type" ports in the 3D time domain simulators suffer from the same problems with multiple conductors as the 3D FEM simulators. We can generally obtain a multimode *S*-parameter matrix, which must then be converted to a single mode, multinode *S*-parameter matrix. This is an additional limitation in the 3D time domain simulators for problems with more than two ports. A multiport problem requires one complete domain solution for each port if the full *S*-parameter matrix is to be found. That is, the simulation at each port fills one row and one column of the *S*-parameter matrix. For two-port problems the simulation time is comparable for the 3D time domain and FEM simulators. As the number of ports goes up, the time domain simulation times become much longer.

8.7 INTERNAL, LUMPED, AND GAP PORTS

So far our discussion of ports and de-embedding has focused on cases where all the ports are on the periphery of our problem space. These are, in general, the easiest type of calibrated port to realize. But there are cases where we may wish to define ports that are "internal" to the problem space. Most of the 2.5D and 3D solvers have

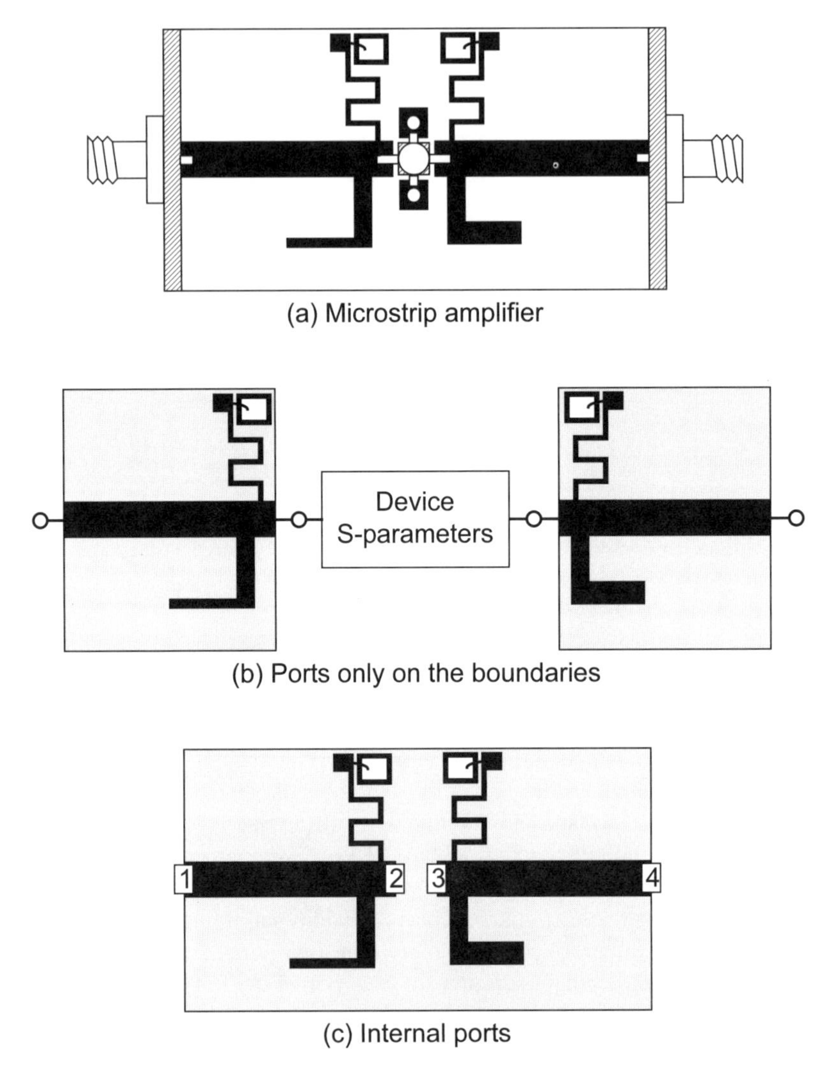

Figure 8.15 (a) Microstrip amplifier with matching and bias structures that may couple to each other. (b) Analysis using only ports on the boundaries; any potential interaction is lost. (c) Analysis using internal ports.

now implemented some type of port that can be placed more arbitrarily in the layout. They go by various names including *internal ports*, *lumped ports*, or *gap ports*. A simple microstrip amplifier problem will be used to demonstrate the concept of internal ports.

Our simple amplifier in Figure 8.15(a) has matching networks and bias networks at both the input and the output. Using only ports on the periphery we would

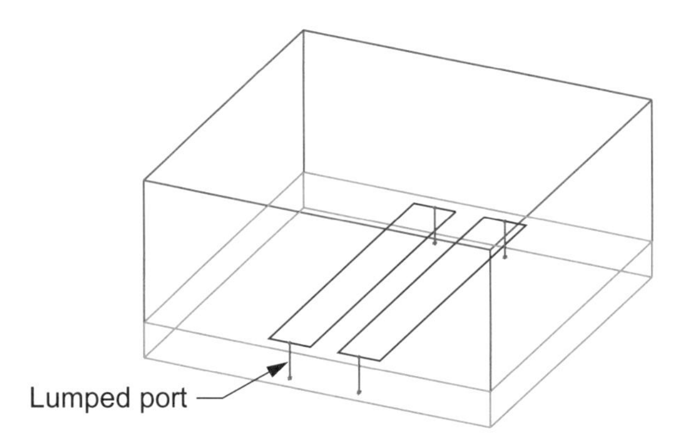

Figure 8.16 3D FEM analysis of microstrip coupled line pair. Lumped or gap ports are used at the ends of the microstrip lines.

subdivide the problem into three pieces for analysis. The input substrate would be one two-port, the output substrate would be a second two-port, and the active device *S*-parameters are the third piece of the problem. We would analyze both substrates using a field-solver, then cascade three *S*-parameter files using our favorite linear simulator (Figure 8.15(b)). This amplifier analysis assumes there are no interactions between the input and output substrates. How would we handle the problem if we suspected there was interaction between the two circuits? What is required are internal ports which provide points to connect our active device *S*-parameters.

Now our field-solver problem is a four-port (Figure 8.15(c)) and we will connect our device parameters to Ports 2 and 3. The distance between Ports 2 and 3 in the analysis matches the physical distance in the actual circuit. We could also expand this analysis to include the grounding structure for the active device. In this case the internal ports are ideal 50-ohm sources connected from the strip to the ground plane.

The internal ports have a physical size in the field-solver and must be small in terms of wavelengths. Also, there is generally no information about the impedance or phase velocity of the line that the port is connected to. Therefore, de-embedding is generally not possible with an internal port. However, if the port is small in terms of wavelengths and located exactly where it is needed, the inability to de-embed may not be a serious limitation.

Internal ports can also help us overcome some of the limitations in the 3D FEM codes when we are interested in multistrip and CPW structures. In Figure 8.16 we consider a pair of coupled microstrips once again. We now have the option of pulling the ends of the strips back from the side walls and placing lumped or gap ports from the ends of the strips to the ground plane or the box walls. The result is a four port *S*-parameter file that can be imported directly into a circuit simulator.

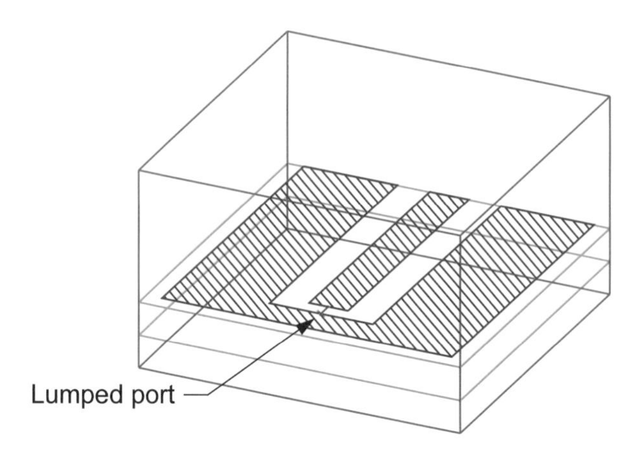

Figure 8.17 A lumped or gap port used to symmetrically excite a CPW port in a 3D FEM solver. De-embedding is not an option and the fringing capacitance across the port gap is not known.

Again, de-embedding is not an option and any fringing capacitance off the open ends will be included in the analysis.

The lumped or gap port also gives us another option for exciting a CPW port in the 3D FEM simulators (Figure 8.17). Again we pull the center strip back from the side wall and wrap the ground connection around the open end. We may or may not connect the ground metal to the box walls. If we do make a connection to the box walls, there is a transverse path for current flow that is perpendicular to the direction of propagation. Placing the gap port on the center line of the strip forces a symmetrical excitation of the desired CPW mode. Again, the capacitance from the strip end to ground is not known at this point. An analysis of a reference structure with a symmetry plane down the center and a conventional wave type port would allow us to find that parasitic capacitance. There is probably some excess inductance in the ground structure as well.

If we place a ground plane below a conventional CPW structure it becomes a coplanar waveguide with ground (CPWG). Questions then arise as to how the three ground planes should be tied together to avoid undesired modes and resonances. Very useful information on grounding and modeling CPWG structures can be found in [26].

8.7.1 Exceptions to the Comments on Internal Ports

One exception to these comments on internal ports for 2.5D solvers is the "auto-grounded" port in Sonnet ***em***. This port is calibrated and supports de-embedding. However, this port must be referenced to the bottom of the simulation box. In a multilayer environment, there must be a clear path from the port location to the bottom of the simulation box with no interfering metal.

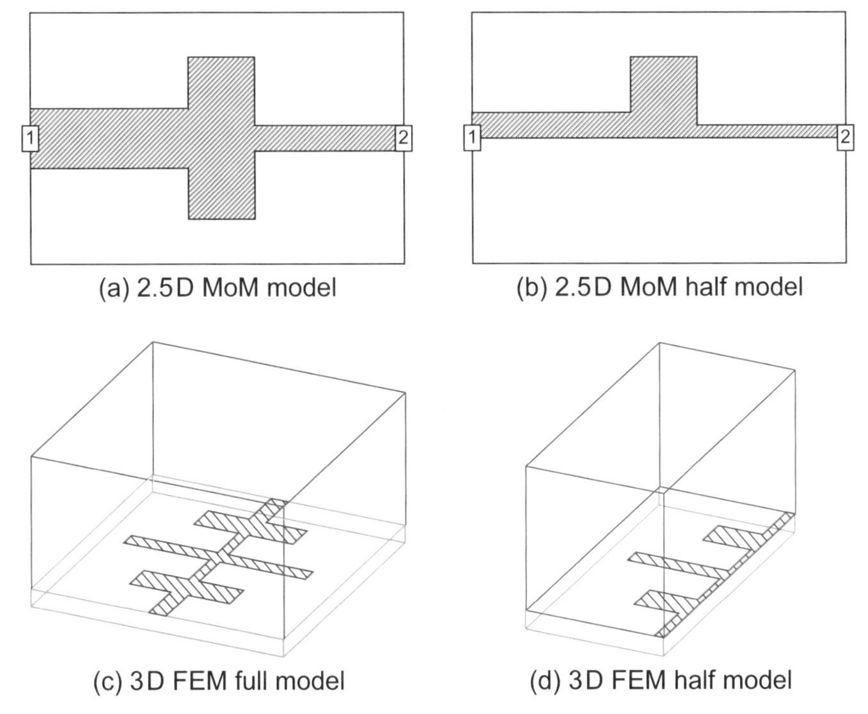

Figure 8.18 Symmetry applied to field-solver models: (a) full model of double step discontinuity using an MoM solver; (b) half model of double step; (c) full model of a microstrip lowpass filter using a 3D FEM solver; and (d) half model of lowpass filter.

8.8 SYMMETRY AND PORTS

Making use of symmetry can be an important strategy to reduce the size of field-solver problems. If we can cut our problem in half using a symmetry plane, solution time may be reduced by a factor of four to eight in a frequency domain solver. If the symmetry plane bisects a single port or a pair of ports, then we only need apply an "impedance multiplier" to recover the *S*-parameters for the full geometry. In Figure 8.18(a) we show the full model of a microstrip double step discontinuity in a 2.5D MoM solver. Figure 8.18(b) shows the half model of the same structure. The same concept applies to the 3D frequency domain and time domain field-solvers. Figure 8.18(c) is the full model of a "complete" circuit, in this case a lowpass filter. We also have the option of analyzing half of the lowpass filter (Figure 8.18(d)). In both cases we place a magnetic wall down the center of the structure, which enforces the symmetry condition.

If one or more ports are deleted in the process of applying symmetry, then we must do additional field-solver problems and some post processing of the *S*-parameters. The simple branch line coupler shown in Figure 8.19(a) is an example ana-

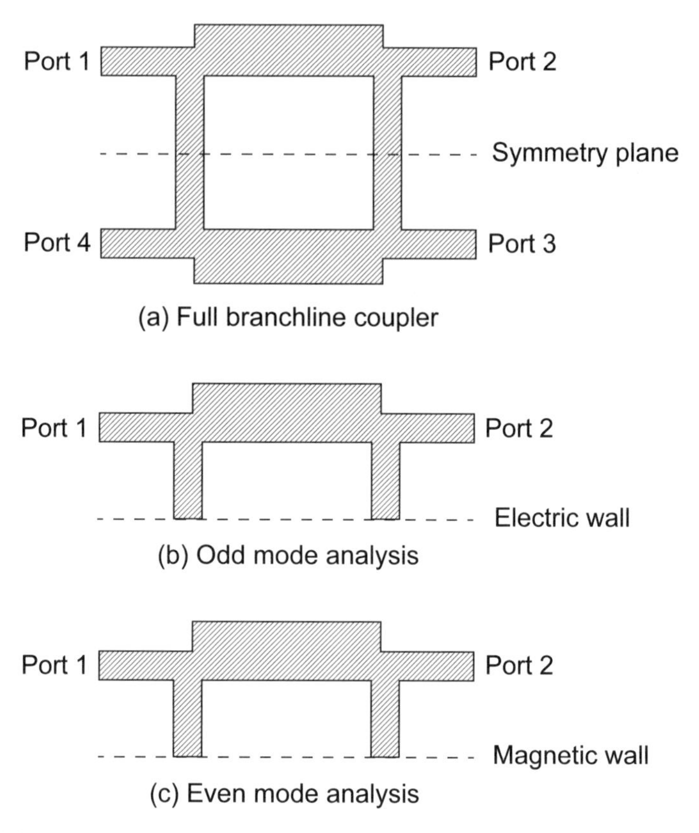

Figure 8.19 Applying symmetry to a branchline coupler: (a) full coupler with horizontal line of symmetry; (b) odd mode analysis with electric wall along the symmetry plane; and (c) even mode analysis with magnetic wall along the symmetry plane.

lyzed in a classic paper by Reed and Wheeler [27]. If we inject 1 W into Port 1, we expect half of the power to appear at Port 2 and half the power at Port 3. Port 4 is the isolated port and ideally does not receive any power.

If we take a symmetry plane in the horizontal axis and discard the bottom half, we have eliminated Ports 3 and 4 from the problem. Now if 1 W is injected into Port 1, it can only appear at Port 2. Simple conservation of energy warns us that a new treatment for the problem may be necessary. We can still compute *S*-parameters for the four-port device, but we must do two analysis runs on the field-solver and some post processing of the results. The two field-solver problems are laid out in Figure 8.19(b, c). One analysis uses an electric wall on the symmetry plane; this is the so called odd-mode. The second analysis uses a magnetic wall on the symmetry plane, which is the even-mode.

The 3D field-solvers allow the most flexible placement of electric and magnetic walls for this type of analysis, but we could also do this analysis with many of

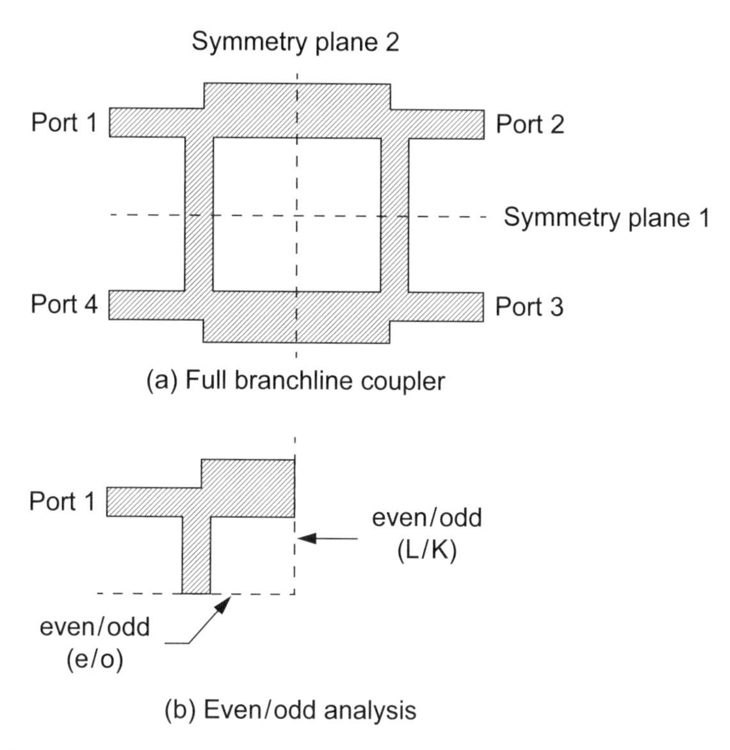

Figure 8.20 Applying symmetry to a branchline coupler: (a) full coupler with horizontal and vertical planes of symmetry, and (b) even/odd analysis on two planes simultaneously.

the 2.5D planar solvers. Once we have the two-port *S*-parameters from the field-solver runs, we can reconstruct the four-port *S*-parameters of the coupler using the following equations:

$$S_{11} = (S_{11}^e + S_{11}^o)/2 \tag{8.7}$$

$$S_{21} = (S_{21}^e + S_{21}^o)/2 \tag{8.8}$$

$$S_{31} = (S_{21}^e - S_{21}^o)/2 \tag{8.9}$$

$$S_{41} = (S_{11}^e - S_{11}^o)/2 \tag{8.10}$$

The additional entries in the matrix can be filled in using the assumed symmetry. If you are willing to do the post-processing, in some cases it may be faster to do two smaller problems. This technique may also turn an impossibly large problem into two smaller problems that can be solved in a finite amount of time.

In the case of the branch line coupler there is a second plane of symmetry that we have not yet exploited (Figure 8.20(a)). This second plane cuts through the original structure vertically. This raises the possibility that we can analyze the four-port coupler using a series of four one-port analyses on the field-solver. At millimeter wave frequencies, these branchline couplers start to resemble an irregular patch of metal with a small, irregular hole in the middle. Tefiku [28] used the analysis technique in Figure 8.20(b) to optimize the shape, port locations, and port feed line angles for millimeter wave branchline couplers. Using Tefiku's notation with port numbers modified to match Wheeler's, the coupler S-parameters can be reconstructed using the following equations:

$$S_{11} = (R_{eL} + R_{eK} + R_{oL} + R_{oK})/4 \tag{8.11}$$

$$S_{21} = (R_{eL} - R_{eK} + R_{oL} - R_{oK})/4 \tag{8.12}$$

$$S_{31} = (R_{eL} - R_{eK} - R_{oL} + R_{oK})/4 \tag{8.13}$$

$$S_{41} = (R_{eL} + R_{eK} - R_{oL} - R_{oK})/4 \tag{8.14}$$

Generally, the size of the field-solver problem we are able to solve is limited by solution time or computer resources. Symmetry is an important concept for reducing the problem size and the solution time. But there are limits to what we can do with one simple symmetry plane. If we have a one-port or a two-port, and the symmetry line bisects the ports, then at most we need to apply an impedance multiplier to obtain the correct S-parameters. In many cases the software does this automatically. On the other hand, if applying a symmetry plane removes ports from the problem we must do extra field-solver problems and some post processing of the results. This extra effort may be justified if we can turn an impossibly large problem into two or more smaller problems that can be solved in an acceptable amount of time. The 3D field-solvers generally offer the most flexibility for defining the symmetry planes needed for this type of analysis.

References

[1] Faché, N., F. Olyslager, and D. De Zutter, *Electromagnetic and Circuit Modelling of Multiconductor Transmission Lines*, New York: Oxford University Press, 1993, pp. 1–15.

[2] Marx, K. D., "Propagation Modes, Equivalent Circuits and Characteristic Terminations for Multiconductor Transmission Lines with Inhomogeneous Dielectrics," *IEEE Trans. Microwave Theory Tech.*, Vol. 21, No. 7, 1973, pp. 450–457.

[3] Jansen, R. H., "Unified User-Oriented Computation of Shielded, Covered and Open Planar Microwave and Millimeter-Wave Transmission-Line Characteristics," *IEE Journal of Microwaves, Optics and Acoustics*, Vol. 3, No. 1, 1979, pp. 14–22.

[4] Lindell, I. V., "On the Quasi-TEM Modes in Inhomogeneous Multiconductor Transmission Lines," *IEEE Trans. Microwave Theory Tech.*, Vol. 29, No. 8, 1981, pp. 812–817.

[5] Jansen, R. H., and M. Kirschning, "Arguments and an Accurate Model for the Power-Current Formulation of Microstrip Characteristic Impedance," *Arch. Elek. Übertragung*, Vol. 37, No. 3/4, 1983, pp. 108–112.

[6] Getsinger, W. J., "Measurement and Modeling of the Apparent Characteristic Impedance of Microstrip," *IEEE Trans. Microwave Theory Tech.*, Vol. 31, No. 8, 1982, pp. 624–632.

[7] Jansen, R. H., "The Spectral Domain Approach for Microwave Integrated Circuits," *IEEE Trans. Microwave Theory Tech.*, Vol. 33, No. 10, 1985, pp. 1043–1056.

[8] Brews, J. R., "Transmission Line Models for Lossy Waveguide Interconnections in VLSI," *IEEE Trans. Electron Devices.*, Vol. 33, No. 9, 1986, pp. 1356–1365.

[9] Brews, J. R., "Characteristic Impedance of Microstrip Lines," *IEEE Trans. Microwave Theory Tech.*, Vol. 35, No. 1, 1987, pp. 30–34.

[10] Tripathi, V. K., and H. Lee, "Spectral-Domain Computation of Characteristic Impedances and Multiport Parameters of Multiple Coupled Microstrip Lines," *IEEE Trans. Microwave Theory Tech.*, Vol. 37, No. 1, 1989, pp. 215–221.

[11] Faché, N., and D. De Zutter, "Circuit Parameters for Single and Coupled Microstrip Lines by Rigorous Full-Wave Space-Domain Analysis," *IEEE Trans. Microwave Theory Tech.*, Vol. 37, No. 2, 1989, pp. 421–425.

[12] Carin, L., and K. J. Webb, "Characteristic Impedance of Multilevel, Multiconductor Hybrid Mode Microstrip," *IEEE Trans. on Magnetics.*, Vol. 25, No. 4, 1989, pp. 2947–2949.

[13] Faché, N., and D. De Zutter, "New High-Frequency Circuit Model for Coupled Lossless and Lossy Waveguide Structures," *IEEE Trans. Microwave Theory Tech.*, Vol. 38, No. 3, 1990, pp. 252–259.

[14] Rautio, J. C., "A New Definition of Characteristic Impedance," *IEEE MTT-S Int. Microwave Symposium Digest*, Boston, MA, June 10–12, 1991, pp. 761–764.

[15] Bauer, R. F., and P. Penfield, Jr., "De-Embedding and Unterminating," *IEEE Trans. Microwave Theory Tech.*, Vol. 22, No. 3, 1974, pp. 282–288.

[16] Glasser, L. A., "An Analysis of Microwave De-Embedding Errors," *IEEE Trans. Microwave Theory Tech.*, Vol. 26, No. 5, 1978, pp. 379–380.

[17] Swanson, Jr., D. G., "Ferret Out Fixture Errors With Careful Calibration," *Microwaves*, Vol. 18, No. 1, 1980, pp. 79–85.

[18] Kruppa, W., "An Explicit Solution for the Scattering Parameters of a Linear Two-Port Measured with an Imperfect Test Set," *IEEE Trans. Microwave Theory Tech.*, Vol. 19, No. 1, 1971, pp. 122–123.

[19] Rehnmark, S., "On the Calibration Process of Automatic Network Analyzer Systems," *IEEE Trans. Microwave Theory Tech.*, Vol. 22, No. 4, 1974, pp. 457–458.

[20] Fitzpatrick, J., "Error Models for Systems Measurement," *Microwave Journal*, Vol. 21, No. 5, pp. 63–66.

[21] "Evaluation of Electromagnetic Microwave Software," Publication EVAL98-01, Sonnet Software, June 1, 1998.

[22] Rautio, J. C., "A De-Embedding Algorithm for Electromagnetics," *Int. Journal of Microwave and Millimeter-Wave Computer-Aided Engineering*, Vol. 1, No. 3, 1991, pp. 282–287.

[23] Wiemer, L., and R. H. Jansen, "Reciprocity Related Definition of Strip Characteristic Impedance for Multiconductor Hybrid-Mode Transmsission Lines," *Microwave and Optical Technology Letters*, Vol. 1, No. 1, 1988, pp. 22–25.

[24] Jansen, R. H., "Some Notes on Hybrid-mode versus Quasi-static Characterization of High Frequency Multistrip Interconnects," *23rd European Microwave Conference Proceedings*, Madrid, Spain, September 1993, pp. 220–222.

[25] Wolff, I., G. Kompa, and R. Mehran, "Calculation Method for Microstrip Discontinuities and T Junctions," *Electronics Letters*, Vol. 8, No. 9, 1972, pp. 177–179.

[26] Haydl, W. H., "On the Use of Vias in Conductor-Backed Coplanar Circuits," *IEEE Trans. Microwave Theory and Tech.*, Vol. 50, No. 6, 2002, pp. 1571–1577.

[27] Reed, J., and G. J. Wheeler, "A Method of Analysis of Symmetrical Four-Port Networks," *IRE Trans. on Microwave Theory Tech.*, Vol. 4, No. 4, 1956, pp. 246–252.

[28] Tefiku, F., E. Yamashita, and J. Funada, "Novel Directional Couplers Using Broadside-Coupled Coplanar Waveguides for Double-Sided Printed Antennas," *IEEE Trans. Microwave Theory Tech.*, Vol. 44, No. 2, 1996, pp. 275–282.

Chapter 9

Numerical Methods Summary

Now that we have introduced the major numerical methods in some detail, this is perhaps a good point to summarize the basic concepts that apply to all these methods.

9.1 MESHING

We will begin with meshing. There is no such thing as an ideal mesh. Meshing is a combination of science and art. Two different software packages that use the same numerical method may generate quite different starting meshes. We must always make trade-offs between accuracy, computation time, and memory required. In any mesh we should avoid elements with high aspect ratios. Remember, the goal is an accurate approximation of the real fields or currents in our problem.

If our structure is symmetric and we expect symmetric S-parameters, then the mesh must be symmetric. How we draw our project can affect the quality of the mesh. Meshing algorithms must start at one edge or surface of the problem and scan across the geometry. Meshing algorithms detect the edges of objects. Some algorithms also detect redundant vertices. The intelligent user can use this behavior to fine tune mesh. Key meshing concepts are summarized below:

- No such thing as ideal mesh;
- Trade-offs:
  - Accuracy;
  - Computation time;
  - Memory required.
- Avoid elements with high aspect ratios;
- Symmetric mesh for symmetric S-parameters;

- How you draw affects the mesh:
  - Mesher detects edges of objects;
  - Some meshers detect redundant vertices.

9.1.1 Surface Meshing

When we use the surface meshing codes, we would like to place small, square cells where the current changes direction rapidly. We should avoid large, complex polygons. Breaking our drawing into smaller units forces the mesher to start over at boundaries and helps us control the mesh quality. If the mesher uses triangles, limit the resolution of arcs.

- Small square cells where current changes direction;
- Avoid large, complex polygons;
- Break big polygons into smaller units;
- When using triangles, limit resolution of arcs.

9.1.2 Volume Meshing

There are also some general guidelines for volume meshing codes. When we use FEM codes we can use dummy objects to control aspect ratio. We can also seed various objects in the model to force a know mesh resolution in a critical region. Again we want limit the resolution of arcs and cylinders. In the FDTD and TLM codes we need to snap the mesh to the edge of key elements, particularly strip type elements.

- Use dummy objects in FEM to control aspect ratio;
- Seed model objects to force known mesh resolution;
- Limit resolution of arcs and cylinders;
- In FDTD and TLM snap mesh to edges.

9.2 CONVERGENCE

Convergence is an issue for any numerical method. We have identified three major elements of the convergence equation. They are guide wavelength, spatial wavelength, and geometrical resolution. When the mesh must be manually refined, we make intelligent estimates of where more mesh is needed and run the analysis again. When automatic mesh refinement is available, the intelligent user will still attempt to "steer" the solver towards the correct solution.

9.2.1 Guide Wavelength

Guide wavelength is the variation of field quantities or voltages and currents in the direction of propagation. It is easy to compute and visualize and is a very familiar circuit theory CAD concept. To capture guide wavelength we generally subdivide our problem using a cell size between $\lambda/10$ and $\lambda/30$ at the highest frequency of interest.

- In direction of propagation;
- Cell size of $\lambda/10$ to $\lambda/30$ at highest frequency of interest;
- Easy to compute and visualize;
- Familiar circuit theory CAD concept.

9.2.2 Spatial Wavelength

Spatial wavelength is the variation of field quantities or currents across the width of a strip transmission line or waveguide. This variation is perpendicular to the direction of propagation and is typically not a function of frequency. It typically requires a finer discretization than guide wavelength. Capturing the edge singularity in strip problems is one important example of spatial wavelength. Spatial wavelength is harder to visualize and is not needed or considered in circuit-theory-based CAD.

- Variation across the width of strip or guide;
- Perpendicular to direction of propagation;
- Capture edge singularity in strip problems;
- Harder to visualize, not part of circuit-theory-based CAD;
- Typically not a function of frequency;
- Typically requires finer discretization than guide wavelength.

9.2.3 Geometrical Resolution

The third aspect of convergence is geometrical resolution. Geometrical resolution must be considered in both surface meshing and volume meshing codes. In general we want just enough resolution for good convergence without over-specifying arcs and cylinders. The variational principle applies to most of the circuit parameters we are trying to compute. This means that we only have to get the average energy right to find accurate *S*-parameters; we do not have to find the exact value of the fields. Experience using the tools and convergence studies lead us to trust coarser geometrical descriptions.

- What resolution do we really need for our geometry?
- Applies to surface and volume meshing;
- Don't over-specify arcs;
- Don't over-specify cylinders;
- Variational principle applies:
  - *S*-parameters are variational;
  - Only need correct average energy, not exact fields.
- Experience helps us trust coarser descriptions.

9.3 VALIDATION STRUCTURES

If we are interested in the absolute accuracy of these tools, we need some kind of validation or canonical structure. For strip type problems microstrip is a poor choice due to the uncertainty in the computation of impedance. A better choice is air-filled stripline because there is an exact analytical formula for impedance and we also know the phase velocity in air. A coaxial standard is useful for the volume meshing codes because it forces us to think about how we approximate curved boundaries. A rectangular waveguide resonator is another valuable standard for volume meshing codes, particularly the time stepping codes.

- Stripline Standard:
  - Applies to all numerical methods;
  - Consider edge singularity.
- Coaxial Standard:
  - Applies to all volume meshing tools;
  - Approximation of curved boundaries.
- Resonator Standard:
  - Applies to volume meshing tools;
  - Particularly useful for time stepping codes.

9.4 CALIBRATION STRUCTURES

The goal of our "calibration" structures is to allow the user to experiment with the various features of the software. We are trying to train or calibrate the user rather than measure or improve the accuracy of the software. There is, however, a strong analogy to the kit of standards we use to calibrate a vector network analyzer in the lab. These simple structures give the user the opportunity to explore the various aspects of meshing, convergence, de-embedding, and visualization. These structures also help the user identify "normal" behavior of our circuits. Later, when we encounter some unusual behavior it will stand out more distinctly.

- Microstrip cal kit for surface meshing:
  - Through, short, open, and load.
- Coaxial cal kit for volume meshing:
  - Through, short, open, and load.
- Others, depending on transmission line type:
  - Must be simple and intuitive.
- Explore basic features of simulator;
- Develop intuition for 1% error mesh;
- Develop intuition for "normal" behavior.

9.5 PORTS AND DE-EMBEDDING

There are differences between the various numerical methods when it comes to ports and de-embedding. In the early days of commercial tools, only ports on the periphery of the problem space were available. Later, various types of internal ports were offered. De-embedding is used in all the 2.5D and 3D codes to remove any port discontinuity and some of the uniform line that leads up to the geometry of interest.

9.5.1 MoM Ports

Ports in MoM codes are "circuit theory" based; they find the total voltage and current (or current and power) at the port without regard to modes. Microstrip, differential, CPW, and multistrip ports are all easy to realize.

- Find total voltage and current at the port;
- No information on modes;
- Easy to realize CPW and differential ports;
- Easy to realize multistrip ports.

9.5.2 FEM, FDTD, and TLM Ports

Ports in the 3D frequency and time domain codes are generally "wave type." A 2D eigenmode solution is used to find the field pattern of the desired mode. Higher order modes can also be specified, so mode conversion can be studied. Multistrip type ports are generally harder to implement. Generally a one node, multimode solution at the port must be converted to multinode solution, the "modes-to-nodes" problem.

- Find field distribution of desired mode at the port;
- Higher modes allowed, can study mode conversion;
- Multistrip modes more difficult;
- "Modes-to-nodes" conversion needed.

9.5.3 Internal, Lumped, and Gap Ports

Various types of internal ports have been implemented in both the 2.5D and 3D simulators. They allow the user to place a port at an arbitrary location in the problem space. However, there are limitations. We assume the port is ideal, but it has physical size in the simulation. So the port must be small in terms of wavelengths for this approximation to hold. There is no information on the impedance of phase velocity of the line we are connected to in this case; so de-embedding is not possible.

- Allow ports at arbitrary locations;
- Port must be small in terms of wavelengths;
- De-embedding is generally not possible.

Chapter 10

Microstrip

While the early days of microwave technology were dominated by waveguide components, various strip-based technologies soon developed and later became dominant. Researchers looking for an easier and cheaper way to integrate multiple functions in a single package developed tri-plate (stripline) and microstrip in the 1950s [1]. Military microwave systems soon evolved into a combination of thin-film circuits on ceramic substrate, stripline components using Teflon-based substrates, and waveguide components—each with their own strengths and weaknesses. Today, many RF and high-speed digital systems depend on multilayer PCBs, which are often a combination of microstrip and stripline type geometries. RFICs also include multilayer strip type geometries as part of their basic design library.

Microstrip is a two conductor transmission line consisting of a strip over a ground plane, both of which are usually supported by a layer of dielectric material. The fields around the strip conductor exist both in the air and in the dielectric. This inhomogeneous dielectric arrangement leads to quasi-TEM behavior and greatly complicates the analysis of microstrip components. By quasi-TEM we mean that there is always some small, but finite longitudinal component to the fields on a microstrip line [2].

10.1 DISCONTINUITIES

If working components could be built solely from straight runs of microstrip transmission lines, life would be relatively easy. But in fact we need bends, tee-junctions, steps in width, vias to ground, and vias between layers to make a real component (Figure 10.1). All of these deviations from a simple straight line geometry we call discontinuities. In the early days of microwave CAD, analytical formulas were developed which linked the dimensions of a given discontinuity to an equivalent lumped element circuit. These models were developed by carefully measuring a set of representative structures, or more often by running a custom field-

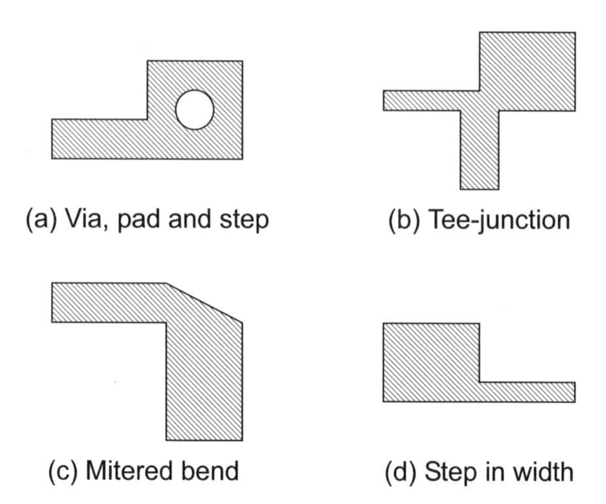

Figure 10.1 Typical microstrip discontinuities: (a) via hole with pad and offset step in width, (b) tee-junction, (c) mitered bend between unequal widths, and (d) offset step in width.

solver code which generated data for the equation fitting process. This basic approach is still the basis for many of the models found in circuit-theory-based microwave simulators.

But all of these formulas and equivalents have some finite range of usable accuracy. Equivalent circuit models are most likely to break down when the aspect ratio between critical dimensions in the discontinuity becomes high. For the step in width, there is some critical ratio of line widths at which error will increase. The tee-junction is a notoriously difficult geometry to model when the ratio of line widths grows large. If you add a broad range of dielectric constants, a broad range of substrate thicknesses, and a broad frequency range to the requirements, the modeling task is difficult indeed.

In the early days of the commercial field-solvers, a single discontinuity was a relatively large problem. But even so, users were thrilled with the ability to analyze an arbitrary geometry and get an accurate solution [3–6]. These field-solver solutions could be used to validate an existing analytical model or as a substitute for that model. And of course the more ambitious users were not prevented from building their own analytical models from the field-solver data [7].

A single discontinuity, a small group of discontinuities or a small matching network are all very easy problems for today's field-solvers. A problem of this type will solve in, at most, a minute or two per frequency point. If the structure is not resonant, and many are not, data points are only needed every few GHz; the linear simulator can easily interpolate between the computed data points.

The field-solver also frees the user from the limitations of the standard library of circuit elements. The creative engineer is free to use geometries that cannot be easily or accurately described using a library of predefined circuit elements.

10.2 MICROSTRIP VIAS AND SLOTS

Vias are used in single layer microstrip, multilayer PCBs, and RFICs to provide a path to ground [8]. Whether we are grounding the source of an FET or terminating a quarter wavelength transmission line, it is important to know the exact inductance of the via structure. Figure 10.2 shows two different via structures and one slot structure. Note that the round holes are approximated as octagons here. There is vertical metal from the edges of the holes and the slot down to the ground plane.

At first glance, there are several variables we need to consider in the single via structure shown in Figure 10.2(a). But in any technology we choose, several of the variables are defined by the manufacturing group, rather than the design group. For a given substrate thickness there is a minimum diameter for the via hole. Given the diameter of the hole, there is a minimum size for the pad around the hole. In this case the via pad is 25 mil square, the hole diameter is 13 mil, and the substrate is 15-mil thick alumina. If we set the reference plane at the step discontinuity, the equivalent inductance of the via barrel, the pad and the step is 121.6 pH at 18 GHz.

If we look closely at the current distribution in Figure 10.2(a), we see the distribution we expect on the microstrip feed line. When the current reaches the via, the leading edge of the via barrel carries the majority of the current. The backside of the via barrel and pad carry very little current. After some contemplation, this makes perfect sense; the current is following the lowest impedance path to ground. However, most engineers have probably never thought about what this current distribution actually looks like.

If we know that most of the current is on the edges of the microstrip, perhaps placing vias closer to the strip edges will lower the equivalent inductance. Figure 10.2(b) shows this double via geometry. The equivalent inductance for this case is 79.6 pH, a decrease of 35% over the single via case. There is probably some mutual inductance between the two vias. Looking at the current distribution, we again see that it is the leading edge of the vias that carry the majority of the current. We can also see that the current is excluded from the outside corners of the step. This is due to the charge singularity at the sharp point [9].

When inductance must be minimized, slots have often been proposed as the best alternative. But slots require a different, more difficult manufacturing process than vias. And the actual current distribution for the slot (Figure 10.2(c)) is quite similar to the double via case. The equivalent inductance is 70.6 pH at 18 GHz, a marginal improvement over the double via. If we look closely at the current distribution for the slot, there is actually a null at the midpoint of the leading edge. So, after fighting with manufacturing to create a geometry with much more surface area, we find that very little of that surface actually carries current. The current scale is 0 to 40 amps/meter for all three plots in Figure 10.2.

It may be possible to further optimize the shape of the metallization pad for the double via geometry [10]. For narrow microstrip feed lines we can make the pads around the vias circular and move the microstrip line feed point closer to the center line between the two vias.

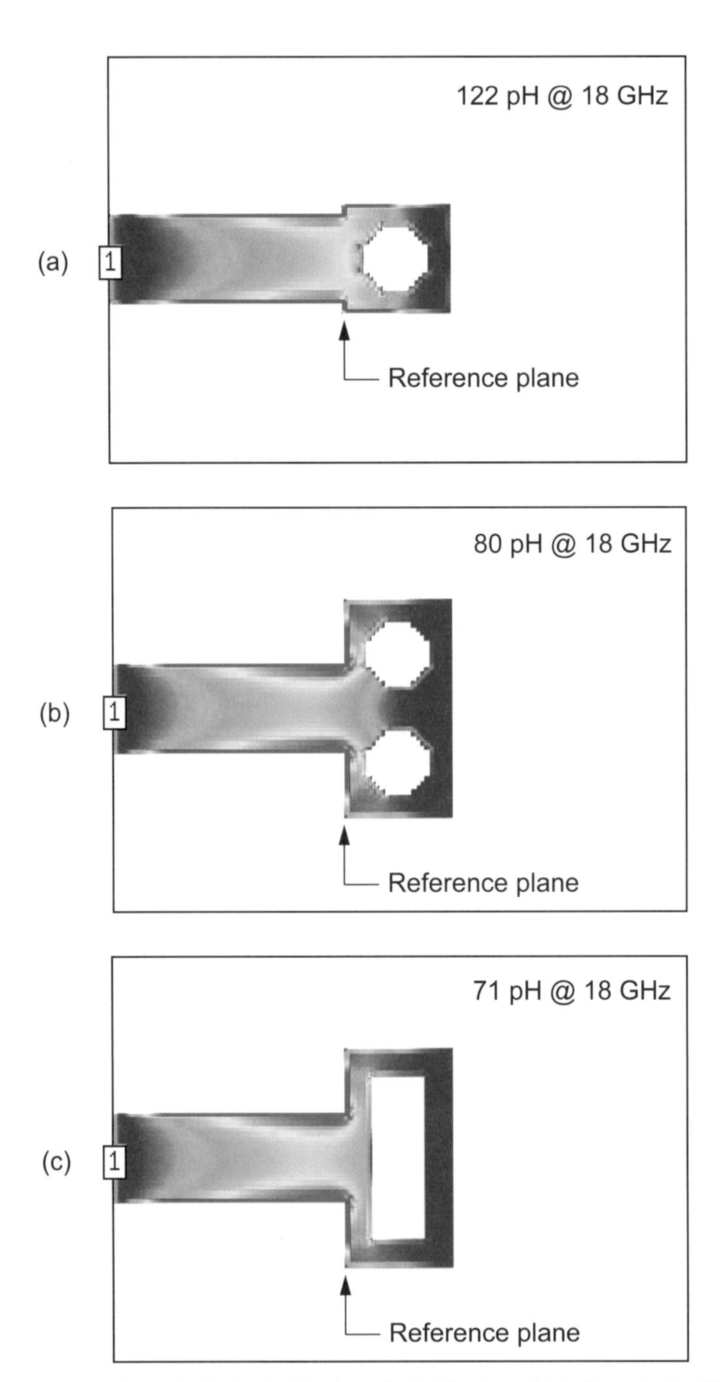

Figure 10.2 (a) Microstrip single via; (b) microstrip double via; and (c) microstrip slot. The scale in all three plots is 0 to 40 amps/meter (Sonnet ***em*** Ver. 7.0).

10.3 MICROSTRIP 3D VIAS

The previous current plots used a 2.5D solver, which assumes infinitely thin conductors. Also, we cannot view the current on the via barrel with most of the 2.5D codes. Perhaps we can get a more complete view of the current distribution using one of the 3D solvers. In Figure 10.3 we present three views of the conduction current at 18 GHz on a single, hollow via using Flomerics Micro-Stripes [11], a 3D TLM code. The via has the same dimensions as before but the strip thickness is now 1 mil. The via barrel is made up of cubic cells, 0.625 mil per side. In Figure 10.3(a) the current is clearly flowing on the near side of the via barrel, as was implied in Figure 10.2(a). There is also a small amount of current that flows down the inside surface of the via barrel. For the first time, we can also see the return current in the ground plane. We tend to forget that microstrip is a two conductor system; the return path is just as important as the signal path. We will see in later chapters that the return path is a critical consideration in modeling and designing multilayer transitions in PCBs.

Figure 10.3(b) shows a top view of the complete structure. Note the nonuniform current distribution across the width of the strip. We can also see the part of the ground plane that is not obscured by the top strip.

Finally, in Figure 10.3(c) we have a view of the bottom side of the top strip. Comparing the last two views we note that there is more current on the bottom side than the top side. At low frequencies, the current would split more or less equally between the top and bottom surfaces of the strip. As frequency increases, the current distribution shifts towards the bottom side of the strip. The scale for all three plots in Figure 10.3 is 0 to 120 nA/m.

At various times, engineers have debated the merits of hollow vias versus solid vias for RF performance. There is a great temptation to believe that making the via solid will somehow reduce the equivalent inductance of the structure. We have already dealt with the hollow via case; what happens when we fill the via with a solid plug of metal? In Figure 10.4 we have again used Flomerics Micro-Stripes to explore the solid via case. In the perspective view (Figure 10.4(a)), the current distribution has not changed significantly. But if we look closely, we can see some small differences on the top strip where the hole used to be.

Figure 10.4(b) is the top view of the solid via case. There is less current on the top of the via pad where the hole used to be. We can explain this if we note there is no longer a path for current down the inside of the via barrel.

Finally, in Figure 10.4(c), we have a view of the bottom side of the top strip. As in the hollow via case, there is much more current on the bottom side of the strip compared to the top side. The scale for all three plots in Figure 10.4 is 0 to 120 nA/m.

If we extract the equivalent inductances for the hollow via and solid via cases, they are virtually identical. But before we draw any conclusions, we need to consider skin effect and the assumptions made in the numerical model. At 18 GHz, skin depth effects alone will force the current to the surface of the solid via. In the

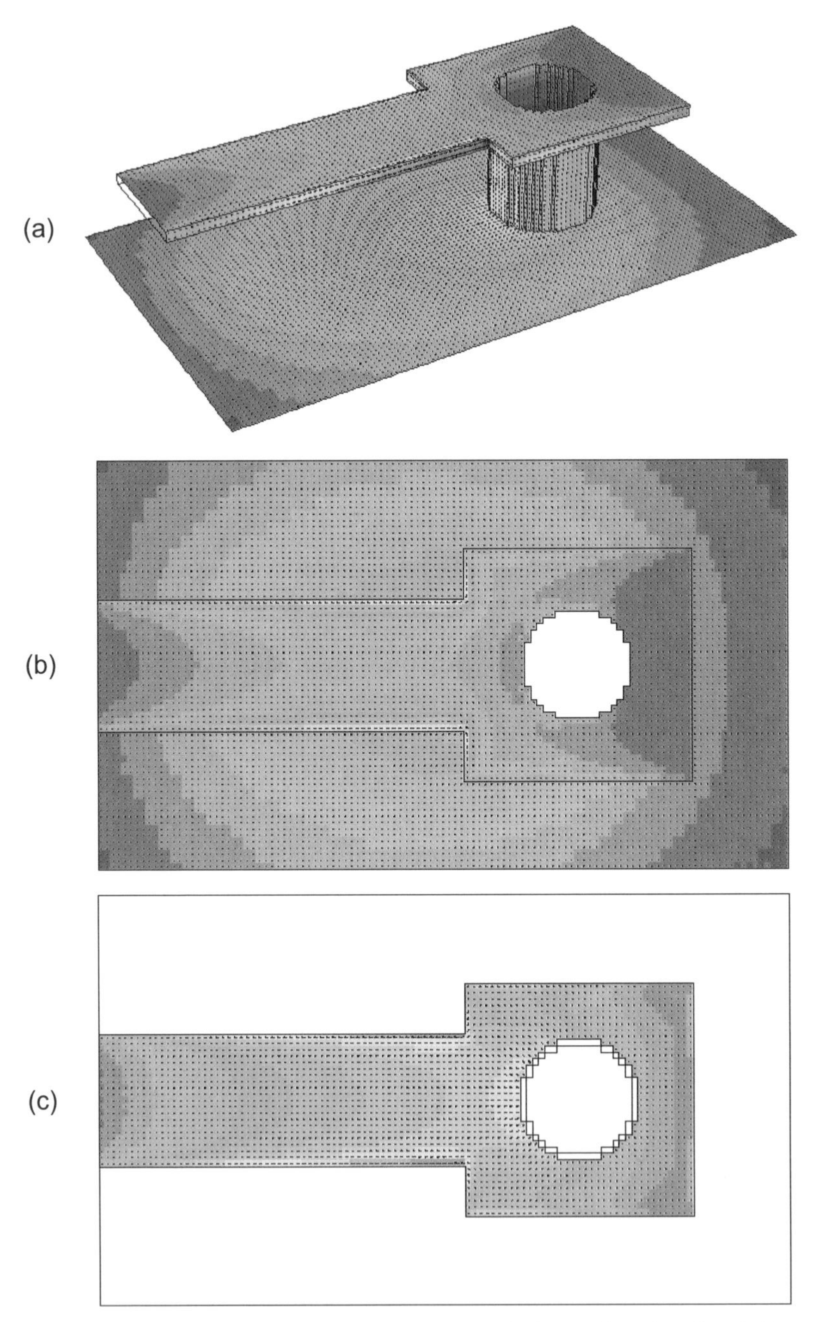

Figure 10.3 Conduction current at 18 GHz on microstrip single via with hollow barrel: (a) perspective view, (b) top view of strip and ground plane, and (c) bottom side of top strip. The scale in all three plots is 0 to 120 nA/m (Flomerics Micro-Stripes Ver. 6.0).

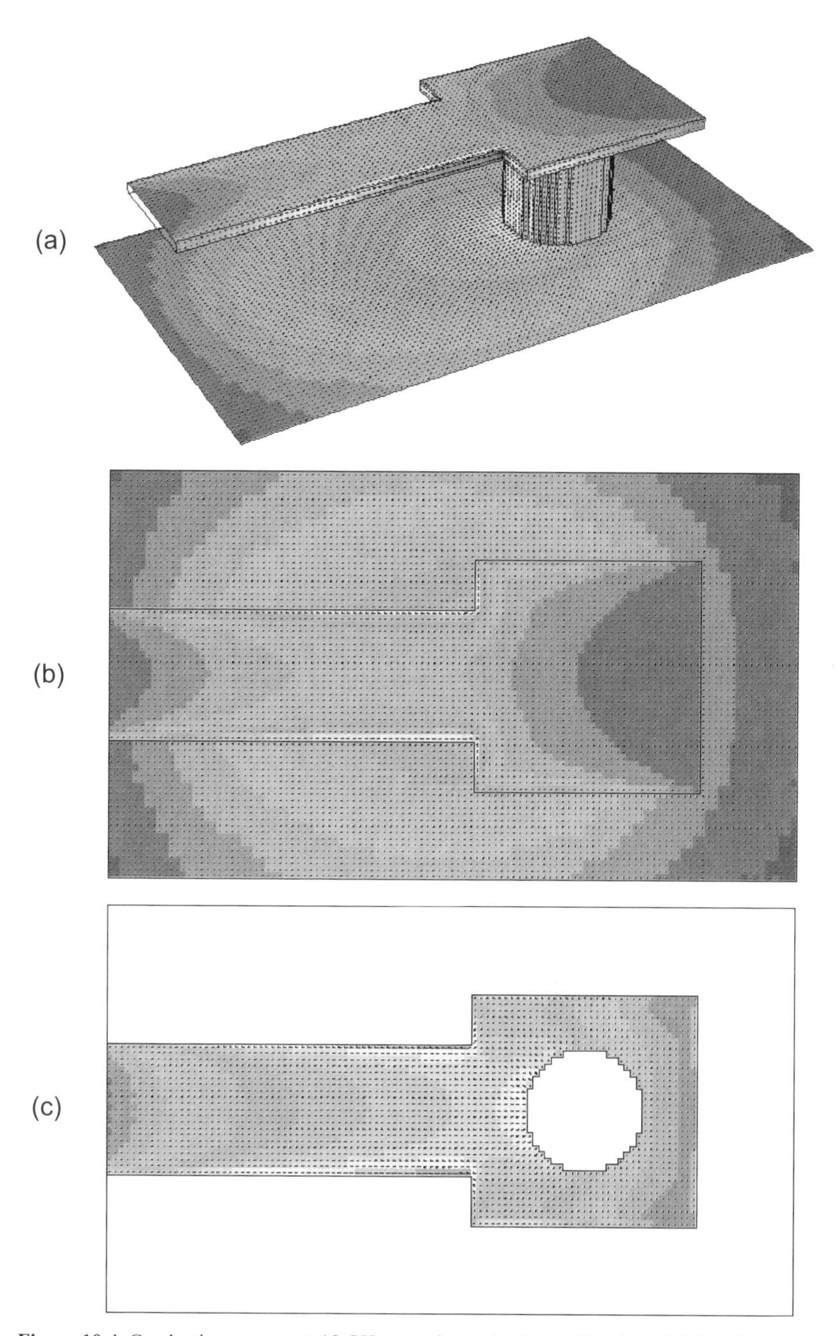

Figure 10.4 Conduction current at 18 GHz on microstrip single via with solid barrel: (a) perspective view, (b) top view of strip and ground plane, and (c) bottom side of top strip. The scale in all three plots is 0 to 120 nA/m (Flomerics Micro-Stripes Ver. 6.0).

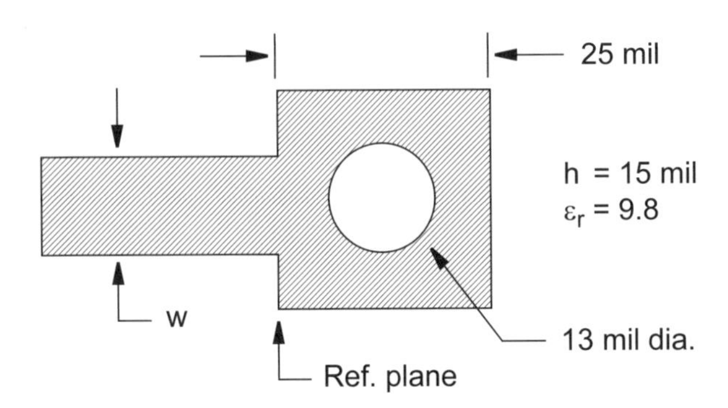

Figure 10.5 Typical via structure in a ceramic substrate. If we assume a reference plane at the edge of the pad, the equivalent circuit must include the via barrel, the pad around the via, and the step discontinuïty. © 1992 IEEE [8].

numerical model, metals with finite conductivity are often modeled as an infinitely thin sheet with an equivalent surface impedance. So our numerical model may not allow currents to actually penetrate the surface of the metal. In some 2D cross-section-solvers and in some 3D solvers it is possible to mesh and solve the interior of metal conductors. However, to capture skin depth effects, an extremely fine mesh resolution is needed and the solution time will be quite long.

10.4 MODELING MICROSTRIP VIAS

The microstrip via is a good example of how a typical user might generate a custom model. Whenever a printed circuit is fabricated, there are design rules that set a minimum diameter for the via barrel and any pads that connect to the barrel. This is true for both ceramic-based substrates and epoxy-glass-based substrates. In any design project, the substrate type, relative dielectric constant, and layer thicknesses are fixed at a very early stage. If we are trying to model a via structure like the one shown in Figure 10.5, the designer really only has to consider the line width, *w,* as a variable. The via diameter and pad dimensions are fixed by the manufacturing process.

Using batch processing it is easy to solve several via problems as a function of frequency and line width, *w*. If *Z*-parameters are computed, rather than *S*-parameters, an equivalent inductance can be extracted from the field-solver data using a pocket calculator.

$$X_L = j\omega L \qquad L = \frac{7.9577 X_L}{f(\text{GHz})}\ \text{nH} \tag{10.1}$$

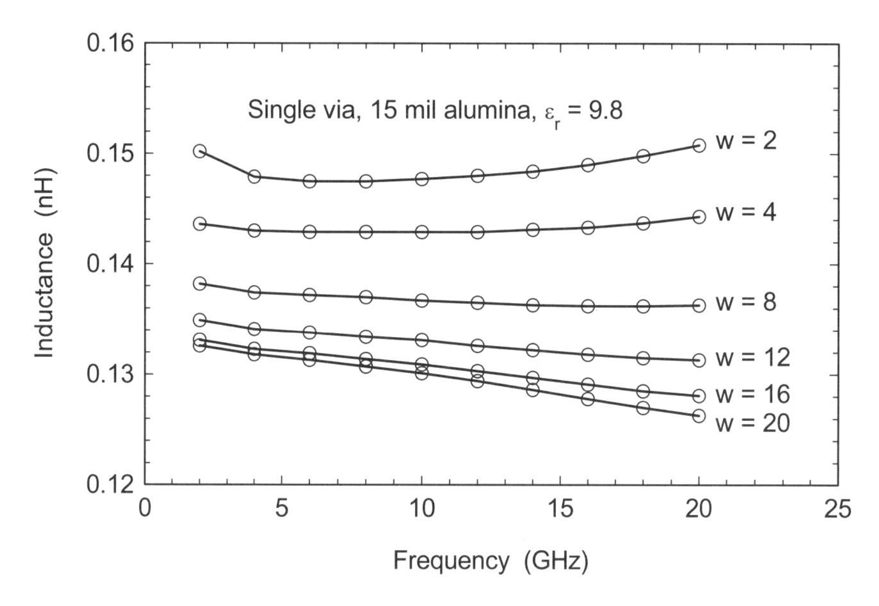

Figure 10.6 Equivalent inductance of the geometry shown in Figure 10.5 as a function of line width and frequency. © 1992 IEEE [8].

This assumes that the combination of the via barrel, via pad, and the step discontinuity can be described as a single lumped inductance to ground. After analyzing the structure in Figure 10.5 at several different line widths, the results are shown in Figure 10.6. The graph indicates that the simple, single inductor model is a function of both line width and frequency. The variation with frequency is nonphysical and indicates that the model is too simple. An ideal lumped model would have elements that are only a function of the physical dimensions. Despite these limitations, the data in Figure 10.6 is still useful over narrow bandwidths. Given a desired line width and frequency an equivalent inductance can be extracted from the chart and used in a circuit simulation. In amplifier design this approach might be useful for a source grounding geometry or emitter grounding geometry that is used over and over again.

The geometry in Figure 10.5 is quite simple; why even bother with the field-solver? It should be possible to construct this same structure using a combination of standard library elements in any microwave linear simulator. After reading the fine print in the element catalogue, it is clear that most via models are for the post or barrel (hollow cylinder) only and the reference point is the center of the barrel. It is tempting to construct a collection of models as shown in Figure 10.7.

Using the dimensions shown in Figure 10.5 and assuming $w = 17$ mil we computed data for the via alone, the collection of analytical models and for the field-solver model of the same geometry using Microwave Office and EMSight [12]. The

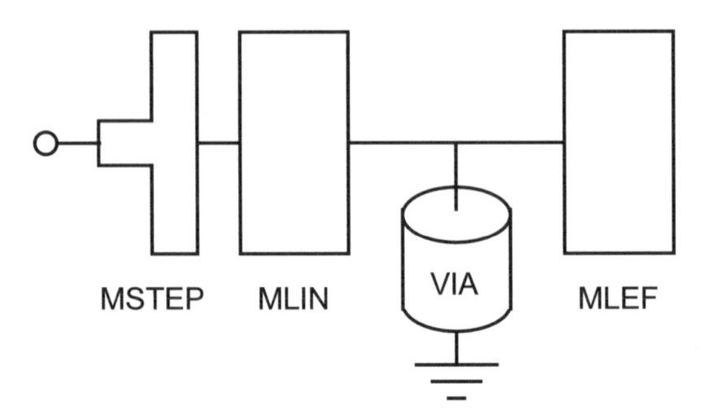

Figure 10.7 A collection of analytical models that describe the geometry shown in Figure 10.5.

results of this experiment are shown in Figure 10.8. Assuming the field-solver data is correct, the via model alone seriously underestimates the inductance and the more complicated analytical model overestimates the inductance. At 10 GHz the spread in phase angle across the three solutions is about 20 degrees. This is a very large uncertainty in a filter or matching network design.

Looking at the current plots in Figure 10.2 we might hypothesize that the MLEF element in Figure 10.7 is not needed. When the MLEF element is removed, the results are very similar to the full analytical model. For the line widths used here the MSTEP model has very little impact on the results. It is possible to make

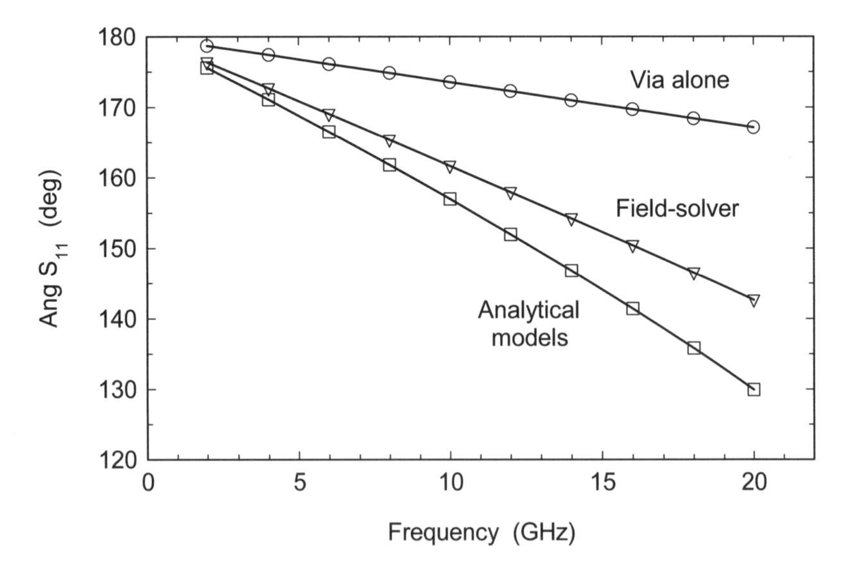

Figure 10.8 The geometry in Figure 10.5 is analyzed three different ways. The phase angle of S_{11} for the via alone, the collection of analytical models, and the field-solver model is shown (AWR Microwave Office and EMSight Ver. 4.0).

the analytical model match the field-solver data by adjusting the length of the MLIN element. However, that model is nonphysical in the sense that the length needed to match the data does not match any of the obvious physical dimensions of the via and pad.

This simple structure is a good example of how the traditional approach of cascading individual discontinuity models can sometimes fail to capture the true behavior of the network. We should emphasize that on their own, the individual elements used in Figure 10.7 are indeed correct, but in combination they fail. However, the field-solver computes the correct current distribution for this geometry without resorting to an arbitrary subdivision of the problem based on strictly visual cues or mathematical convenience.

10.5 MICROSTRIP MITERED BEND

The microstrip bend is a discontinuity where current flow around a corner is critical. Figure 10.9(a) shows a right angle bend in a 15-mil wide line on 15-mil thick alumina substrate. Note the current null at the outer corner and the current maximum on the inner corner. The current null at the outer corner is again due to the charge singularity at the sharp point. If we drive the structure at Port 1 we see what we recognize as the normal microstrip current distribution over most of the feed line. However, at the bend, the current tends to follow the shortest path and bunches up around the inside corner. As the current continues towards Port 2, the normal microstrip current distribution is again established. Note that it takes time and distance to reestablish the normal microstrip current distribution. If we set reference planes at the inside corner, we can extract a simple lumped element model.

In Figure 10.9(b) the bend has been mitered by 50% of the distance from the inside to the outside corner. The equivalent capacitance has decreased almost 50% and the series inductance has increased slightly.

Finally, the "optimum" miter (Figure 10.9(c)) has been computed using a well known formula [13]. The optimum miter is typically 60% to 70% of the distance between the inside and outside corners. The capacitance has been cut in half again and the inductance is two times larger than the full right-angle bend case. The current scale is 5 to 55 amps/meter in all three color plots.

The return loss for all three bends is shown in Figure 10.10. Mitering clearly improves the return loss. From a circuit theory point of view, we are trying to find a well matched lowpass filter with a high cutoff frequency. The current plots help us understand how that is actually achieved. Removing metal simultaneously reduces the capacitance and increases the inductance by pinching the current. Results from an analytical model for the optimum miter are also shown [14]. This analytical model seems to work quite well for these substrate parameters and line width. So we might consider this to be a validation of the analytical model. We could then use the faster analytical model with confidence as long as the bends were relatively isolated in the layout. To do a more careful validation, we should also look at the phase

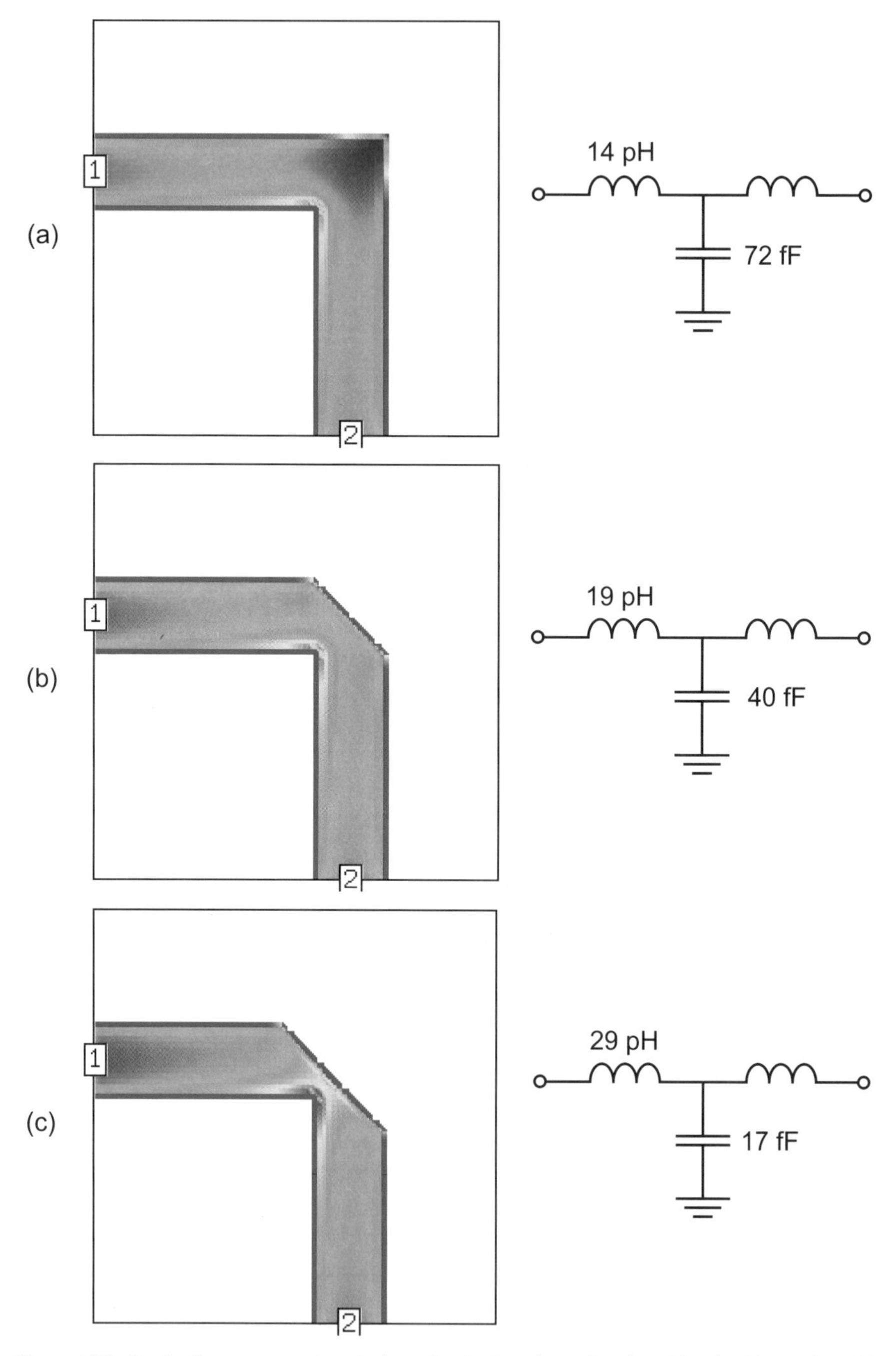

Figure 10.9 Conduction current and equivalent circuits for microstrip mitered bends: (a) no miter, (b) 50% miter, and (c) optimum miter (Sonnet ***em*** Ver. 7.0).

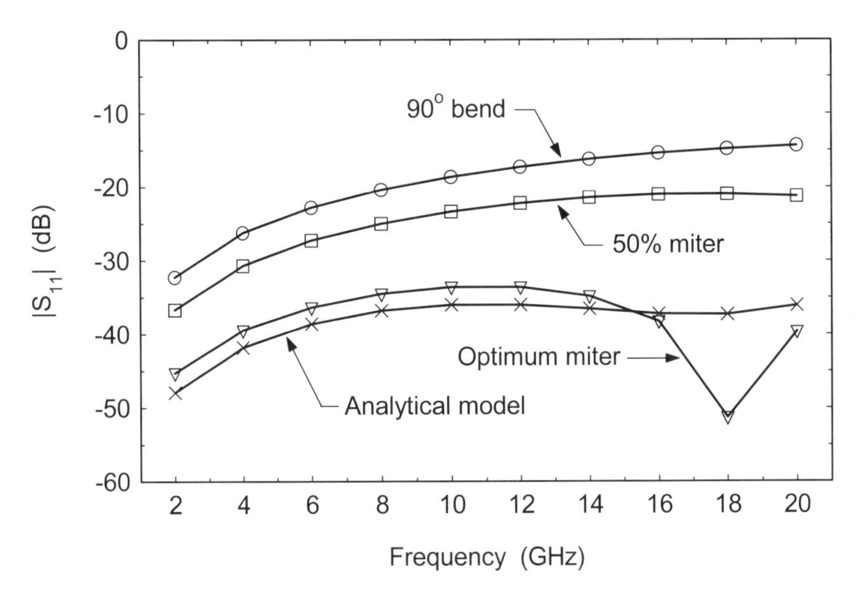

Figure 10.10 Return loss for the three cases shown in Figure 10.9. The "optimum" miter data from the field-solver is also compared to data from an analytical model for the same geometry.

responses. Some recent experimental data on the microstrip bend can be found in [15].

10.6 MICROSTRIP TEE-JUNCTION

Another interesting and frustrating discontinuity is the tee-junction. The example in Figure 10.11(a) is driven at Port 1. Note how the current flows around the corners. It takes considerable time and distance for the normal microstrip current distribution to be reestablished on the left and right arms. Some current from the corner regions must cross over to the top edges of the left and right arms. The result is a considerable area with very little current flow across from the common arm.

The junction in Figure 10.11(a) is unmatched. One published matching technique [16] shapes the "dead" area across from the common arm (Figure 10.11(b)). The amount of compensation shown here is greater than what was recommended in [16]. This tee-junction was optimized manually by making several runs at 12 GHz and using the mitered bend as a guide. Another matching technique [17] modifies the common arm and the transition region around the corners (Figure 10.11(c)). The current scale in all three plots is 0 to 30 amps/meter.

Figure 10.12 contains S-parameter data for the three junctions shown in Figure 10.11. The theoretical simultaneous match condition for any three-port is –9.5 dB return loss. The second compensation technique (Figure 10.11(c)) appears to be

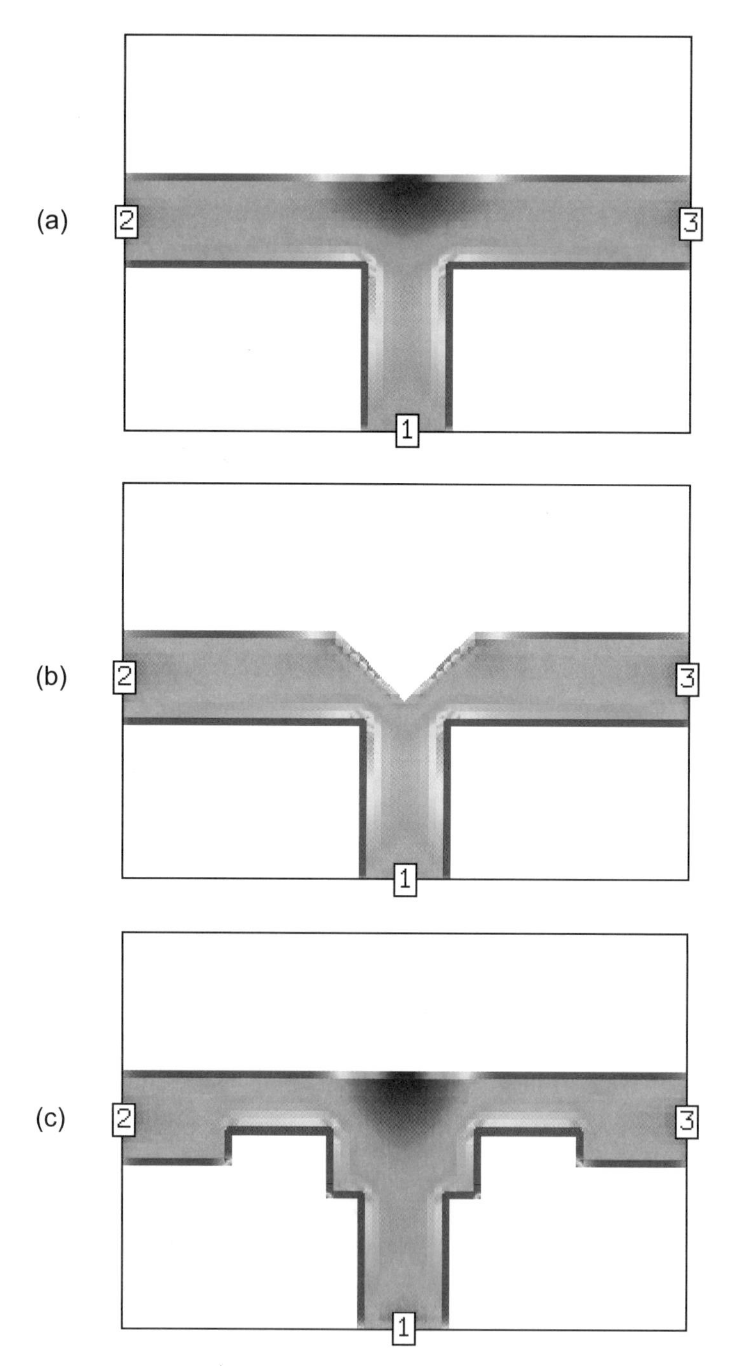

Figure 10.11 Conduction current on microstrip tee junctions: (a) uncompensated junction, (b) one possible compensation, and (c) another possible compensation (Sonnet ***em*** Ver. 7.0).

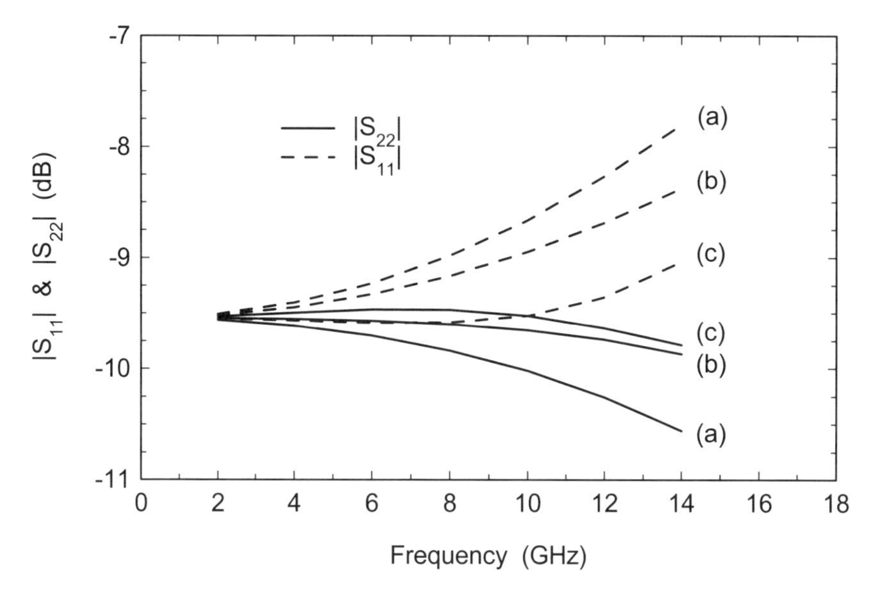

Figure 10.12 Return loss for the tee junctions in Figure 10.11. The theoretical simultaneous match condition for a three-port is –9.5 dB.

superior for these dimensions. Both compensation techniques are candidates for further optimization on the field-solver.

10.7 SUMMARY FOR MICROSTRIP DISCONTINUITIES

Like the "calibration" examples we looked at earlier, these microstrip discontinuities emphasize the highly nonuniform current distributions that can be found on planar circuits. Inside corners tend to create very high current densities while outside corners produce current nulls. Once we have upset the "normal" microstrip line current distribution with a discontinuity, it takes time and distance to reestablish the normal current distribution. If we place a second discontinuity close to the first, there is no time to reestablish the normal current distribution before the current encounters the second discontinuity. In this sense, discontinuities can "interfere" with each other locally. This is distinct from the idea of the fields generated by discontinuities coupling to each other or to modes in housing.

In practice, we often get unexpected results when we try to predict the performance of a tightly packed circuit with standard analytical models. Clearly there will be interactions between planar circuit elements that isolated analytical models cannot predict. In the last section of this chapter we present some very simple examples of compaction of microstrip and stripline circuits.

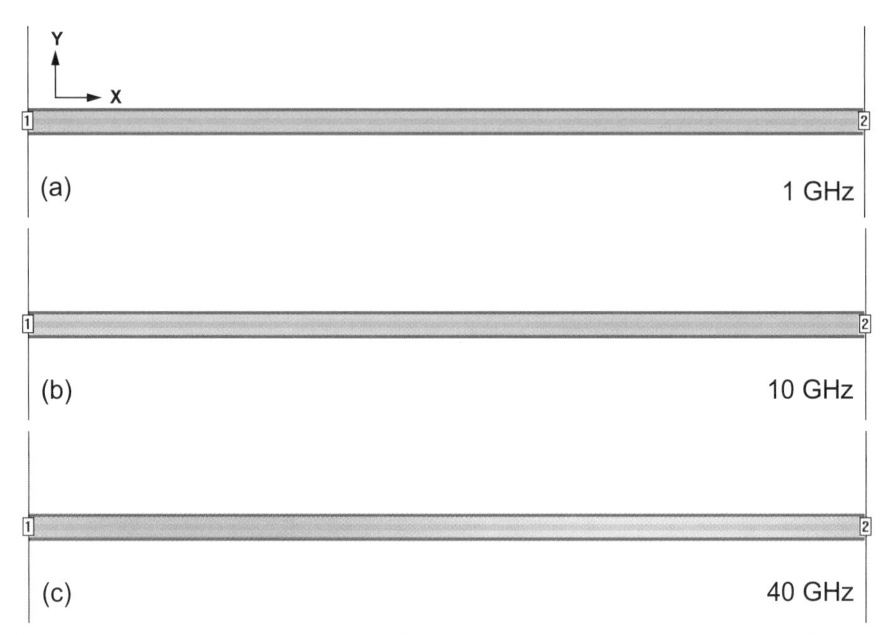

Figure 10.13 Conduction current, mostly X-directed, on a microstrip line on a gallium arsenide substrate: (a) current at 1 GHz, (b) current at 10 GHz, and (c) current at 40 GHz. The scale is 0 to 300 amps/meter (Sonnet ***em*** Ver. 7.0).

10.8 QUASI-TEM NATURE OF MICROSTRIP

The visualization tools in field-solvers give us the opportunity to view first hand some of the more subtle aspects of microstrip behavior. One of these topics is the quasi-TEM nature of microstrip. Any transmission line with a homogeneous cross-section supports a pure transverse electric magnetic mode. The vector direction of both the E-field and H-field is perpendicular to the direction of propagation. The currents will be directed only along the direction of propagation. In an inhomogeneous transmission line, like microstrip, there are longitudinal components to the fields, as well as the transverse components. The result is a small, but finite, transverse current component on the microstrip line.

Figure 10.13 shows various views of a 40 by 1,280 micron microstrip line on 50-micron thick gallium arsenide substrate. The scale is 0 to 300 amps/meter in all three plots. In all three figures we see the normal microstrip current distribution that we are now familiar with. At 40 GHz (Figure 10.13(c)) the line is long enough that we start to see some variation due to wavelength. The visualization software allows us to independently turn the X- and Y-directed currents on and off. If we turn off the Y-directed currents and keep the same scale the plots are visually unchanged. The majority of the current is in the X direction.

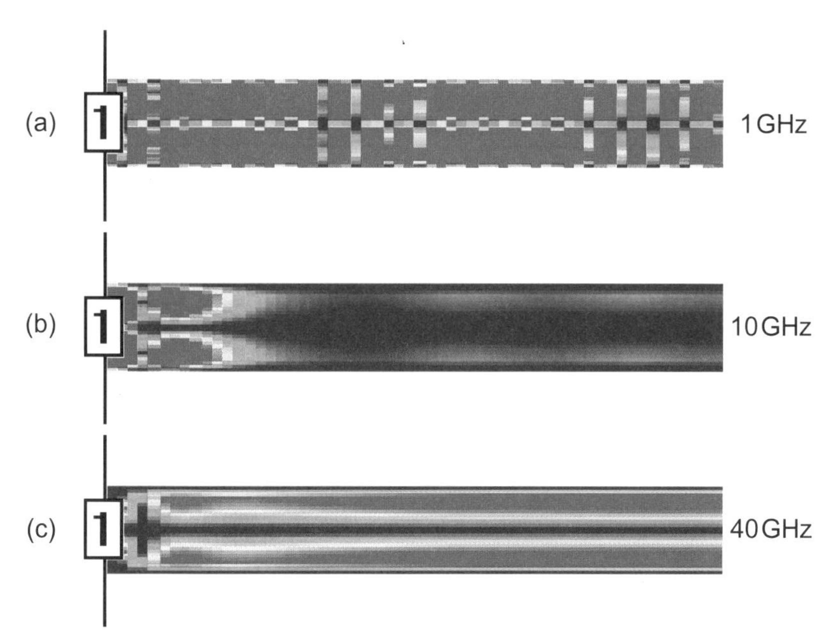

Figure 10.14 Transverse current (Y-directed only) on the microstrip line: (a) current at 1 GHz, (b) current at 10 GHz, and (c) current at 40 GHz. The scale is 0 to 0.035 amps/meter (Sonnet ***em*** Ver. 7.0).

To observe the quasi-TEM nature of microstrip we need to look for small transverse components in the conduction current. We can do this in Sonnet ***emvu*** by turning off the X-directed currents and viewing the Y-directed currents only at a much higher sensitivity. In Figure 10.14 we have zoomed in on a region close to Port 1 and the current scale is now 0 to 0.035 amps/meter in all three plots.

First we should note that the transverse currents we are observing are roughly four orders of magnitude lower than the longitudinal currents. At 1 GHz the "speckled" nature of the plot (Figure 10.14(a)) indicates we are down in the numerical noise and we should disregard this plot. In the 10-GHz plot (Figure 10.14(b)) the transverse currents are evanescent and die out quickly away from the port. At 40 GHz the transverse current component is fully supported along the entire length of the line (Figure 10.14(c)). The transverse component is maximum at the edges of the strip and has a null at the center of the strip. If we repeat this experiment on a substrate that is electrically thicker, the transverse component appears at a lower frequency.

The transverse current distribution we observe actually wraps around the strip. This distribution has been computed and observed by others. Figure 10.15 is from Hoffman [18] and shows a qualitative view of the longitudinal distribution due to the fundamental mode and the transverse distribution which appears at high frequencies.

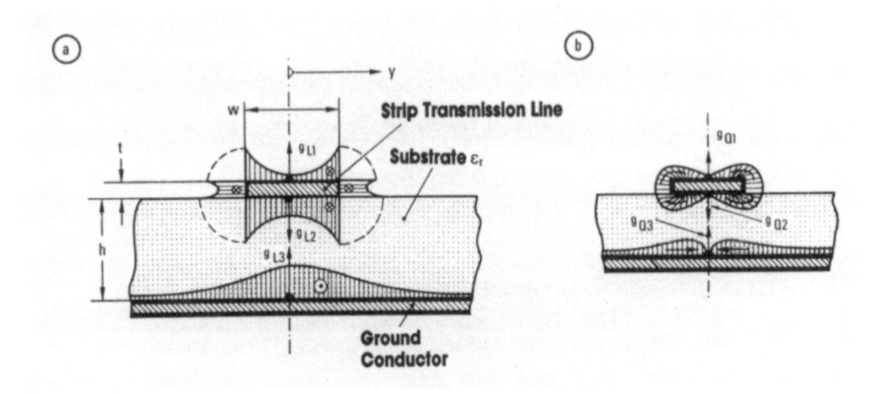

Figure 10.15 Qualitative current distributions for the microstrip fundamental mode: (a) longitudinal distribution at low frequencies with static approximation; and (b) transverse distribution at high frequencies with dynamic analysis. From Hoffman [18].

10.9 EVANESCENT MODES IN MICROSTRIP

Another more subtle aspect of microstrip behavior is the formation of evanescent modes around discontinuities. We have seen that the quasi-TEM mode has conduction currents that flow only in the direction of propagation. Any currents flowing perpendicular to the direction of propagation indicate some kind of energy storage. One common interpretation of this storage mechanism is evanescent modes, modes that are formed around the discontinuity but do not propagate down the transmission line.

Figure 10.16(a) shows a simple test structure to demonstrate this concept. We have a line terminated in a via, a line with a step discontinuity, and a through line for reference. The via is actually a simple ribbon of metal to ground in this case, rather than a circular or square metal post. In this experiment Ports 1, 2, and 4 are excited simultaneously to generate the current plots.

First we should observe the full solution (Figure 10.16(a)). We see the normal microstrip current distribution over most of the structures. If we were to display only the Y-directed currents, the picture would not visibly change. The scale is 0 to 44 amps/meter and the frequency is 1 GHz.

Next we turn off the Y-directed currents and display only the X-directed currents at a higher sensitivity (Figure 10.16(b)). Any currents in the X-direction are not part of the normal quasi-TEM mode. The scale is now 0 to 4.4 amps/meter. We can see X-directed current components in the vicinity of the discontinuities, but they go to zero rather rapidly as we move away from the discontinuity. The reference line shows no X-directed currents. Note the scale for the X-directed currents is only one order of magnitude below the full solution. These X-directed currents represent energy stored in evanescent modes near the discontinuity.

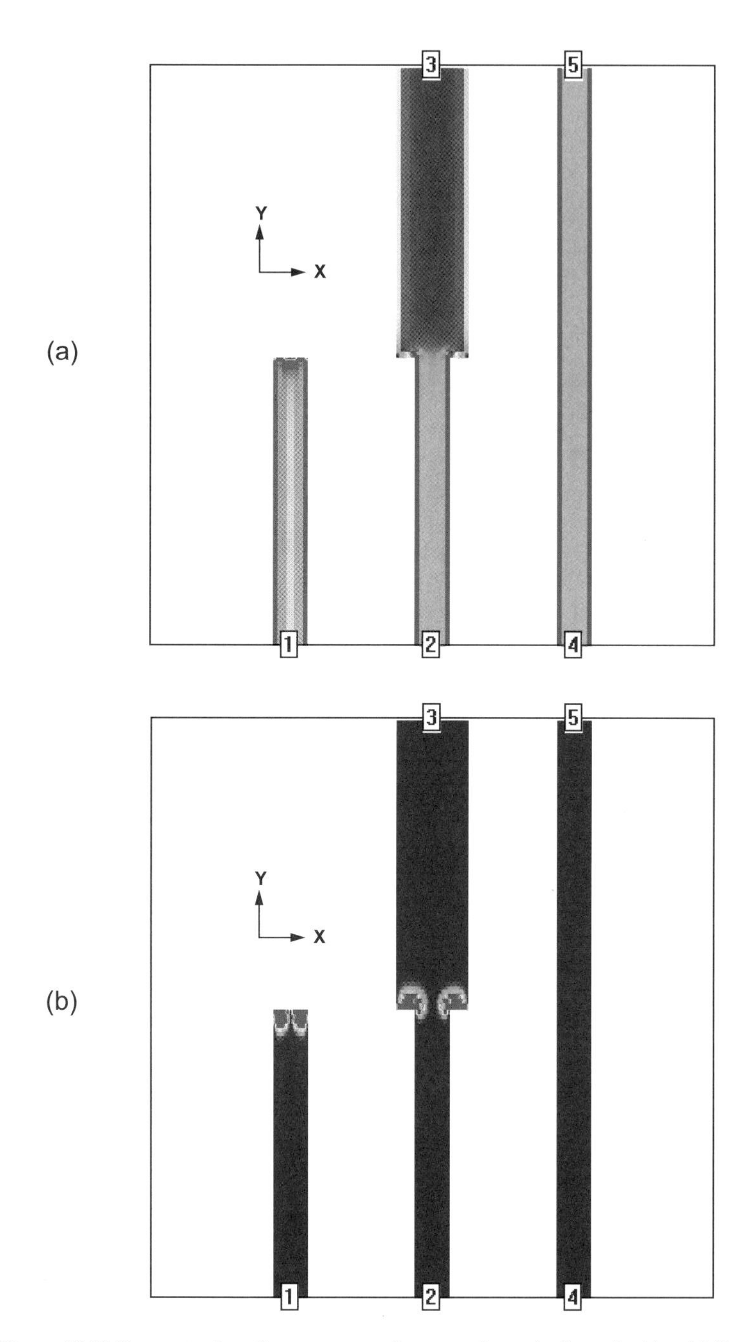

Figure 10.16 Demonstration of evanescent modes near microstrip discontinuities: (a) X- and Y-directed currents at scale of 0 to 44 amps/meter; and (b) X-directed currents at scale of 0 to 4.4 amps/meter (Sonnet ***em*** Ver. 7.0).

Evanescent mode behavior is one more subtle aspect of microstrip technology that we can study using visualization. When we compact our circuits, and discontinuities are very close together, these modes are one of the mechanisms that cause a circuit-theory-based analysis to fail. The standard models assume that each discontinuity is isolated from its neighbors. We can also observe this kind of behavior in 3D objects like waveguide discontinuities. Analyzing groups of discontinuities on the field-solver is one effective strategy for tightly packed circuits.

10.10 MICROSTRIP LOSS

Microstrip loss and predicting loss in general are topics of continued interest [19–26]. Questions revolve around the assumed conductivity for metals, the impact of strip thickness and the potential impact of surface roughness and edge roughness. Many engineers also assume that loss follows the square root of frequency rule. But if we measure the loss of a microstrip line with an ohm meter, we observe a small, but finite resistance. Recently, a very interesting discussion of microstrip loss appeared in [25] and a more detailed analysis in [26]. The first article was prompted by lengthy Internet discussion on loss modeling.

We assume that high-frequency loss increases with the square root of frequency and is dominated by skin effect. Skin effect forces RF currents to flow near the surface of a good conductor. As frequency increases, the skin-effect layer becomes thinner and resistance increases. Skin depth is given by

$$\delta = \frac{1}{\sqrt{\pi\mu\sigma f}} \tag{10.2}$$

where

$$\delta = \text{skin depth (m)}$$

$$\mu = \text{conductor magnetic permeability } (\mu_0 = 4\pi \times 10^{-7}\ \text{H/m})$$

$$\sigma = \text{bulk conductivity (S/m)}$$

$$f = \text{frequency (Hz)}$$

If the frequency quadruples, then skin depth is cut in half. If the current now flows in half the original cross-section, then resistance doubles. The other high-frequency effect we have already discussed in detail is the edge singularity. The combination of skin effect and the edge singularity drives high-frequency loss.

At very low frequencies, the skin depth is much larger than the conductor thickness, and current is uniform through the entire thickness of the line. If we lower the frequency further, the cross-section the current sees remains constant and the loss should remain constant with frequency. The frequency at which the conductor transitions from high-frequency behavior to medium-frequency behavior is

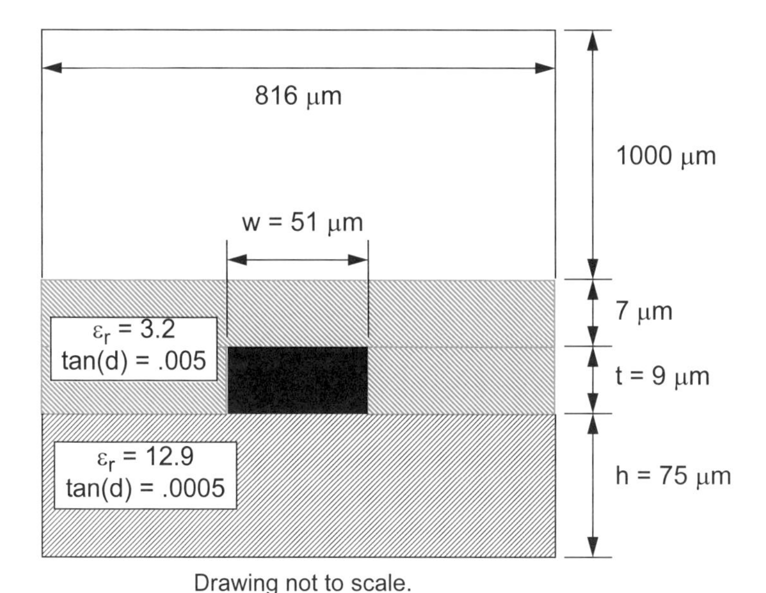

Figure 10.17 Test structure for microstrip loss experiment. The thick strip is modeled by two thin sheets, top and bottom in the simulator. © 2000 IEEE [25].

$$f_{c2} = \frac{4}{\pi \mu \sigma t^2} \tag{10.3}$$

where f_{c2} is the critical frequency (Hz) at which the conductor thickness equals twice the skin depth. Twice the skin depth is used because there is skin effect current on the upper and lower surfaces of the conductor.

A test case on GaAs was fabricated to provide some experimental data for this study (Figure 10.17). The conductor sits on the GaAs substrate and is surrounded by polyimide passivation. The microstrip line is modeled as two infinitely thin sheets of metal. At high frequencies, the two strips represent the two skin depth sheets of current on the top and bottom sides of the strip. At low frequency, the current splits equally between the two sheets and the R_{DC} is adjusted for thickness.

Sonnet ***em*** was used for this loss study. This program uses two terms for loss analysis. R_{DC} is the basic resistive loss at dc in ohms/square for each sheet of the conductor. R_{RF} is the skin effect, square root of frequency loss term. Specifically

$$R_{DC} = \frac{n}{\sigma t} \tag{10.4}$$

and

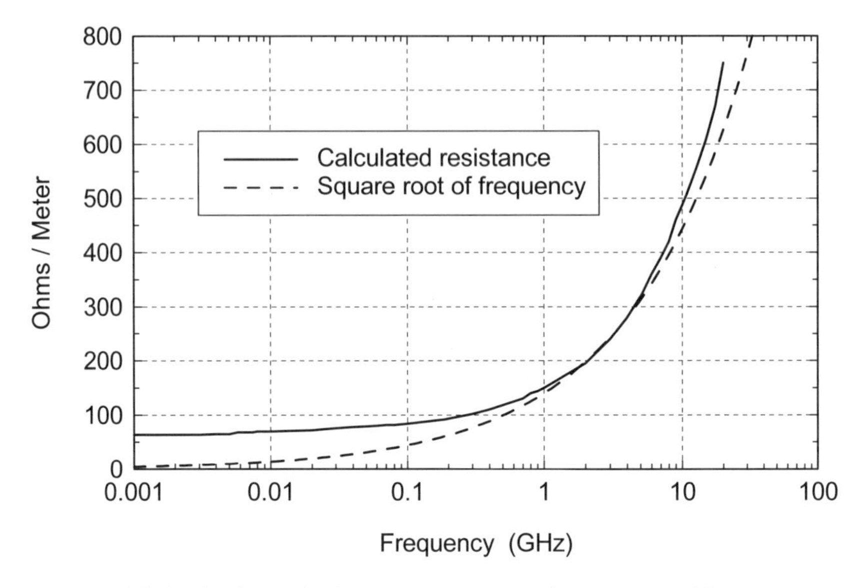

Figure 10.18 Calculated resistance for the test structure compared to square root of frequency curve. © 2000 IEEE [25].

$$R_{RF} = \sqrt{\frac{\pi\mu}{\sigma}} \tag{10.5}$$

where n is the number of sheet conductors used in the model. The key parameter in both terms is the bulk conductivity of the metal. The handbook values for conductivity tend to be optimistic because they do not include porosity and other defects due to industrial manufacturing processes. For the experiment reported in [25], the bulk conductivity of four through lines was measured with an average value of 3.42×10^7 S/m. This compares to 4.09×10^7 S/m for pure gold. From many measurements on the Q of silver plated and copper resonators and filters we have observed that a conductivity that is 80% of the handbook value fits the measured data quite closely. Note that in this case, the measured conductivity for the gold microstrip is also close to 80% of the handbook value. The actual formula for surface impedance used in Sonnet ***em*** is

$$Z_s = \frac{(1+j)R_{RF}\sqrt{f}}{1 - e^{((1+j)R_{RF}\sqrt{f})/R_{DC}}} \tag{10.6}$$

where Z_S is the total surface impedance in ohms/square. At low frequency, $Z_s = R_{DC}$. At high frequencies, $Z_s = (1+j)R_{RF}\sqrt{f}$. When operating more than one

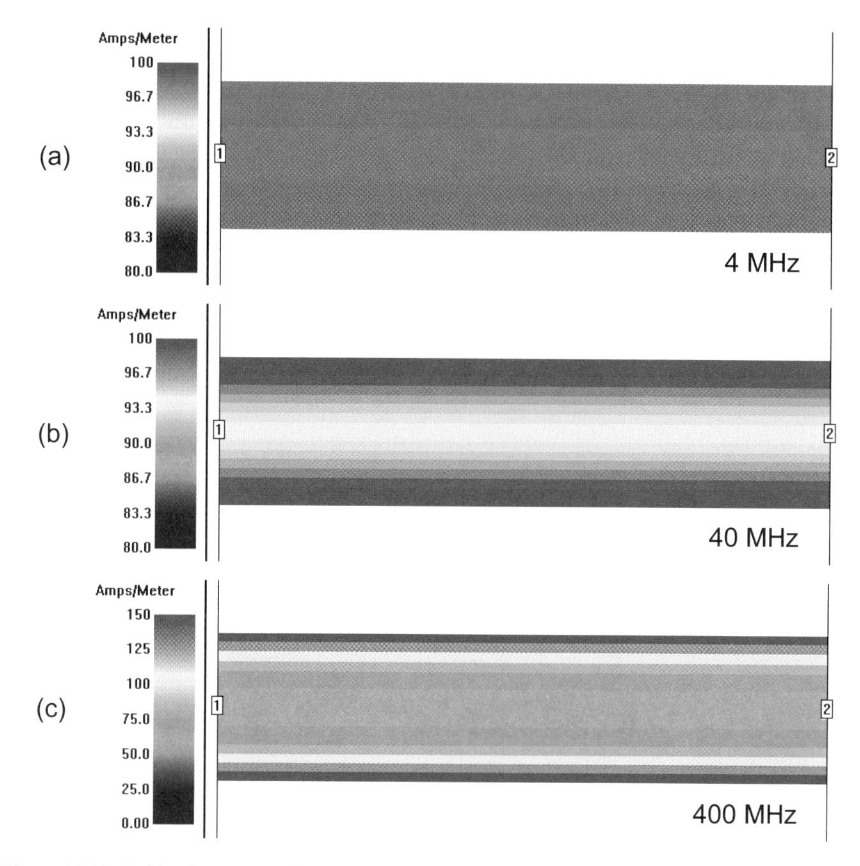

Figure 10.19 Conduction current distributions for the test structure: (a) uniform current at 4 MHz, (b) edge singularity is starting to form at 40 MHz, and (c) edge singularity fully formed at 400 MHz. Note the compressed current scale in (a) and (b) and different current scale in (c) (Sonnet ***em*** Ver. 7.0). © 2000 IEEE [25].

octave above f_{c2}, only R_{RF} is important. When operating more than one octave below f_{c2}, only R_{DC} is important. A sample of microstrip line roughly 200 μm long was analyzed and the resistance per meter for the line was extracted. Figure 10.18 shows the computed resistance plotted against square root of frequency.

At first glance, the computed resistance seems to behave exactly as we expected. Near the critical frequency, f_{c2} = 363 MHz, the resistance is definitely becoming more constant with frequency. But it is still falling, and it apparently does not truly become constant until about 10 MHz. This raises the possibility that there are two critical frequencies for this curve, rather than just one. Above 5 GHz, the resistance is also increasing faster than square root of frequency.

To explore the qualities of the computed resistance curve more fully, we again turn to the current density plots (Figure 10.19). We have chosen three frequencies:

4 MHz will serve for dc, 40 MHz is near the suspected lower critical frequency, and 400 MHz is just above the computed critical frequency of 363 MHz. Of course, these frequencies also fall on a nice decade scale.

At 400 MHz we see the strong edge-singularity that we now associate with normal microstrip behavior. At 40 MHz the edge-singularity is starting to disappear and is finally gone at 4 MHz. Note the compressed current scale for the upper two plots. Apparently, the low-frequency behavior of the strip has two critical frequencies. The first is when the edge-singularity starts to form and the second is when the strip reaches two skin depths in thickness.

The initial hypothesis regarding the first critical frequency assumed it was related to when the width of the strip (not the thickness) was small compared to skin depth. However, this hypothesis was discarded when it was tested for other transmission line geometries. The second hypothesis assumes the transition frequency is that frequency for which the resistance per unit length equals the inductive reactance per unit length.

$$f_{c1} = \frac{R}{2\pi L} \tag{10.7}$$

$$R = \frac{1}{\sigma w t} \tag{10.8}$$

$$L = \frac{Z_0}{v} \tag{10.9}$$

where

f_{c1} = first critical frequency (Hz)

R = resistance per unit length (ohm/cm)

w = width of strip (m)

t = thickness of strip (m)

L = inductance per unit length (H/m)

v = velocity of propagation (m/s)

This formula gives a critical frequency of 20 MHz for this geometry. It also gives reasonable answers for other cases, but it is purely empirical and has not been derived from Maxwell's equations in a rigorous way.

The additional resistance above 5 GHz in Figure 10.18 can be attributed to microstrip dispersion. If characteristic impedance and velocity of propagation change with frequency, then the current distribution must also change. If we probe the current plots for quantitative data, we find that as frequency increases the current shifts to the bottom strip and the edge singularity effect becomes stronger. This conclusion is also supported in [24].

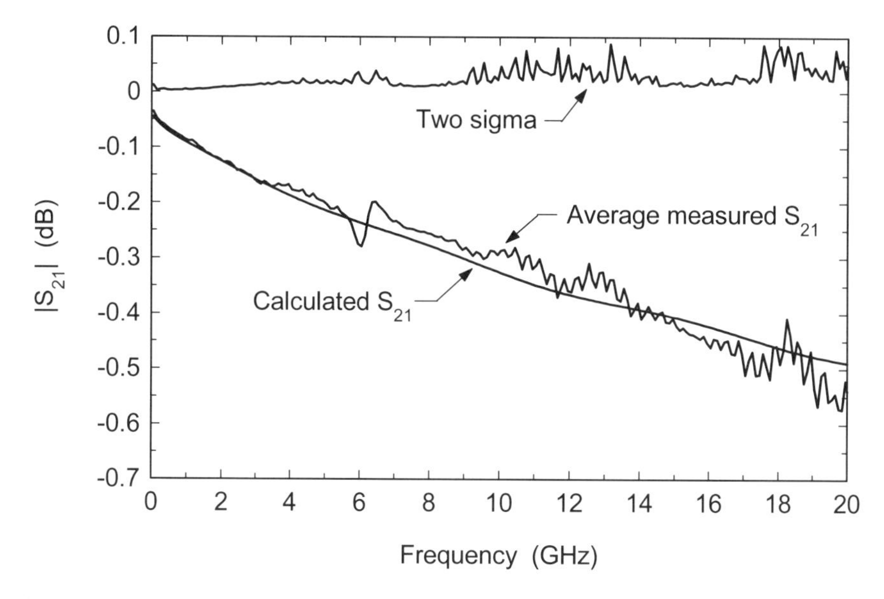

Figure 10.20 Measured versus modeled loss for the 6,888-μm long line on GaAs. The measured trace is the average of eight transmission magnitudes. © 2000 IEEE [25].

Experimental validation was performed on a 6,888-μm long line on GaAs (Figure 10.20). A line with 1/32 of the total length was analyzed for comparison. This result was cascaded with itself 32 times in the final analysis. Data was measured for four separate through lines. The S12 and S21 magnitudes from all four lines (a total of eight values) were averaged and plotted. We can observe that the calculated data is within two sigma of the measured data. Unfortunately, there are only a few data points below 363 MHz and no data points below 50 MHz. The calculated results do not include roughness or loss in the ground plane.

Accurate prediction of loss has been a challenge since the early days of microwave CAD. Before commercial field-solvers came along, we were limited to analytical models that were often derived from a limited set of experimental data. This study seems to confirm our "80% rule" for conductivity of good metals. It also confirms previous understanding of the critical frequency due to skin depth. What is new is the possibility of a second critical frequency related to the formation of the edge-singularity. Again, it is the visualization aspect of the field-solver that often leads to insight when we are faced with unexplained behavior.

10.11 COMPACTION OF MICROSTRIP CIRCUITS

We often use very simple circuits with very simple geometrical layouts to demonstrate CAD principles. But real circuits are highly compact with somewhat arbitrary

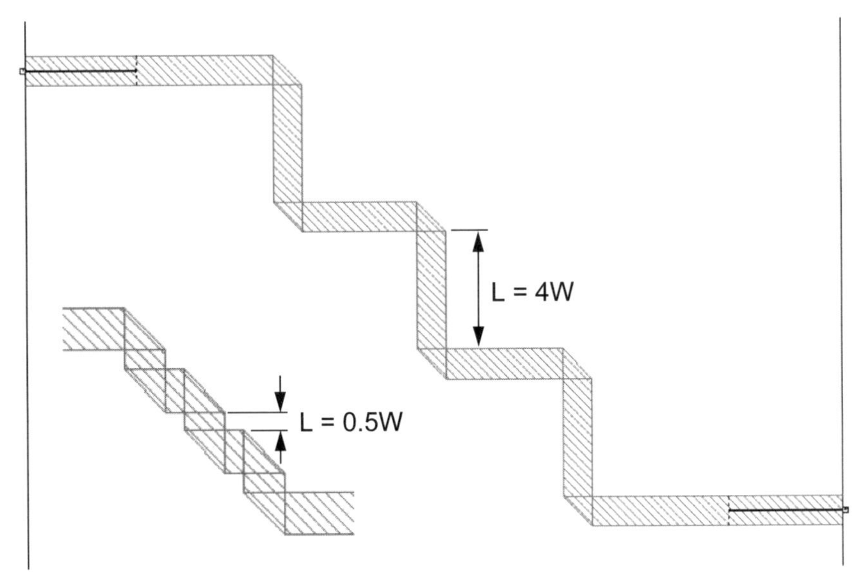

Figure 10.21 Test case with six mitered bends. In the main figure the distance between bends is four times the line width. In the inset the distance between bends is one-half the line width. In all cases the total de-embedded length of line is 2,560 μm (AWR Microwave Office).

shapes. Before field-solvers were available, we often blamed "nonadjacent couplings" when measured results did not match the computer prediction. When the commercial field-solvers arrived, we soon found that a global analysis of the circuit matched the measured results much more closely. Goldfarb and Platzker [27] published an excellent early article on EM analysis of MMIC circuits.

Circuit-theory-based CAD assumes we can cascade a set of independent, isolated models and get the correct answer for our circuit. When the circuit is highly compacted this assumption sometimes fails, but what is the exact failure mechanism? In earlier sections of this chapter, we saw how the normal microstrip current distribution is modified by a discontinuity. Before field-solvers were widely available, this detailed view of actual current distributions could not be visualized by the average design engineer.

10.11.1 Cascade of Mitered Bends

Figure 10.21 is a numerical experiment we devised to explore compaction issues. It is a 50-ohm microstrip line on a 100-μm thick GaAs substrate. There are six mitered bends in cascade along the line. The total length of the line, measured down the centerline, is 2,560 μm. We would like to compare a global EM solution of this structure to a circuit-theory-based simulation. The distance between discontinuities, *L*, will be the variable in the experiments.

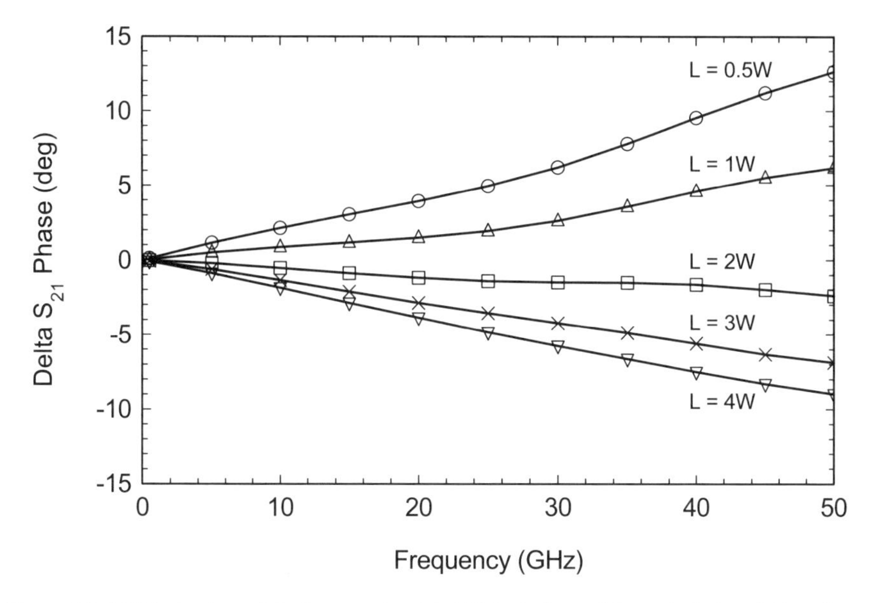

Figure 10.22 Delta in phase of S_{21} between the global EM solution and the analytical model based on a cascade of circuit-theory-based models for the straight line segments and the mitered bends (AWR Microwave Office 2002).

Before we began, we ran a straight through line calibration and compared the field-solver solution to the analytical line model. There was a difference in S_{21} phase of about 5 degrees at 50 GHz. This corresponds to a change in length of about 28 μm. This 28-μm correction was applied to all the analytical models that follow. With the length correction the delta phase error was less than ±0.3 degrees.

Figure 10.22 compares the S_{21} phase for the global EM analysis with a cascade of circuit theory models for the bends and straight lines. When the spacing between bends is large ($L = 4W$), there is apparently some error between the analytical bend model and the EM solution. As the distance decreases, the sign of the error between the two models actually changes. The total uncertainty in the length is about 125 μm.

In the next experiment we created a field-solver model for the mitered bend. In Figure 10.23 we again compare the S_{21} phase for the global EM analysis with a cascade of analytical models for the line segments and an EM-based model for the mitered bend. When the spacing between bends is large ($L = 4W$), the error between the two solutions is relatively low. However, as the spacing between bends decreases, the error increases. For $L = 0.5W$ the error in length is again about 125 μm, almost twice the line width. The EM-based model for the mitered bend is clearly more accurate for large L. But as the circuit is compacted, the total error is about the same as the experiment with all analytical models.

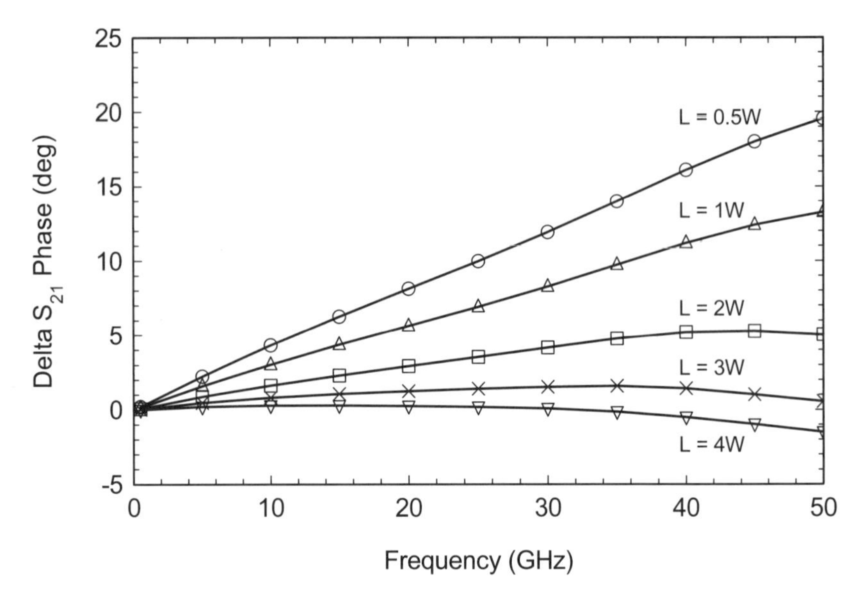

Figure 10.23 Delta in phase of S_{21} between the global EM solution and the analytical model based on a cascade of analytical line models and an EM-based model of the mitered bends (AWR Microwave Office 2002).

If we look the inset in Figure 10.21, the reason for the error between the global EM solution and any cascade of models is rather obvious. As L gets very small, the geometry approaches that of a diagonal line. The actual current is following a path that is not predicted by the individual discontinuity models. Only the global EM solution can capture the actual current distribution.

10.11.2 Stripline Meander Line

As networks get more complicated, it is more difficult to identify the exact mechanism that may cause a circuit-theory-based analysis to fail. One simple network that may be a candidate for further analysis is the stripline meander line (Figure 10.24). This network is commonly used in high-speed digital circuits to adjust the "skew" or relative arrival time of a clock signal. At low frequencies a simple line length computation down the center line of the meander was perhaps good enough. At higher frequencies the path the current takes around the corners and the coupling between line segments both impact the actual delay.

An excellent article by Rubin and Singh [28] has explored the meander line delay problem in some detail. Figure 10.24 shows one of their numerical experiments with 10-mil pitch between the meandered segments. The line width is 3.3 mil and the ground plane spacing is 15.9 mil. In order to separate the effects of the cor-

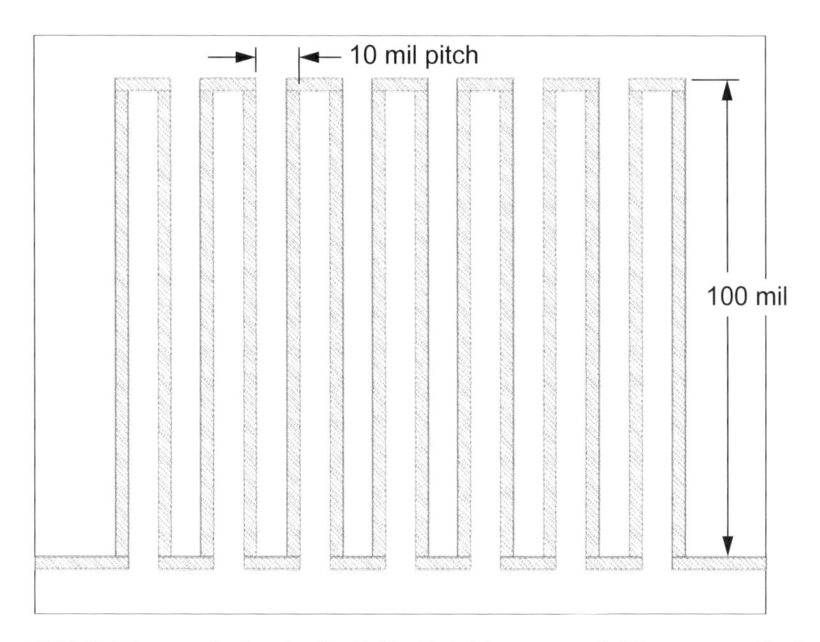

Figure 10.24 Stripline meander line circuit with 10-mil pitch between coupled line segments. The line-width is 3.3 mil and the ground plane spacing is 15.9 mil. © 2000 IEEE [28].

ners from the coupling effects they varied the pitch between line segments from 10 to 60 mils while keeping the overall physical length constant. Their baseline delay for a straight line segment was 927 pS. At 60-mil pitch and negligible coupling the delay was reduced by about 8 pS, which was attributed to the corners. As the pitch was reduced, the total delay continued to decrease; the additional reduction in delay was attributed to coupling between the meandered segments. At 10-mil pitch the total reduction in delay was about 50 pS.

10.11.3 Microstrip Branchline Coupler

The microstrip branchline coupler is another relatively simple geometry that we have used as an example several times. The single section branchline realizes a narrowband, 3 dB coupler without resorting to fine line geometries. However, it does occupy a fair amount of area on the PC board and several workers have attempted to compact it in various ways. Figure 10.25(a) shows the conventional, single section branchline coupler. Figure 10.25(b) shows a coupler compacted by folding the low impedance line sections (NEC Corp., circa 1980). We can probably get a starting point for the design using a cascade of conventional discontinuity models, but the discontinuities are so close together we would expect significant interaction between them. For the final optimization it would be easier and more accurate to

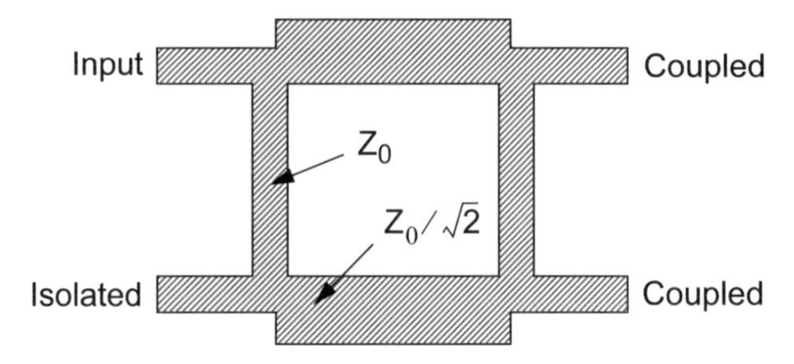

(a) Conventional branchline coupler

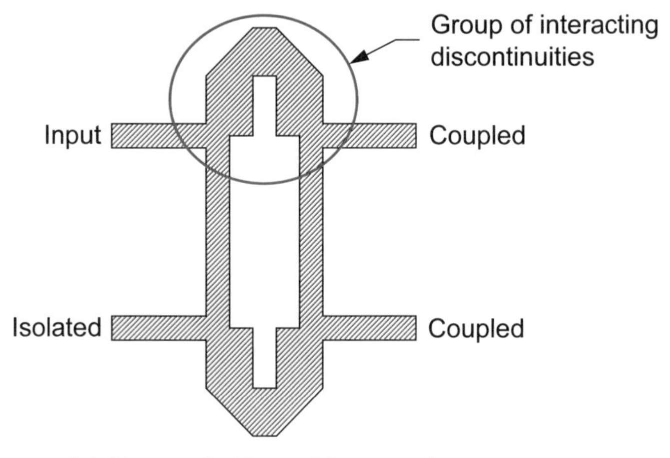

(b) Compacted branchline coupler

Figure 10.25 Microstrip single section branchline couplers: (a) conventional design and (b) compacted design taken from a trade magazine photo, circa 1980. Using a field-solver we would lump the folded low impedance line and the discontinuities into one four-port problem.

lump the discontinuities and the folded low impedance line into one, EM-based, four-port model. Analyzing groups of discontinuities rather than cascading individual models is a key strategy when using field-solvers.

References

[1] Barrett, R. M., "Microwave Printed Circuits–The Early Years," *IEEE Trans. Microwave Theory and Tech.*, Vol. 32, No. 9, 1984, pp. 983–990.

[2] Hoffmann, R. K., *Handbook of Microwave Integrated Circuits*, Norwood, MA: Artech House, Inc., 1987, pp. 95–99.

[3] Oldfield, W., et al., "Simple Microstrip Structures Calculated Versus Measured," *37th ARFTG Conference*, Boston, June 1991, pp. 10–20.

[4] Swanson, Jr., D. G., "Electromagnetic Simulation of Microwave Components," *37th ARFTG Conference*, Boston, June 1991, pp. 3–9.

[5] Swanson, Jr., D. G., "Electromagnetic Simulation Software," *Emerging Microwave Technologies and Applications Conference*, Stanford, CA, March 1991.

[6] Swanson, Jr., D. G., "Designing Microwave Components Using Electromagnetic Field Solvers," *IEEE International Microwave Symposium Workshop WSA Digest*, Albuquerque, NM, June 1992.

[7] Goldfarb, M., and R. Pucel, "Modeling Via Hole Grounds in Microstrip," *IEEE Microwave and Guided Wave Letters*, Vol. 1, No. 6, 1991, pp. 135–137.

[8] Swanson, Jr., D. G., "Grounding Microstrip Lines With Via Holes," *IEEE Trans. Microwave Theory and Tech.*, Vol. 40, No. 8, 1992, pp. 1719–1721.

[9] Marchetti, S., and T. Rozzi, "H-field and J-current Singularities at Sharp Edges in Printed Circuits," *IEEE Trans. Antennas Propagation*, Vol. 39, No. 9, 1991, pp. 1321–1331.

[10] Bonato, P., et al., "A New Shape Reducing Via Hole Inductance," *Int. Conference on Electronics Technology*, Brighton, UK, June 4–5, 1996.

[11] Micro-Stripes, Flomerics Inc., Southborough, MA.

[12] Microwave Office and EMSight, Applied Wave Research, El Segundo, CA.

[13] Douville, R. J. P., and D. S. James, "Experimental Study of Symmetric Microstrip Bends and Their Compensation," *IEEE Trans. Microwave Theory and Tech.*, Vol. 26, No. 3, 1978, pp. 175–181.

[14] Touchstone Ver. 3.5, Agilent EEsof EDA, Santa Rosa, CA.

[15] Slobodnik, Jr., A. J., and R. T. Webster, "Experimental Validation of Microstrip Bend Discontinuity Models from 18 to 60GHz," *IEEE Trans. Microwave Theory and Tech.*, Vol. 42, No. 10, 1994, pp. 1872–1878.

[16] Chadha, R., and K. C. Gupta, "Compensation of Discontinuities in Planar Transmission Lines," *IEEE Trans. Microwave Theory and Tech.*, Vol. 30, No. 12, 1982, pp. 2151–2156.

[17] Wu, S., et al., "A Rigorous Dispersive Characterization of Microstrip Cross and Tee Junctions," *IEEE MTT-S Int. Microwave Symposium Digest*, Dallas, TX, May 8–10, 1990, pp. 1151–1154.

[18] Hoffmann, R. K., *Handbook of Microwave Integrated Circuits*, Norwood, MA: Artech House, Inc., 1987, p. 138.

[19] Wheeler, H. A., "Formulas For the Skin Effect," *Proc. IRE*, Vol. 30, 1942, pp. 412–424.

[20] Pucel, R. A., D. J. Masse, and C. P. Hartwig, "Losses in Microstrip," *IEEE Trans. Microwave Theory Tech.*, Vol. 16, No. 6, 1968, pp. 342–350.

[21] Pucel, R. A., D. J. Masse, and C. P. Hartwig, "Losses in Microstrip (Correction)," *IEEE Trans. Microwave Theory Tech.*, Vol. 16, No. 12, 1968, p. 1064.

[22] Wheeler, H. A., "Transmission-line Properties of a Strip on a Dielectric Sheet on a Plane," *IEEE Trans. Microwave Theory Tech.*, Vol. 25, No. 8, 1977, pp. 631–647.

[23] Hoffmann, R. K., *Handbook of Microwave Integrated Circuits*, Norwood, MA: Artech House, Inc., 1987, Chapter 6, pp. 193–203.

[24] Faraji-Dana, R., and Y. L. Chow, "The Current Distribution and AC Resistance of a Microstrip Structure," *IEEE Trans. Microwave Theory Tech.*, Vol. 38, No. 9, 1990, pp. 1268–1277.

[25] Rautio, J. C., "An Investigation of Microstrip Conductor Loss," *IEEE Microwave Magazine*, Vol. 1, No. 4, 2000, pp. 60–67.

[26] Rautio, J. C., and V. Demir, "Conductor Loss Models for Electromagnetic Analysis," *IEEE Trans. Microwave Theory and Tech.*, Vol. 51, No. 3, 2003, pp. 915–921.

[27] Goldfarb, M., and A. Platzker, "The Effects of Electromagnetic Coupling on MMIC Design," *Int. J. MIMICAE*, Vol. 1, No. 1, 1991, pp. 38–47.

[28] Rubin, B. J., and B. Singh, "Meander Line Delay in Circuit Boards," *IEEE Trans. Microwave Theory and Tech.*, Vol. 48, No. 9, 2000, pp. 1452–1460.

Chapter 11

Computing Impedance

The two most basic parameters that define a transmission line are characteristic impedance, Z_0, and the velocity of propagation, v_p. Analytical models (equations) exist for all the standard geometries like microstrip, stripline, coax, and waveguide. However, there are times when it is useful to be able to compute the impedance of some arbitrary cross-section. This is particularly true in multilayer environments like PCBs, LTCC modules, and RFICs. In these multilayer environments we have the freedom to place a signal conductor in a rather arbitrary orientation to its reference plane or planes. The fastest and most inexpensive tool for many of these calculations is a 2D cross-section solver. The most inexpensive tools are often general purpose partial differential equation (PDE) solvers which must be programmed for the specific problem we are trying to solve.

11.1 SINGLE STRIP IMPEDANCE AND PHASE VELOCITY

For single strip transmission lines it is easy to compute characteristic impedance, Z_0, and the velocity of propagation, v_p, using a stand-alone, 2D cross-section solver. Characteristic impedance and velocity of propagation are defined in terms of inductance per unit length, L, and capacitance per unit length, C.

$$Z_0 = \sqrt{\frac{L}{C}} \qquad v_p = \frac{1}{\sqrt{LC}} \tag{11.1}$$

These equations are for lossless lines and ignore skin depth effects [1]. The equations imply that we need two types of 2D cross-section-solvers, an electrostatic solver to compute C, and a magnetostatic solver to calculate L. We can simplify the problem if we take advantage of one special case where we know the phase velocity in advance. This special case is any air dielectric transmission line where the velocity must be the speed of light,

$$c = 2.998 \times 10^8 \text{ m/s} \tag{11.2}$$

Substituting into the equation for velocity of propagation we get

$$c = \frac{1}{\sqrt{LC_0}} \tag{11.3}$$

where C_0 is the capacitance per unit length of the line when all of the dielectrics are air ($\varepsilon_r = 1$). If all the materials are also nonmagnetic ($\mu_r = 1$) we can solve for L and substitute back into the equations for Z_0 and v_p.

$$L = \frac{1}{c^2 C_0} \tag{11.4}$$

$$Z_0 = \frac{1}{c\sqrt{CC_0}} \qquad \left(\frac{c}{v_p}\right)^2 = \frac{C}{C_0} \tag{11.5}$$

For mixed dielectric problems, like microstrip, we can define an effective relative dielectric constant, ε_{eff}, which is related to the actual velocity of propagation in the medium.

$$\varepsilon_{eff} = \left(\frac{c}{v_p}\right)^2 \tag{11.6}$$

Substituting in the equation for v_p as a function of C and C_0 we get

$$\varepsilon_{eff} = \frac{C}{C_0} \tag{11.7}$$

Now we can solve for impedance and effective relative dielectric constant by using only an electrostatic solver to find the two values of capacitance per unit length. The first computation is for C, the capacitance per unit length with all dielectrics present. The second computation is for C_0, the capacitance per unit length with all dielectrics removed. With these two values in hand, Z_0 and ε_{eff} are simply

$$Z_0 = \frac{1}{c\sqrt{CC_0}} \qquad \varepsilon_{eff} = \frac{C}{C_0} \tag{11.8}$$

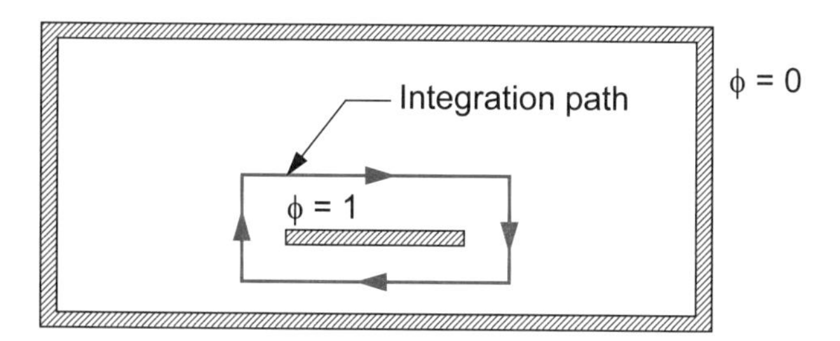

Figure 11.1 Generic strip in a box problem governed by Laplace's equation. The strip is set to 1 V and the outer conductor to 0 V. © 2001 IEEE [2].

For a transmission line with a homogeneous dielectric region, like standard coax or stripline, the impedance equation simplifies to

$$Z_0 = \frac{\sqrt{\varepsilon_r}}{cC} \tag{11.9}$$

We will look at two stand-alone electrostatic solvers suitable for finding C and C_0. These tools are actually general-purpose solvers for PDEs. They can be used for electrostatic and magnetostatic problems, thermal analysis, and stress/strain problems, among others. In our case, we are interested in solving Laplace's equation in two dimensions

$$\frac{\partial^2 \phi}{\partial x^2} + \frac{\partial^2 \phi}{\partial y^2} = 0 \tag{11.10}$$

where $\phi(x, y)$ is the potential in the region of interest. The classic paper by Green [3] outlines the solution for several geometries using the finite difference method. This formulation is so simple it can also be implemented as a spreadsheet program [4].

If we are solving for a strip in a box (Figure 11.1), we typically set the strip potential to 1 V, the boundary to 0 V and solve for the potential at a number of points inside the box. If we then integrate along a closed path around the strip, we get the total charge on the strip in Coulombs per meter, which is what the software will typically report. If we remember that charge equals capacitance times voltage,

$$Q = CV \tag{11.11}$$

then we can interpret the results from the software directly as capacitance per meter.

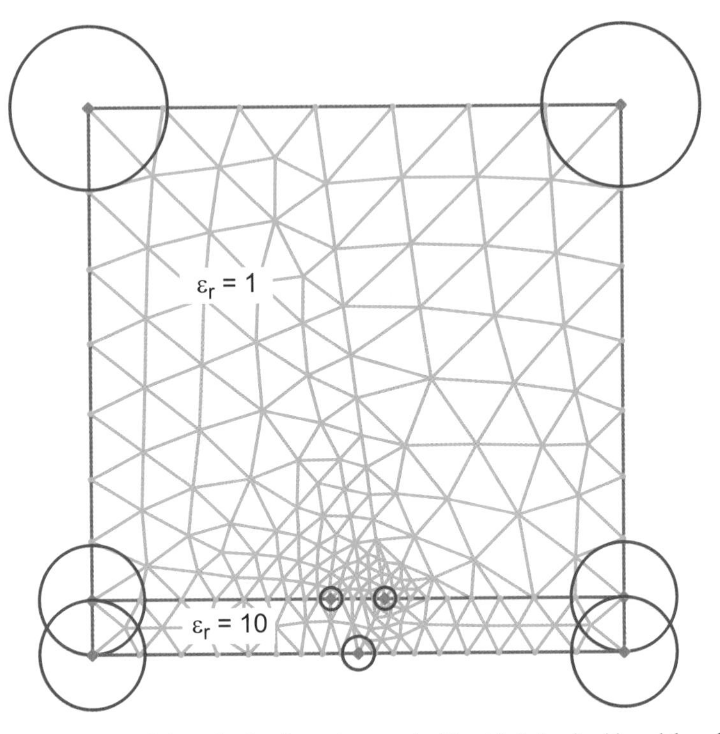

Figure 11.2 2D FEM mesh for a simple microstrip example. The strip is 1 unit wide and the substrate is 1 unit thick. The substrate relative dielectric constant is 10 (QuickField Ver. 4.2T).

Note that we are performing a static ($f = 0$) analysis of our transmission line structure. For pure TEM structures like stripline and coaxial lines, a static analysis will be accurate right up to the first higher order mode frequency. Quasi-TEM structures, like microstrip, will have frequency dependent impedance and effective dielectric constant. However, if the substrate is thin in terms of wavelength, and if the strip conductor is very narrow in terms of wavelength, then a static analysis should be perfectly adequate.

Using QuickField we can compute a simple microstrip example. QuickField is a popular 2D PDE solver that uses the finite element method [5]. The finite element mesh for a simple microstrip example is shown in Figure 11.2. It is always prudent to start with a simple test case where the answer is known. The strip is 1 unit wide and is centered on a substrate that is 1 unit thick. We will assume that εr = 10. An electrostatic FEM code subdivides the region of interest with triangles (the mesh) and solves for the potential at the vertices (the nodes). QuickField allows the user to draw the desired geometry using a graphical user interface (GUI). Although this program does not have automatic mesh refinement, the circles at several vertices indicate the desired mesh size at these points chosen by the user. The vertex cen-

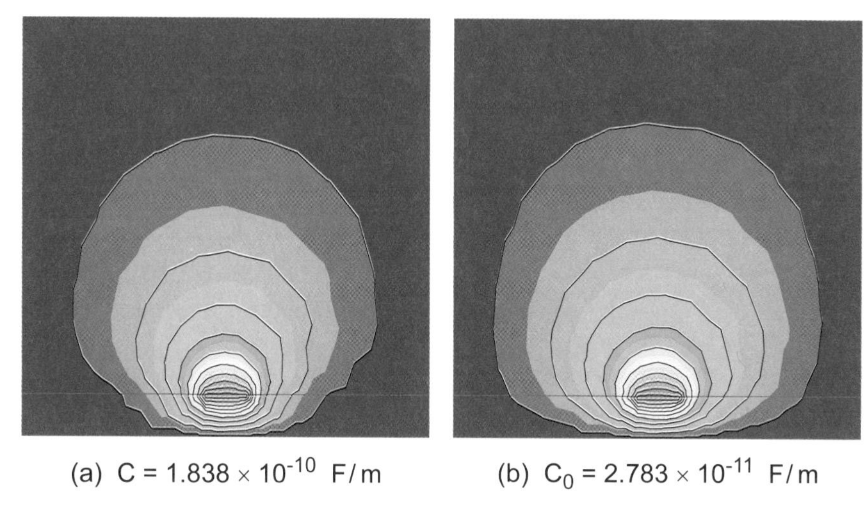

Figure 11.3 Results for the geometry shown in Figure 11.2: (a) the potential lines and capacitance per unit length with all dielectrics present, and (b) the potential lines and capacitance per unit lengths with $\varepsilon_r = 1$ for all dielectrics (QuickField Ver. 4.2T).

tered at the bottom of the box is a "dummy" vertex, which forces more mesh in the high field region under the strip.

Figure 11.3 shows the results of the two analysis runs. The strip potential was set to 1 V and the shield set to 0 V. We have plotted lines of constant potential with an increment of 0.1 V. To compute the capacitance we must integrate the charge along a closed path around the strip. The path for integration should completely enclose the center strip and theoretically its size does not matter. In practice we will see small differences in the result for different integration paths due to differences in the mesh density. QuickField will report the charge per unit length in Coulombs per meter. Because we have set the center strip to 1 V and $Q = CV$, the charge per unit length converts directly to capacitance per unit length in Farads per meter. With the capacitance values in hand we only need to do a quick calculation with a pocket calculator:

$$Z_0 = \frac{1}{c\sqrt{CC_0}} = 46.9\ \Omega \tag{11.12}$$

$$\varepsilon_{eff} = \frac{C}{C_0} = 6.6 \tag{11.13}$$

When using any numerical solution of a partial differential equation we have to constantly test the convergence of our solution. Is the mesh in Figure 11.2 "good

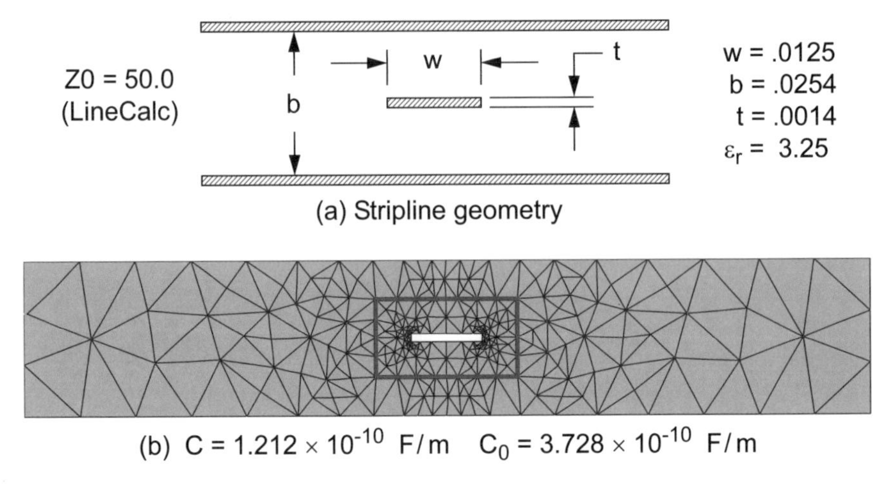

Figure 11.4 Stripline test geometry: (a) dimensions in inches, and (b) FEM mesh and computation results. The integration path is highlighted in red (FlexPDE Ver. 3.01). © 2001 IEEE [2].

enough" for 1% to 2% accuracy in impedance? Unfortunately, the only way to tell for sure is to increase the density of the mesh and solve the problem again. There are clearly any number of tools that will compute this simple microstrip case. There are also analytical equations for microstrip that we can use to check our results. Later we will look at some unusual transmission line geometries that have no simple analytical solution.

Another useful stand-alone 2D cross-section solver for partial differential equations is FlexPDE. This software also uses FEM and offers the additional feature of automatic mesh refinement. FlexPDE uses a text file input rather than a graphical user interface. At first, this may seem cumbersome, but the advantage is that you can "program" your geometry using variables and then make changes very rapidly. Or you can generate analysis files dynamically from some other program.

For demonstration purposes, we will use a simple stripline geometry that should be very close to 50 ohms. Because stripline has an exact analytical solution, it is very useful as a test case [6]. The test case geometry, dimensions, and results from FlexPDE are shown in Figure 11.4. To minimize the influence of the sidewalls, we typically place them several strip widths away from the center conductor. We could also make the sidewalls perfect magnetic conductors (PMCs) to approximate a laterally open structure.

The input file for the stripline problem is shown in Figure 11.5. A "feature" called *test* is used to set up the path for the contour integral that is needed to compute total charge. In FlexPDE the meshing algorithm also detects this feature and uses it to further refine the mesh around the center strip. Therefore, we should probably define *test* to be fairly close to the center strip and symmetrical around the center strip. We can compute Z_0 and ε_{eff} as before:

```
TITLE 'Stripline'
VARIABLES
  V                         {solve for potential}
DEFINITIONS                 {units are inches}
  w = 0.0125                {width of center conductor}
  a = 0.150                 {width of box}
  b = 0.0254                {ground plane spacing}
  th = 0.0014               {thickness of metal}
  eps0 = 8.854e-12          {epsilon sub zero, F/m}
  epsr                      {epsilon sub r, must be defined below}
  k = epsr*eps0

  E = -k*grad(V)            {definition of E field}

EQUATIONS
  div(-k*grad(V)) = 0       {Laplace's equation}

BOUNDARIES
 region 1
   epsr = 3.25
   value(V) = 0             {set outer boundary to zero volts}
   start 'box' (-a/2, -b/2)  line to (-a/2, b/2)  to (a/2, b/2)  to (a/2, -b/2)  to finish
   value(V) = 1             {set center conductor to one volt}
   start 'center' (-w/2, -th/2)  line to (-w/2, th/2)  to (w/2, th/2)  to (w/2, -th/2)  to finish
   feature                  {set path for contour integral}
   start "test" (-w, -b/4)  line to (-w, b/4)  to (w, b/4)  to (w, -b/4)  to (-w, -b/4)
MONITORS
PLOTS
  grid(x, y)                {the finite element mesh}
  contour(V)                {potential plot}
  vector(E)                 {E-field vector plot}
  elevation (normal(E)) on 'test'   {contour integral to find total charge}
END
```

Figure 11.5 Input file for the geometry shown in Figure 11.4(a) (FlexPDE Ver. 3.01).

$$Z_0 = \frac{1}{c\sqrt{CC_0}} = 49.93\ \Omega \tag{11.14}$$

$$\varepsilon_{eff} = \frac{C}{C_0} = 3.25 \tag{11.15}$$

The impedance is very close to the analytical result and $\varepsilon_{eff} = \varepsilon_r$, which is exactly the right result for a homogeneous dielectric.

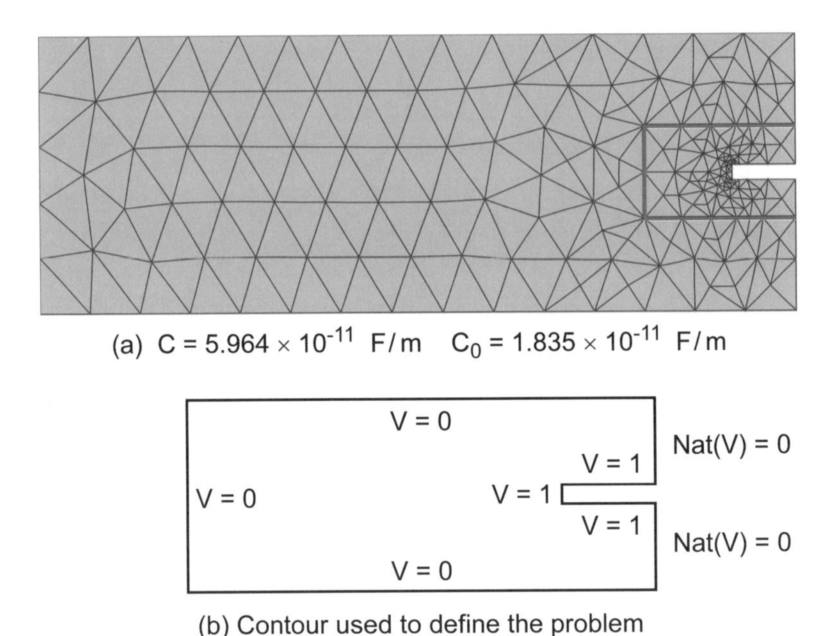

Figure 11.6 Single strip solution using symmetry: (a) FEM mesh and computed results (the integration path is shown in red); and (b) contour used to define the problem (FlexPDE Ver. 3.01).

11.2 SINGLE STRIP IMPEDANCE USING SYMMETRY

If our geometry has one or more planes of symmetry, we can speed up the computation by taking advantage of that symmetry. Computation time is usually not an issue for the 2D cross-section solvers but modeling time may be an issue. It may be quicker to define half the problem if the geometry is complex. Sometimes shareware or student versions of these solvers put a limit on the maximum number of available nodes. We can stretch the usefulness of these programs by applying symmetry to our problems. When we apply symmetry, we also have to apply a correction factor to our impedance calculations. Some vendors call this correction factor an "impedance multiplier." In Figure 11.6(a) we have used a vertical magnetic wall to divide the geometry of Figure 11.4. In this case the "feature" called *test* that sets the path for the line integral only has three segments.

There are a number of different ways this geometry can be described to FlexPDE. If you modify the original file and use a box for the outer conductor and a box for the center conductor, the software can get confused when it comes to the vertical symmetry plane. The most unambiguous way to describe this geometry is shown in Figure 11.6(b). The entire geometry is one closed polygon with the appropriate boundary condition on each segment. The magnetic wall is specified as a "natural" boundary (Figure 11.7). The computed impedance is

```
TITLE  'Stripline - using vertical magnetic wall symmetry'
VARIABLES
   V                           {solve for potential}
DEFINITIONS                    {units are inches}
   w = 0.0125                  {width of center conductor}
   a = 0.150                   {width of box}
   b = 0.0254                  {ground plane spacing}
   th = 0.0014                 {thickness of metal}
   eps0 = 8.854e-12            {epsilon sub zero, F/m}
   epsr                        {epsilon sub r, must be defined below}
   k = epsr*eps0

   E = -k*grad(V)              {definition of E field}

EQUATIONS
   div(-k*grad(V)) = 0         {Laplace's equation}

BOUNDARIES
  region 1
    epsr = 3.25
    value(V) = 0                                              {set outer boundary to zero}
    start 'box' (0, -b/2)  line to (-a/2, -b/2)  to (-a/2, b/2)  to (0, b/2)
    natural(V)=0 line to (0, th/2)                            {upper magnetic wall}
    value(V)=1 line to (-w/2, th/2) to (-w/2, -th/2) to (0, -th/2)   {set strip to 1 volt}
    natural(V)=0 line to (0, -b/2)                            {lower magnetic wall}
    finish

  feature
    start  "test" (0, -b/4)  line to (-w, -b/4)  to (-w, b/4)  to (0, b/4)  {three sides, not four}
MONITORS
PLOTS
   grid(x, y)                  {the finite element mesh}
   contour(V)                  {potential plot}
   vector(E)                   {E-field vector plot}
   elevation (normal(E)) on 'test'   {contour integral to find total charge}
END
```

Figure 11.7 Input file for the geometry shown in Figure 11.6(a) (FlexPDE Ver. 3.01).

$$Z_0 = \frac{0.5}{c\sqrt{CC_0}} = 50.71\ \Omega \tag{11.16}$$

Because we cut the geometry in half, we now have half the capacitance per unit length of the full structure. To correct this an "impedance multiplier" of 0.5 appears in the numerator of the impedance equation.

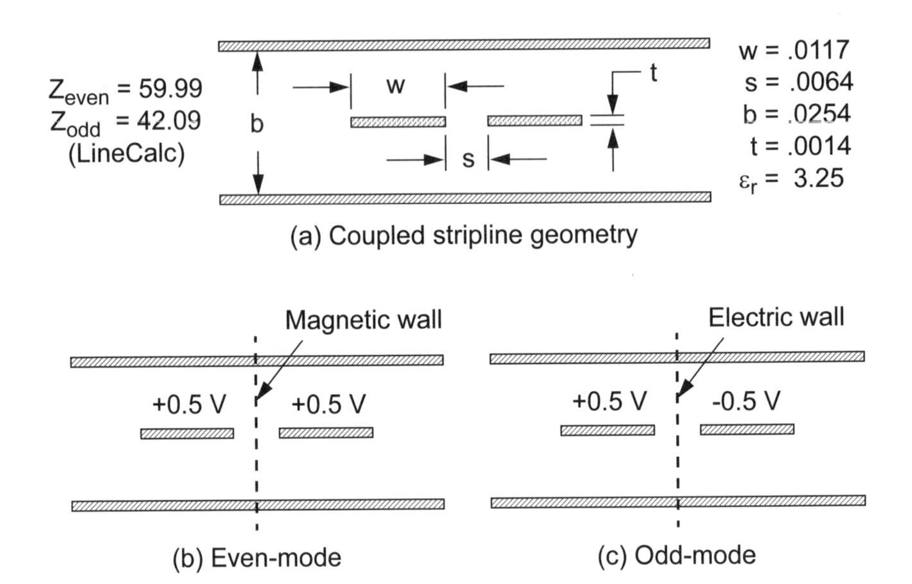

Figure 11.8 Coupled line parameters using symmetry: (a) coupled stripline geometry, (b) even-mode analysis with magnetic wall down the center, and (c) odd-mode analysis with electric wall down the center. © 2001 IEEE [2].

11.3 COUPLED LINE PARAMETERS USING SYMMETRY

Symmetry is also a useful concept when we compute coupled line parameters. We could compute the complete two strip cross-section, or we can make use of the vertical line of symmetry in most coupled strip problems. Applying symmetry to the coupled strips reduces a three conductor problem to a simpler two conductor problem. Figure 11.8(a) shows a coupled stripline example with electrical parameters computed by LineCalc.

The even-mode has equal potentials on both strips with the same sign (Figure 11.8(b)). We can place a vertical magnetic wall between the two strips without modifying the pattern of electric field lines. The odd-mode has equal potentials with opposite signs on the two strips (Figure 11.8(c)). A vertical electric wall between the two strips will not modify the pattern of electric field lines. The superposition of the even-mode and odd-mode solutions is equivalent to driving one of the strips with a 1 V source.

Like the single strip case, we need to compute C and C_0 for the even-mode and for the odd-mode, a total of four capacitance calculations. For more than two strips, we can develop a similar procedure that excites one strip at a time. But this procedure can get very tedious and you may want to consider a more sophisticated tool for more than two strips.

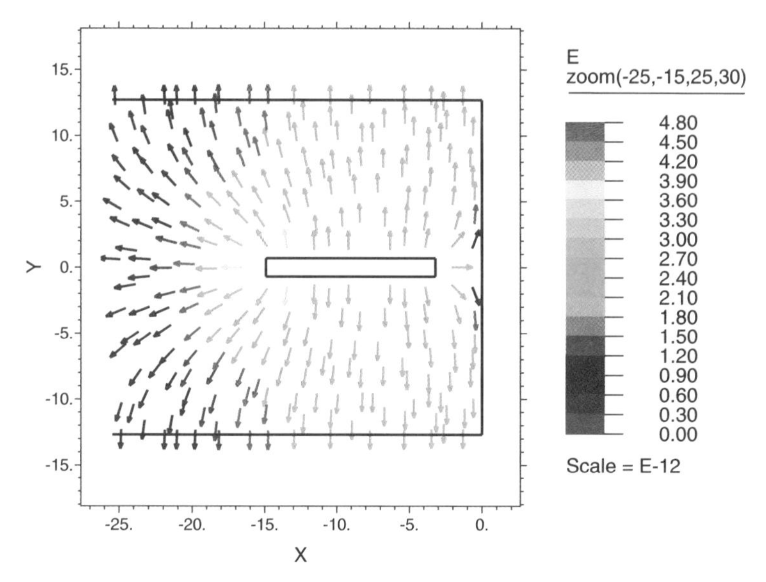

Figure 11.9 E-field vectors for the even-mode, with all lengths normalized to one value and magnitude denoted by color (FlexPDE Ver. 3.01). © 2001 IEEE [2].

The even-mode analysis proceeds as expected, computing C then C_0. We put 0.5 V on the strip, 0 V on the outer conductor, and a magnetic wall down the symmetry plane. The "feature" used to define the integral should completely enclose the strip and not touch the center symmetry plane. Figure 11.9 is a plot of the E-field vectors for the even-mode, with all the lengths normalized to one value. The color scale chosen by the software is rather confusing.

The computed electrical parameters for the even-mode are shown below. We must remember to apply the impedance multiplier, which appears in the numerator of the equation for Z_{even}.

$$C = 4.965 \times 10^{-11} \text{ F/m} \tag{11.17}$$

$$C_0 = 1.528 \times 10^{-11} \text{ F/m} \tag{11.18}$$

$$Z_{even} = \frac{0.5}{c\sqrt{CC_0}} = 60.56\ \Omega \tag{11.19}$$

The computed even-mode impedance is within 1% of the analytical value computed by LineCalc. For microstrip we would expect the even-mode ε_{eff} to be lower than ε_r.

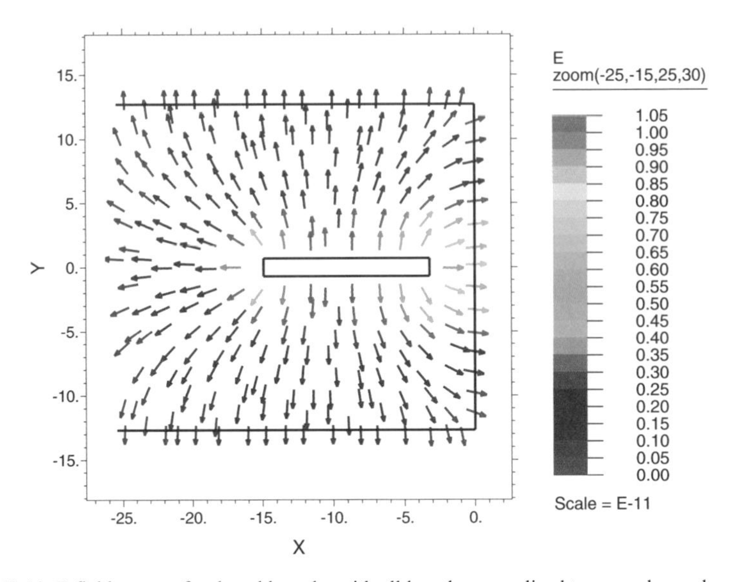

Figure 11.10 E-field vectors for the odd-mode, with all lengths normalized to one value and magnitude denoted by color (FlexPDE Ver. 3.01).

The odd-mode analysis puts 0.5 V on the strip, 0 V on the outer conductor, and an electric wall at the symmetry plane. The "feature" used to define the integral should completely enclose the strip and not touch the center symmetry plane. Figure 11.10 is a plot of E-field vectors for the odd-mode, with all the vector lengths normalized to one value. The computed electrical parameters for the odd-mode are shown below. Again, we must apply the impedance multiplier to the equation for Z_{odd}.

$$C = 7.324 \times 10^{-11} \text{ F/m} \tag{11.20}$$

$$C_0 = 2.254 \times 10^{-11} \text{ F/m} \tag{11.21}$$

$$Z_{odd} = \frac{0.5}{c\sqrt{CC_0}} = 41.05 \ \Omega \tag{11.22}$$

The odd-mode impedance computed by FlexPDE is within 2.5% of the analytical value computed by LineCalc. Because the dielectric is homogeneous, there is no need to compute ε_{eff}.

For more than two strips above a ground plane, or for any number of arbitrarily shaped conductors referenced to a ground plane, we can devise a procedure to build

Table 11.1

Partial List of 2D Field-Solvers

Stand-Alone Software - PDE Solvers
FlexPDE – PDE Solutions
QuickField – Tera Analysis
FEMLAB – COMSOL
Stand-Alone Software – Dedicated Static Solvers
Maxwell SI 2D – Ansoft
LINPAR and MULTLIN – Artech House
ApsimRLGC – Applied Simulation Technology
ELECTRO – Integrated Engineering Software
ElecNet – Infolytica
Opera-2D – Vector Fields
Integrated with Linear/Nonlinear Simulator
MCPL model – Ansoft Designer
VUSTLS model – AC Microwave LINMIC+/N
MLnCTL model – Agilent ADS

the C and C_0 matrices using a general purpose PDE solver. The static capacitances to ground and between conductors are the coefficients in a set of linear equations that relate charge to voltage.

$$\begin{aligned} q_1 &= c_{11}v_1 + c_{12}v_2 + \ldots + c_{1N}v_N \\ q_2 &= c_{21}v_1 + c_{22}v_2 + \ldots + c_{2N}v_N \\ &\ldots \\ q_N &= c_{N1}v_1 + c_{N2}v_2 + \ldots + c_{NN}v_N \end{aligned} \tag{11.23}$$

where $q_1, q_2, \ldots, q_N$ are the per-unit-length charges on the signal conductors, $v_1, v_2, \ldots, v_N$ are the voltages on the signal conductors with respect to the reference, and $c_1, c_2, \ldots, c_N$ are the per-unit-length capacitance coefficients in F/m.

If we set the first signal conductor to 1 V and all the other conductors to 0 V and solve for the charge on each conductor, we can find the first column of capacitance coefficients. Next we set the second signal conductor to 1 V, solve for the charges, and compute the second column of capacitance coefficients. All the capacitance terms on the diagonal, the so-called "self terms" are positive and all the off diagonal terms are negative. A more detailed treatment of multiconductor systems can be found in [7–9].

Although these multiconductor calculations are possible with a general purpose solver, at some point this becomes a very tedious process. For more than two conductors there are dedicated electrostatic solvers listed in Table 11.1, which will easily find L and C in matrix form. The solvers are often used to define the parameters of a high-speed digital bus. If you need to optimize multiple microstrip or strip-

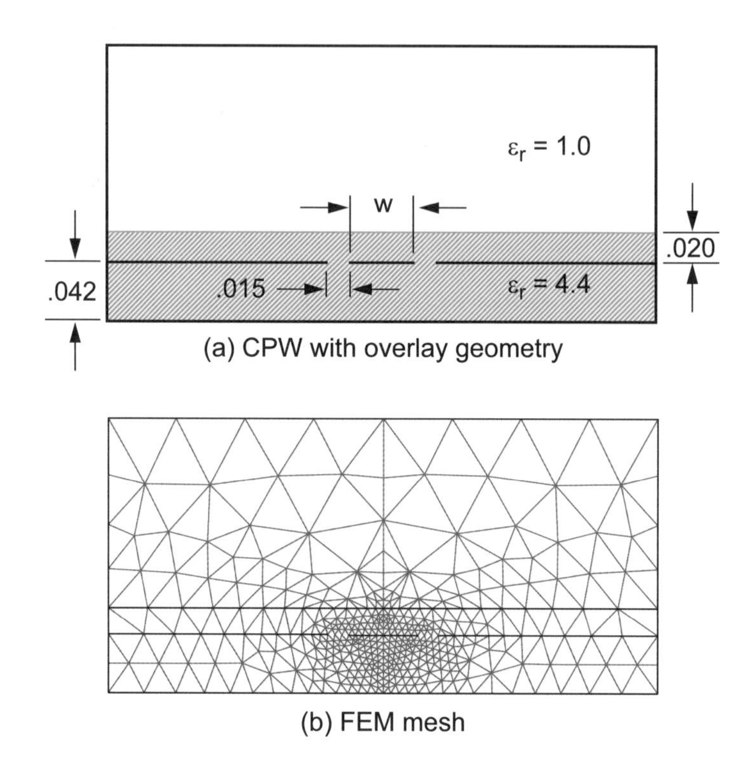

Figure 11.11 CPW with dielectric overlay: (a) geometry and dimensions (inch), and (b) FEM mesh. The mesh was fine-tuned at the edges of the strips (QuickField). © 2001 IEEE [2].

line conductors, the integrated solvers found in some circuit simulators (Table 11.1) are probably the best choice. In addition, MCPL in Ansoft Designer and VUSTLS in LINMIC+/N are full-wave solvers that include impedance and phase velocity dispersions. Both of these models have been successfully used to design microstrip filters at millimeter-wave frequencies (Section 16.2).

All of our microstrip and stripline examples so far have been quite simple, even trivial, with well-known solutions. But this is by design. These simple problems are quite useful for getting started with a new software tool before we attempt a more complicated problem.

11.4 CPW WITH DIELECTRIC OVERLAY

In multilayer PC boards we can dream up many transmission line configurations that are not addressed by standard analytical models. But if we can compute an impedance and phase velocity for the structure, we can include these nonstandard cross-sections in our circuit designs. The geometry in Figure 11.11(a) was brought

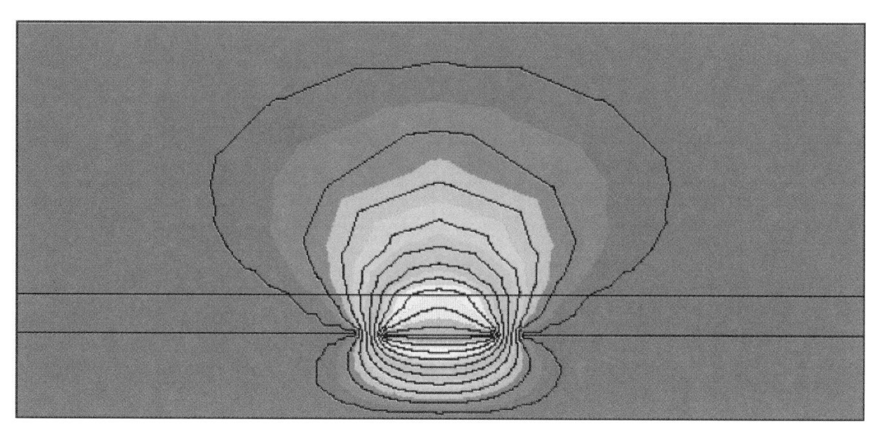

Figure 11.12 Lines and shaded contours of constant potential. The center strip is at 1 V and contour lines have an increment of 0.1 V (QuickField Ver. 4.2). © 2001 IEEE [2].

to me (D.S.) by a co-worker one afternoon. He wanted to use this CPW-like structure with dielectric overlay to route some signals on his board.

We used QuickField to compute impedance as a function of the center line width, w, while holding the gap constant. In this case we chose to ignore metal thickness. Figure 11.11(b) shows the mesh we developed for this problem. The mesh was fine-tuned at the edges of the strip to maximize accuracy.

After setting up our problem and doing the first solution, we can look at the lines of constant potential to make sure they make sense (Figure 11.12). The center strip is set to 1 V and the contour lines are plotted with an increment of 0.1 V. We computed results for several line widths with the gap held constant at 0.015 inch. The results can be found in Table 11.2.

Once we have the impedance and effective dielectric constant data, we can plug those into an ideal transmission line element in our favorite linear simulator. With a little more work, we could fit curves to the Z_0 and ε_{eff} data and program those equations into our linear simulator using the equation block feature.

Table 11.2

Results for CPW with Dielectric Overlay, Figure 11.11(a)

w (inch)	C (F/m)	C_0 (F/m)	Z_0 (ohms)	ε_{eff}
0.030	1.376e-10	3.373e-11	49.26	4.08
0.040	1.490e-10	3.720e-11	45.07	4.01
0.050	1.583e-10	3.991e-11	42.22	3.97
0.060	1.711e-10	4.333e-11	38.97	3.95

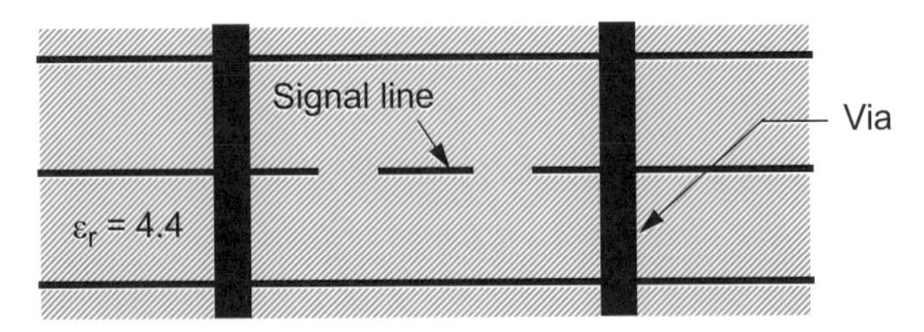

Figure 11.13 Typical buried, shielded transmission line. Only three layers of multilayer board are shown. © 2001 IEEE [2].

11.5 BURIED TRANSMISSION LINES

In multilayer PC boards we often use buried transmission lines to route RF and high-speed digital signals. To increase the isolation between lines in the same layer, we sometimes bring metal close to the signal line and connect it to the ground planes above and below with vias. Depending on the relative dimensions, we might

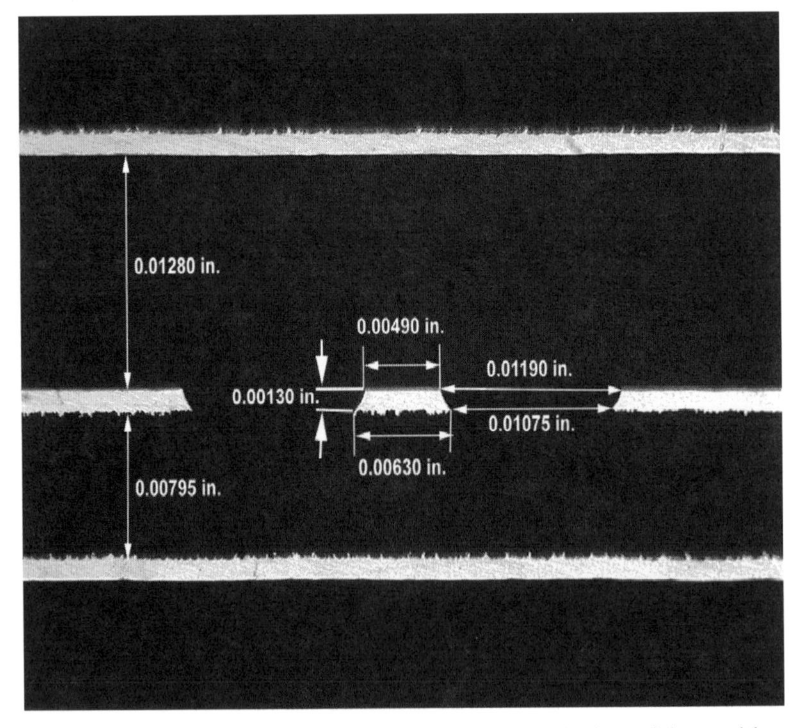

Figure 11.14 Actual cross-section from an eight-layer PC board, only three of the metal layers are shown. The trapezoidal edges are due to the etch process. The "tooth" or roughness added to the metal planes for adhesion can also be seen. Photo courtesy of M/A-COM.

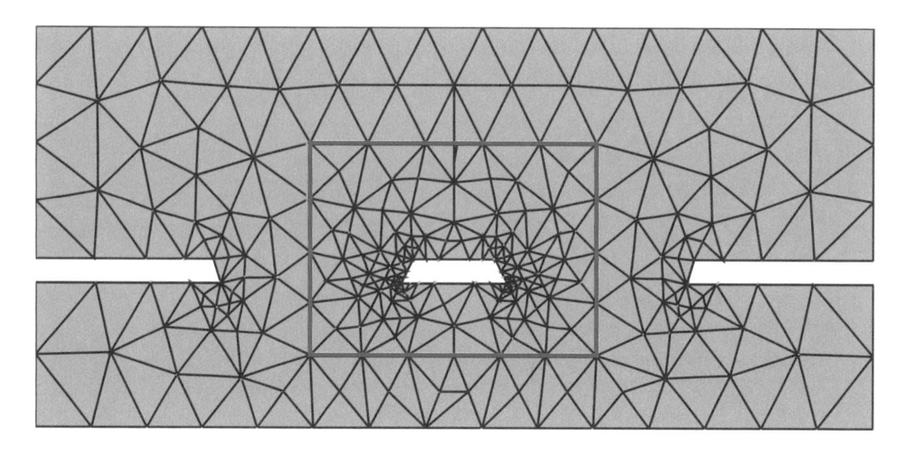

Figure 11.15 The finite element mesh for the geometry shown in Figure 11.14. The actual cross-section of the conductors are included in the analysis (FlexPDE Ver. 3.01). © 2001 IEEE [2].

call this CPW or we might call it stripline. The label we put on the geometry is less important than our ability to analyze any structure than can be manufactured at a reasonable cost. A typical cross-section is shown in Figure 11.13.

This may represent three metal layers out of a board with eight layers or more. There may or may not be vias close to the edge of what we intend to be buried ground planes. The thickness of the dielectric layers might be anywhere from 0.005 to 0.032 inch. In addition, the layers may have different thickness.

Figure 11.14 is a cross-section from an actual eight-layer PC board, but only three of the metal layers are shown. We can clearly see the trapezoidal cross-section of the etched conductors. Also, note that the distance to the upper and lower ground planes is not equal. Using FlexPDE we can easily compute the impedance of this structure for the given dimensions. The finite element mesh for this problem is shown in Figure 11.15.

We can also explore the effects of the trapezoidal cross-section on impedance. Alternatively, we might investigate the effects of various ground planes on the computed impedance. Once we are comfortable with these software tools, we are only limited by our imagination.

11.6 OTHER APPLICATIONS OF 2D CROSS-SECTION SOLVERS

The are other applications of 2D PDE solvers that do not involve impedance calculations, but are still quite interesting. Many engineers who design active devices use these solvers to calculate the parasitic capacitances of the bonding pads and other metal structures. One very sophisticated application is a combined parasitic and thermal analysis of a GaAs FET [10]. If you use complicated thin-film or thick-film resistor geometries, you can solve for the resistance and current distribution. We can also find the waveguide cutoff of a complex package cross-section with inhomogeneous, layered dielectric layers.

References

[1] Edwards, T. C., *Foundations for Microstrip Circuit Design*, Chichester, UK: John Wiley & Sons, 1981, pp. 38–40.

[2] Swanson, Jr., D. G., "What's My Impedance?" *IEEE Microwave Magazine*, Vol. 2, No. 4, 2001, pp. 72–82.

[3] Green, H. E., "The Numerical Solution of Some Important Transmission-Line Problems," *IEEE Trans. Microwave Theory and Tech.*, Vol. 13, No. 5, 1965, pp. 676–692.

[4] Lockyear, W. H., "Spreadsheets Cut Finite-Difference Computing Costs," *Microwaves & RF*, Vol. 26, No. 11, 1988, pp. 99–108.

[5] Sadiku, M., *Numerical Techniques in Electromagnetics*, Boca Raton, FL: CRC Press, 2001, pp. 334–340.

[6] Rautio, J. C., "An Ultra-High Precision Benchmark for Validation of Planar Electromagnetic Analysis," *IEEE Trans. Microwave Theory and Tech.*, Vol. 42, No. 11, 1994, pp. 2046–2050.

[7] Paul, C. R., *Analysis of Multiconductor Transmission Lines*, New York: John Wiley & Sons, 1994.

[8] Faché, N., F. Olyslager, and D. De Zutter, *Electromagnetic and Circuit Modelling of Multiconductor Transmission Lines*, New York: Oxford University Press, 1993, pp. 1–15.

[9] Brandao Faria, J. A., *Multiconductor Transmission-Line Structures, Modal Analysis Techniques*, New York: John Wiley & Sons, 1993.

[10] Budka, T. P., "Simultaneous Electrical and Thermal Modeling of GaAs FETs Using a Two-Dimensional Partial Differential Equation Solver," *Microwave Journal*, Vol. 40, No. 11, 1997, pp. 142–146.

Chapter 12

Vias, Via Fences, and Grounding Pads

In Section 10.2 we looked at various microstrip via structures. The emphasis at that point was on gaining insight into the actual current distribution and how it affects the RF performance. In this chapter we will look at some simple, single layer PCB vias and develop a more sophisticated model for the single layer via. We will also study via isolation fences and look at grounding pads for larger devices.

12.1 VIAS IN FR4

In our discussion of microstrip discontinuities we looked at vias and potential modeling errors using standard analytical models. We noted significant modeling errors at 10 GHz for vias in alumina substrates. But what about vias in FR4 at lower frequencies? Will we find the same kind of problems? Figure 12.1 shows a single layer via in a fairly thick, RF PCB. The pad geometry and via hole geometry are approximated as octagons in the closed box MoM solver. There is a plane of symmetry down the center of the geometry that we can exploit.

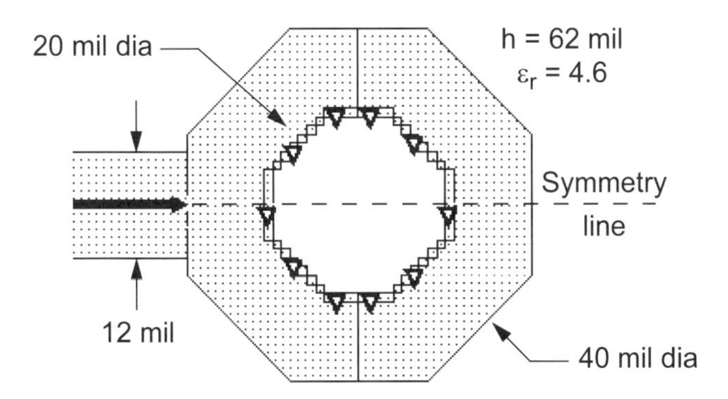

Figure 12.1 Single via to ground in FR4. The heavy black triangles indicate vertical metal down to the ground plane (Sonnet ***em*** Ver. 7.0).

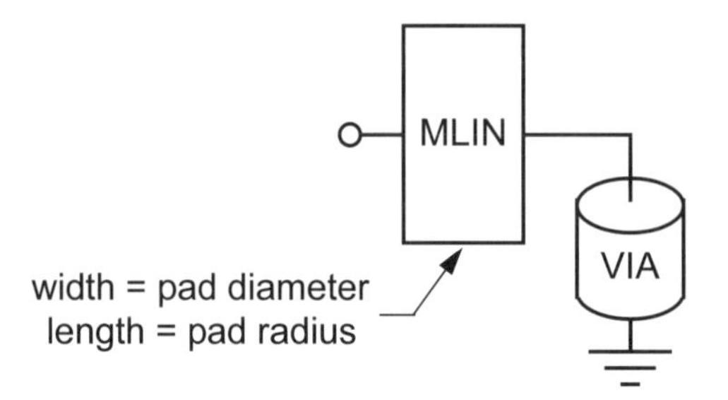

Figure 12.2 Potential analytical model for the FR4 via in Figure 12.1.

We again try to model the via with a cascade of standard analytical models, Figure 12.2. In this case we use have used a single microstrip line and the provided via model. We choose the width of the line to be the pad diameter and the length of the line to be the pad radius.

In Figure 12.3 we compare the results for this via modeling experiment. As with the vias in alumina, the analytical via model alone underestimates the inductance of the post and pad combination. However if we add a section of microstrip line with width of 40 mil and length of 20 mil to the via model, the match to the field-solver data is quite good. The thicker board material, lower dielectric con-

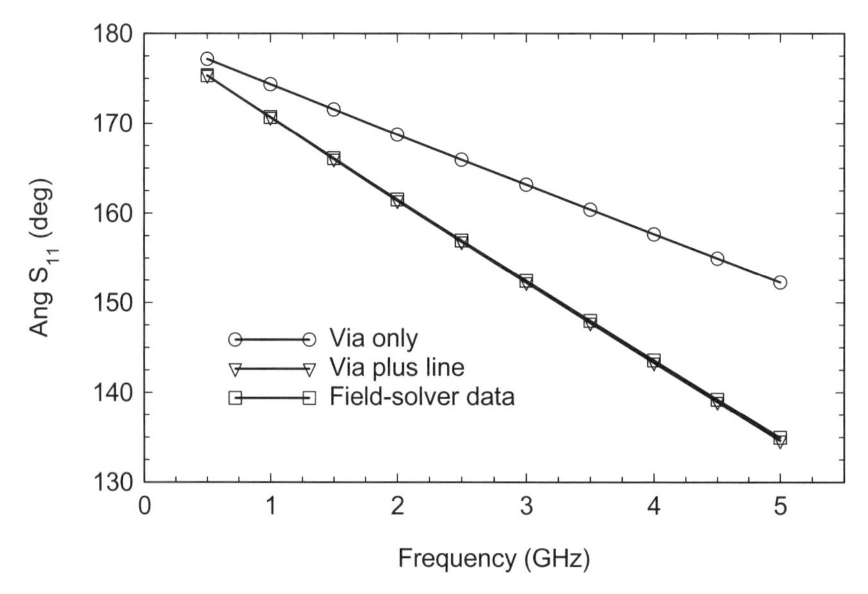

Figure 12.3 Phase angle of S_{11} for the via model alone, the combination of analytical models, and the field-solver data for the geometry shown in Figure 12.1.

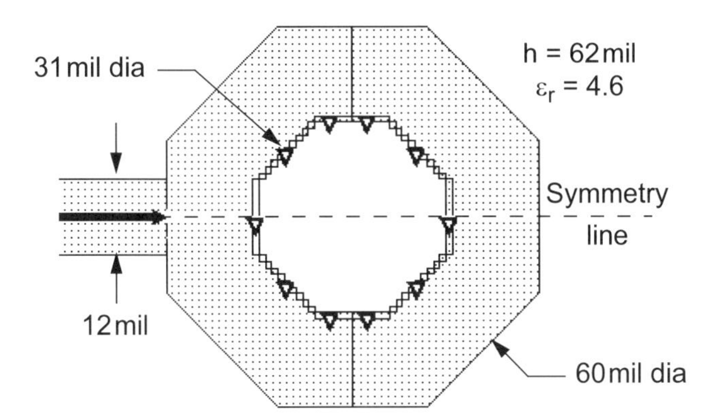

Figure 12.4 Second via to ground in FR4 experiment. The heavy black triangles indicate vertical metal down to the ground plane (Sonnet ***em*** Ver. 7.0).

stant, and lower frequency range appear to favor a cascade of analytical models compared to the vias in alumina we looked at earlier.

Before we rush to any conclusions it would be prudent to test several other similar cases. In Figure 12.4 we show a similar via problem with a larger via diameter and pad diameter. The same comments on symmetry and layout made for the previous example apply here as well. Figure 12.5 shows the comparison between

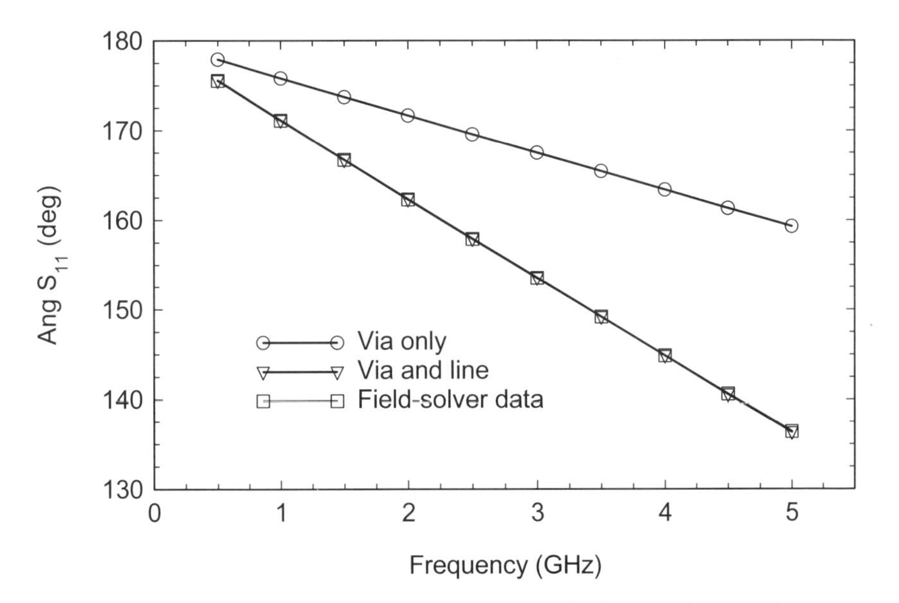

Figure 12.5 Phase angle of S_{11} for the via model alone, the combination of analytical models, and the field-solver data for the geometry shown in Figure 12.4.

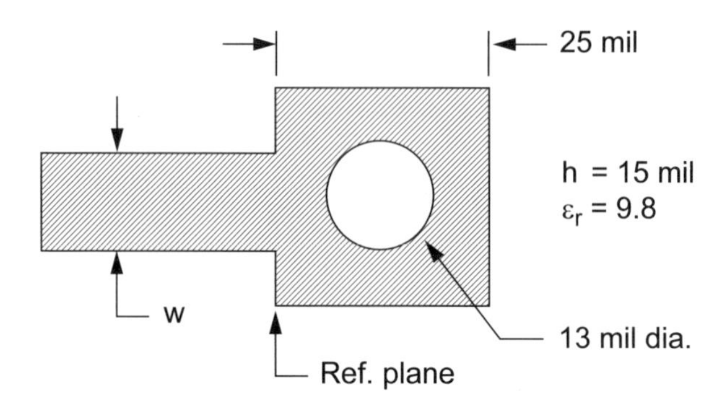

Figure 12.6 Typical via structure in a ceramic substrate. If we assume a reference plane at the edge of the pad, the equivalent circuit must include the via barrel, the pad around the via, and the step discontinuity. © 1992 IEEE [1].

the various models. Again the via model alone underestimates the inductance of the post and pad combination. However, if we add a section of microstrip line with width of 60 mil and length of 31 mil to the via model, the match to the field-solver data is again quite good.

12.2 A MORE ADVANCED VIA MODEL

In Section 10.2 we looked at several via geometries in an alumina substrate (Figure 12.6). We used a single inductor to model the via hole, the pad around the hole, and the step discontinuity into the pad. In this section we will develop a more sophisticated model of the via.

We can use any of the 2.5D or 3D field-solvers to compute the electrical parameters of this network. If we ask the solver to output *Z*-parameters we can compute the equivalent inductance of the structure with a pocket calculator.

$$X_L = j\omega l \qquad L = \frac{7.9577 X_L}{f(\text{GHz})}\ \text{nH} \tag{12.1}$$

We could also try to obtain the equivalent inductance by optimizing to a set of *S*-parameters, but we may get different results depending on the starting point we choose. In Figures 12.7 and 12.8 we show two sets of curves for a single via in 15 and 25-mil thick alumina. These are useful design charts if you are willing to interpolate a value off them over a narrow frequency range. Note the equivalent inductance varies with line width and frequency. The variation with frequency indicates that we need a more sophisticated model.

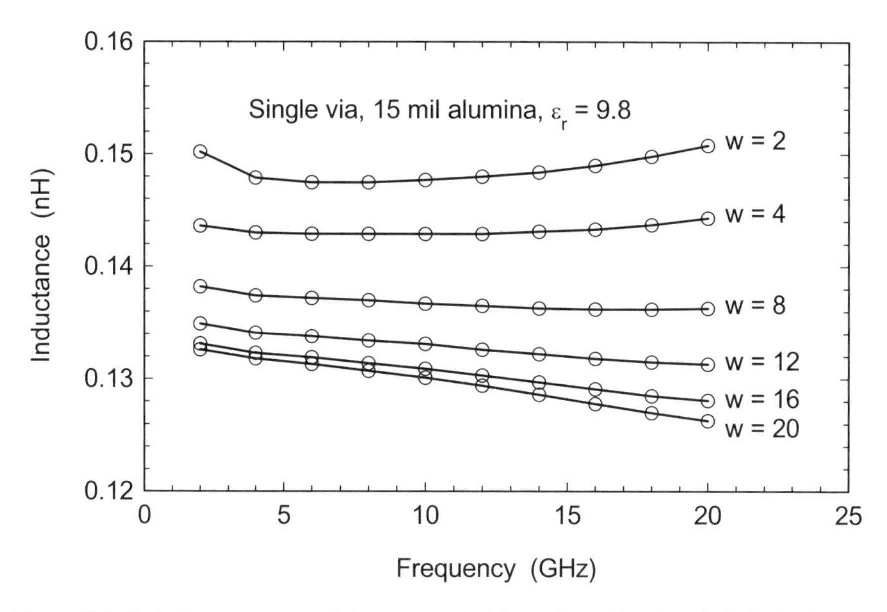

Figure 12.7 Equivalent inductance of the geometry in Figure 12.6 with a 15-mil thick alumina substrate, $\varepsilon_r = 9.8$. © 1992 IEEE [1].

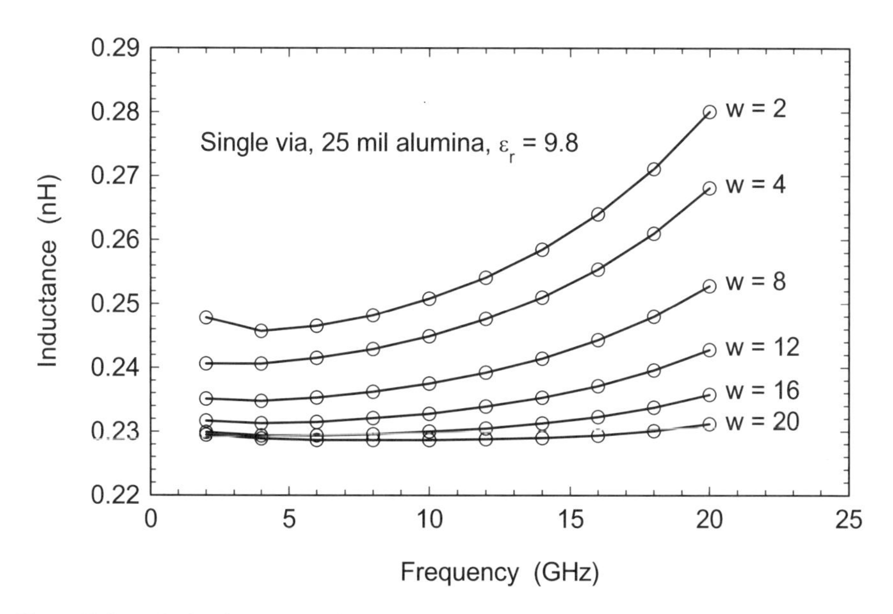

Figure 12.8 Equivalent inductance of the geometry in Figure 12.6 with a 25-mil thick alumina substrate, $\varepsilon_r = 9.8$. This thickness is typically not used above 10 GHz due to dispersion in Z_0. © 1992 IEEE [1].

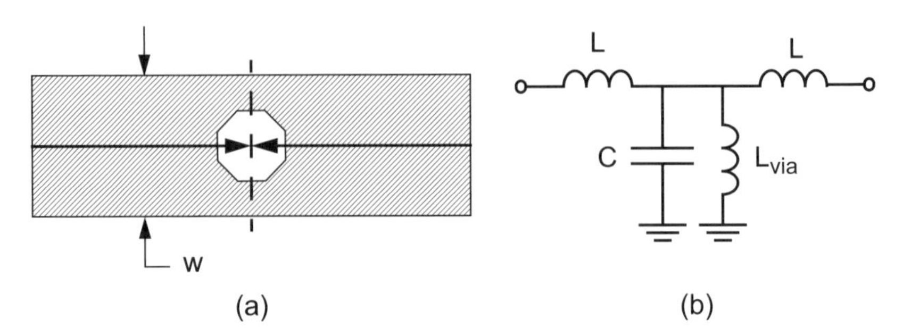

Figure 12.9 Geometry used to extract via post inductance: (a) field-solver problem, and (b) assumed equivalent circuit.

Using the field solver we can develop a more sophisticated lumped element model for the via. First let's try to determine the inductance of the post alone using a technique proposed by Goldfarb [2]. In Figure 12.9(a) we have a via embedded in a two-port microstrip line of some width, *w*. We can use a 2.5D or 3D solver for the analysis; in either case we must approximate the cylindrical via barrel with a finite number of segments. Eight to 12 segments should offer more than enough accuracy, and there is a symmetry plane down the center of the line that could be used as well.

The reference plane for de-embedding is set to the center of the hole; we are looking for the inductance of the barrel only and wish to ignore any of the planar feed structure at this stage. This structure was analyzed for two different substrate heights and two different line widths. The data was fit to the lumped element model in Figure 12.9(b) and is summarized in Table 12.1.

The good news here is that what we have identified as the via inductance only depends on the d/h ratio as we would hope. The negative series L's represent the distance from the center of the hole, back to the edge of the hole where the current

Table 12.1

Results of Via Inductance Calculation, Figure 12.9

	$w = 17$ mil	$w = 25$ mil
$h = 15$ mil	$L = -0.044$ nH $C = -0.594$ pF $L_{via} = 0.053$ nH	$L = -0.023$ nH $C = -0.711$ pF $L_{via} = 0.055$ nH
$h = 25$ mil	$L = -0.046$ nH $C = -0.231$ pF $L_{via} = 0.123$ nH	$L = -0.024$ nH $C = -0.312$ pF $L_{via} = 0.126$ nH

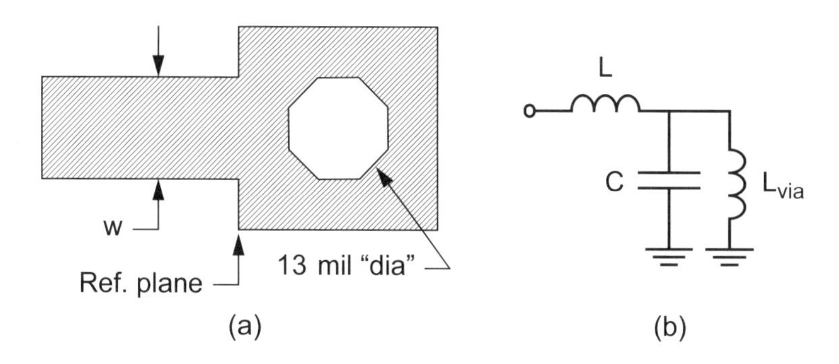

Figure 12.10 Single via model: (a) geometry analyzed with field-solver includes via barrel, pad, and a variable line width; and (b) lumped element model.

is actually going to ground. These series L's only depend on the line width. The negative shunt C is a fitting parameter that helps to adjust the slope of the reactance curve for the via. We'll concentrate on the shunt C in the next modeling step.

Now we can return to the original modeling problem. In this step we'll fit a lumped element model to the field-solver generated data for the via barrel, the surrounding pad, and a line width that varies (Figure 12.10(a)). The proposed lumped element model is shown in Figure 12.10(b). During the fitting process L_{via} is held constant while the series L and shunt C are allowed to vary. We are hoping that the remaining parameters will only be a function of the line width, w. If we had skipped the previous step which identified L_{via} uniquely, fitting the complete model by optimization would be very unstable.

In Table 12.2 we have the results for six different line widths. The series L is a very weak function of the line width; it mostly has to do with the distance from the edge of the hole to the reference plane, which is fixed in this case. Notice, however,

Table 12.2

Results of Model Fitting, Figure 12.10

w (mil)	L (nH)	C (pF)	L_{via} (nH)
5.0	0.077	0.142	0.054
9.0	0.072	0.123	0.054
13.0	0.066	0.109	0.054
17.0	0.065	0.097	0.054
21.0	0.064	0.087	0.054
25.0	0.064	0.055	0.054

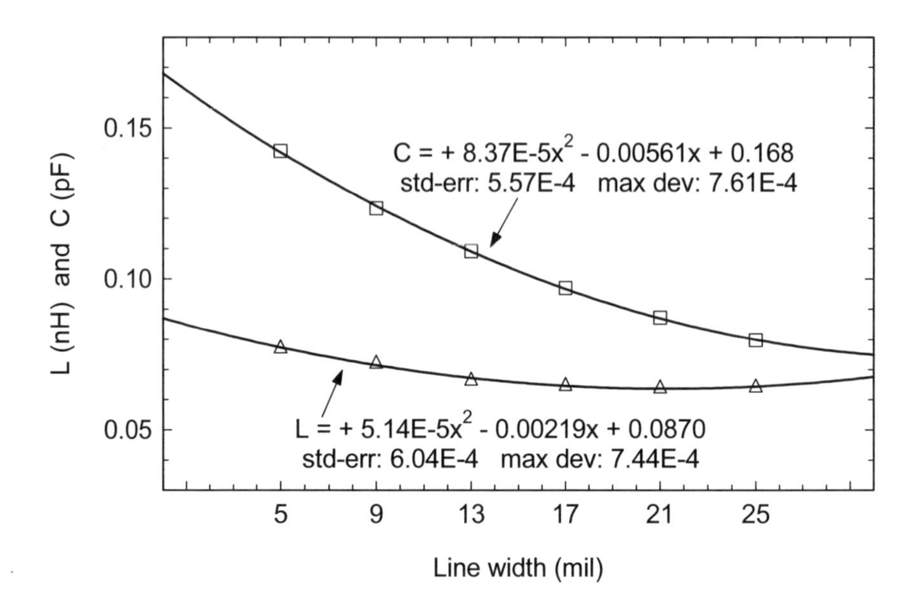

Figure 12.11 The series L and shunt C from the via model are plotted along with second order polynomial curves fit to the data.

that this series L is slightly larger than L_{via}. The shunt C represents the total pad capacitance; it decreases as the line gets wider because the fringing capacitance is reduced.

In Figure 12.11 we have plotted the series L and shunt C data as a function of the line width and also fit both data sets to a second order polynomial. If a second or third order polynomial does not fit the data well, the lumped model probably has the wrong topology or is again somehow too simple. With the resulting equations in hand, we now have a model that can easily be entered into any linear simulator using the equation block capability.

12.3 SUMMARY FOR MICROSTRIP SINGLE LAYER VIAS

We have touched on microstrip single layer vias to ground several times now. This is perhaps a good opportunity to summarize some observations. In Section 12.2 we extracted data on the via barrel alone and noted that its inductance depends only on its diameter to height ratio, as we would hope. In the same modeling study we observed that the reactance of the via pad and feed network is typically greater than or equal to the reactance of the via post. At microwave frequencies in alumina substrates we were unable to find a combination of analytical models that accurately describe the via post and pad combination (Section 10.4). However, at lower fre-

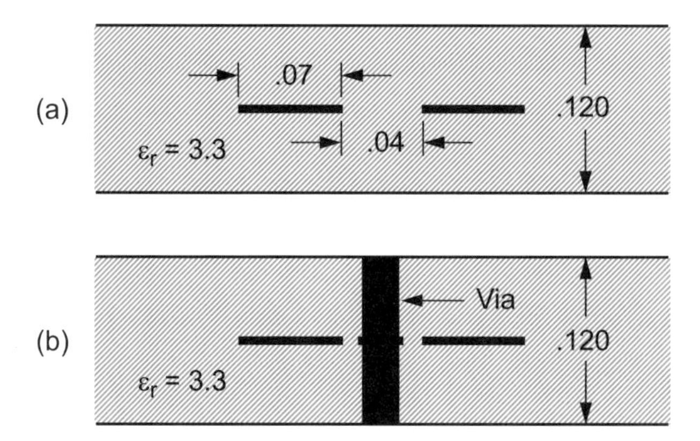

Figure 12.12 Coupled stripline isolation problem: (a) the original geometry, and (b) vias added between the strips. Dimensions are inches.

quencies and lower dielectric constant substrates a combination of via post and microstrip line analytical models seems to work well (Section 12.1).

In any given design project the geometry used for vias to ground is determined by manufacturing design rules and is relatively static. Once we obtain the field-solver data for this geometry we can use it in at least three ways. First, we can use the *S*-parameter file directly as a black box. Second, we can compute a simple equivalent inductance, which is a narrowband solution. Finally, we can develop a more sophisticated lumped element model and program it into our preferred linear simulator using the equation block feature or some other technique. This same methodology can obviously be applied to many other types of discontinuities in single layer and multilayer environments.

12.4 VIA ISOLATION FENCES—PART I

Vias are often used in multilayer environments to improve the isolation between neighboring signal conductors and to create isolated pockets within the board. Via "fences" have also been used in multilayer ceramic modules to form waveguide-like structures [3].

12.4.1 2.5D MoM Simulation

The geometry in Figure 12.12 was brought to me (D.S.) by a student in 1996. It is a simple pair of coupled striplines. The question was, would a row of vias between the lines improve the isolation and by how much? The ground plane spacing in this example is rather large by today's standards so we will also look at an example with reduced spacing in the next section.

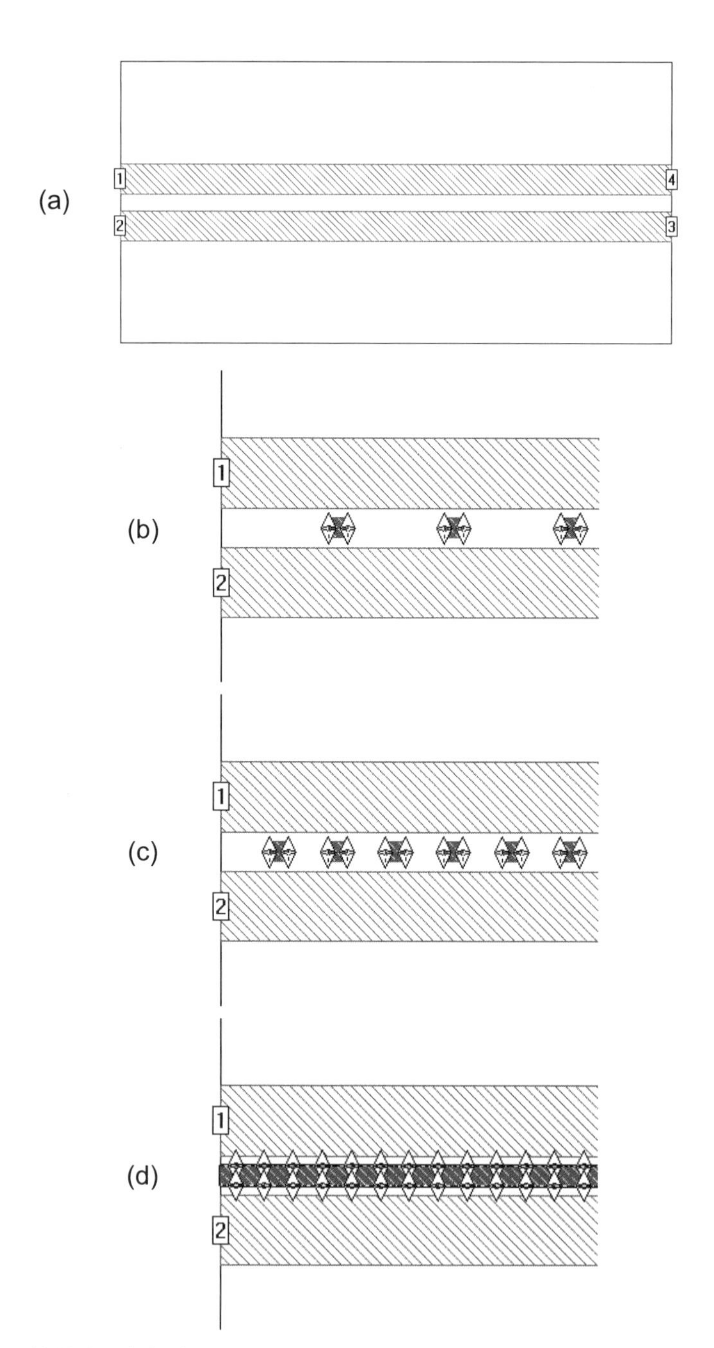

Figure 12.13 Coupled strip geometries: (a) initial geometry; (b) vias between strips, 0.12 in on center; (c) vias between strips, 0.06 in on center; and (d) wall of via metal between strips. The arrows indicate vertical via metal connecting the ground planes (Sonnet ***em*** Ver. 7.0).

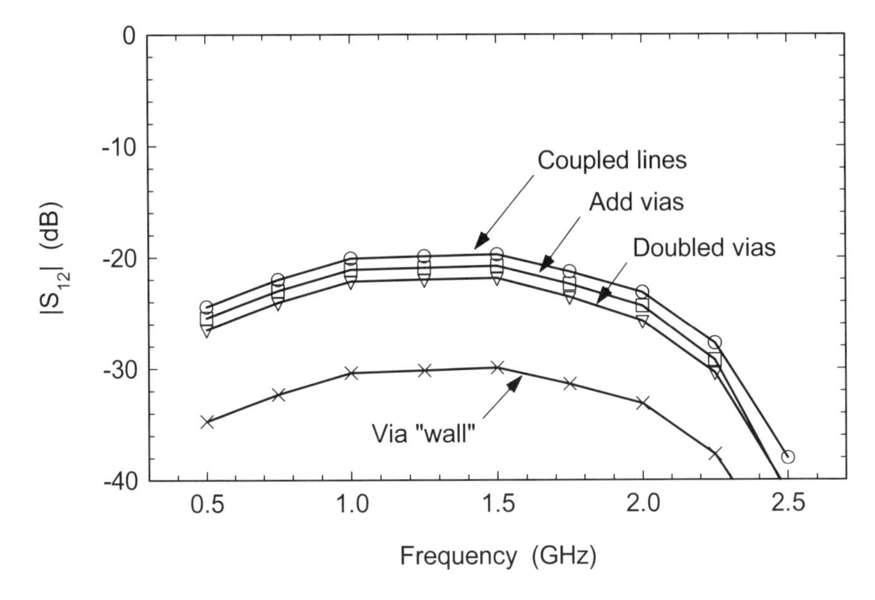

Figure 12.14 Results of the 2.5D MoM via isolation experiments in Figure 12.13.

The dimensions in Figure 12.12(a) yield a matched, 18-dB stripline coupler at 1.5 GHz. The baseline geometry is shown again in Figure 12.13(a) and the RF performance in Figure 12.14. Next we added a line of vias between the two strips. The vias were 0.02 inch square on 0.12 inch centers. In the 2.5D simulator, we put a small square of metal in the plane of the strips. We then connected that square to the cover and the ground plane using via metal. The triangles in Figure 12.13(b) indicate via metal going up and down from the plane of the strips. The surprising result of this experiment is that the isolation improves very little.

Next we doubled the number of vias between the two strips (Figure 12.13(c)). The vias were 0.02 inch square on 0.06 inch centers. Again, the improvement in isolation between the lines was very small. There may be several reasons why the vias are so ineffective. Most of the current is directed along the length of the strips. We have put very little metal in the plane of the strips to couple to this current. Also, because the vias have finite inductance, the mutual inductance they create between the strips may negate some of the shielding effects.

The improvement in isolation was so poor we were prompted to explore this problem a little further. As we add more vias, in the limit we will create a solid "wall" of vias. We would expect a solid wall to have perfect isolation. So this was the next case we tested on the field-solver (Figure 12.13(d)). Initially we expected our "wall of vias" to support much higher isolation between the two strips. But in Sonnet ***em***, the vias only support Z-directed currents, so they cannot intercept the mostly X-directed currents on the coupled strips. In effect, we have a strip of metal

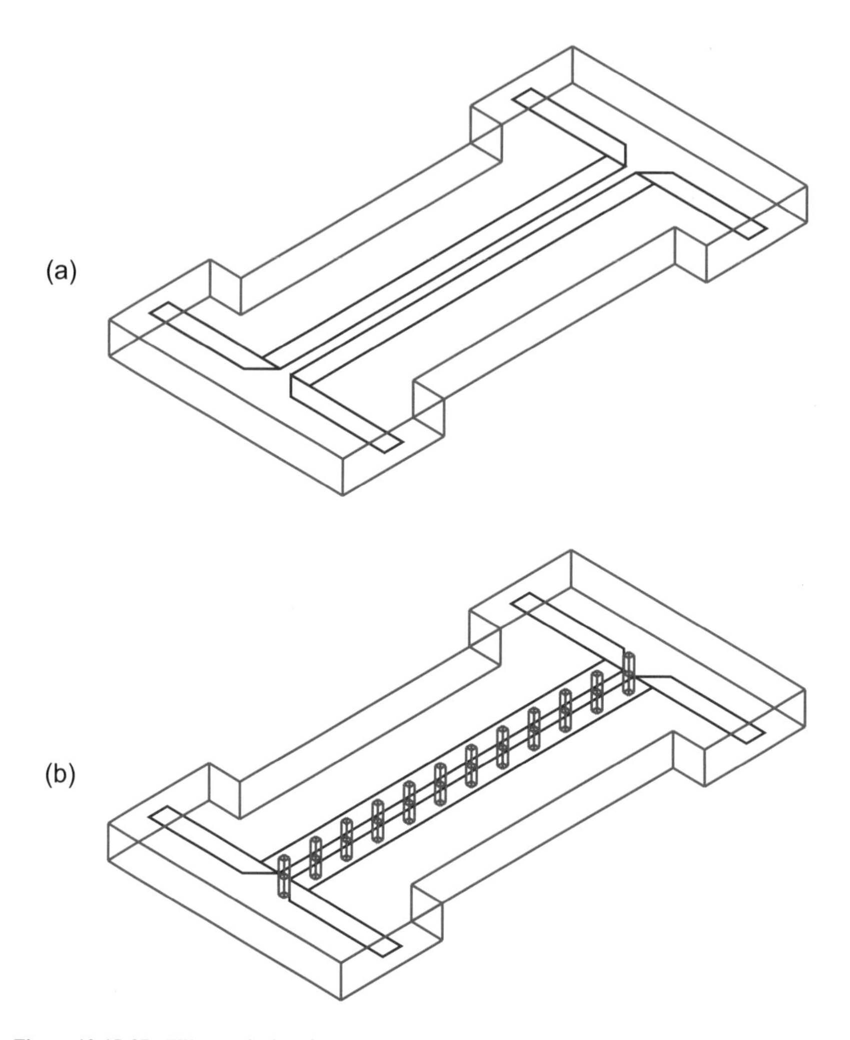

Figure 12.15 3D FEM analysis of the via isolation experiment: (a) initial geometry, and (b) vias between strips, 0.02 inch square on 0.12 inch centers (Ansoft HFSS Ver. 8.0).

in the plane of the coupled strips that is connected to the upper and lower ground planes by the via metal. What we have computed here is the noise floor of our previous experiments. This same noise floor has been observed by other users who attempt to build isolation walls within packages using this type of software.

The results for the "wall" of vias in the 2.5D simulator underline the importance of the background information on numerical methods in Chapter 3. The lack of isolation with the wall of via metal is not a bug in the code, it is simply a limitation of the formulation.

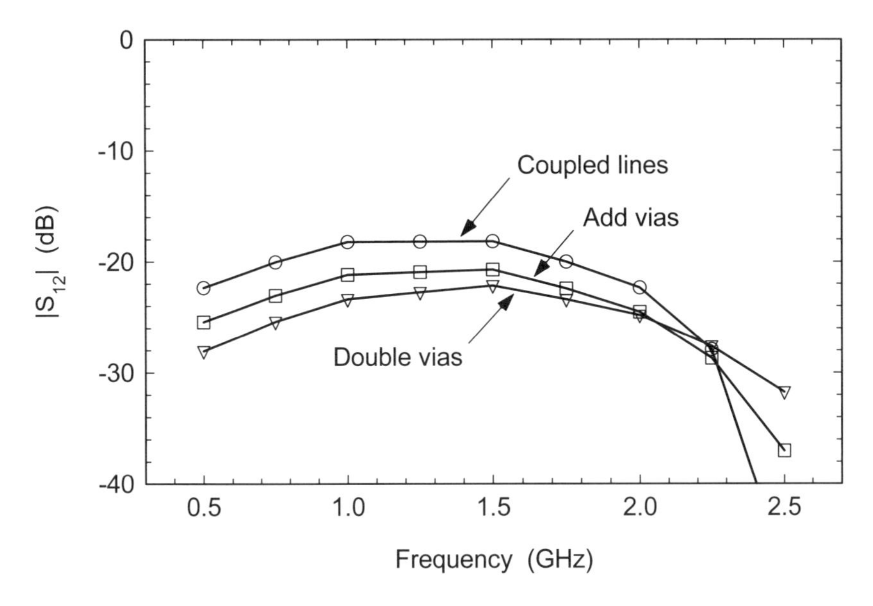

Figure 12.16 Results of the 3D FEM via isolation experiments in Figure 12.15.

12.4.2 3D FEM Simulation

The numerical noise floor for the 2.5D MoM set of experiments was fairly high. Just to check our results we repeated this set of experiments using a 3D FEM field-solver. Using the 3D solver we expect fewer uncertainties regarding the interaction of the vias and the strips. On the negative side, the time required to create the models is higher and the solution time is higher. Figure 12.15 shows two of the geometries we created using Ansoft HFSS.

As before, we get the coupled response we expect from the initial geometry (Figure 12.16). Now we add a row of vias to the structure (Figure 12.15(b)). As before, the vias add very little isolation to the structure. Now we double the number of vias between the two strips. The improvement in isolation is similar to what we computed with the 2.5D solver (Figure 12.14). Overall, we get a very similar result using the 3D solver compared to the 2.5D solver. The computed improvement in isolation is slightly greater using the 3D solver. If we add a metal wall between the two strips using the 3D solver we get a computed isolation of greater than 200 dB. The numerical noise floor in this case is not an issue.

This example has several interesting aspects. First, we explored the idea of using vias to isolate a pair of coupled lines. The improvement in isolation is surprisingly small. But doing our experiments on the computer probably saved us some expensive mistakes in the lab. Second, we found that using one of the 2.5D solvers, we had a fairly high "noise floor" for our experiments on the computer. If we

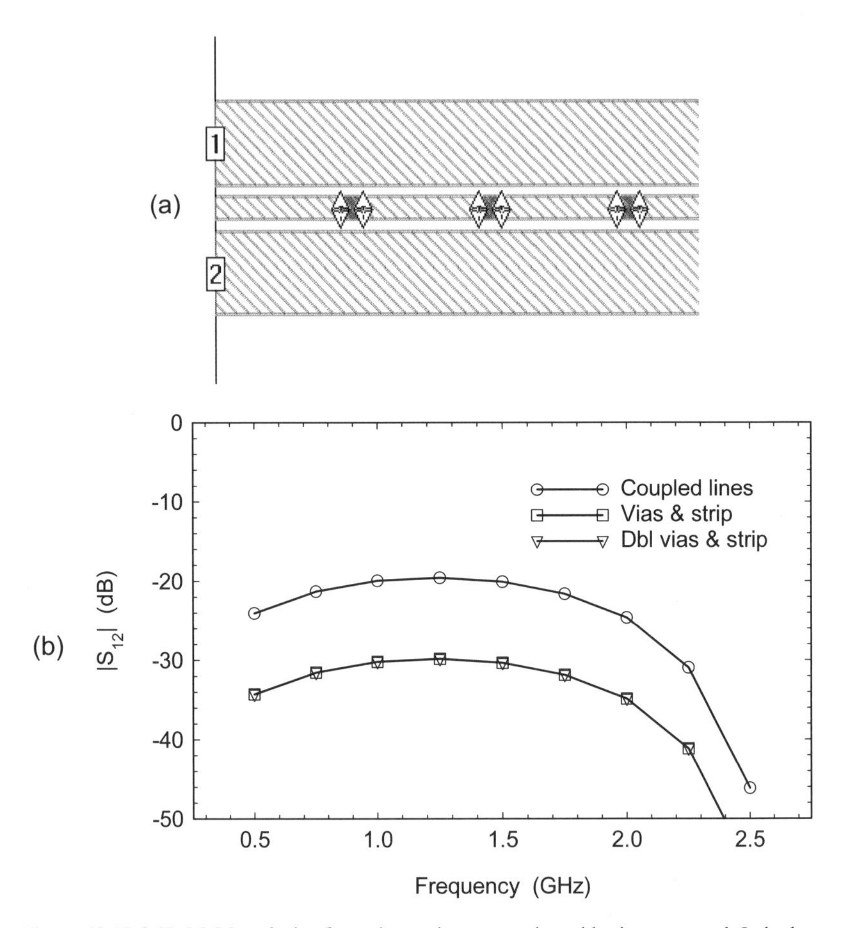

Figure 12.17 2.5D MoM analysis of guard trace between strips with vias to ground. In both cases the analysis reaches the noise floor of the analysis and stops (Sonnet ***em*** Ver. 7.0).

expected results below this noise floor we would clearly be misled. We can explain the results we obtained by understanding how the software works. Finally, we "checked our work" by using a 3D field-solver. We computed comparable isolation numbers with the 3D solver where the noise floor was much lower.

12.5 VIA ISOLATION FENCES—PART II

The previous example has generated a fair amount of discussion and comment since it was first presented. Several students and colleagues have suggested that the vias should be connected together in the plane of the strips. Recently, we had the opportunity to run these new experiments.

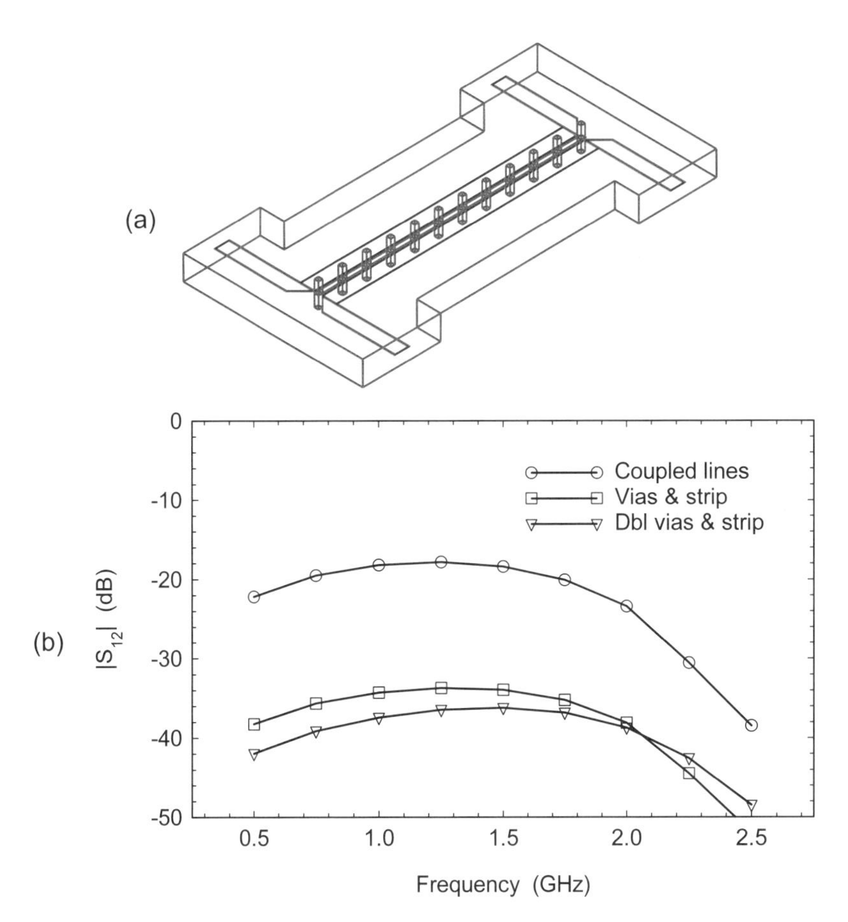

Figure 12.18 3D FEM analysis of guard trace between strips with vias to ground. The addition of the guard trace has improved the isolation (Agilent HFSS Ver. 5.6).

The first experiment was to add the strip to the Sonnet ***em*** analysis. Three runs were done: the initial geometry, adding 10 vias with a shorting line, and adding 21 vias with the shorting line. Figure 12.17(a) shows a detail of the geometry with 10 vias and the shorting strip. As before, the pair of lines alone has about 20 dB of coupling at 1.5 GHz. When the vias and shorting line are added, in both cases the isolation drops to the noise floor of the analysis and stops (Figure 12.17(b)).

In the next experiment we repeated the analysis using 3D FEM. In this case we used Agilent HFSS. The same three runs were done with the coupled strips alone, then vias and grounding strip. Figure 12.18(a) shows the complete geometry for the case with 10 vias and a shorting strip plus the RF results for all three cases. Again, the coupling for the coupled pair alone is around 18 dB at 1.5 GHz. However, now we can see a more realistic view of what the isolation structure might be doing. We

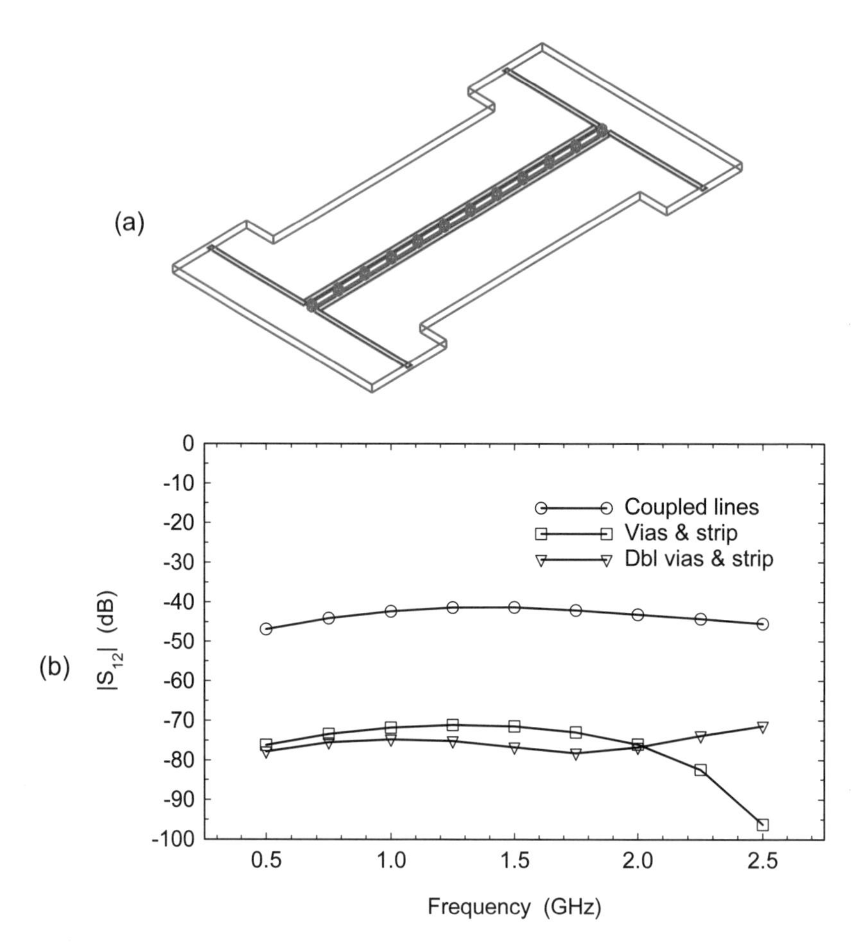

Figure 12.19 3D FEM analysis of isolation experiment with reduced ground plane spacing (Agilent HFSS Ver. 5.6).

have at least 10 dB more isolation with the shorting strip. And we get a little more isolation at most frequencies when we double the number of vias.

In the original via isolation example the ground plane spacing was quite large. Layer thicknesses for typical RF and high-speed digital PCBs today tend to be much thinner. The next analysis reduces the ground plane spacing from 0.120 to 0.030 inch (Figure 12.19(a)). The strip widths we changed to 0.018 inch to maintain a Z_0 of 50 ohms. The original gap between the strips was maintained, as was the dimensions of the vias and the shorting strip. The starting point for coupling is now around –40 dB and we can improve the isolation by about 30 dB (Figure 12.19(b)). Note there is little difference between the two different via densities. In a real PCB achieving this level of isolation consistently is quite difficult.

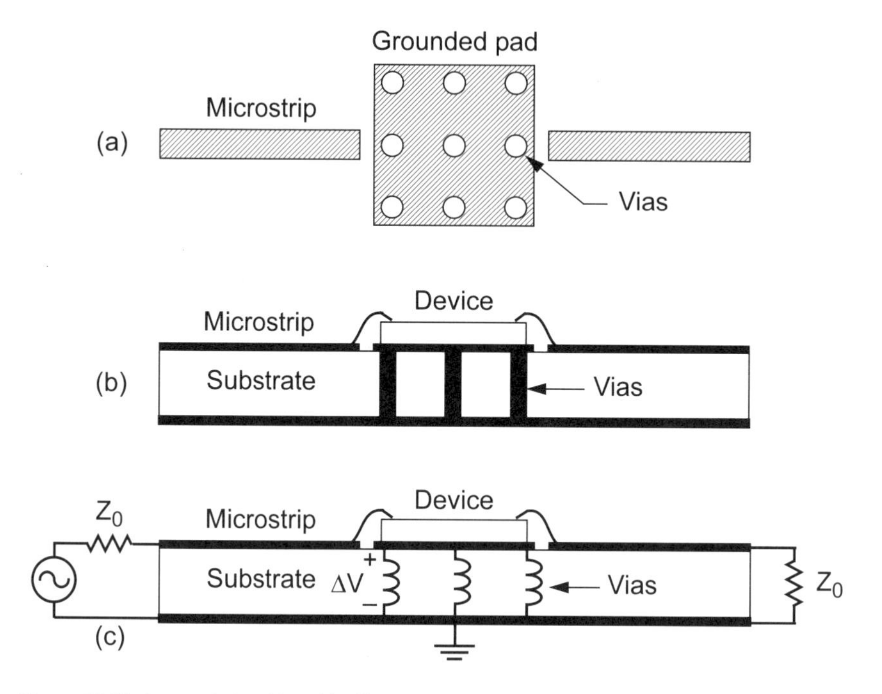

Figure 12.20 A grounded pad in a thin-film substrate: (a) top view of geometry, (b) side view of device mounted on pad, and (c) schematic of device and grounding structure.

This last set of examples is typical of what really happens when we attack a new problem. The first results are often not exactly what we expected and we begin to design "what if" experiments. Adding the shorting strip between the vias did improve the isolation. We also saw the same "noise floor" in the 2.5D analysis. Reducing the ground plane spacing is much more effective for improving isolation. Basically, as the spacing is reduced, the field lines terminate on the ground planes more quickly. Achieving 60 to 80 dB isolation in real PCBs can be quite difficult. Any signal via in the board can potentially launch parallel plate modes between the ground planes. Once this energy gets into the board it goes everywhere and it is very difficult to eliminate.

12.6 GROUNDING PADS

When we need to connect a device to ground in a thin-film hybrid circuit or a multilayer PCB, we often use a metal pad with several vias to the nearest ground plane. For small diode and transistor chips this may be a small pad with just one or two vias. For MMIC chips the metal pad may be larger with several vias to case ground. We use the same techniques in multilayer surface mount boards to connect pack-

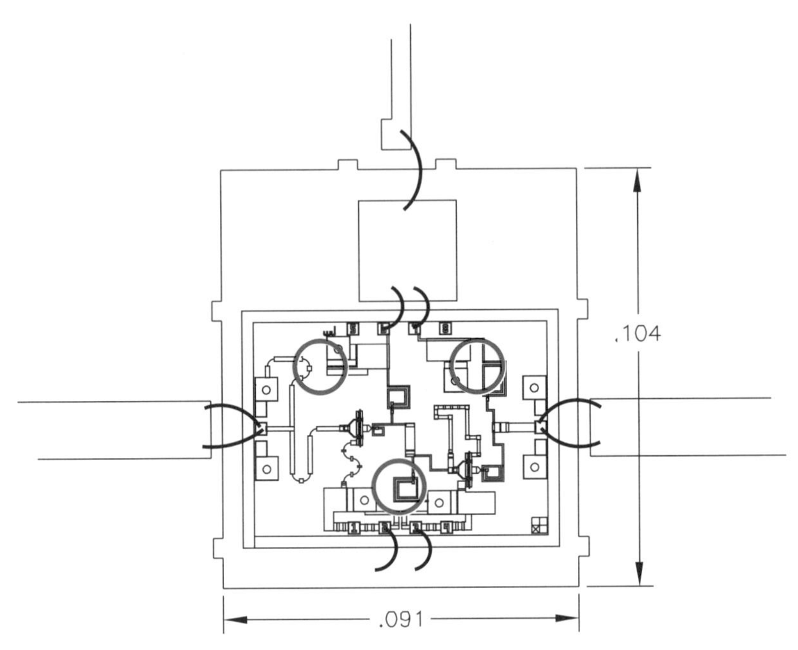

Figure 12.21 A two-stage MMIC amplifier with bypass capacitor mounted on a grounded pad. The heavy red circles indicate the locations of the vias from the pad to case ground. © 1993 Horizon House Publications [4].

aged active and passive devices to case ground. Smaller active devices need connections to ground in amplifiers and oscillators. Large MMIC devices, like converter chips, also need connections to case ground. And there are some fairly large surface mount devices like mixers, filters, and couplers that also need connections to ground. A very simple example of this type of problem is shown in Figure 12.20(a, b).

The problem is that as soon as we throw down a few vias in our layout, we assume we have an ideal connection to case ground. In fact, the vias have a finite reactance and we never have an ideal connection to case ground. The vias are a "common mode" element and can very effectively couple the input to the output in this case (Figure 12.20(c)). The input signal travels down the input microstrip line and drives the MMIC chip. The return current for signal passes through the chip and through the grounding structure near the input of the chip. Because the vias have a finite resistance and reactance the return current creates a voltage across the grounding structure. This voltage can now induce a current in the output loop. An oscillator designer would recognize this as a series feedback network.

The following is a case study [4] that demonstrates how a "simple" grounding problem can impact a design. Figure 12.21 shows a two-stage MMIC amplifier and bypass capacitor mounted on a 91 by 104-mil microstrip pad. The substrate is 15-

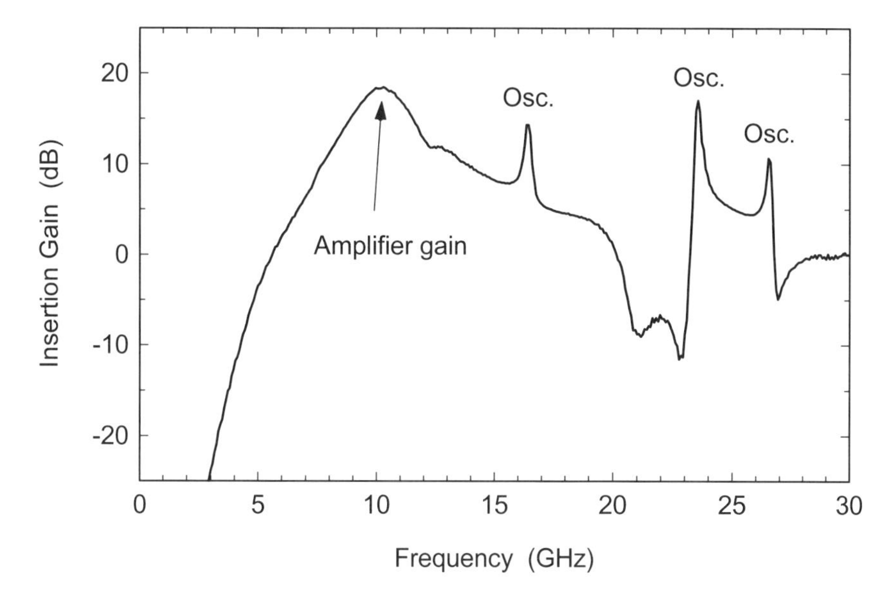

Figure 12.22 Broadband response of the MMIC amplifier shown in Figure 12.21. The desired gain region is a narrow band near 10 GHz. The circuit also oscillated at 16.5, 23.5, and 26.5 GHz. © 1993 Horizon House Publications [4].

mil thick alumina and the nominal center frequency of the amplifier is 10 GHz. There are three solid vias, 13 mil in diameter (heavy red circles), connecting the mounting pad to case ground. The substrate via locations were chosen to fall close to the vias on the MMIC chip. This was a special filled and planarized via process that allowed the MMIC chip to be mounted over the via holes. If hollow via technology is used, there is typically build-up of metal around the edges of the holes, which occurs during plating. Hollow vias are typically placed along the upper and lower edges of the mounting pad and the MMIC chip would not overlap any via holes for reliability reasons. The circuit designer felt this via hole placement would guarantee a good connection to ground.

When this amplifier was tested in an open test fixture (no cover), spurious oscillations were found at 16.5, 23.5, and 26.5 GHz (Figure 12.22). Wafer probe measurements of this chip did not reveal any of these oscillations. Nor could oscillations be found with the MMIC amplifier mounted directly to a metal carrier with microstrip lines at the input and output. So attention was focused on the microstrip grounding structure.

A new substrate with no components attached was measured as a two-port network by connecting the input and output microstrip lines directly to the grounding pad. Because the input and output lines of the MMIC chip induce return currents in the microstrip pad at these points, driving the pad itself at these points seems justi-

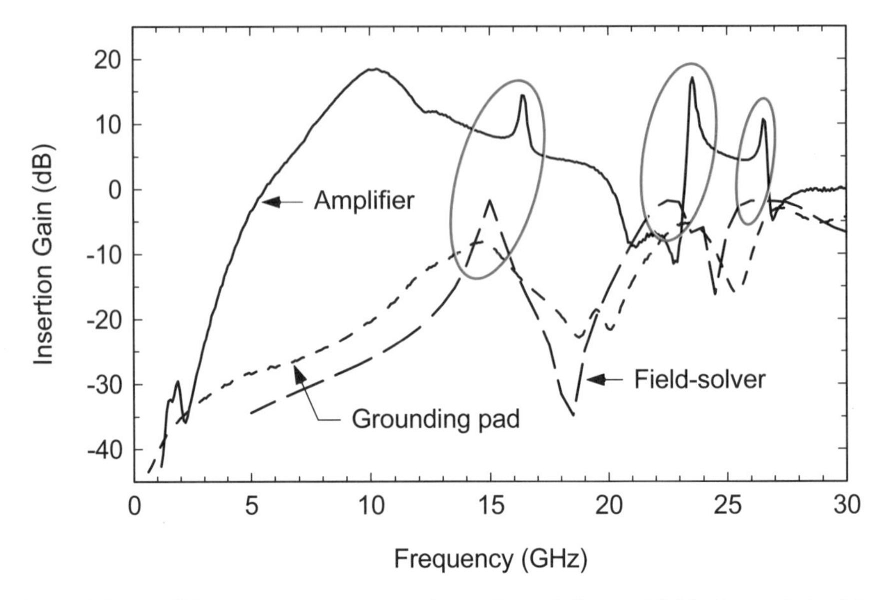

Figure 12.23 Amplifier response, measurement of grounding pad alone, and field-solver analysis of the grounding pad geometry. The peaks in grounding pad response correspond to the amplifier oscillation frequencies. © 1993 Horizon House Publications [4].

fied. However, most circuit designers would probably not make this measurement, assuming that the pad is a short-circuit and there is nothing of interest to measure. Although the correlation is not exact, the resonances measured under these conditions, shown in Figure 12.23, closely match the oscillation frequencies seen in the MMIC amplifier. The measured passive structure has a strong resemblance to a microstrip patch antenna modified by the presence of via holes.

It became obvious that the passive structure alone could be analyzed and maybe even optimized using an electromagnetic field-solver. A very coarse approximation was used, the grid size was set to 5 mil and the vias were approximated by 10 by 10-mil square posts. The results of this first analysis are also shown in Figure 12.23. The correlation with the passive circuit measurement and the active circuit oscillations was good enough to merit further investigation.

Plots were made of the current distribution on the microstrip pad for this initial case (Figure 12.24). At 10 GHz, the currents are effectively terminated by the vias and the grounding structure performs as we would expect. At 15 GHz, the upper half of the grounding pad behaves like an open-circuit shunt stub. The grounding structure resonates and raises the ground reference of the MMIC chip. The resonance at 23 GHz is similar to a patch antenna type mode where the edges of the patch antenna are radiating. The resonance at 26 GHz is another patch antenna type mode. At the time, seeing these current distributions was truly a revelation. Without

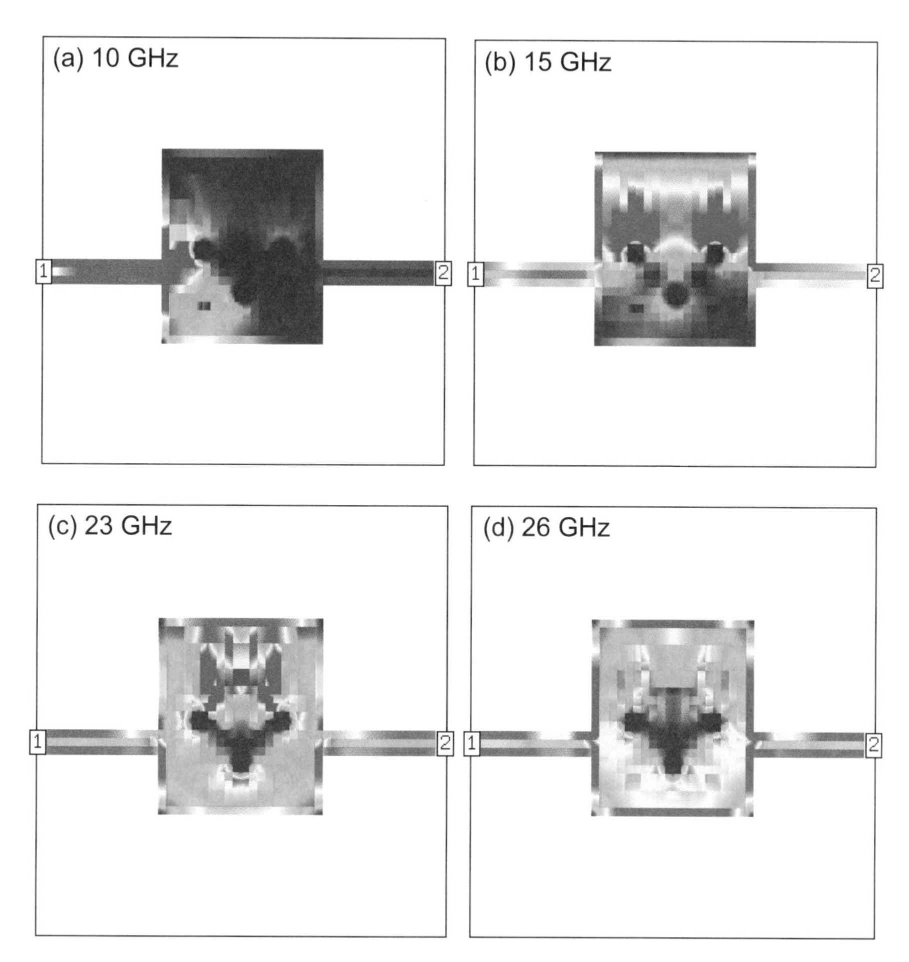

Figure 12.24 Conduction current distributions on the microstrip grounding pad: (a) the current at 10 GHz is effectively terminated by the vias; (b) an open-circuit stub type resonance at 15 GHz; (c) a patch antenna type mode at 23 GHz; and (d) another patch antenna type mode at 26 GHz (Sonnet ***em*** Ver. 3.0). © 1993 Horizon House Publications [4].

the field-solver it would have been impossible to draw these current distributions and predict the resulting resonant frequencies. After studying the various modes on the microstrip pad, it was hypothesized that an optimum set of via hole locations could be found.

A systematic search began with a via in each of the four corners of the pad. Figure 12.25(a) shows the new via hole locations. The insertion gain predicted by the field-solver can be seen in Figure 12.25(e). There is a weak mode at 17 GHz where some current is coupled along the bottom edge of the pad (Figure 12.25(b)).

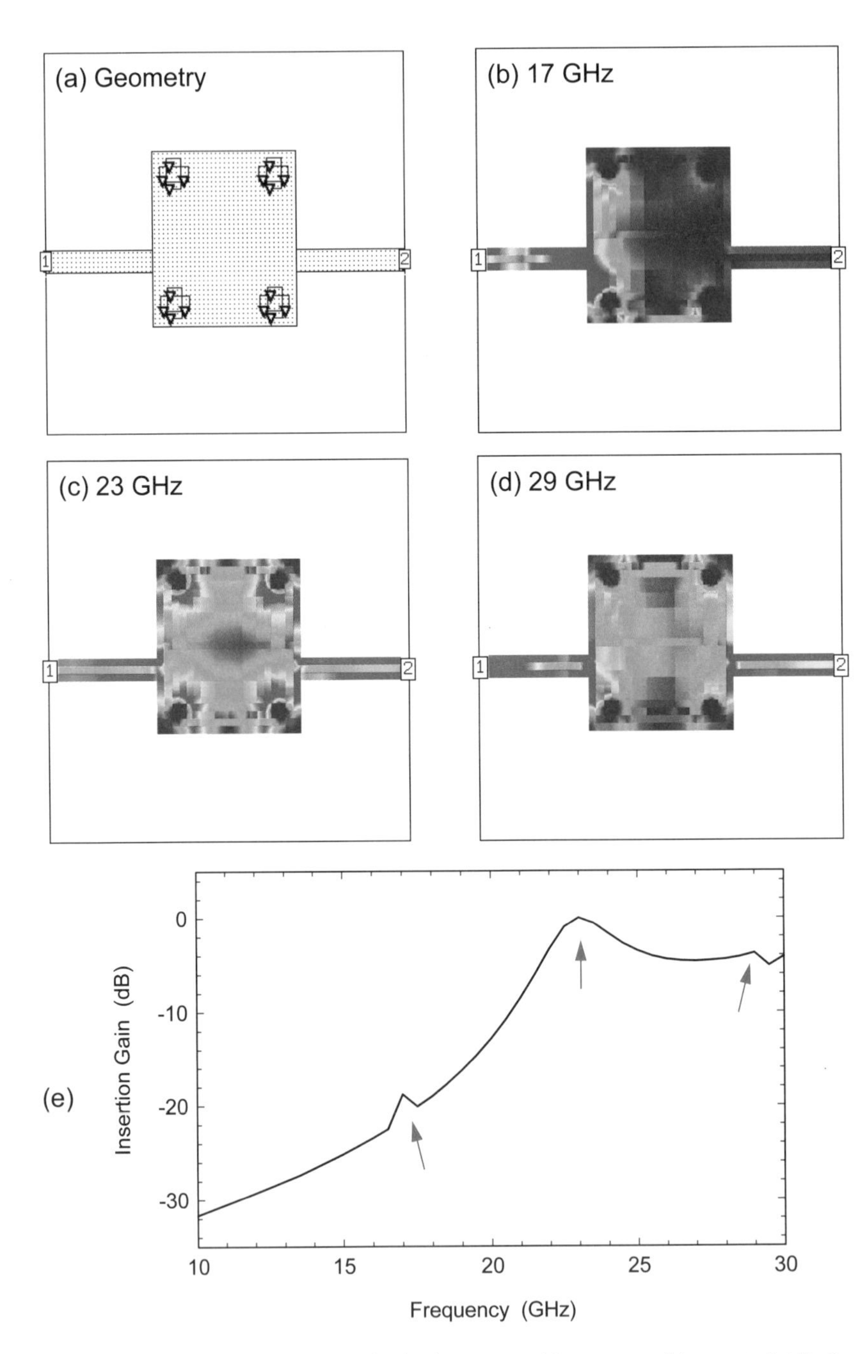

Figure 12.25 Grounding pad with via holes in the corners: (a) geometry; (b) current distribution at 17 GHz; (c) current distribution at 23 GHz; (d) current distribution at 29 GHz; and (e) insertion gain of this configuration. © 1993 Horizon House Publications [4].

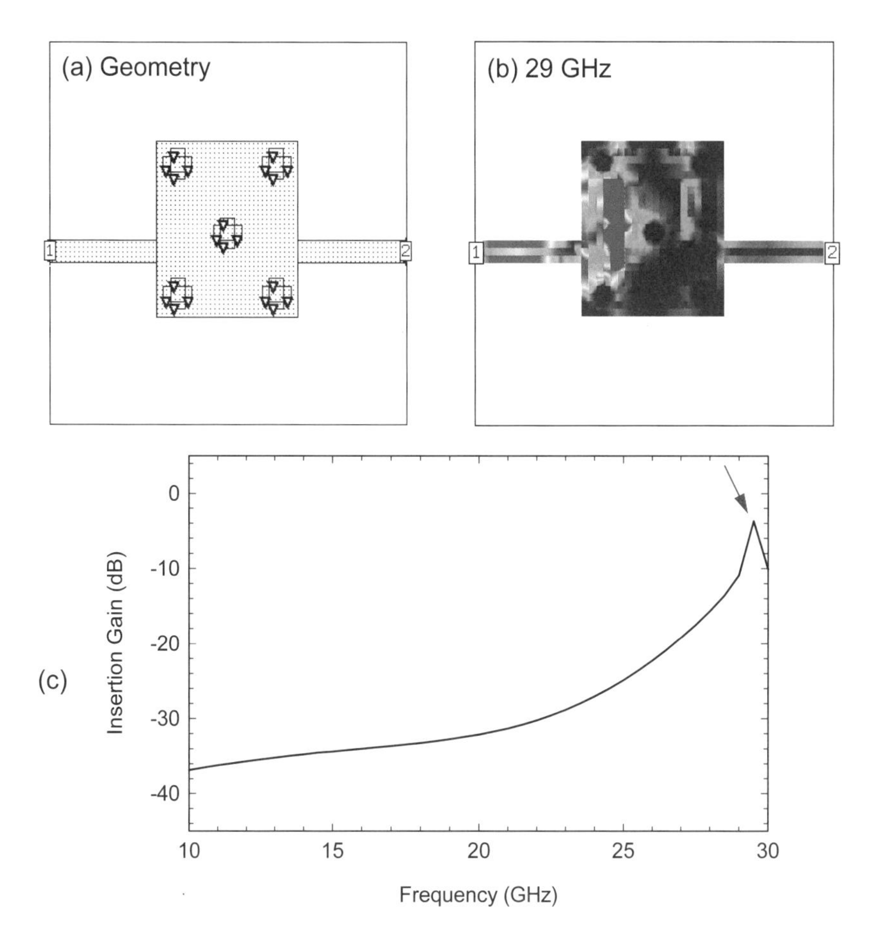

Figure 12.26 Grounding pad with five hole pattern: (a) geometry; (b) current distribution at 29 GHz; and (c) insertion gain of this configuration. © 1993 Horizon House Publications [4].

At 23 and 29 GHz, there are still substantial currents on some edges of the pad, and the input and output are strongly coupled (Figures 12.25(c, d)). At this stage, a voltage maximum was assumed to exist in the center of the microstrip patch. Therefore, the next via was placed in the center of the pad.

Figure 12.26(a) shows the new via locations, and Figure 12.26(c) shows the insertion gain predicted by the field-solver. All but one resonance at 30 GHz has been eliminated. The current plot for this final resonance is shown in Figure 12.26(b). The current maximizes in a region between the via holes near the input and couples to the output side of the microstrip patch radiator. The final two vias were placed to short-circuit these current maxima.

The final via hole configuration, which eliminates all resonances up to 30 GHz, is shown in Figure 12.27 with its predicted insertion gain. The field-solver

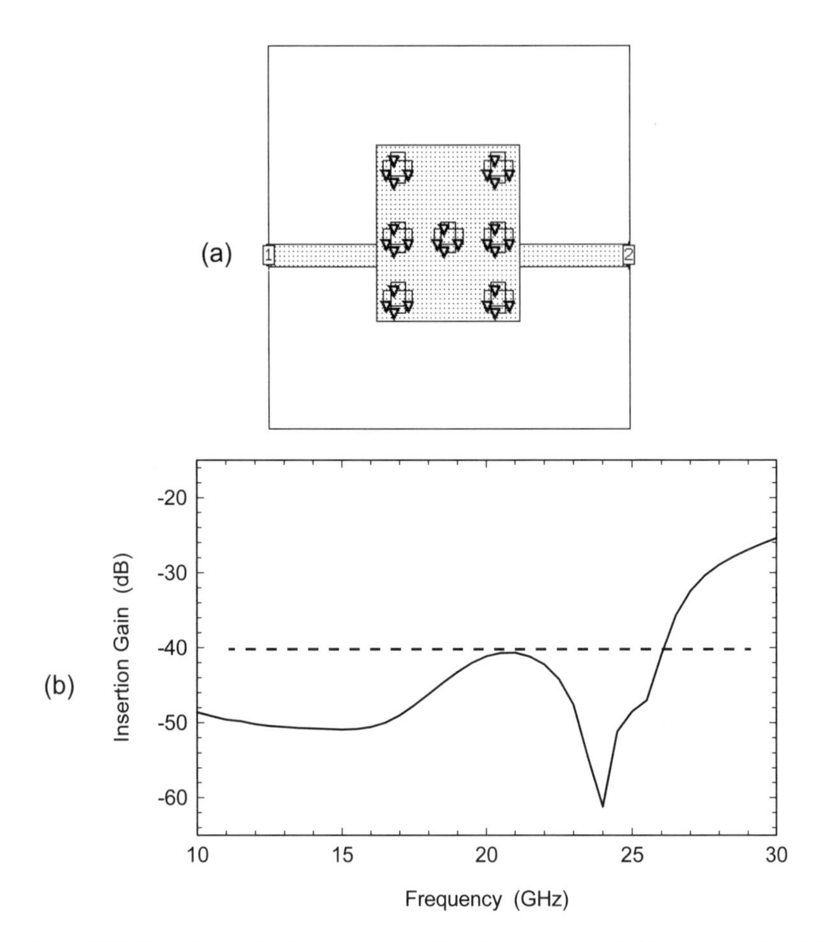

Figure 12.27 Grounding pad with six via hole pattern: (a) geometry, and (b) insertion gain of this configuration (Sonnet ***em*** Ver. 3.0). © 1993 Horizon House Publications [4].

predicts greater than 40-dB isolation below 20 GHz for the passive structure from input to output. The reduced isolation between 20 and 30 GHz is due to the width of the solution region, which is not cut off at these frequencies. When the current plots were examined at several frequencies, it was found that the current terminates on the vias very close to the input.

To verify the final result, one of the original substrates was modified by drilling four new via holes in the corners of the microstrip pad and filling them with conductive epoxy. This configuration was analyzed on the field-solver and found to be similar to the final via hole topology. A MMIC amplifier chip and bypass capacitor were mounted to the modified grounding pad. The measured results for this new configuration are shown in Figure 12.28. No spurious oscillations were observed over a temperature range of –55° to +85°C and several bias conditions.

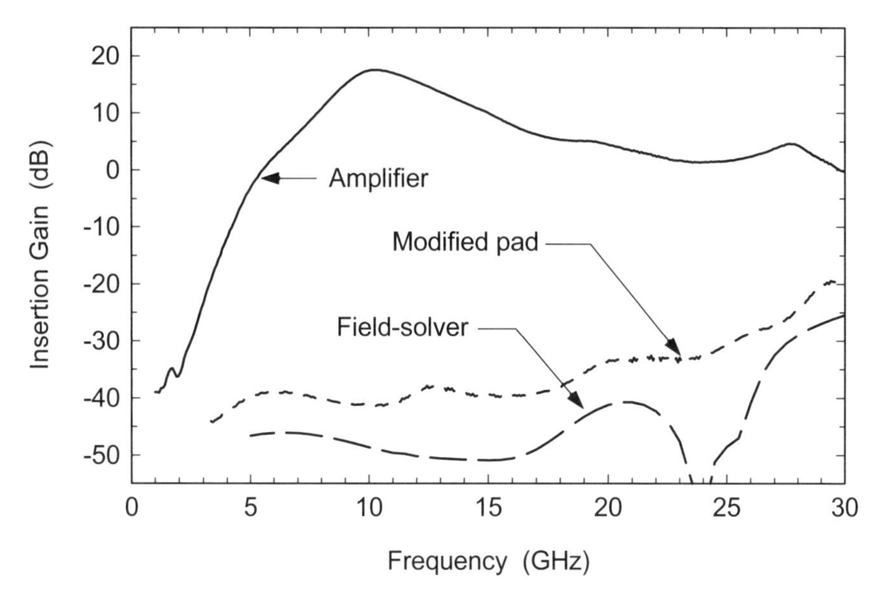

Figure 12.28 Results for final via hole pattern: amplifier response, measured response of the grounding pad, and field-solver prediction for the grounding pad (Sonnet ***em*** Ver. 3.0). © 1993 Horizon House Publications [4].

Although this circuit operates near 10 GHz, designers using GaAs devices in the 1 to 2 GHz range should also remember that these devices have significant gain out to 20 GHz and beyond. The quality of the grounding system at higher frequencies is equally important for these designs.

These results were first presented at a small workshop. At the same workshop, a presentation was made which closely paralleled this work [5]. In that presentation several rules of thumb were offered. The first recommended that vias should first be placed near the I/O locations of the MMIC chip. The second recommended that an

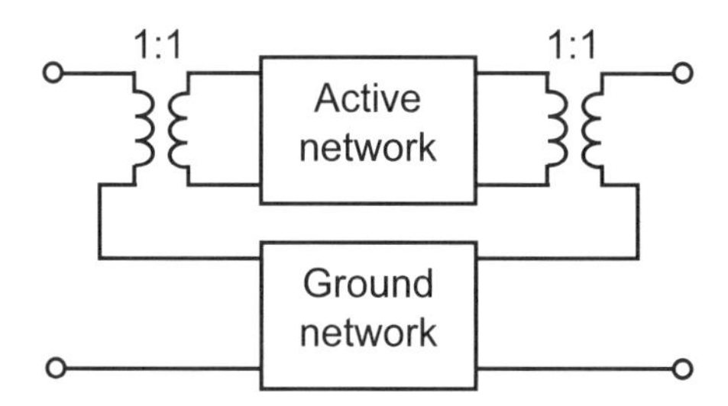

Figure 12.29 Computed or measured grounding system performance can be included in a circuit analysis by forming a series connected network.

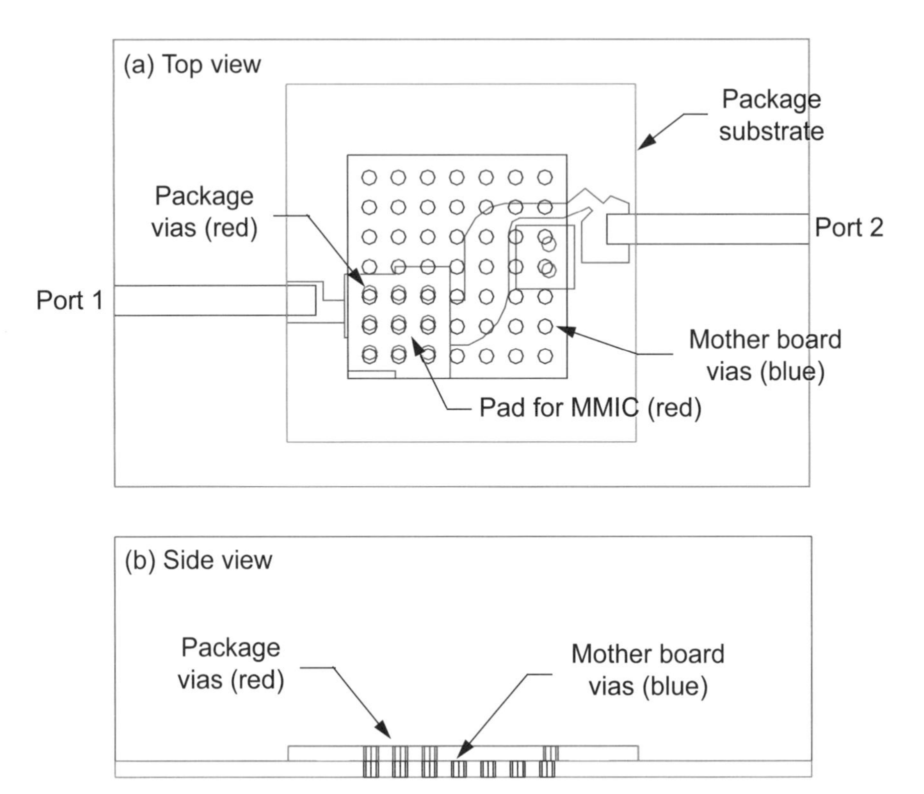

Figure 12.30 A PCB-based package for an RF MMIC mounted on a mother board. The mother board metallization is in blue. The package metallization is in red. There is a through connection from Port 1 to Port 2 to compute the isolation of the grounding structure.

isolation of 40 dB was needed for the grounding pad. This is in good agreement with the results in Figure 12.27(b). Another contribution of [5] was a simple way to analytically include the effect of the grounding structure in a circuit simulation (Figure 12.29). The interested reader should also consult [6, 7] for a very interesting discussion of flipped-chip and BGA package modeling and the resulting circuit theory models.

Figure 12.30 shows another interesting packaging problem that hinges on the isolation of the grounding structure. It is a PCB-based package for an RF MMIC amplifier mounted on top of a mother board. The mother board metallization is in blue, the package metallization is in red. The frequency range is 1 to 2 GHz. The MMIC chip is mounted on the package PCB with nine via holes to the backside of the package. The package is mounted on the mother board with 49 vias to the mother board ground plane (mostly for thermal performance). In Figure 12.30 there is a through connection from Port 1 to Port 2 to compute the isolation of the grounding structure. With the MMIC biased off there was a specification on isola-

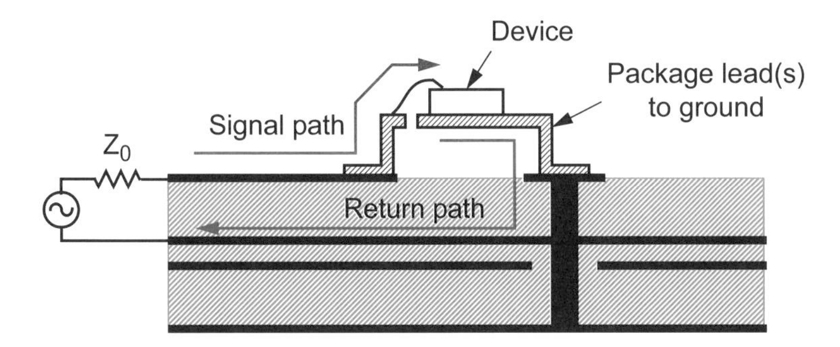

Figure 12.31 A leaded package mounted to a PCB. One or more leads of the package are connected to the microstrip ground plane with vias. The signal path and the return path must be optimized for RF performance. The plastic body of the package has been omitted for clarity.

tion from input to output. An analysis of the grounding structure alone showed that the desired isolation could not be met. Despite the large number of vias used, the grounding structure could not support that level of isolation.

Finally, although the major example of this section was centered at 10 GHz the concepts presented also apply to lower frequency RF and high-speed digital problems. Figure 12.31 shows a gull wing type leaded package mounted on a multilayer PCB. The active die rests on the package paddle. The basic problem in any RF or digital system is to bring the signal onto the mother board, through the package to the die, process the signal and bring it back onto the mother board. While it is easy to focus on the signal path, the path the return current takes (the grounding circuit) deserves equal attention. From this point of view, any RF or digital system is a multitier problem of translating signal paths and return current paths from one physical layer to another. The interested reader is encouraged to study [8, 9], which outline an elegant technique for developing circuit models for packages like the one in Figure 12.31 from field-solver data. These articles also highlight the impact of the chosen return path on RF performance. Additional information on via fences in LTCC packages can be found in [10, 11].

12.7 SUMMARY FOR GROUNDING PADS

Hopefully, we have developed a strong case for paying more attention to the grounding structure and return current path in thin-film hybrid circuits and multilayer PCB environments. Unfortunately, some good MMIC designs have probably been thrown away due to lack of understanding of these issues. Putting down a random pattern of vias clearly does not guarantee a good connection to the system ground. Fortunately, we now have a technique to predict whether a particular geometry will provide the performance needed. Using a field-solver we can place a minimum number of vias in an optimum pattern. Why not just flood the available area with vias? We might be able to do this in some glass-epoxy PCB applications, but vias do have a finite cost. And in any ceramic substrate environment extra holes will impact yield. In a ceramic package, extra vias may also compromise hermeticity or mechanical integrity.

References

[1] Swanson, Jr., D. G., "Grounding Microstrip Lines with Via Holes," *IEEE Trans. Microwave Theory and Tech.*, Vol. 40, No. 8, 1992, pp. 1719–1721.

[2] Goldfarb, M., and R. Pucel, "Modeling Via Hole Grounds in Microstrip," *IEEE Microwave and Guided Wave Letters*, Vol. 1, No. 6, 1991, pp. 135–137.

[3] Uchimura, H., T. Takenoshita, and M. Fujii, "Development of a Laminated Waveguide," *IEEE Trans. Microwave Theory Tech.*, Vol. 46, No. 12, 1998, pp. 2438–2443.

[4] Swanson, Jr., D. G., D. Baker, and M. O'Mahoney, "Connecting MMIC Chips to Ground in a Microstrip Environment," *Microwave Journal*, Vol. 34, No. 12, 1993, pp. 58–64.

[5] Gipprich, J., and S. Grice, "Effective Grounding for High Density Surface Mount Technology," *Microwave Hybrid Circuits Conference*, Wickenburg, AZ, October 17–20, 1993.

[6] Jackson, R. W., and R. Ito, "Modeling Millimeter-wave IC Behavior for Flipped-chip Mounting Schemes," *IEEE Trans. Microwave Theory Tech.*, Vol. 45, No. 10, 1997, pp. 1919–1925.

[7] Ito, R., R. W. Jackson, and T. Hongsmatip, "Modeling of Interconnections and Isolation Within a Multilayered Ball Grid Array Package," *IEEE Trans. Microwave Theory Tech.*, Vol. 47, No. 9, 1999, pp. 1819–1825.

[8] Jackson, R. W., "A Circuit Topology for Microwave Modeling of Plastic Surface Mount Packages," *IEEE Trans. Microwave Theory Tech.*, Vol. 44, No. 7, 1996, pp. 1140–1146.

[9] Jackson, R. W., and S. Rakshit, "Microwave-Circuit Modeling of High Lead-Count Plastic Packages," *IEEE Trans. Microwave Theory Tech.*, Vol. 45, No. 10, 1997, pp. 1926–1933.

[10] Ponchak, G. E., et al., "Characterization of Plated Via Hole Fences for Isolation Between Stripline Circuits in LTCC Packages," *IEEE MTT-S Int. Microwave Symposium Digest*, Baltimore, MD, June 7–12, 1998, pp. 1831–1834.

[11] Mizoe, J., et al., "Miniature 60 GHz Transmitter/Receiver Modules on AIN Multi-layer High Temperature Co-fired Ceramic," *IEEE MTT-S Int. Microwave Symposium Digest*, Anaheim, CA, June 13–19, 1999, pp. 475–478.

Chapter 13

Multilayer Printed Circuit Boards

We have progressed from the basics of microstrip to the behavior of discontinuities and single layer vias. In this chapter we are ready to tackle transitions between layers in multilayer PCBs. When the microwave community shifted its focus from military to commercial applications they were quick to adopt multilayer PCB technology. At the same time, the high-speed digital community was pushing to higher clock speeds and higher bus speeds for data. In the end, the problems faced by both communities are exactly the same, but are still defined using different languages. The RF community tends to define its problems in the frequency domain while the digital community uses the time domain. But both groups are challenged with maintaining signal integrity while signals move between layers and through connectors to other boards or to the outside world.

This chapter is basically a three-part story that spans almost a decade. It demonstrates how my thinking and understanding of multilayer transitions has evolved. The general design procedure presented in the final section has been successfully applied to many RF and high-speed digital transition problems.

13.1 A MULTILAYER TRANSITION IN FR4

One of the freedoms of multilayer boards is the opportunity to use buried layers for shielded runs of RF transmission lines. If we normally use microstrip for the topmost traces, then these buried traces are either stripline or CPW. After working at microwave frequencies for many years, when we consider transitions between layers, we are tempted to think we can "get away with anything" because the frequencies are so low, roughly 0.5 to 2.5 GHz in this case. The frequencies are relatively low, but the parasitics are much higher than we are used to in single layer, ceramic structures at microwave frequencies, so we end up with some interesting problems to solve. Figure 13.1 is a cross-section of a fairly typical scenario in a wireless system, circa 1995, on a multilayer FR4 board. The microstrip line transitions to a buried stripline layer in order to pass under a shielding wall and connect to a packaged

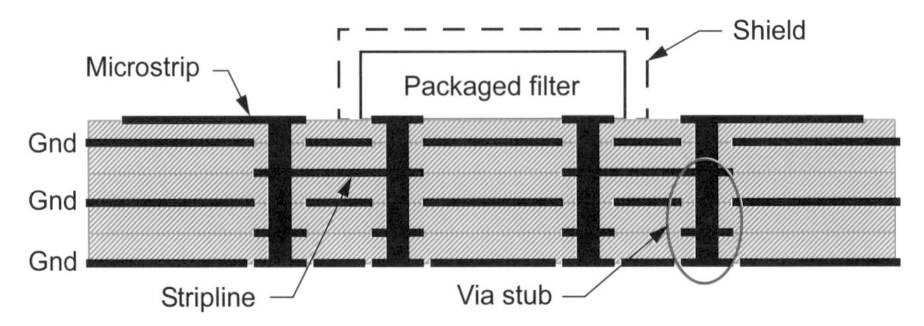

Figure 13.1 Transition from microstrip to stripline in a multilayer PCB. The signal line passes under a shielding wall to connect to a packaged filter with through hole pins. The via stubs below the stripline layer have a negative impact on the performance of the transition.

filter centered near 1.5 GHz. In this design, the vias extend all the way through the board (through hole vias). Through hole vias are generally less expensive than buried or blind vias. However, one disadvantage of through hole vias is that they leave *via stubs* connected below the desired transition signal path. These via stubs add extra parasitic loading to the transition and can resonate when their electrical length becomes a quarter of a wavelength. Back drilling the board to remove the stubs is one option if higher performance is required.

The packaged filter has large diameter through hole pins, so there are actually two different diameter transitions to consider here. The default layout rules were used in the PCB software to define the pad sizes and antipad sizes in the intermediate layers. The "antipad" is the space between a metal pad and the adjacent ground plane metal, if any. Traditionally, via hole, pad, and antipad diameters are chosen

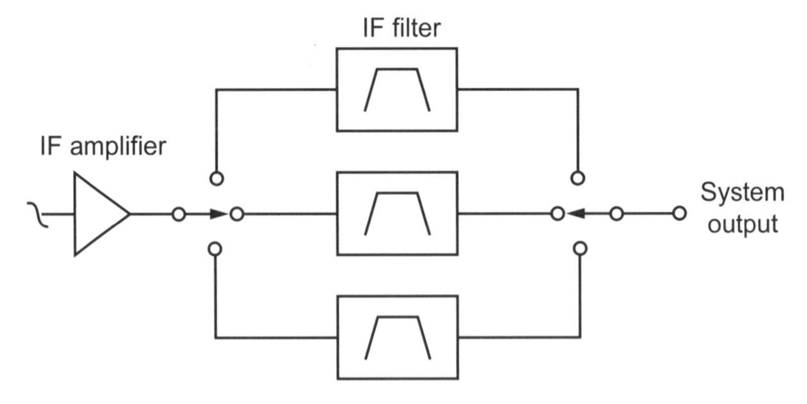

Figure 13.2 Block diagram of the switched filter bank application. Measurements at the system output connector showed poor return loss and large ripples in the S_{21} response.

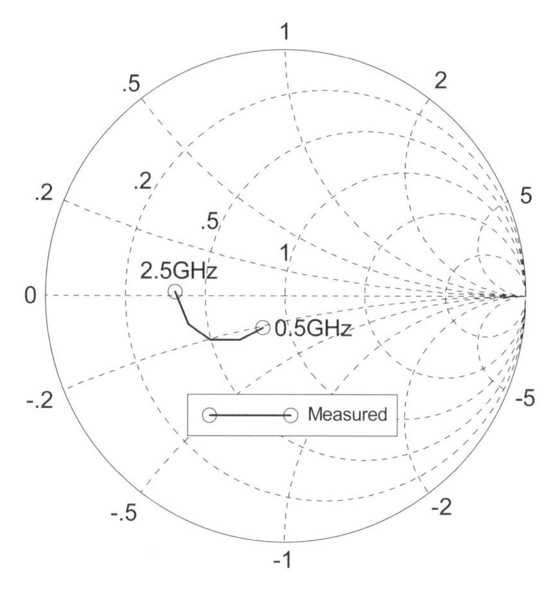

Figure 13.3 Measured S_{11} for a pair of transitions, microstrip to stripline to microstrip. The return loss is about 6 dB at 1.5 GHz and the transition suffers from excess capacitance.

early in the project and applied uniformly throughout the board. Unused pads may or may not be removed depending on the layout rules.

This structure was part of a large, mixed signal, analog/digital PCB. The application was an early implementation of a local multipoint distribution service (LMDS) system. This switched filter bank (Figure 13.2) was located at the IF output of the board. Luckily, we could make measurements at the system output connector, which gave us a good idea of the magnitude of the problem. The return loss at the system output connector was very poor and there was a large amount of ripple in the S_{21} response. When faced with this situation, microwave engineers have historically been comfortable with "tweaking" the design on the test bench. Using intuition, small pieces of metal are added or taken away from the layout. Of course with multilayer technology, there are now large parts of the structure that cannot be experimentally probed or modified on the bench.

When tweaking the design on the bench failed to provide any significant improvement, it was suspected that the microstrip to stripline transitions were the problem. Rather than build a new test board, a pair of transitions, microstrip to stripline to microstrip, were physically cut out of the board and connections were made using small diameter coaxial cables. The measurement for this test case is shown in Figure 13.3. The return loss for the pair of transitions (down and back up) is only about 7 dB at 1.5 GHz. When we cascade passive components we would like to see at least 15 to 20 dB return loss. This helps to minimize interactions between components and minimize insertion loss.

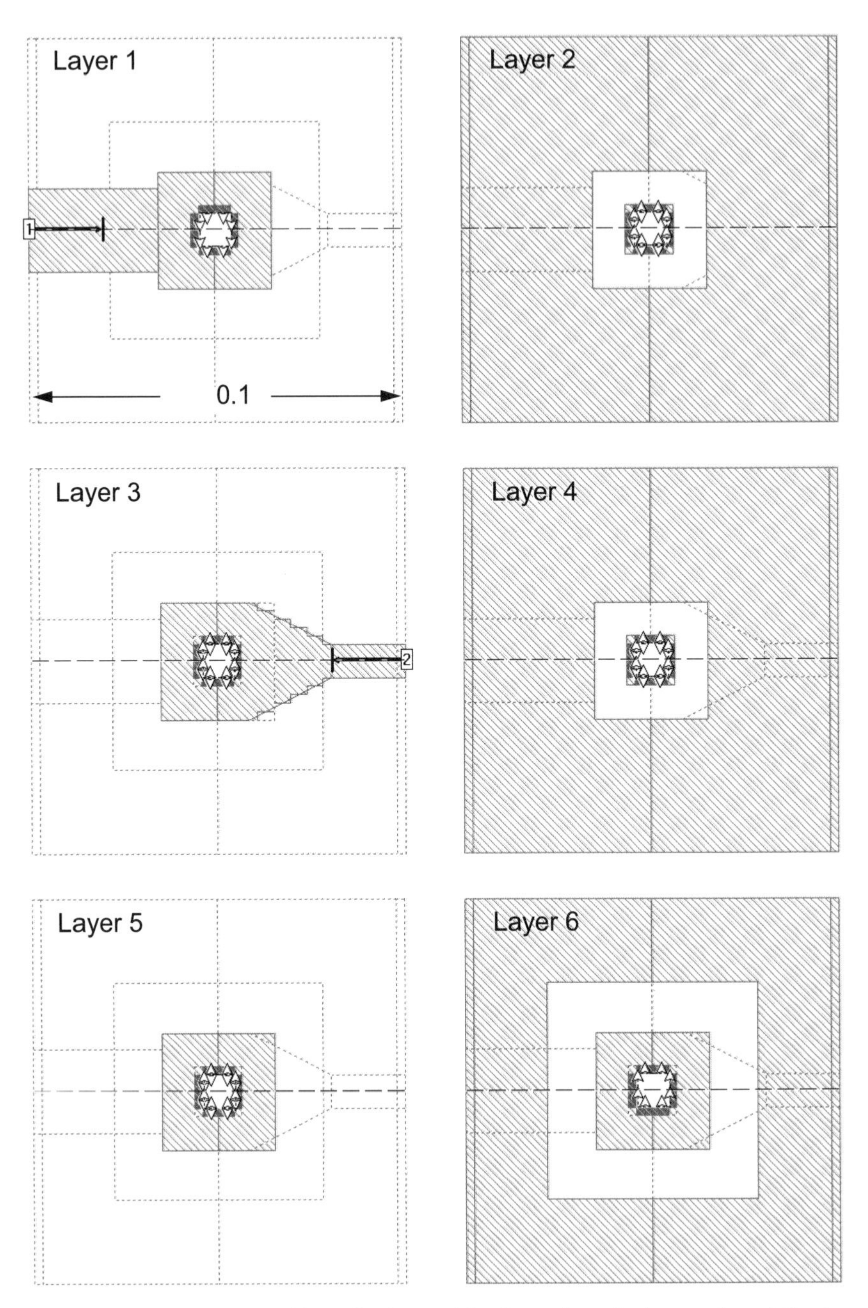

Figure 13.4 Layout of each metal layer for the transition from microstrip down to the stripline layer. Note that the via holes, pads, and antipads have been approximated as squares. The size of the analysis region has been reduced for clarity. The solution time on a SUN SPARC-10 (circa 1995) was 5 min, 46 sec per frequency point (Sonnet ***em*** Ver. 3.0a).

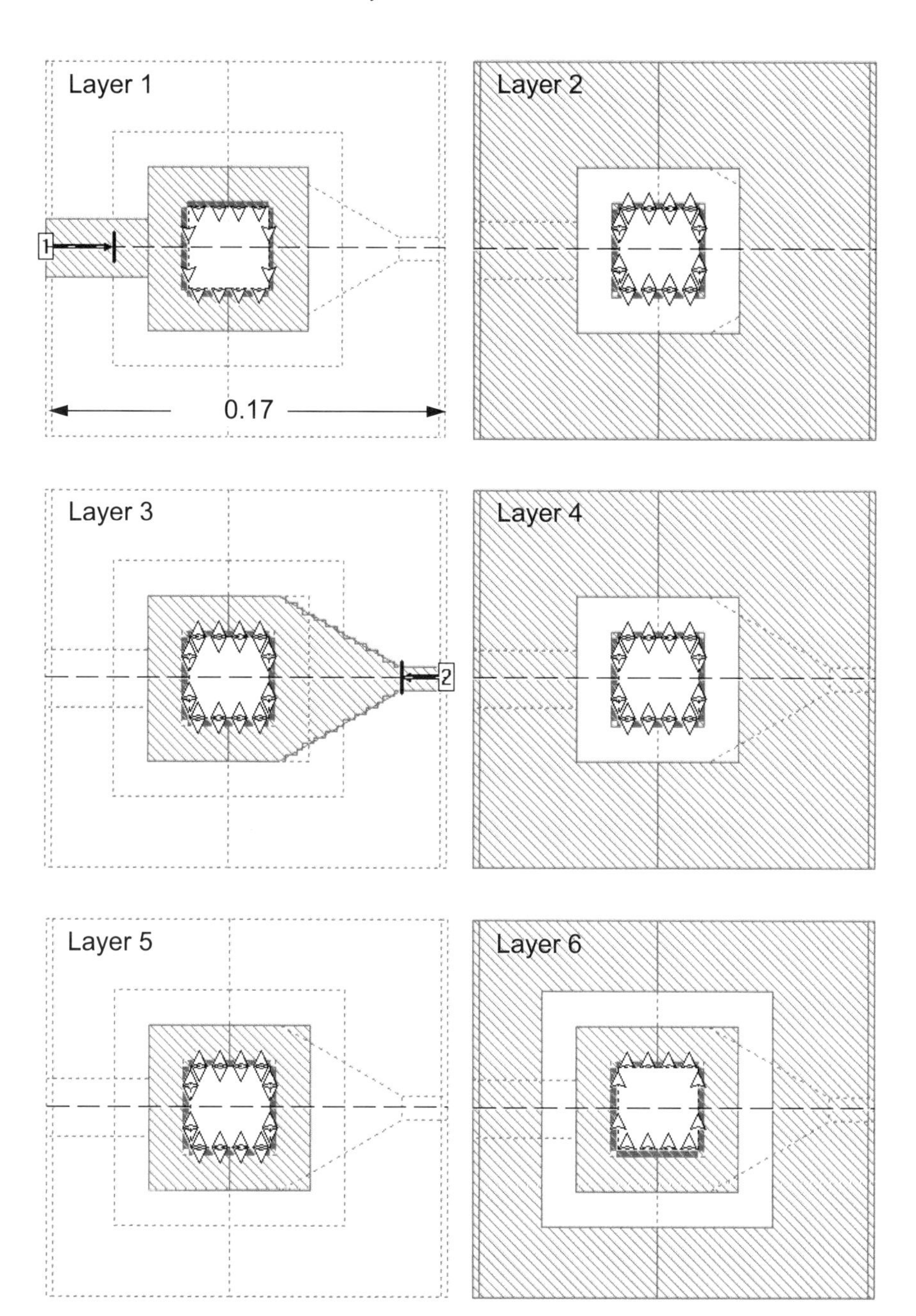

Figure 13.5 Layout of each metal layer for the transition from the stripline layer back up to the microstrip layer. The size of the analysis region has been reduced for clarity. The solution time on a SUN SPARC-10 (circa 1995) was 6 min, 37 sec per frequency point (Sonnet ***em*** Ver. 3.0a).

At this point we brought the problem to the field-solver. At first, we tried to analyze the complete problem, the transition pair plus the stripline between them. But the solution time was too long to rapidly perform “what if” experiments on the geometry, so the transition pair was modeled in two steps. First, we considered the transition from microstrip down to the stripline layer, Figure 13.4. We have reduced the size of the analysis region for clarity. Note that the circular via holes, pads, and antipads have all been approximated as squares. We did this because it is easier and faster to build the model this way and we assumed this approximation would not significantly alter the results at 1.5 GHz. We could approximate all the circular geometries as octagons, but this takes longer to set up and computing the diagonal elements takes longer. One disadvantage of the MoM codes for this type of problem is the need to discretize all the ground plane metal, which has a significant impact on solution time. The analysis time on a SUN SPARC-10 (circa 1995) was 5 min, 46 sec per frequency point.

However, before starting the transition analysis we analyzed a simpler problem that included all the dielectric layers, a 50-ohm microstrip through line and a 50-ohm stripline through line. To our surprise the computed *S*-parameters did not indicate we had well-matched, low-loss transmission lines. After checking the problem setup several times we contacted the software vendor and discovered there was a bug in the meshing algorithm that prevented us from getting the correct answer. The simple work around at that time was to place a one cell wide strip of metal at the left and right edges of the ground plane layers (Figure 13.4). Luckily, we discovered this before we spent a lot of time on the transition analysis. But it also encouraged us to add this simple test to our solution strategy. We have often found mistakes in our own problem setup and occasionally bugs in the software by analyzing a 50-ohm line before we begin the more complicated problem.

The second transition, from the stripline layer back up to the packaged device is shown in Figure 13.5. The size of the analysis region has been reduced for clarity. The layout philosophy is the same, but the dimensions are larger due to the diameter of the package pins. Note that unused pads have been removed in Layer 2 and Layer 4. However, an unused pad was left in the layout in Layer 5. The taper in Layer 3 is standard practice to avoid voids in the metal if the pad and via hole are

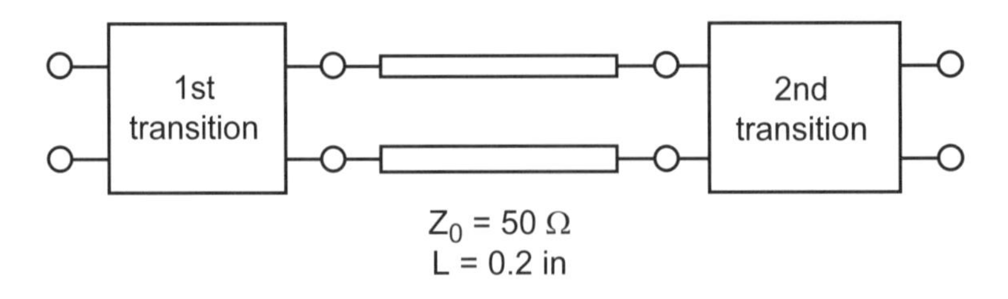

Figure 13.6 Cascade of first transition, a length of stripline and the second transition. Analyzing the complete geometry on the field-solver would take much more time.

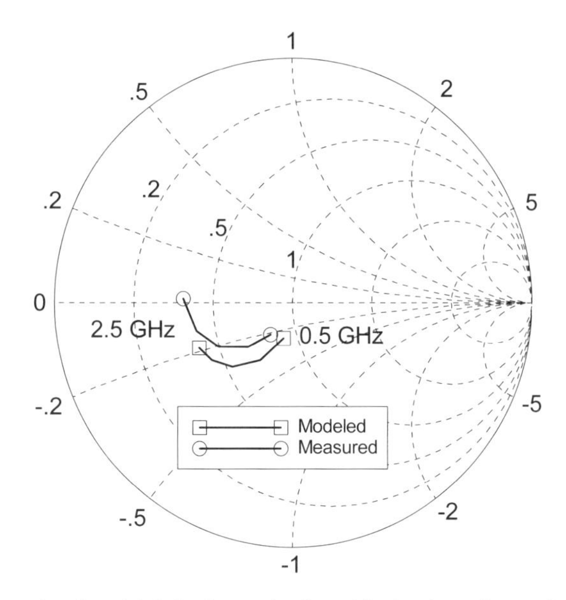

Figure 13.7 Measured and modeled S_{11} for a pair of transitions, microstrip to stripline to microstrip. Although the agreement is not perfect, it was considered good enough to continue the modeling and optimization effort.

misaligned in manufacturing. The analysis time on a SUN SPARC-10 (circa 1995) was 6 min, 37 sec per frequency point.

The initial analysis of the cascade of transitions was completed on the linear simulator (Figure 13.6). The first and second vias are connected using an ideal stripline model. The results for this initial analysis and the measurement that was made are shown in Figure 13.7. The agreement between measured and modeled is certainly not perfect. The discrepancy could be in the measurement or in our rather crude approximation of the geometry. But the correlation is good enough to give us confidence that we can predict what is going on.

The next step was to perform a series of experiments on the field-solver in an attempt to improve on these results. If we examine both the measured and predicted results in Figure 13.7, we note that the transitions suffer from excess capacitance. So using intuition, the effect of unused pads, the large tapers in the stripline layers and the antipad diameters were examined in a series of experiments on the field-solver. Three to four experiments were done on each transition in the course of one afternoon. In the end, all unused pads were removed, the tapers in the stripline layers were removed and the antipad diameters were increased 0.010 inch in the first transition and 0.015 inch in the second.

After updating the field-solver generated data the cascade analysis (Figure 13.6) was performed again. The analysis now predicts greater than 20 dB return loss for the cascade of transitions. This new set of transition dimensions was

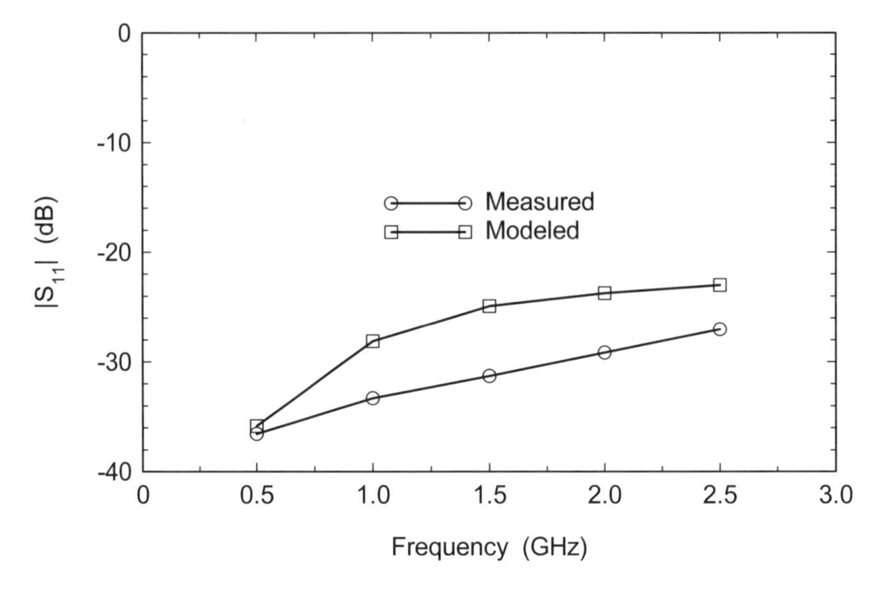

Figure 13.8 Measured and modeled results for the optimized pair of transitions. The new transition pair was cut out of a finished board in the same fashion as the original experiment.

included in a new revision of the multilayer PCB. The system performance was now quite acceptable with the new design. But just to complete the experiment, a pair of transitions was cut out of the new board and measured in the same way as the original pair of transitions. A return loss plot of the measured and modeled data is shown in Figure 13.8. The measured performance is actually better than the prediction. But in this case, as long as the measured performance is greater than 20 dB return loss, exact agreement between measured and modeled data is not needed.

This project was somewhat of a wake up call for a group of engineers used to working in single layer ceramic technology at microwave and millimeter wave frequencies. They found that the standard catalogue of circuit models were not much help in this multilayer environment. They also learned to work more closely with the PCB layout contractor and the PCB fabricator. The traditional, default rules used for PCB layout were not necessarily optimum for RF performance.

13.2 CONTROLLED IMPEDANCE TRANSITIONS

The previous example demonstrates the field-solver's ability to analyze multilayer structures. But in fact, the previous analysis may be somewhat optimistic for a via in a random location on a large printed circuit board. We tend to forget that all our transmission lines are two conductor systems. If we have a signal current on a microstrip, there must be a return current in the ground plane. Similarly, for a signal

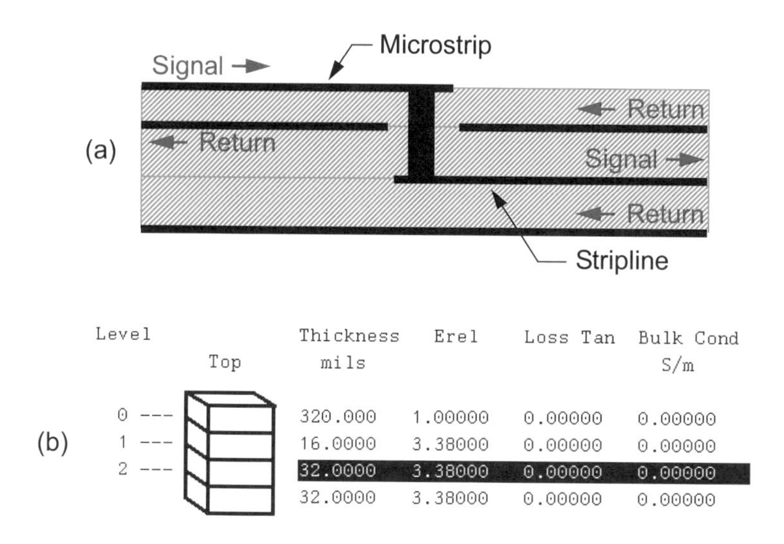

Figure 13.9 A microstrip to stripline transition: (a) cross-section view showing signal and return currents; and (b) layout stack up for the numerical experiments.

current on a stripline, there must be *equal* return currents in the upper and lower ground planes (Figure 13.9(a)). However, in Figure 13.9(a) this is no explicit connection between the upper and lower ground planes. In a typical PC board, the connection between the lower and upper ground planes will be made at several random points with vias. In the previous example, the connection between ground plane layers was made in the box walls provided by the simulator, but these walls are not present in the actual board. And the simulator walls probably have a lower impedance than a few randomly located vias. So in that sense, the analysis can easily be overly optimistic when trying to predict the performance of an individual transition at some random location on a PC board.

13.2.1 Analysis Using Closed Box MoM

In order to clarify these points, we will perform a series of numerical experiments using several different field-solvers. We will use the simple four-layer geometry shown in Figure 13.9(a) with the layer stack up shown in Figure 13.9(b). The first set of experiments will use a closed box MoM simulator.

The layer by layer geometry of our first test case is shown in Figure 13.10. Because we are using symmetry, and to save space, only the top half of the metal pattern is shown. The bottom of the analysis box forms the bottom ground plane. The connection between the ground planes is through the simulator box walls. This structure was actually the starting point for the project in Section 13.3 with performance requirements up to 10 GHz.

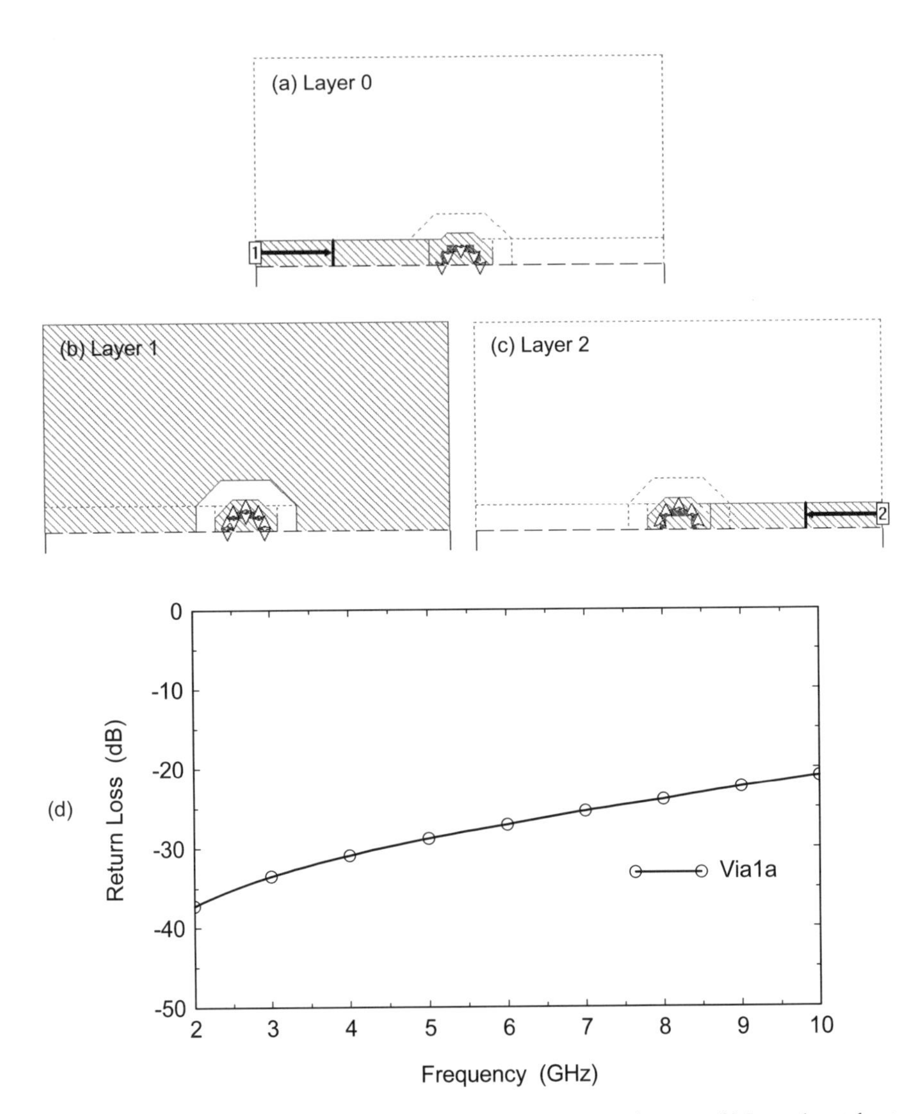

Figure 13.10 Transition Via1a numerical experiment: (a) Layer 0 metal pattern; (b) Layer 1 metal pattern; (c) Layer 2 metal pattern; and (d) return loss for this geometry. Only the metal above the symmetry plane is shown (Sonnet ***em*** Ver. 7.0).

The results for this initial geometry, labeled Via1a, are shown in Figure 13.10(d). The return loss looks quite good up to 10 GHz, in fact too good. Because the analysis box is relatively small, and all the metal planes connect to the ideal box walls, there is a low impedance path for the return current. We instinctively tend to make the analysis region as small as possible to keep the solution time down.

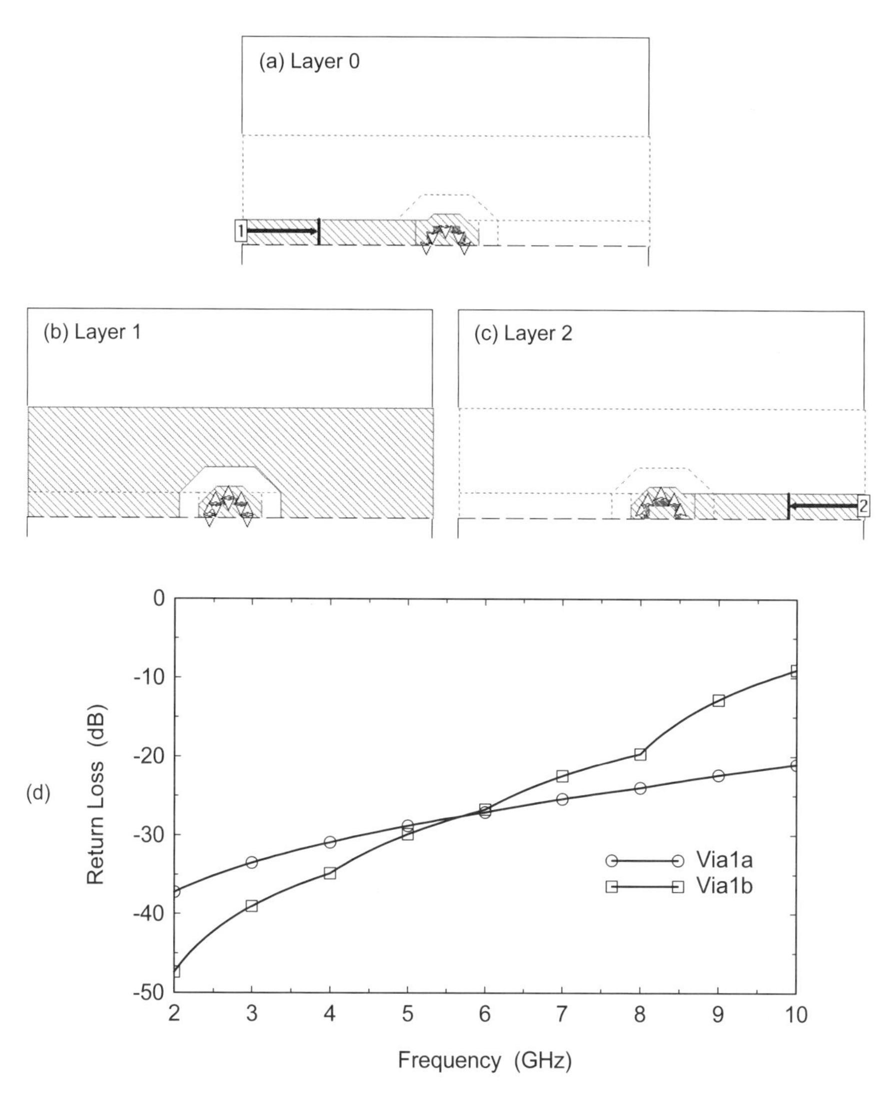

Figure 13.11 Transition Via1b numerical experiment: (a) Layer 0 metal pattern; (b) Layer 1 metal pattern; (c) Layer 2 metal pattern; and (d) return loss for this geometry. Only the metal above the symmetry plane is shown (Sonnet ***em*** Ver. 7.0).

So let's try to make this problem more realistic. The first step is to disconnect the microstrip ground plane from the upper and lower sidewalls. The results for this experiment, labeled Via1b, are shown in Figure 13.11(d). The transition now looks a little less ideal. The next step is to float the bottom stripline ground plane above the floor of the analysis box. We will also force the correct modes at the ports by

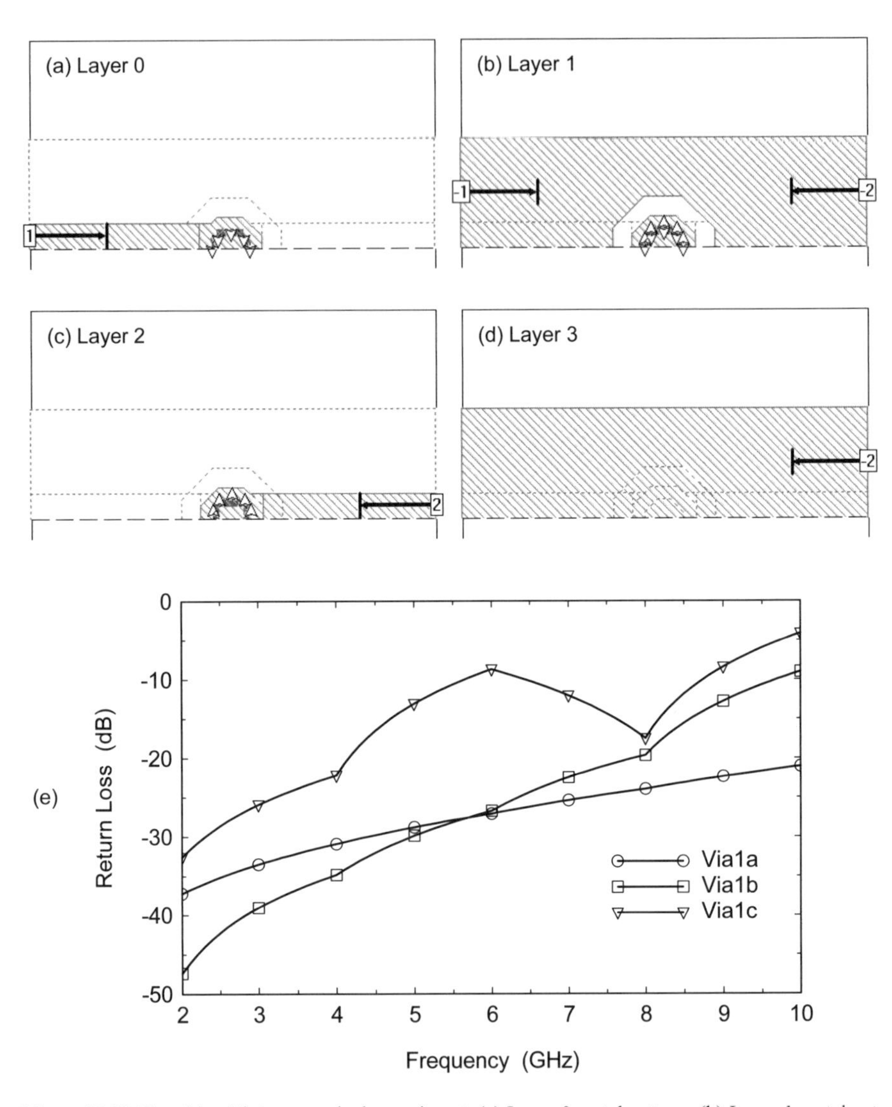

Figure 13.12 Transition Via1c numerical experiment: (a) Layer 0 metal pattern; (b) Layer 1 metal pattern; (c) Layer 2 metal pattern; (d) Layer 3 metal pattern; and (e) return loss for this geometry. Only the metal above the symmetry plane is shown (Sonnet ***em*** Ver. 7.0).

driving the microstrip port plus–minus and the stripline port, minus–plus–minus. Driving the ports in this way also forces the current in the left and right sidewalls at the ports to be zero. This will increase the decoupling of the ground plane layers. The results of this analysis, labeled Via1c, in Figure 13.12(e) show a very poor transition at the higher frequencies and we begin to see wavelength related structure

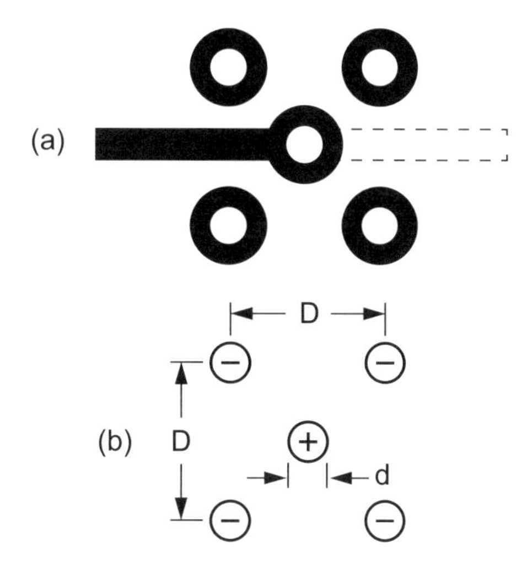

Figure 13.13 Five-wire transmission line: (a) top view of layout, and (b) dimensions.

in the data. Again, we should emphasize that we are only changing the path for the return current in all of these experiments; we have not modified the geometry of the via itself. I believe this is a more realistic analysis of an isolated signal via in a printed circuit board.

Now that we have created a poor transition on the computer, can we find a technique to fix it? By placing several vias that connect the ground planes around the signal via (Figure 13.13(a)), we can do two things [1]. We can provide a low impedance path for the return current that is independent of other structures on the PC board. And we can form a short length of transmission line that is close to 50 ohms through the board. In effect we have created a five-wire transmission line (Figure 13.13(b)) vertically through the board, and there is a simple analytical formula [2] for its impedance

$$Z_0 = (173/\sqrt{\varepsilon_r}) \cdot \log(D/0.933d) \text{ for } d \ll D \tag{13.1}$$

The layout for this new structure with one signal via and four ground vias can be found in Figure 13.14. A pattern of four ground vias may be nearly optimum. With only two ground vias they would have to be much closer to the center via to achieve a 50-ohm impedance. With more than four vias, there would be less space for the signal lines to exit the pattern. The results for the controlled impedance transition, labeled Via1d, are shown in Figure 13.14(e). The electrical performance is now closer to the original, more ideal analysis. There are several opportunities to further optimize this structure. Changing the spacing D and modifications to the microstrip

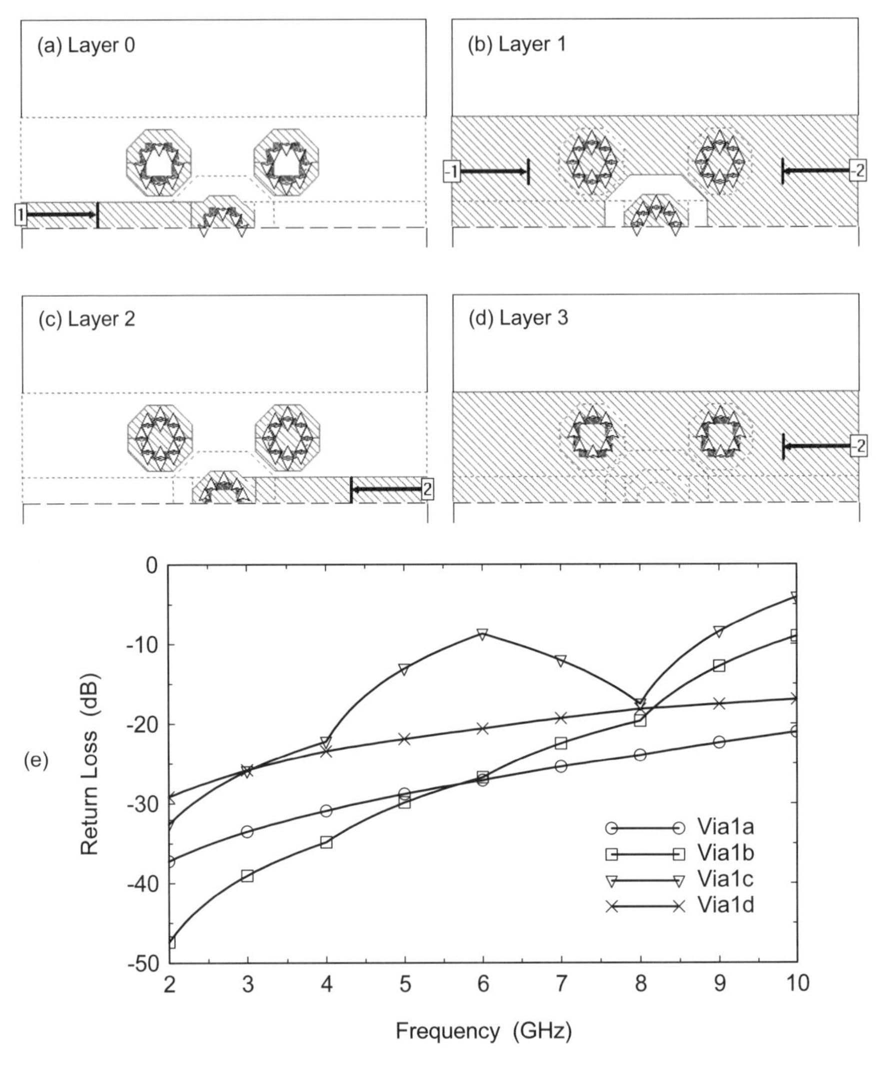

Figure 13.14 Transition Via1d numerical experiment: (a) Layer 0 metal pattern; (b) Layer 1 metal pattern; (c) Layer 2 metal pattern; (d) Layer 3 metal pattern; and (e) return loss for this geometry. Only the metal above the symmetry plane is shown (Sonnet ***em*** Ver. 7.0).

and stripline transitions are only some of the possibilities. The placement of pads in unused layers and antipad diameters are other optimizable parameters.

Looking at the conduction currents is another way to gain an appreciation for happening in the multilayer transition. Figure 13.15 shows the currents in the ground planes at 10 GHz for the Via1c numerical experiment. As we rotate the phase of the signal at the microstrip port a current maxima passes through the tran-

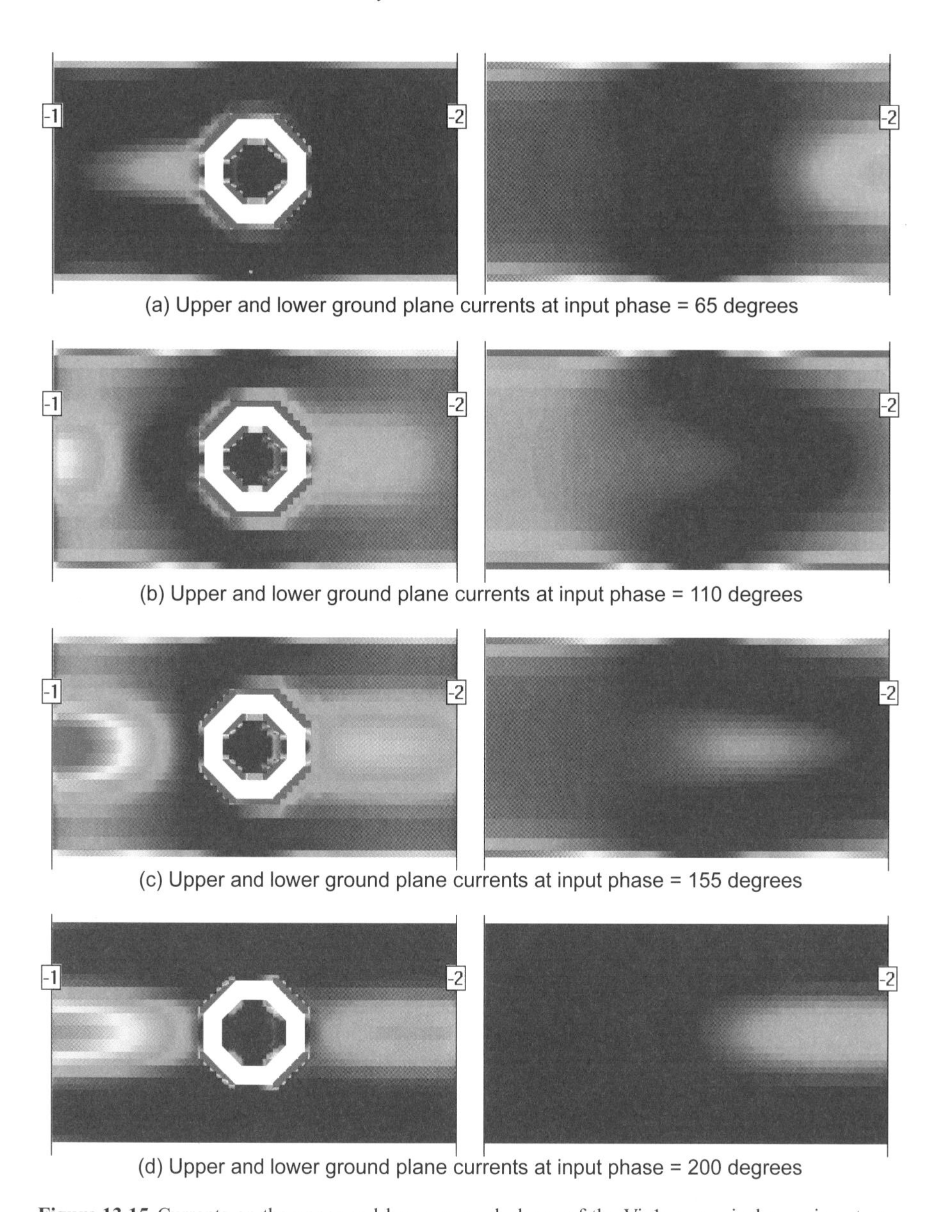

Figure 13.15 Currents on the upper and lower ground planes of the Via1c numerical experiment as a function of the input phase at 10 GHz. The upper ground plane is on the left, the lower ground plane is on the right, and the scale is 0 to 8 A/m (Sonnet ***em*** Ver. 7.0).

sition region. Note that in many of the frames the edges of the ground planes are carrying significant current. Current is being pulled in from the simulator box walls to support the current needed on the ground planes. In Figure 13.15(c) we can see

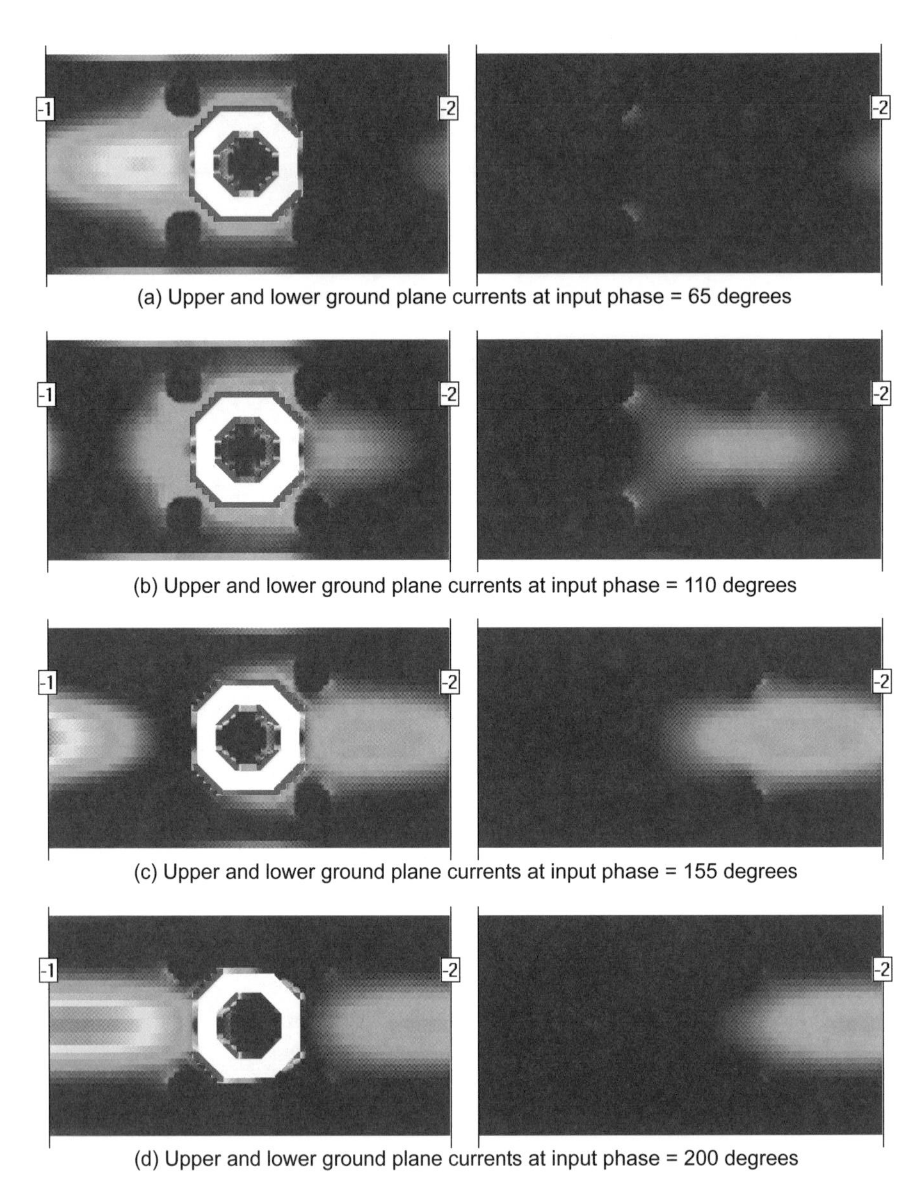

(a) Upper and lower ground plane currents at input phase = 65 degrees

(b) Upper and lower ground plane currents at input phase = 110 degrees

(c) Upper and lower ground plane currents at input phase = 155 degrees

(d) Upper and lower ground plane currents at input phase = 200 degrees

Figure 13.16 Currents on the upper and lower ground planes of the Via1d numerical experiment as a function of the input phase at 10 GHz. The upper ground plane is on the left, the lower ground plane is on the right, and the scale is 0 to 8 A/m (Sonnet ***em*** Ver. 7.0).

that the stripline ground plane currents are not equal in magnitude and phase. For a pure stripline mode we would expect in phase currents in both ground planes. Figure 13.16 shows the ground plane currents at 10 GHz for the Via1d numerical

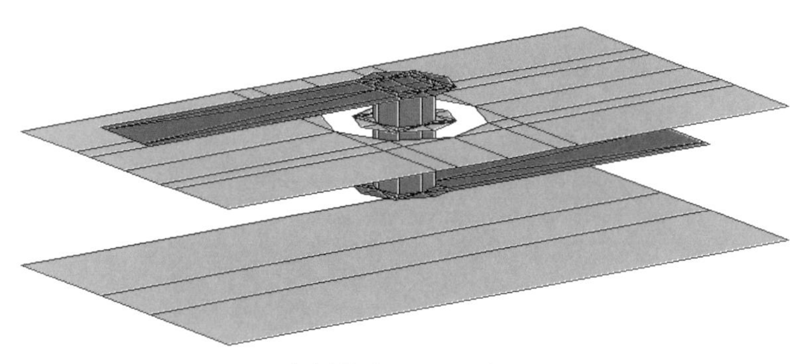

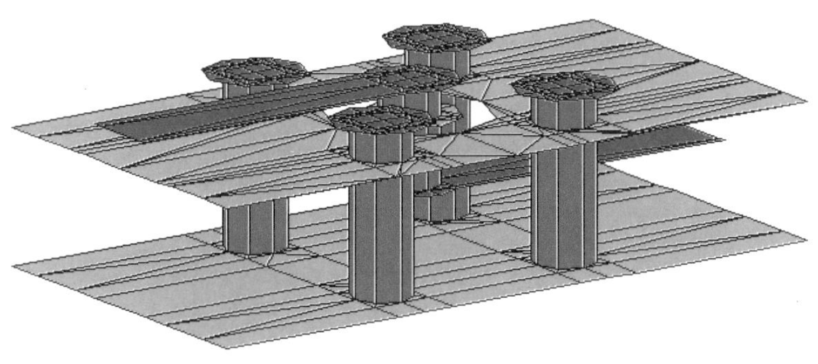

Figure 13.17 Using a laterally open MoM code to solve the transition problem: (a) the Via1c geometry; and (b) the Via1d geometry. The ground planes extend to infinity in the analysis and the actual analysis mesh is not shown (Zeland IE3D Ver. 9.0).

experiment. Now the edges of the ground plane layers carry very little current. The transfer of current between the ground plane layers is being handled by the added ground vias. In Figure 13.16(c) we note that the stripline ground plane currents appear to be equal in magnitude and phase. When we probe the current magnitudes at the same X-Y location on the two plots, we find that they differ by less than 1%. Thus the current plots tend to support our earlier assumptions on the effectiveness of the added ground vias.

13.2.2 Analysis Using Laterally Open MoM

Unfortunately we do not have any measured data for the geometry shown in Figure 13.14. However, in the absence of measured data we can solve the same problem using a different numerical method. If we achieve similar results using a different numerical method, we gain some confidence that our assumptions about the behav-

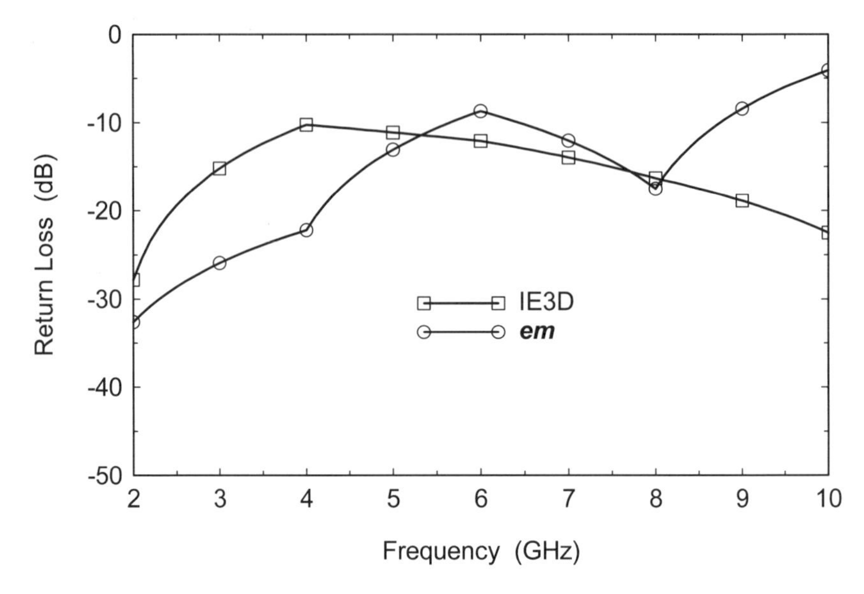

Figure 13.18 Results from the closed box and the laterally open MoM analyses of Via1c. Both solvers indicate poor high-frequency performance for the transition, although the details differ due to the different boundary conditions in the simulators.

ior of the multilayer transition are correct. The next tool we looked at was IE3D from Zeland Software; a code that uses the laterally open MoM formulation.

The geometry for our starting point is shown in Figure 13.17(a). Although the ground planes we have created here are finite, the analysis domain extends to infinity and there are no box walls. The ports are driven differentially, the same configuration we used in the Via1c version of the Sonnet ***em*** analysis. Figure 13.17(a) only expresses the captured geometry; it is not displaying the meshing used for analysis.

The results from IE3D and the analysis from Sonnet ***em*** for the Via1c numerical experiment are compared in Figure 13.18. Although they are not exactly the same, they both indicate that our via structure does not have good high-frequency performance. The differences between these two analyses again has to do with the boundary conditions. We still have a finite size box in the Sonnet ***em*** analysis; the IE3D analysis planes extend to infinity.

Next we added the ground vias around the signal via. Nothing else changes at the boundaries of the problem. The new geometry, which is equivalent to Via1d from the ***em*** analysis, is shown in Figure 13.17(b). As before, Figure 13.17(b) only expresses the captured geometry; it does not show the mesh used for analysis.

Now we can compare the IE3D and Sonnet ***em*** analysis runs for the controlled impedance transition (Figure 13.19). As you can see, there is close agreement between the two. Once we put the ground vias in place, the boundary conditions

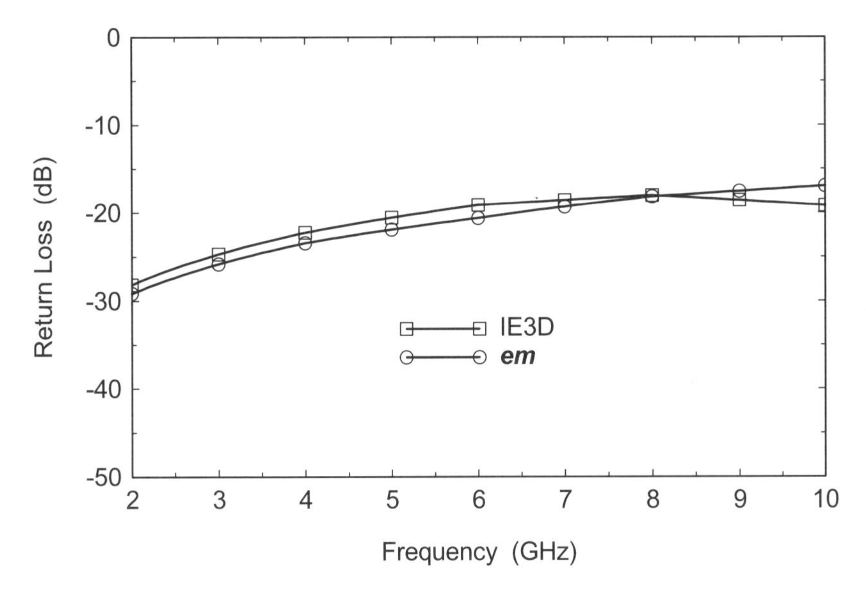

Figure 13.19 Results from the closed box and the laterally open MoM analyses of Via1d. Both solvers agree that the transition performance is much improved. In both cases the ground vias control the boundary conditions, rather than the simulator.

applied by the field-solver are less important. With four ground vias, the solution is not very sensitive to changes in the field-solver boundaries. With only two ground vias, the sensitivity to changes in the numerical problem is much higher, but still better than having no ground vias at all. The geometry of the transition can still be optimized for better high-frequency performance.

13.2.3 Analysis Using 3D FEM

A third analysis of our multilayer transition might seem quite unnecessary at this point. The basic formulation of the 3D FEM codes assumes we have a conducting box enclosing the solution space. We can probably assume that we will have similar problems "decoupling" the simulator walls as we saw in the closed box MoM case.

However, there is an additional aspect to the transition problem that we have not yet considered. With no vias in place (Figure 13.17(a)), there are two possible modes at the stripline port. There is the desired stripline mode which assumes that the two ground planes are at the same potential. And there is a parallel plate mode that runs from the lower ground plane to the upper ground plane. In the parallel plate mode the stripline center conductor floats at some undefined potential between the two ground planes.

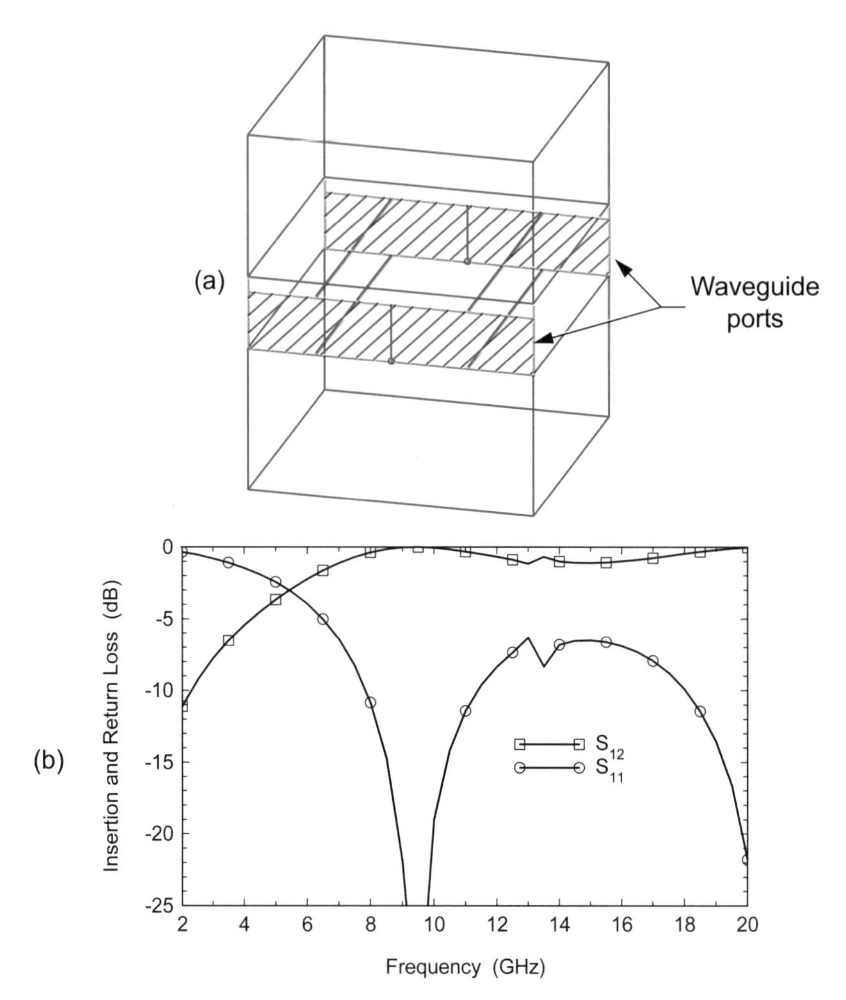

Figure 13.20 Test structure for waveguide ports: (a) the geometry includes the ports, the PCB and the stripline ground planes; and (b) the computed results. The –3 dB frequency is about 5 GHz (Agilent HFSS Ver. 5.6).

So the question arises, is mode conversion possible in the transition region and can we detect it? The MoM simulators use circuit theory type ports which consider only total voltage and current; no information on modes is available. However, with the wave type ports in a 3D FEM simulator we can specify two modes at the stripline port and look for possible mode conversion into the parallel plate mode.

The PCB with stripline ground planes is shown again in Figure 13.20(a). We have already described in detail our attempts to decouple the problem from the simulator walls. There are some additional problems using wave ports in a 3D simula-

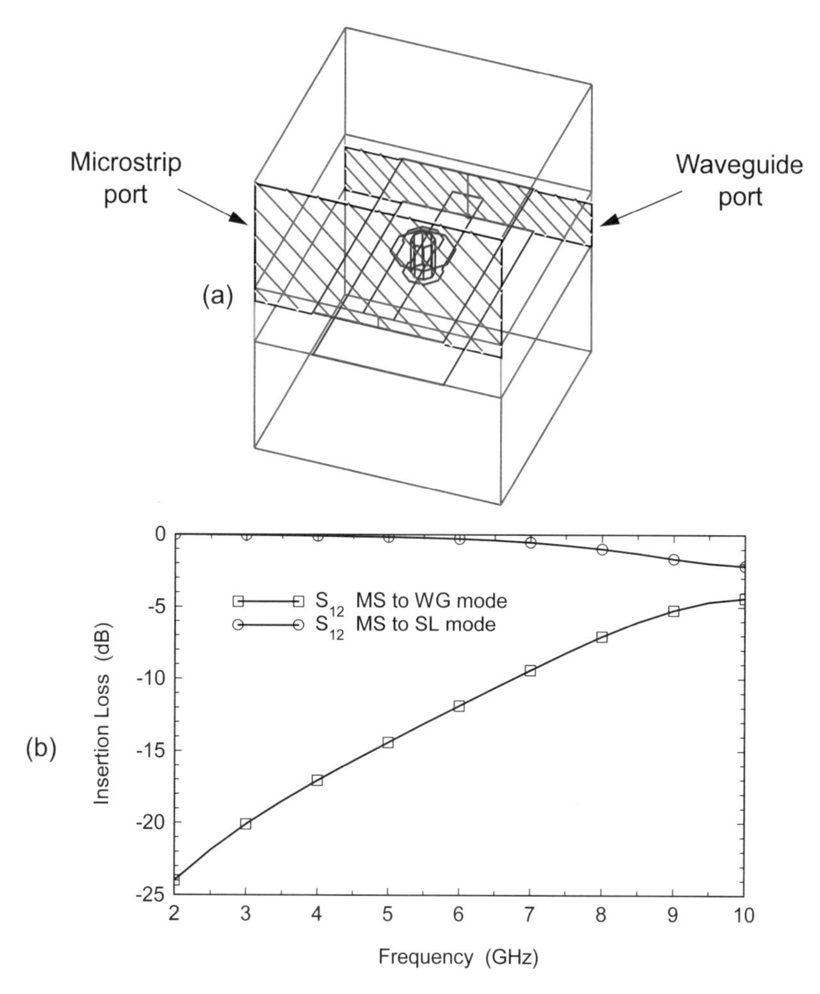

Figure 13.21 The Via1c numerical experiment: (a) the geometry, and (b) the computed results for the two modes at the stripline port. The conversion from the microstrip mode to the waveguide mode is quite high (Agilent HFSS Ver. 5.6).

tor. If we leave the ground planes connected to the box walls, the second mode at the stripline port becomes a waveguide mode with a cutoff frequency determined by the width of the solution box. Any calculation we make may be influenced by the fundamental transmission properties of the waveguide mode. If we pull the ground planes back from the simulator walls, the second mode is now a parallel plate mode which has no cutoff frequency. However, there may be problems defining the stripline mode to the simulator. We are only allowed one reference vector from one of the ground planes to the stripline center conductor. The simulator typi-

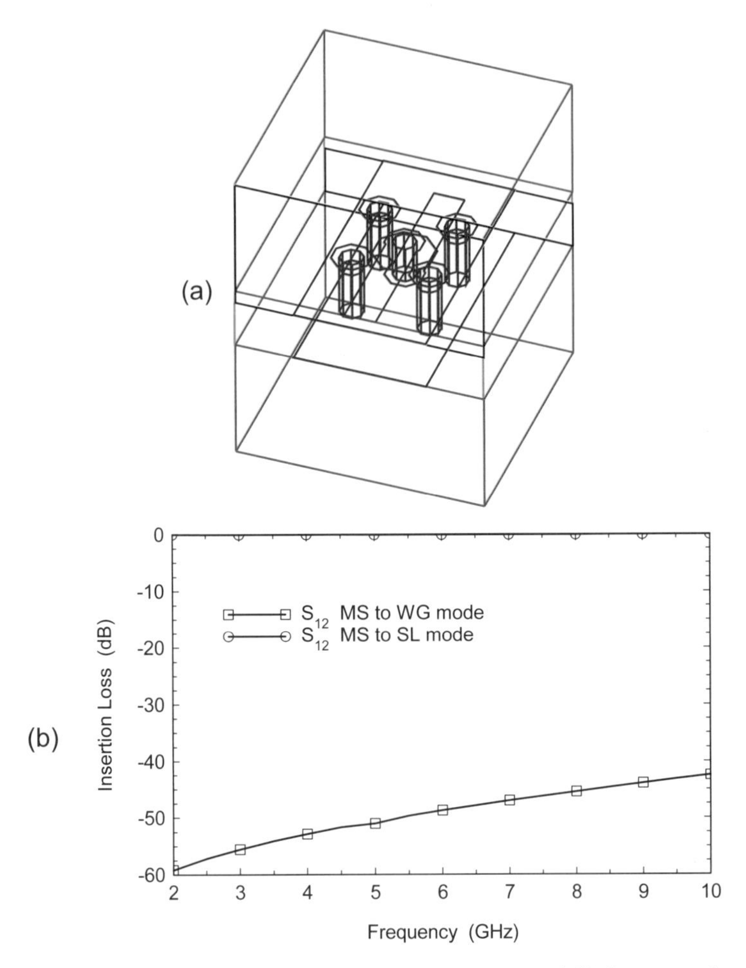

Figure 13.22 The Via1d numerical experiment: (a) the geometry, and (b) the computed results for the two modes at the stripline port. The added ground vias effectively short out the parallel plate mode at the transition (Agilent HFSS Ver. 5.6).

cally cannot figure out that the second ground plane should be connected to the first one. One clue that there is a problem can be found in the reported stripline impedance at the port. If the impedance is not correct, there is a problem and we cannot trust the results. We can also examine the field plot in the plane of the port.

As a compromise, we decided to use a full width, waveguide style port at the stripline end of the transition. To get a feel for the transmission properties of this port and the ground planes on the PCB we made a test structure (Figure 13.20(a)). We have waveguide type ports (hatched) at both ends of the box, the multilayer PCB, and the two stripline ground planes. The ports are essentially dielectric filled

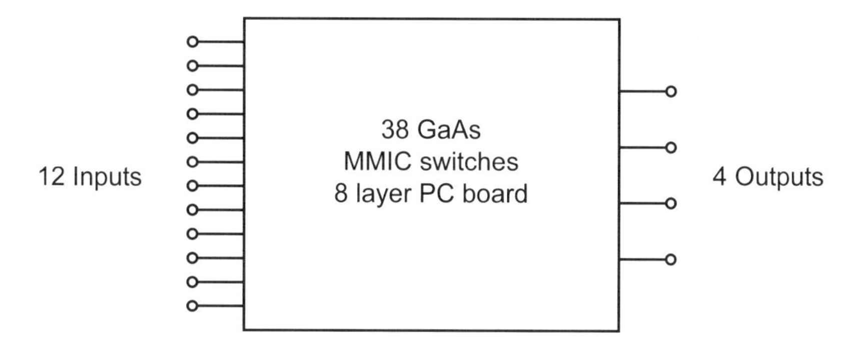

Figure 13.23 Schematic of the 10-GHz switch matrix. Any one of 12 inputs must be switched to any one of four outputs.

waveguides which transition to a pair of finite width, parallel strips supported by the PCB. We expect this structure to support transmission above some critical frequency that is hard to predict due to the complexity of the geometry. The results for this test case are shown in Figure 13.20(b). We find that the –3 dB point for the test structure is about 5 GHz.

Now we can look at the full transition structure. Figure 13.21(a) shows the Via1c geometry in Agilent HFSS. The microstrip and stripline ports have been hatched so they can easily be identified. The are two modes defined at the stripline port: one is the desired stripline mode, and the second is the undesired parallel plate/waveguide mode. The results for this case are found in Figure 13.21(b). With no ground vias, the coupling to the undesired mode is quite strong. In a real PCB, this energy can then easily propagate to other areas of the board.

In Figure 13.22(a) we have added the ground vias. Again we define two modes at the stripline port. The results for this case are found in Figure 13.22(b). With the ground vias in place, the coupling to the undesired parallel plate mode is greatly reduced and is well below the level of the test structure. The vias effectively short out this mode at the source, which is the multilayer transition itself.

13.3 A 10-GHZ SWITCH MATRIX

The third part of the multilayer transition story is the 10-GHz switch matrix shown schematically in Figure 13.23. This project was in fact the original motivation for the development of the controlled impedance transition. This device connects any one of 12 inputs to one of four outputs. Internally there are 38 single pole double throw (SPDT) GaAs MMIC switches and their associated driver circuits. There are no amplifiers internally to isolate the passive components from one another. So

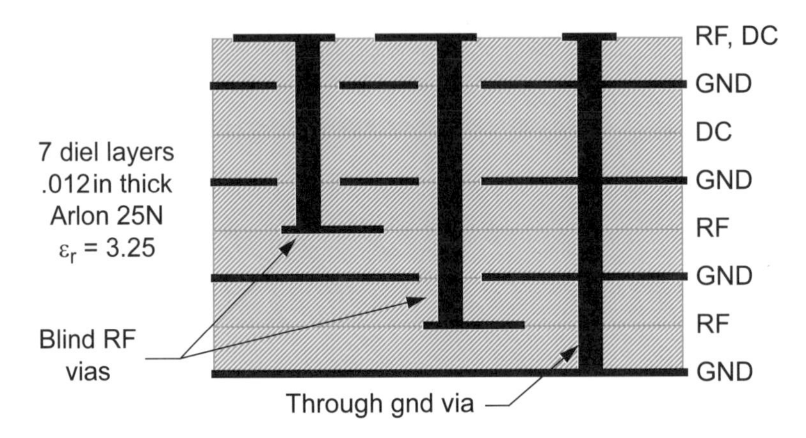

Figure 13.24 Board stackup for the 10-GHz switch matrix. Blind vias transfer the RF signals to one of two buried stripline layers. Through vias tie the ground plane layers together.

each signal path is a cascade of essentially passive components, each of which must have at least 15 dB return loss from dc to 10 GHz.

An eight layer PC board was proposed to mount the MMIC switches, the driver chips, and support the necessary RF and dc routing. Arlon 25N was chosen as a high-performance board material. A simplified view of the board stackup is shown in Figure 13.24. The design challenge here was a microstrip-to-stripline transition from the top of the board to one of two buried stripline layers. Blind vias were used for the signal path, and through vias tie the ground planes together. The transition also has to compensate the bondwire inductance from the MMIC switch chip to the board.

We have already discussed how the ground vias around the signal via form a "controlled impedance transition." The first step in the design process (Figure 13.25) is to lay out the ground via pattern for something close to 50 ohms. Next we will design an ideal back-to-back microstrip transition that incorporates the five hole via pattern. In the third step we design an ideal back-to-back stripline transi-

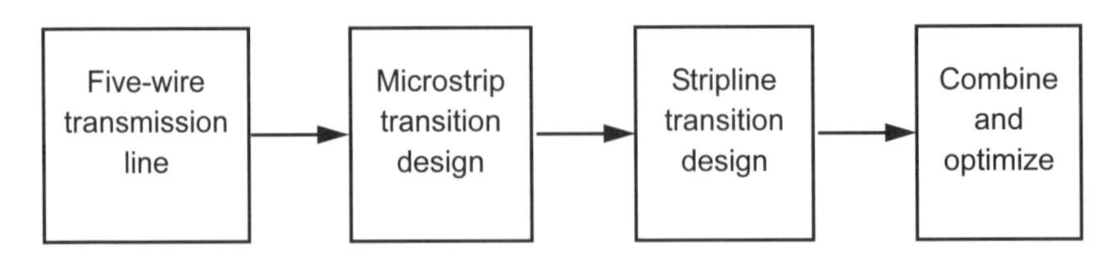

Figure 13.25 Multilayer transition design procedure. After choosing a five-wire line geometry, individual microstrip and stripline transitions are designed. The individual components are then combined and optimized, if necessary.

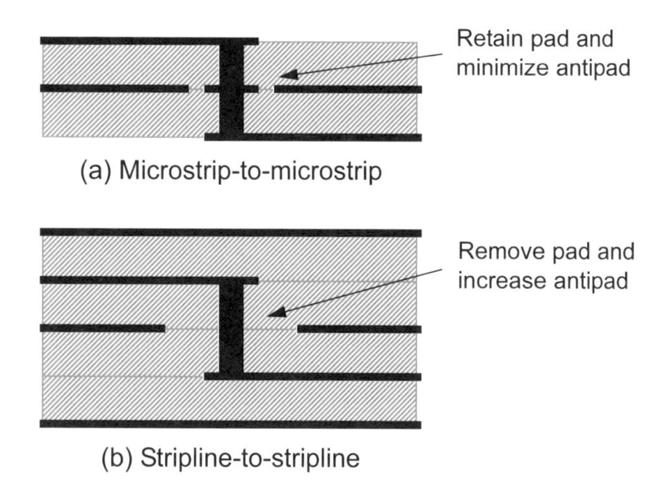

Figure 13.26 Back-to-back transition designs: (a) microstrip, and (b) stripline. Pads are retained or removed and antipad diameters adjusted for maximum return loss. The ground vias are present in the analysis, but omitted here for clarity.

tion which also incorporates the five hole via pattern. The final step combines the two ideal back-to-back structures to form the complete transition. Some optimization of the final structure may be needed, but it is typically only two or three variables. If the ground via spacing is allowed to vary during optimization, the final five-wire impedance is typically 60 or 65 ohms rather than 50 ohms.

One frustration of multilayer board technology is the number of variables. For low-speed digital circuits or very low RF frequencies we can use some simple rules of thumb for pad diameters and antipad diameters where vias pass through ground planes. At microwave frequencies we need to carefully tune pad diameters, antipad diameters, and even add or delete pads in unused layers. If we try to do the whole transition at once, we find there are too many variables and it is difficult to find an optimum solution. The procedure outlined in Figure 13.25 breaks the problem down into more manageable subtasks.

After deciding on via hole diameters and spacings, the next step is to design an optimum microstrip-to-microstrip transition (Figure 13.26(a)). This transition tends to be inductive, so we add extra shunt capacitance on the microstrip layers. We also retain the pad in the ground plane layer and minimize the antipad around that pad. Next, we design a back-to-back stripline transition (Figure 13.26(b)). This transition suffers from excess capacitance, so now we want to minimize the capacitance of the via barrel to the middle ground plane by removing the pad and increasing the antipad diameter. We can also add some series inductance to the signal traces in the two stripline layers. This can be accomplished by narrowing the trace width for a short distance. These two idealized back-to-back transitions can now be connected with the five-wire line to form the complete transition.

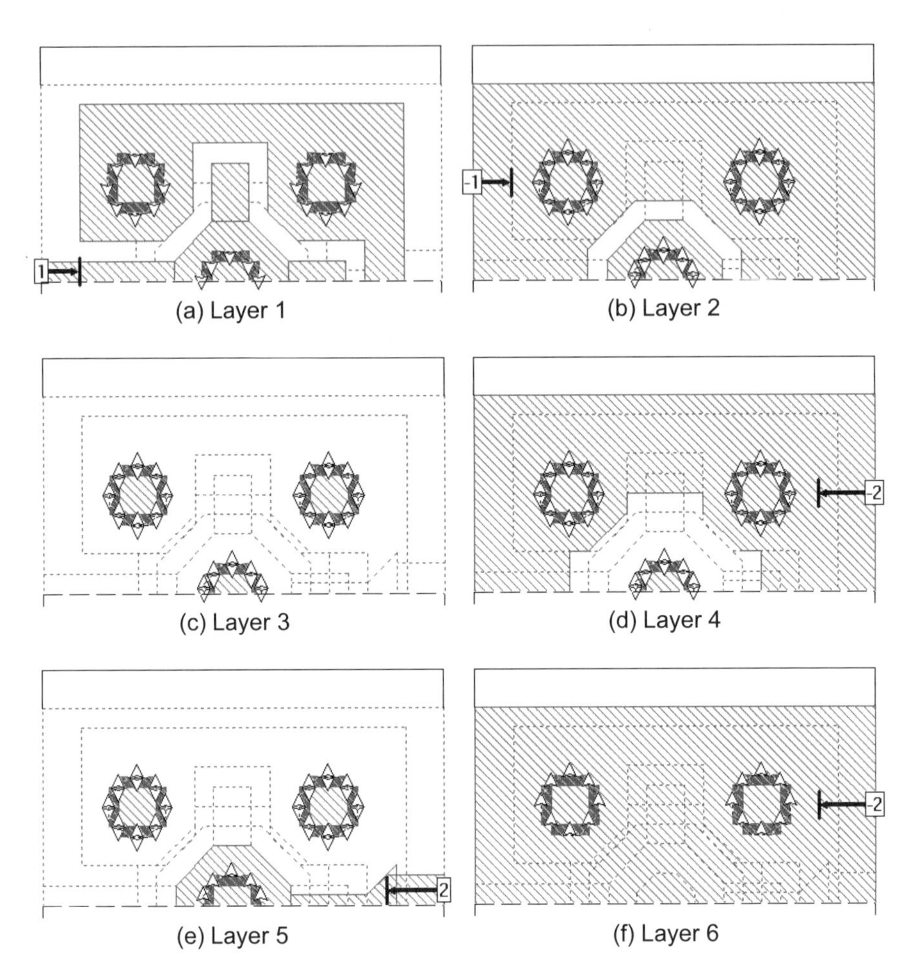

Figure 13.27 The optimized transition from microstrip in Layer 1 to stripline in Layer 5. The vias are 0.015 in diameter, the pads are 0.03 in diameters and the spacing between ground vias is 0.05 in center-to-center (Sonnet ***em*** Ver. 7.0).

The various layers for the optimized transition from microstrip to the first buried stripline layer are shown in Figure 13.27. The vias are all 0.015 in diameter, the pads are 0.03 in diameter and the spacing between ground vias is 0.05 in. Only the top half of the metallization pattern is shown because we have invoked symmetry in the field-solver analysis. Pulling back the ground plane metal from the box walls and driving the ports differentially minimizes the current flow in the box walls of the simulator. The size of the simulation box has also been reduced for clarity.

Our analysis of a simple back-to-back microstrip transition has shown that this transition tends to be inductive. In Layer 1 the ground vias have been tied together

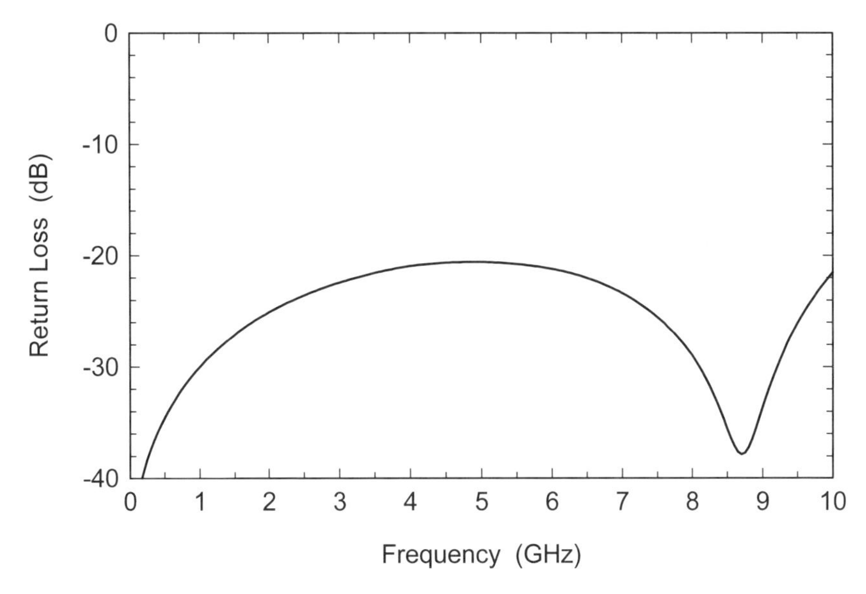

Figure 13.28 Predicted performance for the microstrip to stripline transition shown in Figure 13.27. The goal was 20 dB return loss from dc to 10 GHz.

and the antipad is minimized in order to maximize the shunt capacitance. Stubs were added to the signal pad to tune the transition. In Layer 2 we pass through the microstrip ground plane. We place a pad on the signal via and minimize the antipad to maximize the shunt capacitance. Layer 3 is a dc layer, all five vias pass through this layer, and there are no pads. In the actual analysis this layer can be deleted and the two adjacent dielectric layers merged.

Our analysis of a simple back-to-back stripline transition revealed that it tends to suffer from excess capacitance. Layer 4 is the upper ground plane for the stripline. In this layer we remove the pad from the signal via and maximize the antipad in order to minimize the shunt capacitance. In Layer 5 we have the stripline center conductor. There are no pads on the ground vias and we have added a length of high impedance line in series to tune the transition. Layer 6 is the lower ground plane for the stripline. If necessary, we can open a hole in the ground plane under the signal via which will further decrease the shunt capacitance.

When the complete transition was optimized, only the length of the stubs in Layer 1 and the length of the high impedance line in Layer 5 were allowed to vary. Because the microstrip and stripline transitions are electrically very different, there is no top-to-bottom symmetry in the final design. Also note that we have aggressively added and removed pads in the various layers.

The via transition was designed to include the bondwire inductance from the board up to the MMIC SPDT switch. The return loss for one transition with the

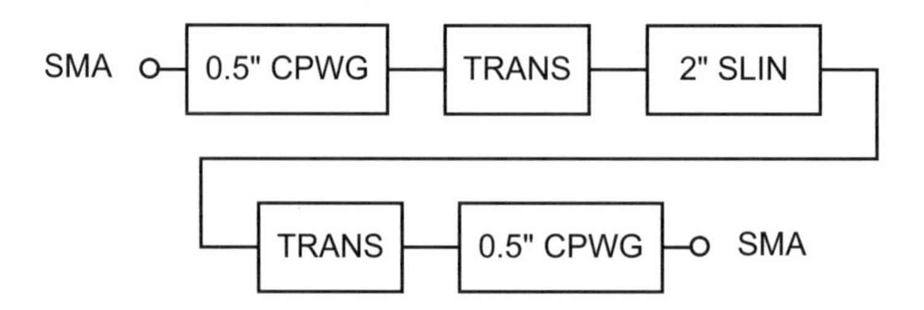

Figure 13.29 Test board for the transition shown in Figure 13.27. The I/O lines are coplanar waveguide with ground plane. The back-to-back transitions are separated by a 2-inch length of stripline.

bondwire inductance (0.25 nH) is shown in Figure 13.28. The goal was 20 dB return loss from dc to 10 GHz.

A test board was made with two transitions back-to-back separated by a 2-inch length of stripline (SLIN). The input and output lines are coplanar waveguide with ground plane (CPWG). We did not redesign the transitions for this test case, so the bondwire compensation is still in place even though the bondwires themselves are not. A schematic of the test board is shown in Figure 13.29. Measured versus modeled data for the test board are presented in Figure 13.30. The return loss is degraded at higher frequencies because the transition is now overcompensated.

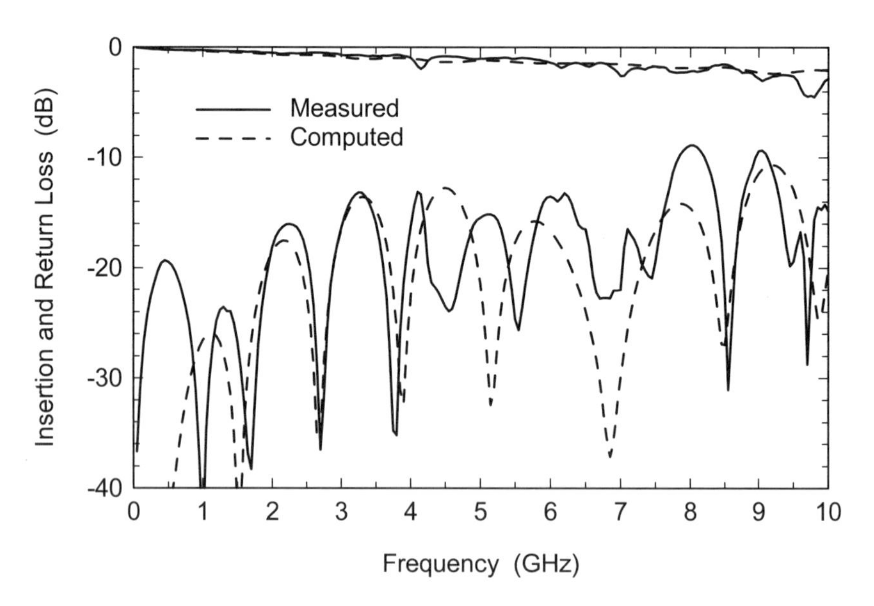

Figure 13.30 Results for the transition test board. The return loss is degraded at higher frequencies because the transitions are overcompensated. Data courtesy of M/A-COM.

Figure 13.31 One of the GaAs SPDT switches surrounded by three multilayer transitions. A group of three transitions was also analyzed on the field-solver, which indicated that the desired isolation could be achieved. Photograph courtesy of M/A-COM.

The photograph in Figure 13.31 is a close-up view of one of the GaAs SPDT switches surrounded by three of the multilayer transitions. The additional tuning stubs on the microstrip layer can clearly be seen. There was also a specification on isolation between ports for this hardware. Before the design was completed, a group of three transitions similar to the group above was also analyzed on the field-solver. The analysis indicated that the desired isolation could be achieved.

A complete switch matrix module was fabricated, tested, and shipped to the customer. Only a small amount of tuning for return loss was needed at the I/Os and no revisions were made to the PCB. The two transition designs are used many times in PCB. The design of the transitions was one key element in achieving first pass success on a fairly complex, high-frequency multilayer PC board.

13.4 SUMMARY

The first example in this chapter was a fairly simple six-layer PCB transition from microstrip to stripline at 1.5 GHz. At this relatively low frequency we were able make some fairly crude approximations to the actual geometry. This simplified modeling process allowed us to perform several numerical experiments on each transition in a single afternoon. It was this process that allowed us to develop some intuition for the effect of unused pads, the impact of the antipad diameter, and the impact of large tapers in buried layers. In the end, this problem was simple enough that intuition alone was enough to find a usable solution. We also learned the value of a simple 50-ohm line analysis to find errors in the problem setup or even bugs in the software. Obviously, the traditional microwave debugging process used on single layer circuits was not much help in this new multilayer environment. However, the field-solver could "get inside" the problem and provide very valuable design data.

We used the second microstrip-to-stripline transition example to point out some of the potential errors in the numerical analysis of a multilayer transition. By paying attention to return currents as well as signal currents we can modify our analysis to more accurately depict the true situation in a multilayer PC board. Any numerical analysis that allows significant return currents to flow in the "walls" of the numerical domain is probably suspect, this includes TLM and FDTD analyses with absorbing boundaries. It is easy to find examples in the technical literature and even in application notes from software vendors where the return current path in a multilayer structure is ignored and by default is in fact controlled by the boundary conditions of the simulator.

If we compare the closed box and laterally open MoM simulators for this problem, the laterally open simulator is perhaps the more "natural." If we are interested in the performance of an isolated via with no grounding vias, the laterally open tool is easier to apply. Once we put the grounding vias in place, we would expect any 2.5D or 3D solver to give us the correct solution. One advantage of the 3D solvers over the 2.5D solvers is the ability to measure the potential mode conversion into unwanted parallel plate modes. At some point, a 3D solution is often more efficient than a 2.5D solution for a multilayer problem. The 2.5D solvers are forced to mesh a lot of metal in the ground plane layers, which increases the solution time dramatically. An RF multilayer board with maximum ground plane fill is mostly metal with small gaps between the traces and ground planes. Recently, one research group has proposed an MoM analysis for this situation that meshes the gaps only [3]. This "magnetic current" solution is potentially more efficient than the traditional electric current formulation for this type of problem.

No matter how we do the analysis, the controlled impedance transition is one of those rare cases where there are many advantages and few disadvantages. The added ground vias provide a low impedance path for the return current, they form a vertical transmission line through the board, and they short out parallel plate modes that might otherwise be launched by the transition itself. The obvious disadvantage is the extra physical space needed for the ground vias.

The final example in this chapter is the 10-GHz switch matrix. This project was a good test case for the controlled impedance concept and forced us to develop a smarter design procedure for these transitions. Placing the ground vias gives us control over the return currents. Placing or removing pads and adjusting antipad diameters gives us control over the parasitics. By studying the microstrip and stripline transitions individually we learn what each one typically requires to optimize performance. The old approach to board layout that uniformly applies a given pad diameter and antipad diameter to every layer simply does not work for high performance RF or digital transitions. A successful design also requires a dialogue with the board fabricator in order to understand how the board will be made and what is allowed in each layer. The biggest limitation on performance is currently the via stub, which can be addressed using blind vias or back drilling if the cost can be justified by the improvement in performance.

References

[1] Pillai, E., "Coax Via—A Technique to Reduce Crosstalk and Enhance Impedance Match at Vias in High-Frequency Multilayer packages Verified by FDTD and MoM Modeling," *IEEE Trans. Microwave Theory and Tech.*, Vol. 45, No. 10, 1997, pp. 1981–1985.

[2] *Reference Data for Radio Engineers*, Fifth Edition, New York: Howard W. Sams & Co., Inc., 1973, p. 22–22.

[3] Abdul-Gaffoor, M. R., et al., "Simple and Efficient Full-Wave Modeling of Electromagnetic Coupling in Realistic RF Multilayer PCB Layouts," *IEEE Trans. Microwave Theory and Tech.*, Vol. 50, No. 6, 2002, pp. 1445–1457.

Chapter 14

Connectors

Many RF and high-speed digital systems rely on multilayer printed circuit board technology. In the last chapter we looked at how we transition signals between layers in a multilayer environment. In this chapter we will look at the transition on and off the board, or in other words, connectors. The fundamental problem is the same in both the RF and digital domains; we have to effectively transition the signal currents and the return currents over some frequency range. In the RF world we often are interested in thicker, single layer boards, although multilayer boards are used as well. In the digital world the individual layers are much thinner but the complete stackup may actually be quite thick, sometimes 0.25 in or more for a backplane application.

In this chapter we will consider single-ended edge launch and surface mount connectors. Through hole connectors are another topic of interest. The controlled impedance transition concept outlined in Chapter 13 also applies to through hole connectors. The only difficulty is we cannot vary the spacing between the pins and thus we have less control over the impedance of the vertical transition. Other problems of interest include larger, multipin digital connectors and differential pairs. Although we will not specifically look at those cases, the concepts we present here should be useful for them as well.

14.1 RF EDGE-LAUNCH CONNECTORS

Various types of edge-mounted or edge-launch connectors are available in several styles from several manufacturers. They generally have "fingers" that fit over the edge of the board and provide mechanical integrity and the connection for the return path. The connector center conductor transitions to a round pin that is 0.030 to 0.050-in diameter or to a rectangular tab.

Figure 14.1(a) shows a perspective view of a typical edge-launch SMA connector mounted on a 0.031-in thick single layer PCB. The board material is Getek II and $\varepsilon_r = 4.2$. In this case the center pin is 0.050-in diameter and 0.155-in

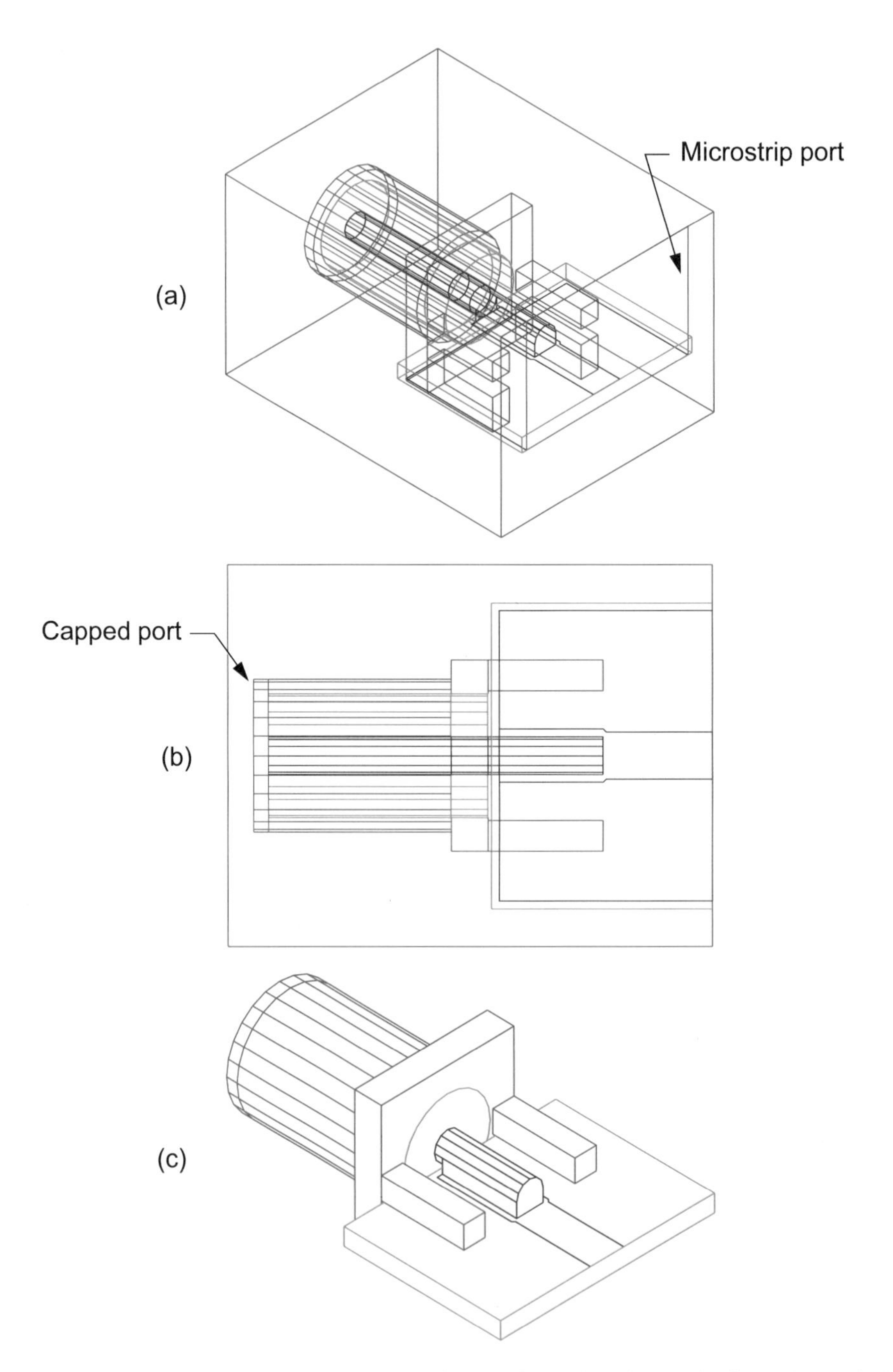

Figure 14.1 Edge-launch connector: (a) perspective wire-frame view, (b) top wire-frame view, and (c) perspective hidden line view (Agilent HFSS Ver. 5.6).

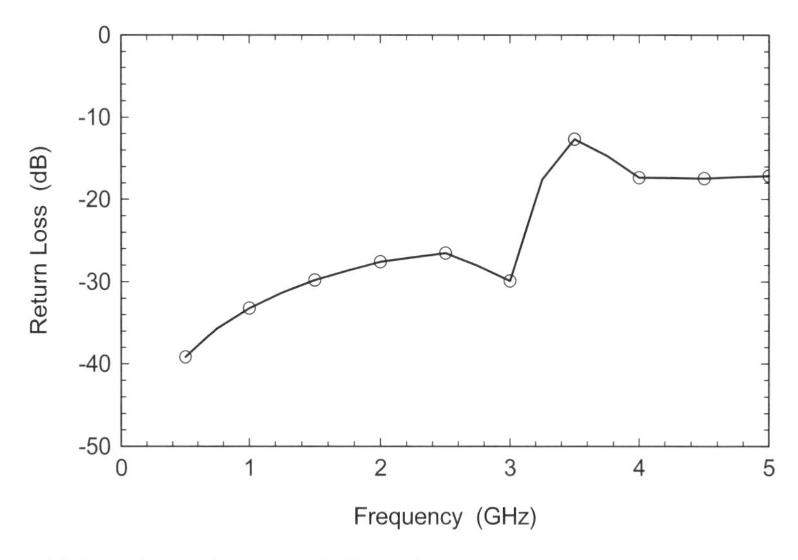

Figure 14.2 Return loss for the geometry in Figure 14.1.

long. Notice we have greatly simplified the geometry on the SMA side and focused our efforts on the PCB side of the connector. We tried to make the PCB side as realistic as possible. There is a 5-mil air gap between the edge of PCB and the connector, and all the metal traces on the PCB are pulled back 10 mil from the edge of the board.

As in the multilayer transition, we need to stop and think for a moment how any potential return path in the simulator might influence the results. Ideally we would like the only return path to be in the connector geometry itself and not in the walls of the simulator. With this in mind we chose to isolate the SMA side of the connector from the walls of the simulator by using a so called "capped port" (Figure 14.1(b)). The SMA port is pulled in from the simulator wall and is literally covered with a conducting metal cap, which we would normally view as a short circuit. However, in HFSS this is viewed by the software as a special case. The software knows it must launch energy down the coaxial line and not let any energy leak out into the surrounding geometry. At the other end of the problem the ground plane for the microstrip port is referenced to the box wall of the simulator. A perspective view with hidden lines removed is shown in Figure 14.1(c).

Although the approach we just outlined sounds good in theory, in this case we ran into some difficulty. Figure 14.2 shows the return loss of the geometry in Figure 14.1. We obviously have some kind of resonance. In Figure 14.3 we have plotted the magnitude of the E-field down the center line of the geometry at 3.5 GHz. The microstrip ground plane and the connector shell together form a resonant stub sitting in solution box. The only solution to this problem might be to shorten the overall length of the structure and move the resonance higher.

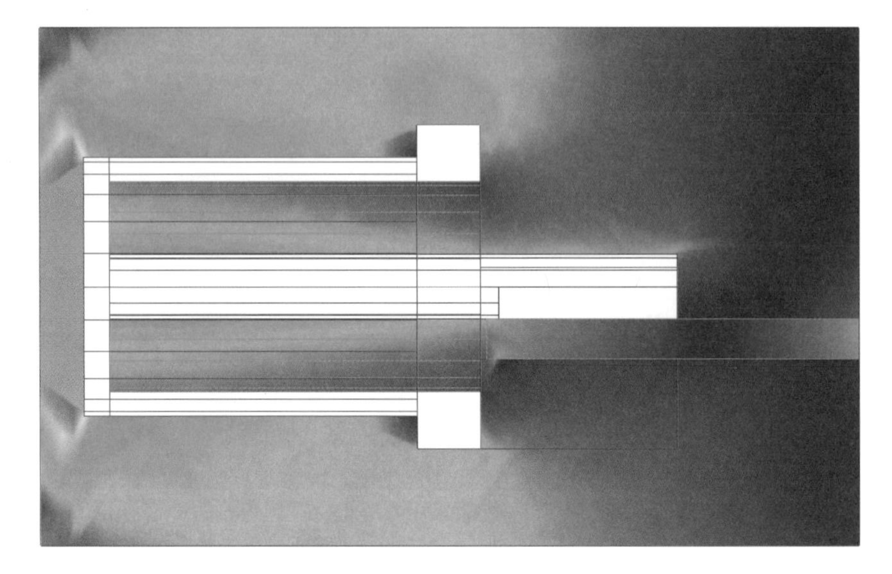

Figure 14.3 Magnitude of the E-field down the center line of the connector at 3.5 GHz. The microstrip ground plane and the connector shell form a resonant stub inside the solution box (Agilent HFSS Ver. 5.6).

In this case we decided to connect both the SMA end and the microstrip end of the problem to the simulator walls (Figure 14.4(a)). The return loss for this case and the case in Figure 14.1 are virtually the same up to 2 GHz, so we can assume that the connector geometry is controlling the return path and not the simulator walls. Now we can begin to explore ways to improve the RF performance of the connector. In our original problem, the fingers over the top side of the board have no connection to the microstrip ground plane. So in Figure 14.4(b) we have added vias from the top fingers to the microstrip ground plane. Surprisingly, this actually hurts the performance of the connector (Figure 14.5(a)). We will offer a theory for this behavior at the conclusion of this section.

Experimentally, we know that shortening the center pin of the connector usually improves performance. In Figure 14.4(c) the center pin is now only 65 mil long. The return loss plot in Figure 14.5(a) shows that this is a significant improvement. We also know from experiments that adding a solder fillet or some copper foil from the microstrip ground plane to the connector body sometimes improves performance. The results for this experiment are also plotted in Figure 14.5(a). The change is not dramatic but it looks like there might be benefits at higher frequencies. The high frequency performance would also improve if we selected a connector with a smaller diameter center pin or tab style pin.

So far we have relied on intuition or previous experimental work to guide our optimization of the transition region. But how can we proceed intelligently from

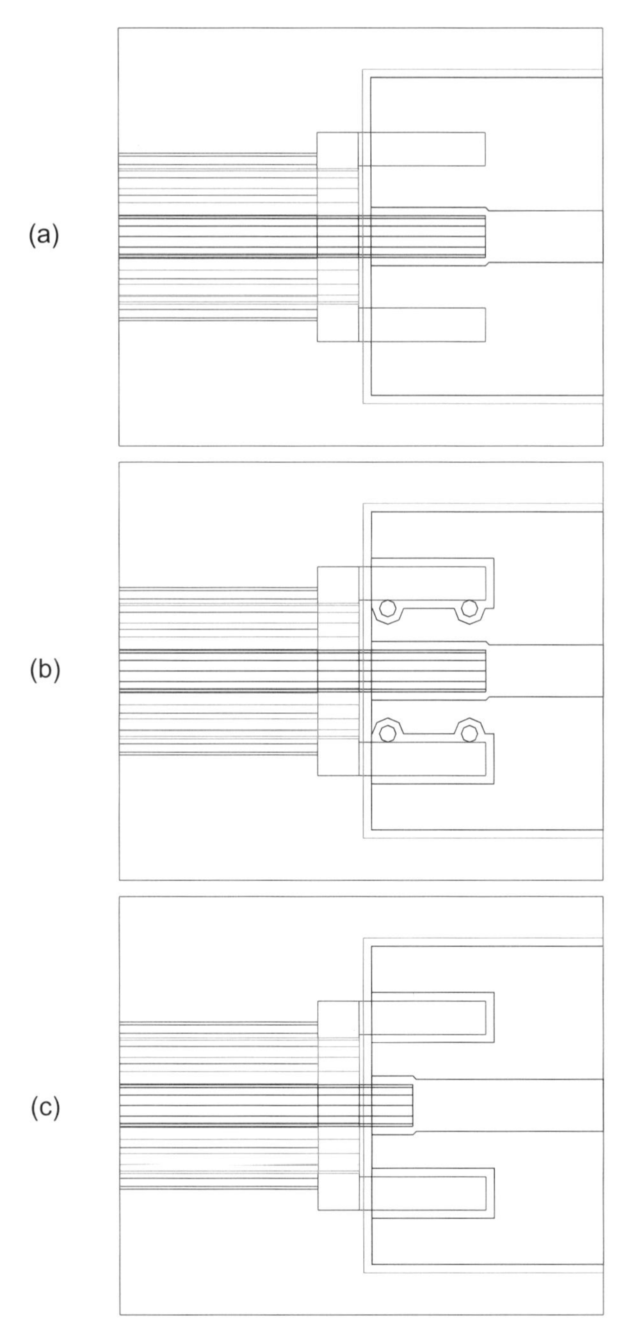

Figure 14.4 Top views of the edge-launch connector optimization: (a) starting point; (b) vias are added to connect the top fingers to the ground plane; and (c) the center pin is shortened (Agilent HFSS Ver. 5.6).

(a)

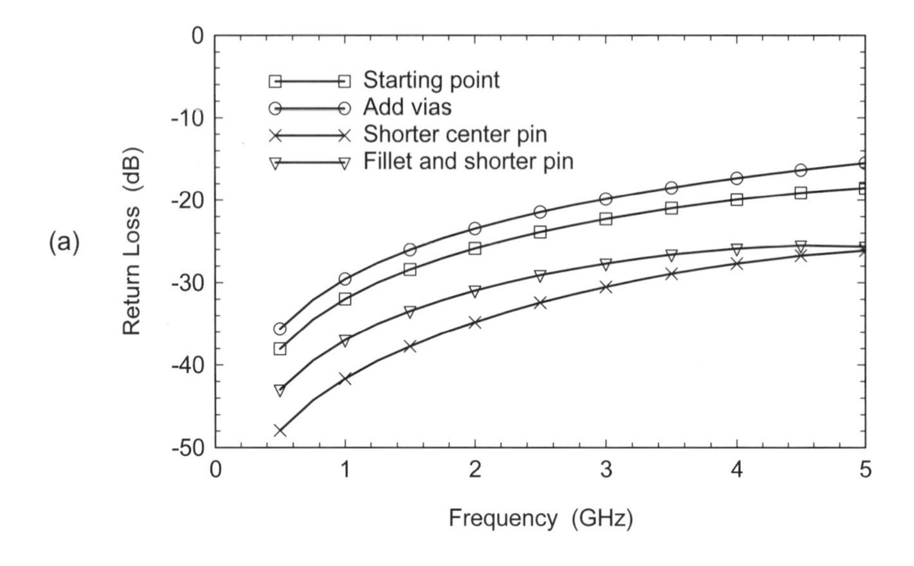

(b)

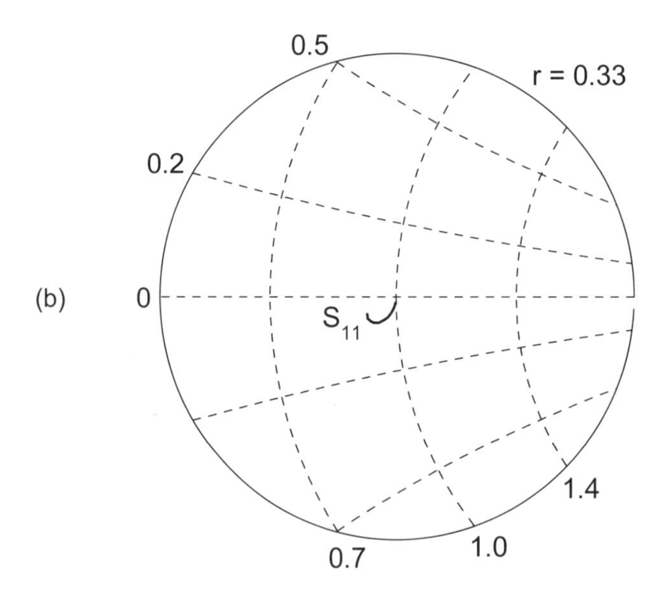

Figure 14.5 Results of connector optimization experiments; (a) return loss results for the geometries in Figure 14.4, and (b) de-embedded reflection coefficient at the edge of the board.

this point? If we de-embed our best result down to the edge of the board on the SMA side, we can plot S_{11} on a Smith chart (Figure 14.5(b)). The location of the S_{11} trace indicates that the connector looks capacitive. This perhaps explains why

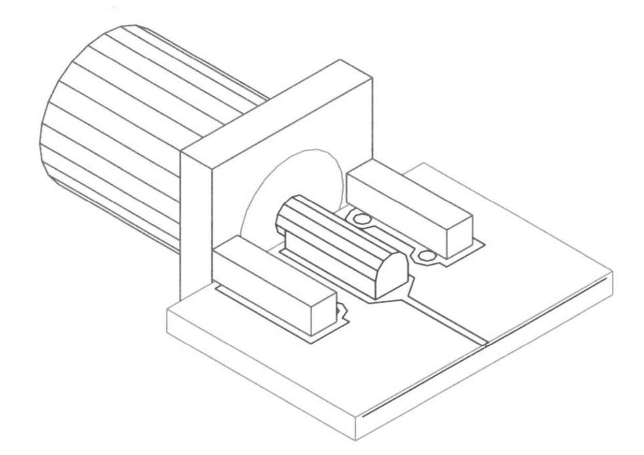

Figure 14.6 Perspective view of the digital edge-launch connector (Agilent HFSS Ver. 5.6).

adding the vias actually hurts the performance; the excess inductance in the return path without the vias in place actually helps to compensate the excess capacitance of the connector. We can take the de-embedded *S*-parameters to our favorite circuit simulator and explore ways to improve the performance. Hanging positive or negative shunt capacitors at either port or placing positive or negative inductors in series will give us ideas on how the geometry could be modified. In this case we would like to remove some capacitance at the launch point onto the board. In the next example we will show how that can be done.

14.2 DIGITAL EDGE-LAUNCH CONNECTORS

The next example was developed for a digital application; the connector is the same but the board material and stackup are now different. We are now interested in a six-layer board made from Nelco 4003. The topmost microstrip layer is 5-mil thick and has $\varepsilon_r = 3.7$. The overall thickness of the board is 32 mil. A perspective view of this new example is shown in Figure 14.6.

To simplify and speed up the modeling we are only including the top two metal layers of the PCB in the analysis. We need some vias in place initially to make the connection from the buried microstrip ground plane to the fingers of the connector. The initial geometry for this example is shown in Figure 14.7(a). If the RF transition had trouble with excess capacitance, we can guess that the digital version will be even worse with a much closer ground plane. The return loss for the starting geometry is shown in Figure 14.8.

If the problem is excess capacitance near the launch point, perhaps we can reduce that capacitance by removing ground plane metal in that region. Figure 14.7(b) shows the next geometry that we tried. We arbitrarily choose to use the out-

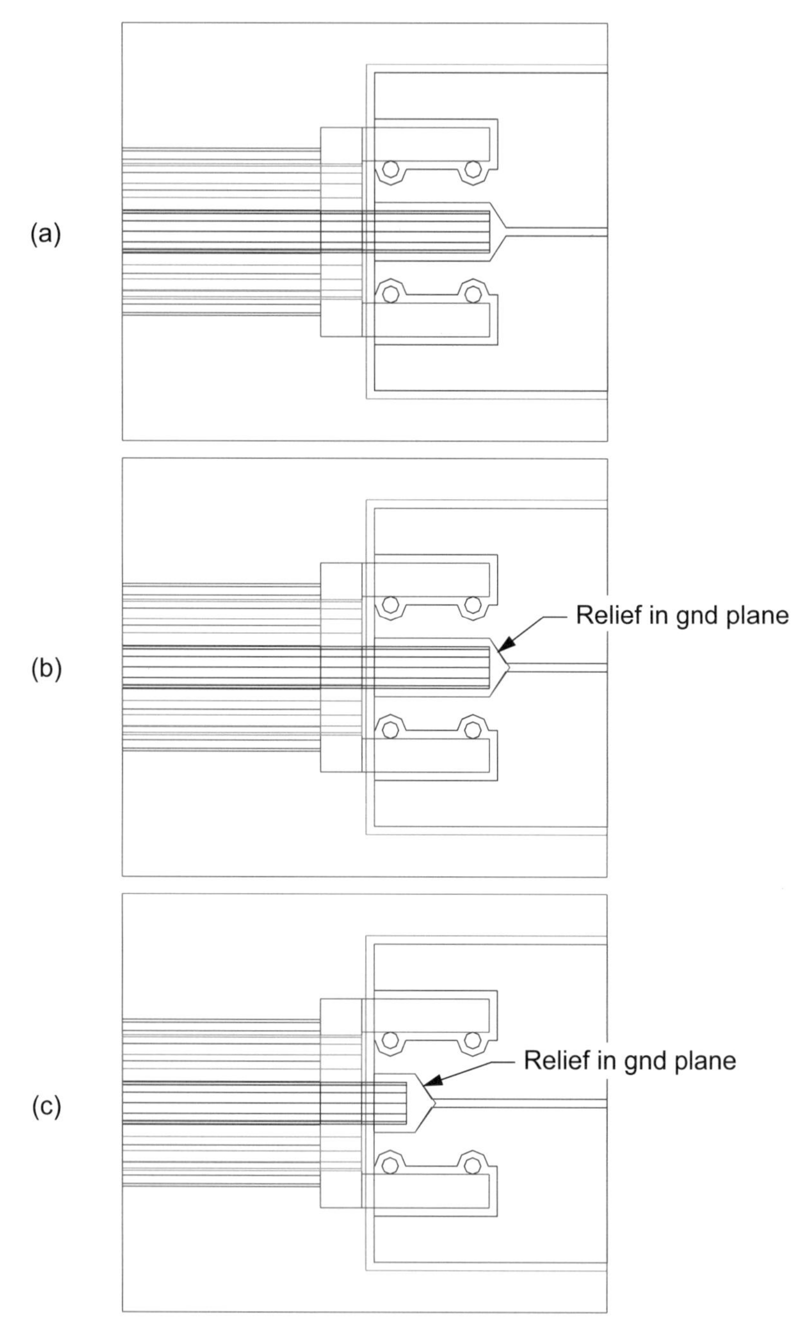

Figure 14.7 Top views of the edge-launch connector optimization: (a) starting point includes vias to the ground plane; (b) ground plane removed under the center pin; and (c) the center pin is shortened and ground plane is removed (Agilent HFSS Ver. 5.6).

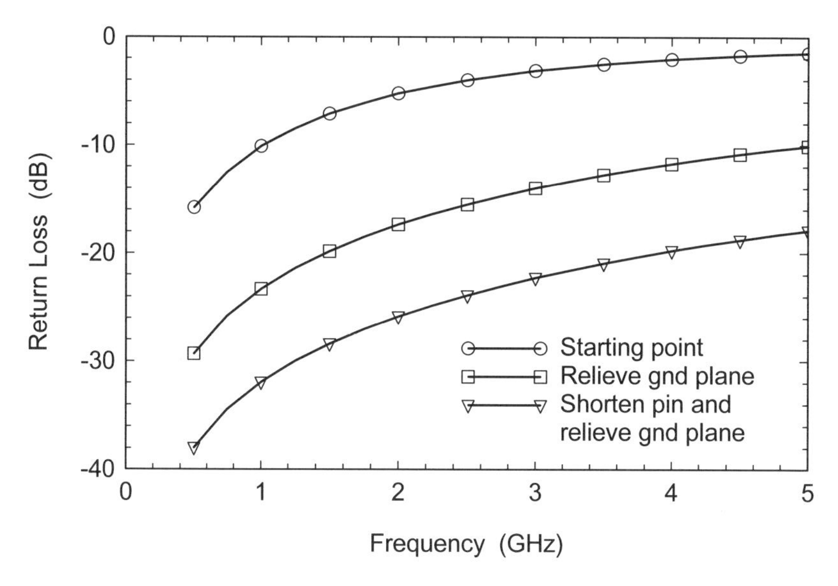

Figure 14.8 Return loss results for the geometries in Figure 14.7.

line of top surface pad to remove the ground plane metal. In Figure 14.8 we can see that the return loss is dramatically improved. We could also provide a continuous taper in the signal line and the ground plane relief [1]. This is analogous to the tapered coax to microstrip transition found in [2]. While this approach may provide superior high-frequency performance, it may also take up a fair amount of real estate on the board.

Now we can apply some of the other tricks that we learned in the previous example. In the third experiment, Figure 14.7(c), have shortened the center pin and removed the ground plane metal under the pad. The return loss for this case is also shown in Figure 14.8. We could now optimize the geometry of the ground plane relief and possibly get better performance. We could also choose a connector with a smaller diameter center pin or a small tab instead of a pin.

14.3 ANOTHER DIGITAL EDGE-LAUNCH EXAMPLE

To push the digital edge-launch case to a lower return loss level or to higher frequencies we need a slightly different style of connector and more aggressive tuning of the PCB metal patterns. Figure 14.9(a) shows a new edge-launch example using a connector with a small tab for the signal connection. The tab dimensions are 10 by 20 by 75 mil. In this case the PCB is 90-mil thick FR4 with an assumed $\varepsilon_r = 4.3$. The thickness of the topmost dielectric layer is 7 mil. The lower fingers of the connector were machined to fit over the board.

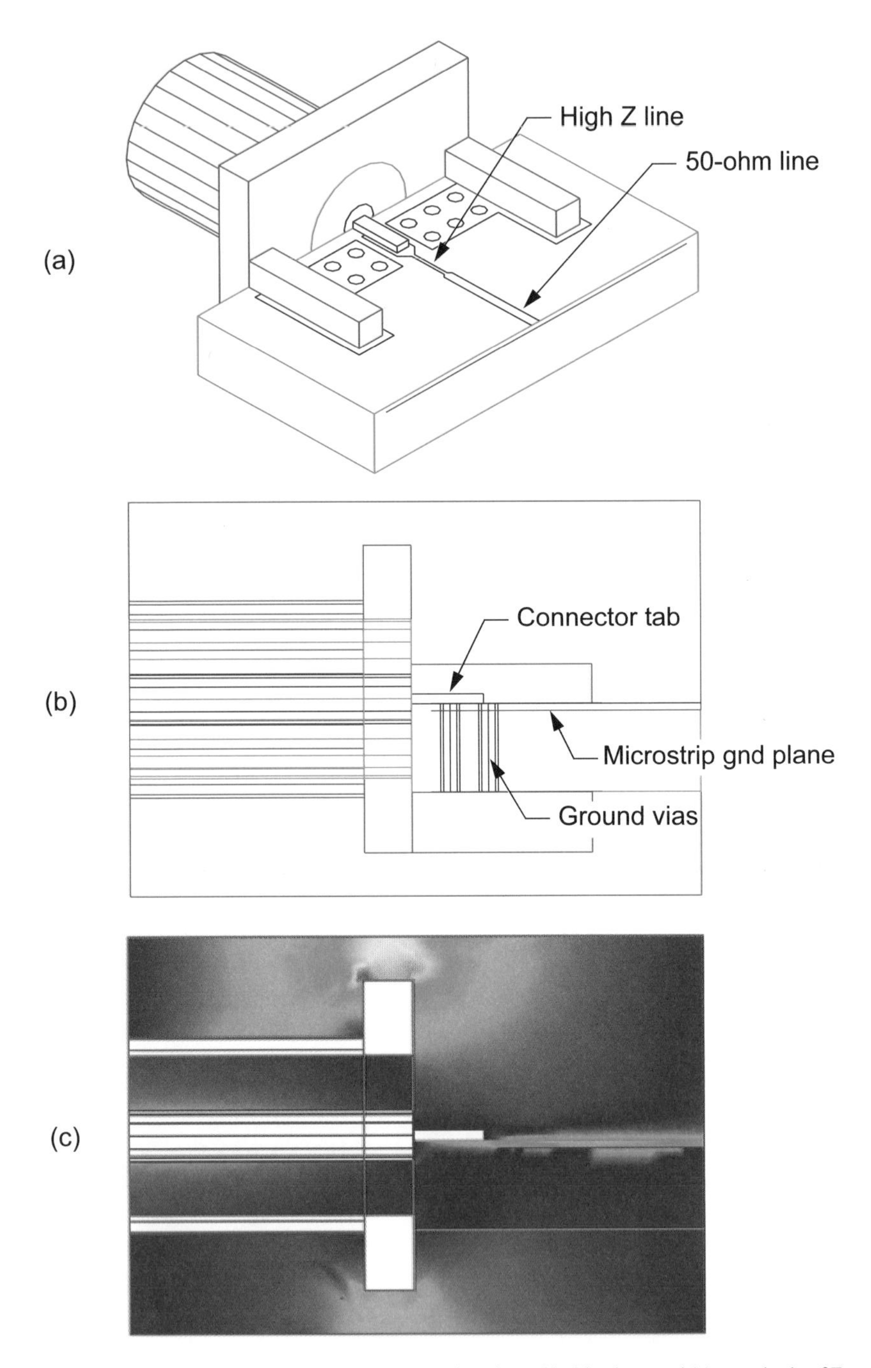

Figure 14.9 Digital edge-launch connector: (a) perspective view; (b) side view; and (c) magnitude of E-field down the center line of the analysis box (Agilent HFSS Ver. 5.6).

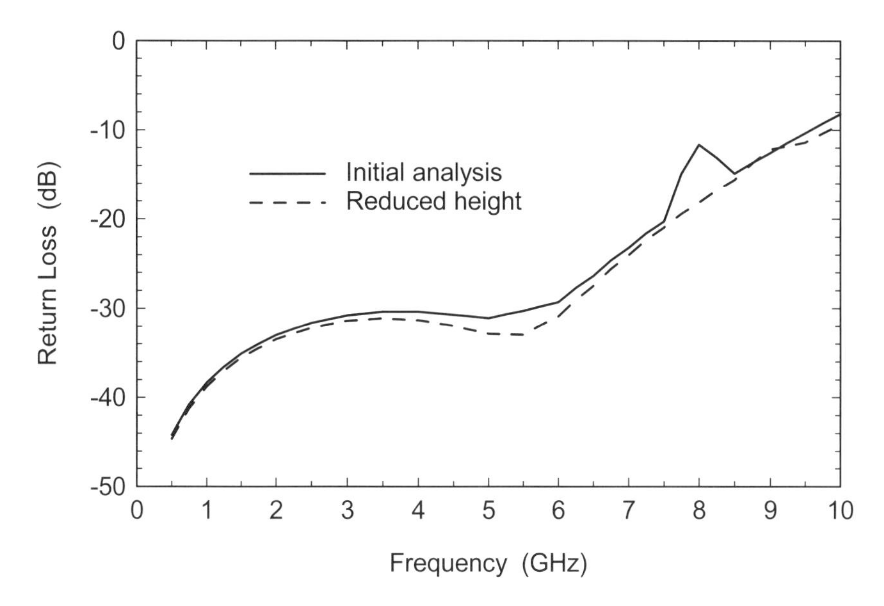

Figure 14.10 Return loss results for the geometry in Figure 14.9.

As in the previous examples, the ground plane under the center contact has been relieved. In addition, some series inductance has been added on the PCB side of the connector. We realized the inductance as a short length of 5-mil wide line between the connector and the 50-ohm trace. The width of the ground plane relief and the length of the high impedance line were optimized on the field-solver for return loss performance from 0 to 5 GHz.

For this demonstration we decided to push the analysis frequency higher to see what would happen. The return loss for this case is plotted in Figure 14.10. First notice that the return loss is below 20 dB to 7.5 GHz. Second, note the spike in the return loss near 8 GHz. If we look at the magnitude of the E-field at 8 GHz (Figure 14.9(c)), we notice a hot spot along the length of the analysis region. We can run some quick numbers with our pocket calculator and conclude this is probably a waveguide type resonance along the length of the analysis region.

One possible way to kill this resonance is to shrink the analysis box vertically so the upper and lower walls touch the main body of the connector. Now the main body of the connector should effectively short out the waveguide resonance. The results for this second experiment are shown in Figure 14.10. The strong resonance is gone but there are still some discontinuities between 8 and 10 GHz. If we compute the cutoff frequency of the cross-section of the analysis box, we find that it falls in the 9 to 10-GHz frequency range also. So if we want to push our analysis of the connector into this region we need to make the analysis box smaller, or find some other way to kill the waveguide modes.

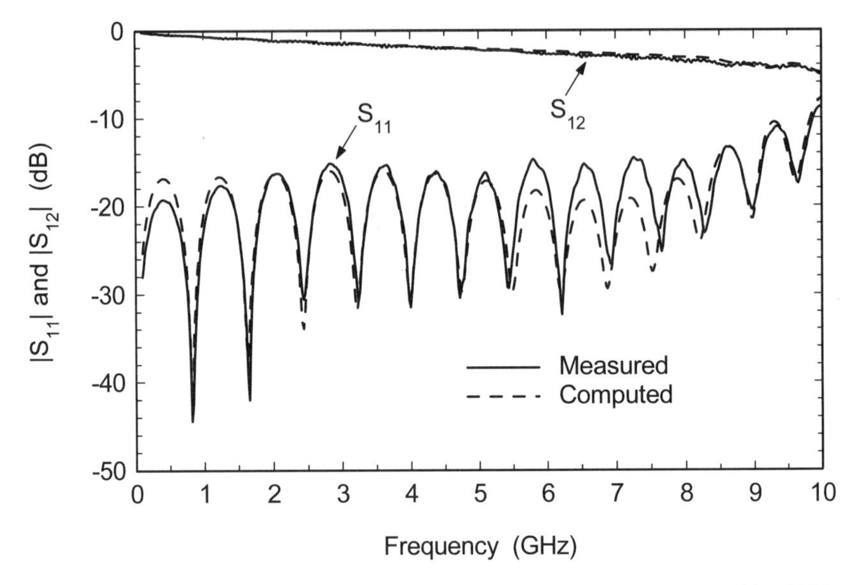

Figure 14.11 Measured versus modeled data for two edge-launch connectors separated by 4.15 in of microstrip line. The impedance of the fabricated microstrip line is about 10% too low.

This edge-launch model was used in a test board and data was taken on a hairpin shaped microstrip through line about 4.15-in long (Figure 14.11). The return loss is not as good as the single connector analysis, primarily because the microstrip line impedance on the test board is about 10% to low. This was verified with careful physical measurements and a TDR analysis. Still, when we put the as-built dimensions in the computer model, the agreement with the measured data is fairly good. For reference, the lossless analysis of the connector using symmetry took about 2 min per frequency point on a 1.13-GHz Pentium notebook, circa 2002. When loss was added the solution time tripled to about 6 min per frequency point.

14.4 THROUGH HOLE SMA CONNECTORS

There are some engineers in both the RF and digital communities who believe it is impossible to obtain good high-frequency performance with a through hole connector. However, we believe that through hole SMA connectors can be treated as a logical extension of the controlled impedance transition concept in Chapter 13. Although we cannot change the position of the connector ground pins, we can still tune the pad and antipad diameters to optimize the performance. There are a few other tricks we can apply as well.

Figure 14.12(a) shows a perspective view of a through hole SMA connector mounted in a high-speed digital multilayer PCB. We are using symmetry so only

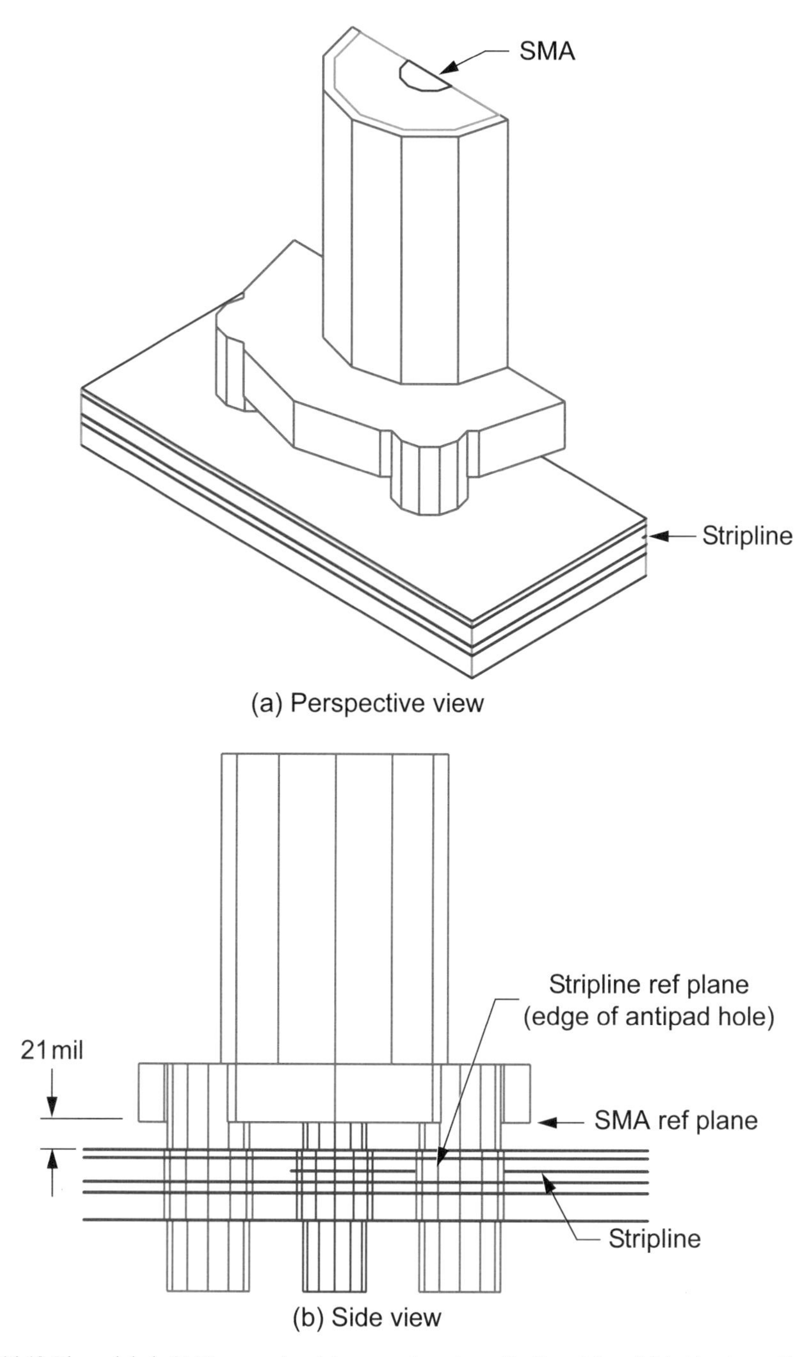

Figure 14.12 Through hole SMA connector: (a) perspective view of half model, and (b) side view with substrate omitted for clarity (Agilent HFSS Ver. 5.6).

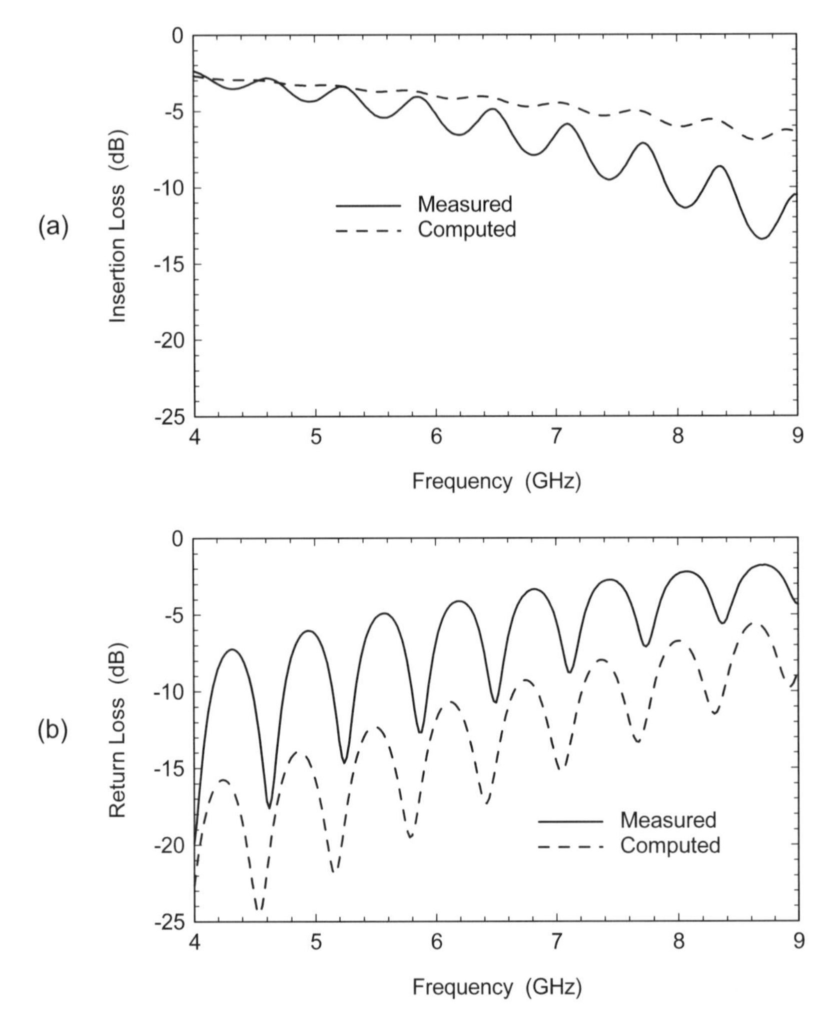

Figure 14.13 Results for the first field-solver model compared to the measured data: (a) insertion loss, and (b) return loss (Agilent HFSS Ver. 5.6). Data courtesy of Bayside Design, Inc.

half the physical geometry is shown. The total FR4 board thickness is 55 mil with an assumed $\varepsilon_r = 4.1$. The metallization is half-ounce copper (0.7 mil) and in this case the metallization thickness was included in all the models. The connector sits 21 mil above the board (Figure 14.12(b)). The connector transitions to a buried stripline layer with the stripline center conductor in the third metal layer down. The vertical location of the stripline center conductor is slightly asymmetrical between the ground planes. There are 70-mil diameter pads on the signal via in every metallization layer. But the antipad diameter is also quite large, 160 mil.

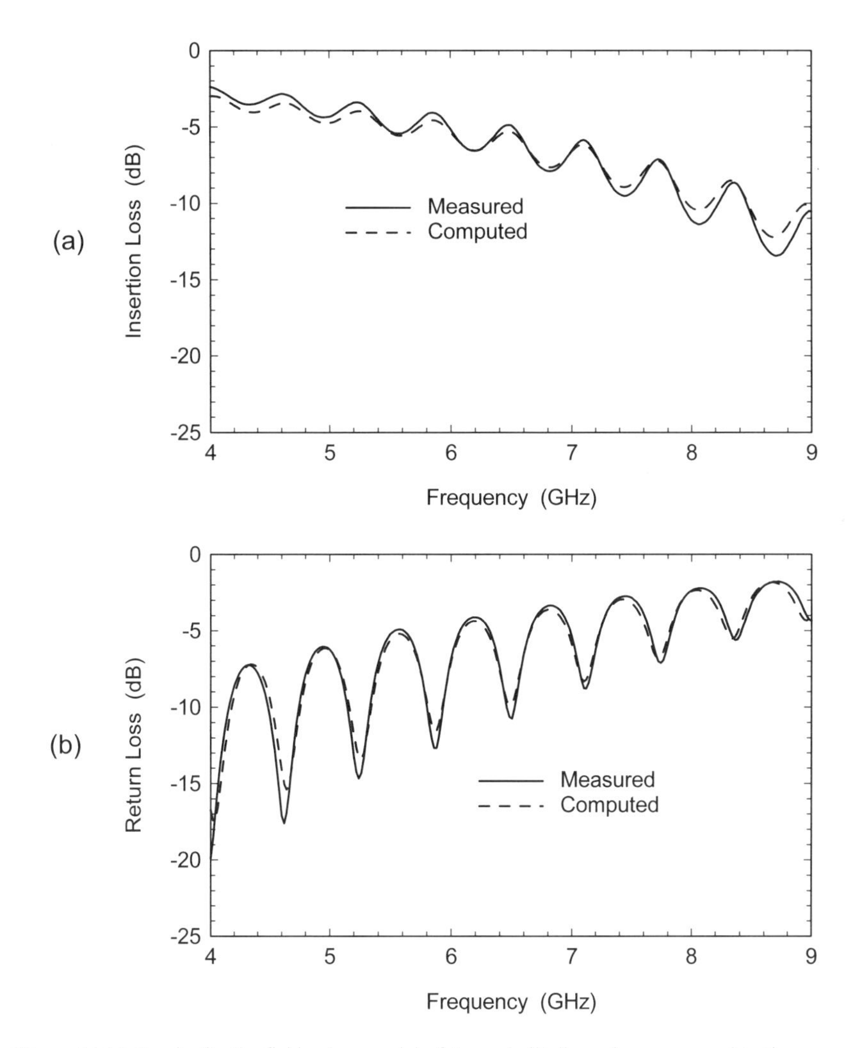

Figure 14.14 Results for the field-solver model of the as-built dimensions compared to the measured data: (a) insertion loss, and (b) return loss (Agilent HFSS Ver. 5.6).

When this project was presented to me (D.S.) for the first time, a test board had already been built and very carefully measured. The spacing between the two connectors was roughly 4.5 in. The measured versus modeled data for this first experiment is shown in Figure 14.13. Obviously the agreement is quite poor and the computer prediction is actually much better than the measurement. The measurement procedure was examined very carefully and the board was remeasured several times with the same result [3].

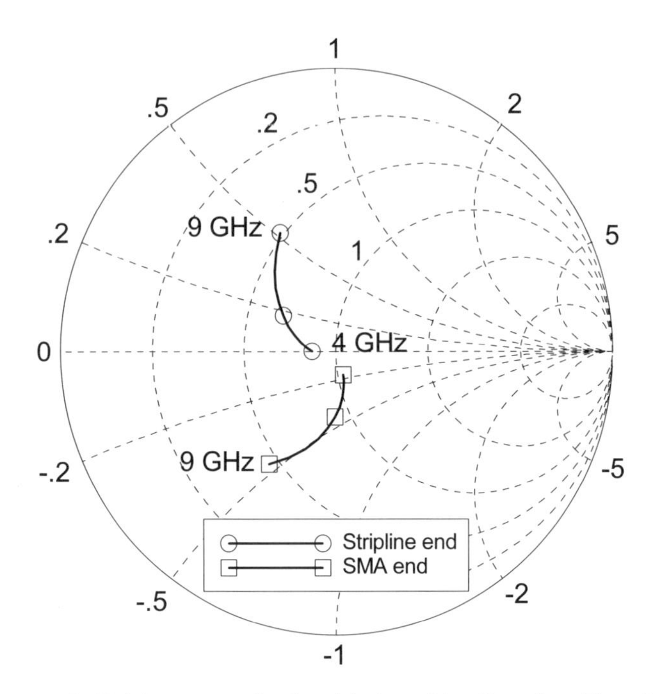

Figure 14.15 De-embedded *S*-parameters for the original transition dimensions. The reference planes are the bottom edge of the Teflon and the edge of the antipad hole.

Finally, I suggested we build a computer model with all the antipad diameters reduced to 80 mil. The model now predicted much worse performance than the measurement. I then encouraged the project engineer to go back and carefully examine the board and the Gerber plots for every layer. He found that the antipad diameter in the top and bottom layers was 90 mil rather the specified 160 mil. The results for the corrected model and the measurement are compared in Figure 14.14. Of course, this is not the first or the last time that a board will come back with an error in it.

Now that we had confidence in our model, we then decided to see if the RF performance of the connector to stripline transition could be improved. Again, the best way to start is to examine the de-embedded results on a Smith chart. We de-embedded the original result file down to the bottom edge of the Teflon on the SMA side and to the edge of the antipad hole on the stripline side (Figure 14.15). The SMA end is suffering from excess capacitance (below the horizontal axis) while the stripline end is suffering from excess inductance (above the horizontal axis).

We could probably reduce the excess inductance by reducing the antipad diameter in only the stripline ground plane layers. But this will also cause the parasitic capacitance to these layers to increase. Instead, we opted to put a small tuning pad

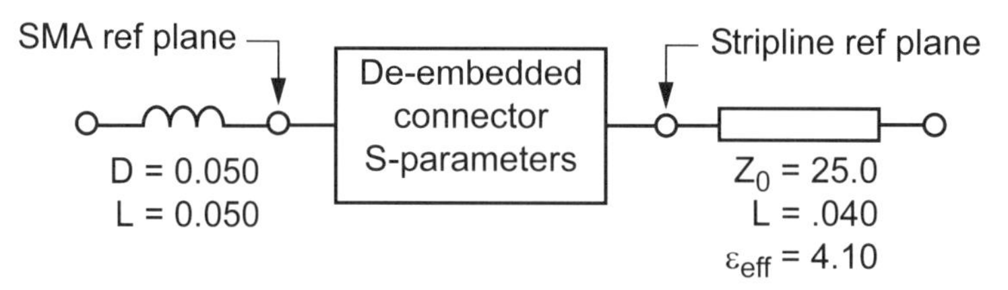

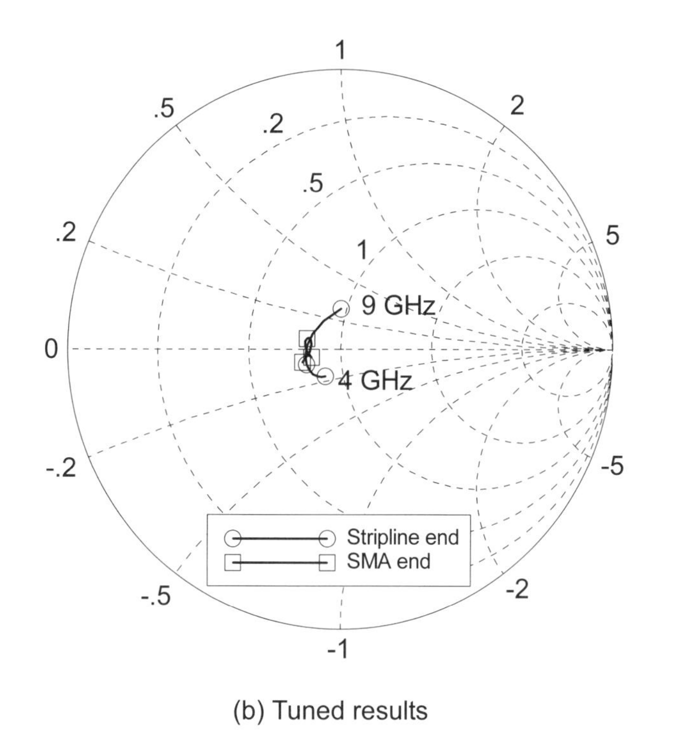

Figure 14.16 Tuning the first field-solver connector model: (a) extra center pin length is added on the SMA end and a small, capacitive pad is added on the stripline end. (b) The *S*-parameters after tuning.

on the stripline center conductor at the edge of the antipad hole. In our circuit model this is a short length of low impedance stripline (Figure 14.16(a)). We picked a line width and length by manually tuning the stripline transmission line model in our circuit simulator. On the SMA end we need some excess inductance in series to tune out the excess shunt capacitance. In our circuit model, we added extra center pin length using a simple "wire" model, which only requires a diameter and a length. We also could have computed the equivalent five-wire impedance of the

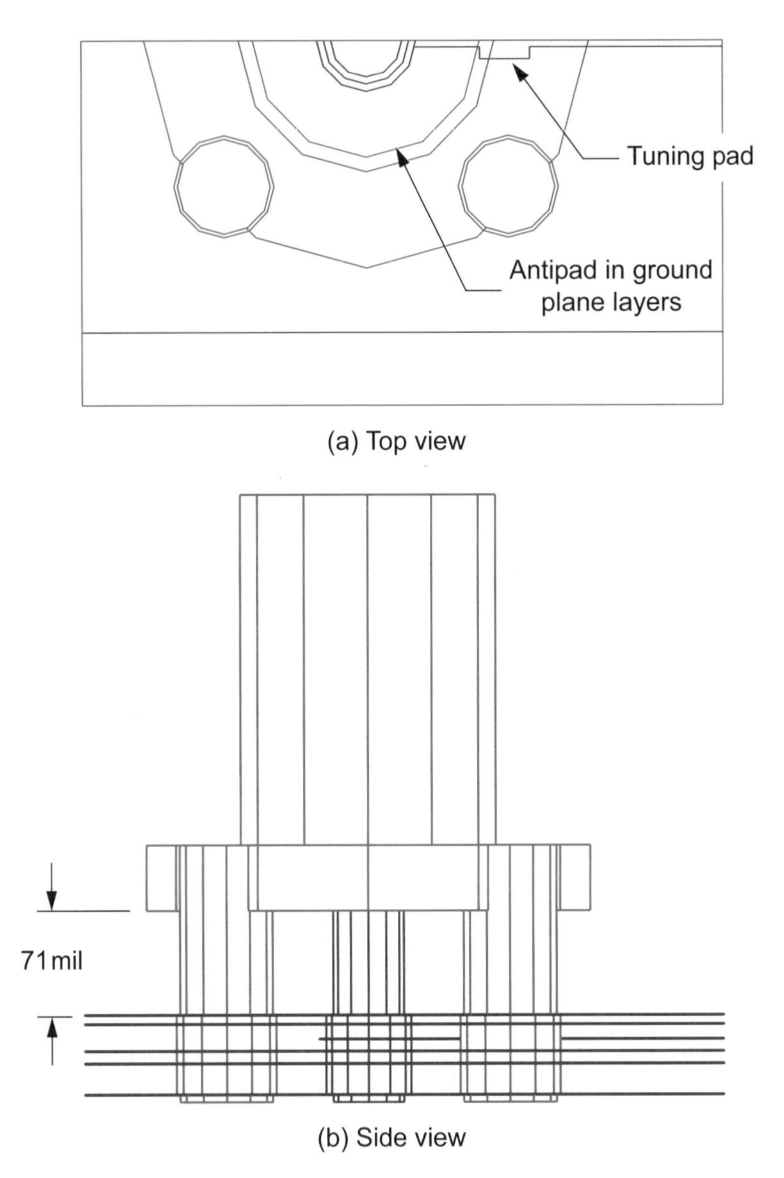

Figure 14.17 Modified transition design showing: (a) capacitive tuning pad, and (b) increased height of the connector above the board (Agilent HFSS Ver. 5.6).

center pin and ground pins and then used a transmission line model in the circuit simulator. The results of our tuning are shown in Figure 14.16(b).

The new field-solver model for the through hole SMA is shown in Figure 14.17. The added tuning pad in the stripline layer can be seen in Figure 14.17(a). In

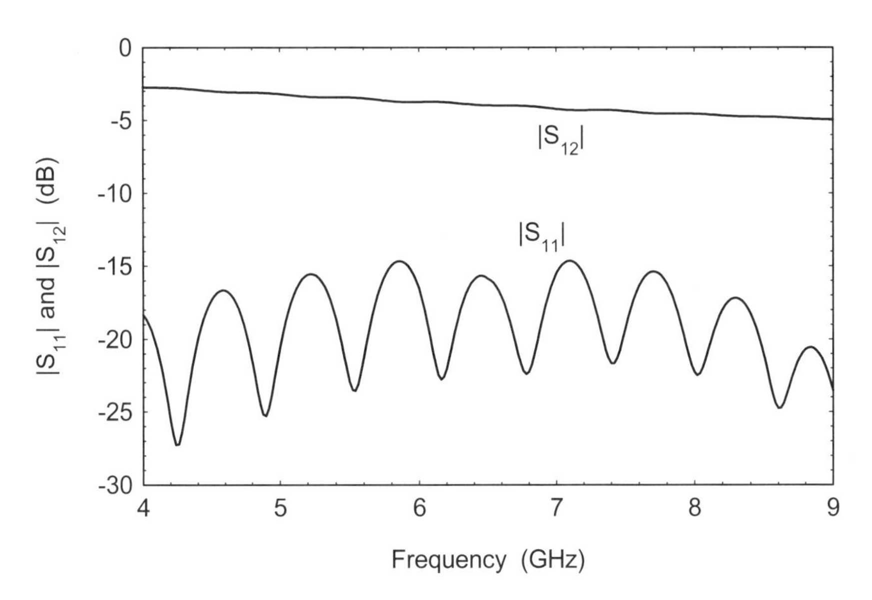

Figure 14.18 Computed results for the transition geometry in Figure 14.17. Two transition models are cascaded with 4.5 in of lossy stripline (Agilent HFSS Ver. 5.6).

Figure 14.17(b) the connector is now 50 mil higher off the board. This has the added benefit of reducing the length of the unused pins that extend through the board. On the center pin in particular, this extra length can only hurt high frequency performance.

In analyzing the new field-solver model we cascaded two transition models with 4.5 in of lossy stripline as before. The computed results for the optimized transition pair can be found in Figure 14.18. Now that the performance is much improved, we might consider doing some further optimization with the field-solver model in the optimization loop.

14.5 SURFACE MOUNT SMA CONNECTORS

The final connector category we will consider is the surface mount SMA. Compared to the through hole connectors, these connectors offer much more flexibility in the placement of the ground vias. Their disadvantages include reduced mechanical integrity and hidden solder connections that are difficult to inspect.

The example in Figure 14.19 is a transition to the top layer of a multilayer board like the one in Section 14.2. The board material is Nelco 4003 and the overall board thickness is 32 mil. The topmost microstrip layer is 5-mil thick with an assumed $\varepsilon_r = 3.7$. In Figure 14.19 we have not placed any ground vias yet so we can check for sneak return paths in the simulator. Also note that we have further simpli-

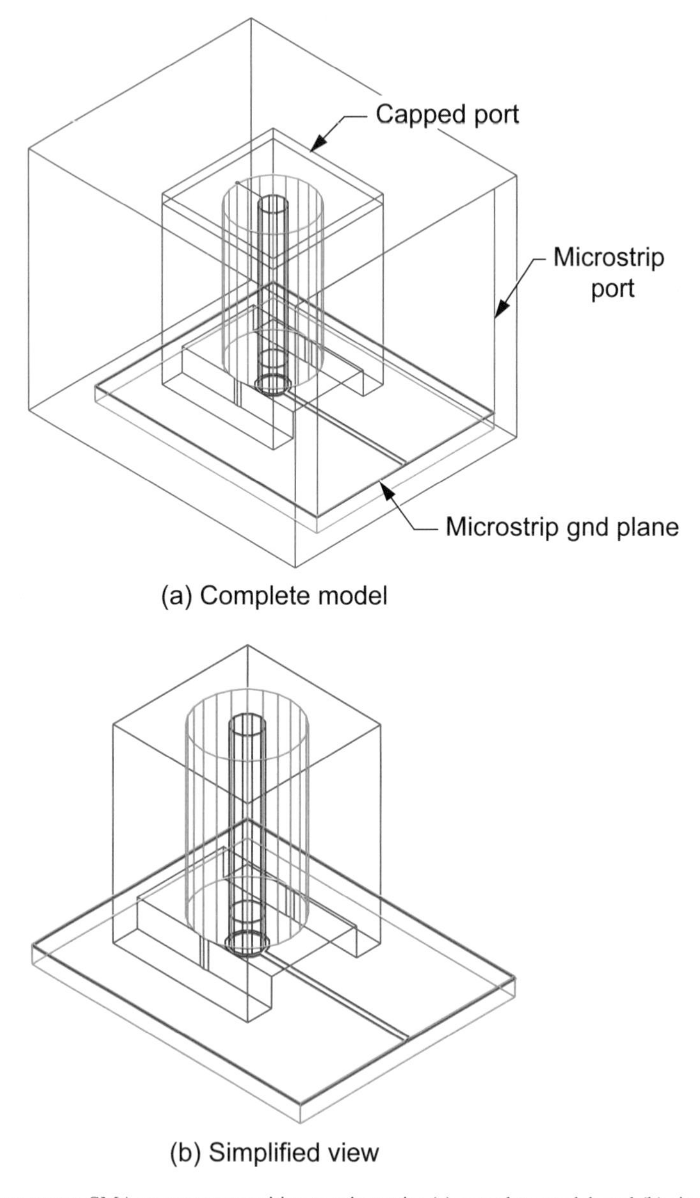

Figure 14.19 Surface mount SMA connector transition to microstrip: (a) complete model; and (b) simplified view of model with ports and outer simulation box hidden. There is no explicit connection from the connector to the microstrip ground plane (Agilent HFSS Ver. 5.6).

fied the SMA end of the model. The upper part of the connector is just a rectangular block of metal. The results for our initial analysis are shown in Figure 14.21.

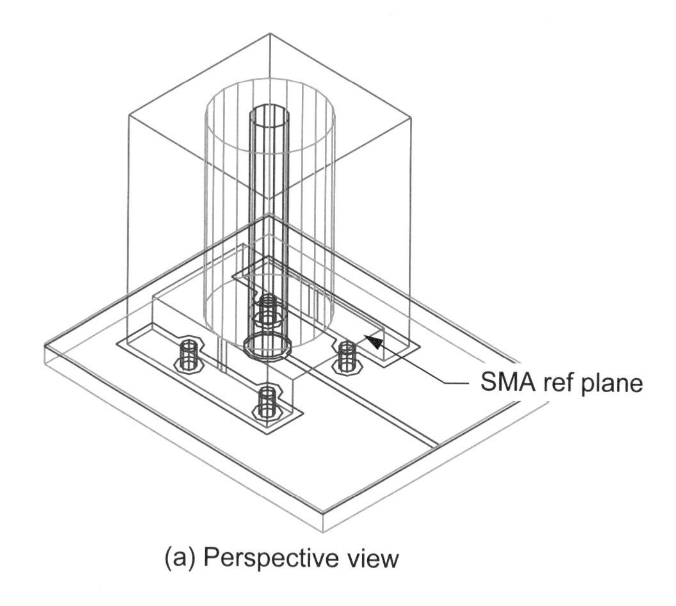

(a) Perspective view

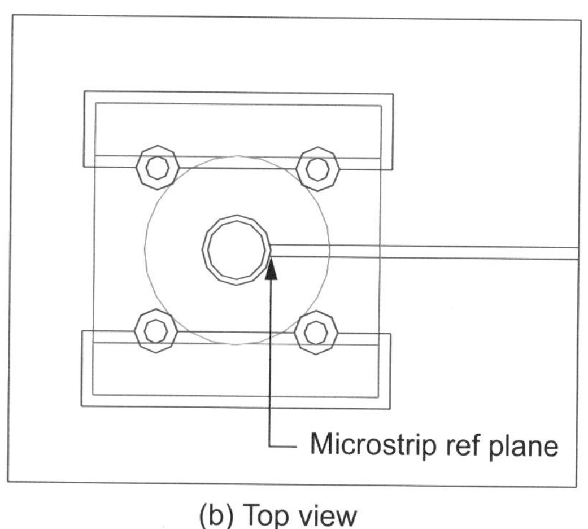

(b) Top view

Figure 14.20 Surface mount SMA connector transition with ground vias and relief in the microstrip ground plane: (a) perspective view, and (b) top view (Agilent HFSS Ver. 5.6).

Next we added four through ground vias from the connector body to the microstrip ground plane layer. Because we are going a very short distance in the board, we arbitrarily chose a spacing of 140 mil center to center. For a transition to a buried layer in a thicker board, we could set the impedance of the vertical section using the five-wire transmission line formula. The results for this model are shown in Figure 14.21.

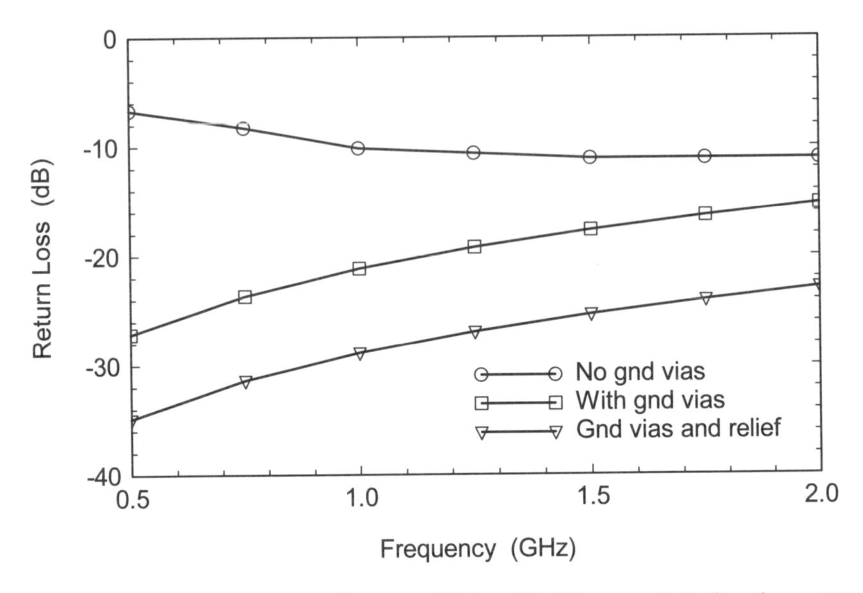

Figure 14.21 Results for the various surface mount SMA models: (a) no ground vias from the connector to the microstrip ground plane; (b) ground vias added; and (c) ground vias plus relief in the microstrip ground plane below the center pin (Agilent HFSS Ver. 5.6).

Finally, we de-embed into the connector model to see what is limiting the performance. In this case we observed excess capacitance at the edge of the microstrip pad. So we decided to relieve the microstrip ground plane under the pad (Figure 14.20). The diameter of the relief is the same as the diameter of the center pin pad (60 mil). The results for this final model are shown in Figure 14.21. The performance with the ground vias and the ground plane relief was "good enough" for this particular application. I am certain that this geometry could be further optimized and that transitions to buried layers can easily be realized.

14.6 SUMMARY

Like the transitions between layers we studied in Chapter 13, transitions from a connector onto the PCB must be designed for good high-frequency performance. Once we have successfully completed a few designs, some common trends and fixes become apparent. Like the multilayer transitions, in most cases we have the freedom to move the ground vias around and aggressively modify the placement of interior pads and antipads. A transition onto a relatively thick RF board is fairly easy. Transitions onto thinner, high-speed digital layers are more difficult because of the larger parasitics. But the basic approach for RF and high-speed digital applications is the same.

References

[1] Johnson, H., "Tapered Transitions," *EDN*, October 11, 2001, p. 34.

[2] Eisenhart, R. L., "A Better Microstrip Connector," *IEEE MTT-S Int. Microwave Symposium Digest*, Ottawa, Ontario, Canada, June 27–29, 1978, pp. 318–320.

[3] Private conversation, Dan Lambalot and Kevin Rosellel, Bayside Design, Inc.

Chapter 15

Backward Wave Couplers

Backward wave couplers [1] are one of the basic distributed components found in many types of systems. Weak couplers, 30 to 50 dB, are used to pick off a sample of a signal in many systems. Couplers with equal power split, 3-dB couplers, are used to create balanced amplifiers, balanced attenuators, and can also be found in some mixer circuits.

While the coupler is useful in its own right, it is also useful as vehicle to test the accuracy of our field-solver modeling. The coupler is still a relatively simple structure to model and it is easy to measure. In this chapter we will study three different couplers and look at our first optimization example. Predicting absolute coupling and directivity and the impact of metal thickness are among the topics to be explored.

15.1 PCS BAND CPW COUPLER

If we wish to build a balanced amplifier, we need a design for a 3-dB coupler. If we flood the empty space on the surface of an RF printed circuit board with metal, then the working environment is basically CPW or CPWG. A simple edge-coupled, two-strip coupler (Figure 15.1(a)) generally cannot achieve the necessary coupling level; the gap between the strips is too small to fabricate. The arrangement of the ports is also inconvenient for our purposes; we would like the coupled and direct ports on the same side.

To get tighter coupling we can subdivide the strips into narrower strips, interleave the narrower strips and rearrange ports in a more convenient configuration. This was the innovation that Lange [2] introduced in 1969. Since then, there have been many variations on the Lange coupler theme. The coupler in Figure 15.1(b) is not a true Lange coupler, but it does use interleaved strips and it gets the ports into the desired configuration. The disadvantages are: we have to fabricate lines with smaller dimensions, the terminations at each end are not symmetrical, and we need cross-over connections between strips.

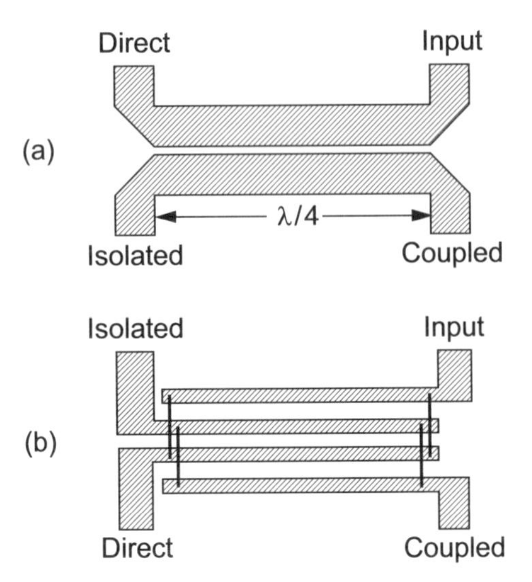

Figure 15.1 3-dB couplers: (a) two-strip edge-coupled, and (b) interdigital. Bondwires or some other form of connections between strips are required in the interdigital coupler.

We can fabricate the coupler in Figure 15.1(b) using microstrip or CPW technology. Microstrip concentrates the fields in the lossy PC board and is sensitive to the dielectric layer thickness. Using CPW technology allows us to reduce the concentration of fields in the lossy PC board and perhaps improve the performance of the coupler. CPW also gives us more control over the even- and odd-mode impedances by varying only the widths and gaps.

Figure 15.2(a) shows the first iteration of a CPW coupler for PCS band (1.8 GHz) applications, circa 1995 [3]. One goal was to minimize the overall length of the coupler. Another goal was to use standard multilayer board technology for any cross-overs. The input port (1) is at the upper right, the coupled port (2) is at the lower right, and the direct port (3) is at the lower left. The isolated port and its 50-ohm termination are tucked in next to direct port. The cross-overs are buried one layer down in the multilayer board. All metal is removed below the strips.

When the performance of this design did not meet expectations, we attempted to back model the results using a field-solver. We could model the complete structure using a 2.5D or 3D solver. But this would only validate the problem we already measured and would be too slow to use for design. Instead we can divide the problem into three smaller problems. The geometry of the center section determines the coupling, the center frequency, and the characteristic impedance. In this case we used a 2.5D planar solver on the center section of the coupler (Figure 15.2(b)). It would be more efficient to use a 2D cross-section-solver.

The parasitics of the end terminations (Figure 15.2(c)) will influence the return loss at all the ports and the directivity of the coupler. At the time, the only solver

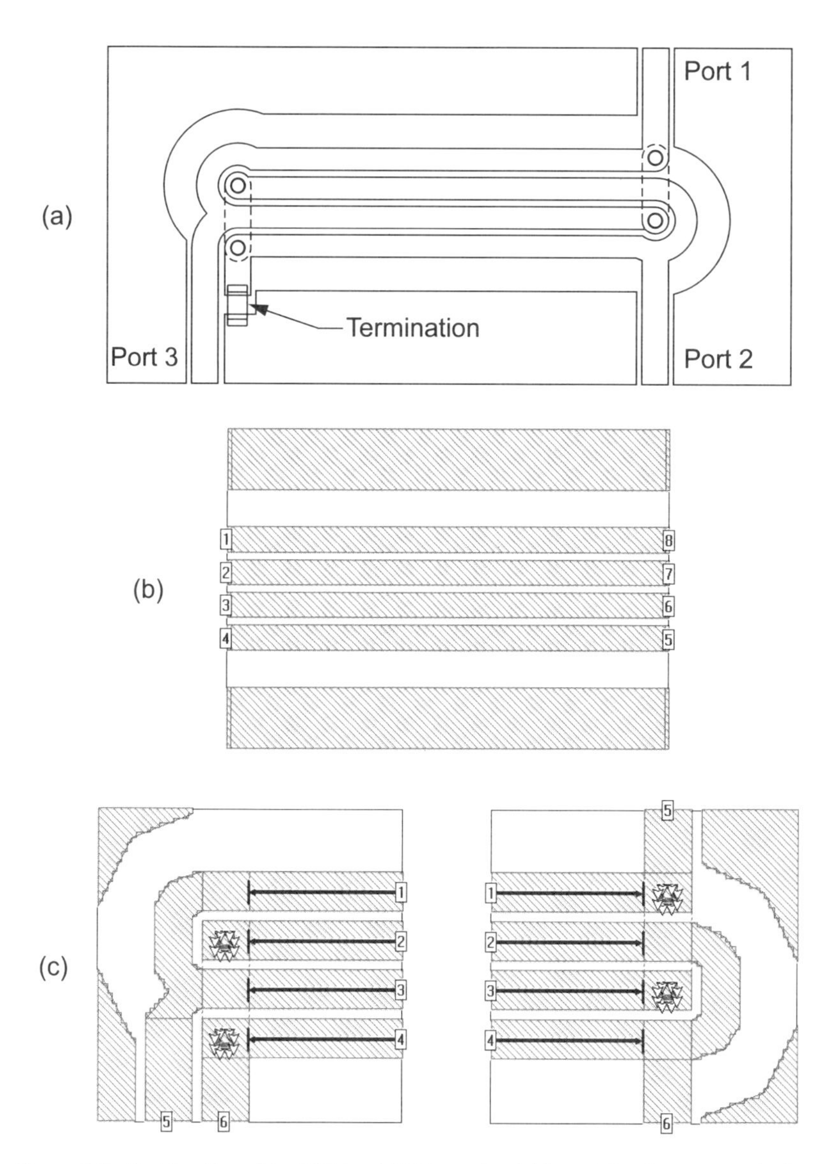

Figure 15.2 First iteration of the PCS band CPW coupler: (a) layout showing port numbering and the termination resistor; (b) center section analysis; and (c) end termination analysis problems (Sonnet ***em*** Ver. 7.0). © WJ Communications. Reprinted with permission.

available to us was the closed box MoM type, so we did the best we could to approximate the smooth curves of the layout. One of the laterally open MoM solvers would allow us to approximate the curves more exactly. We spent most of our time studying the end terminations of the coupler.

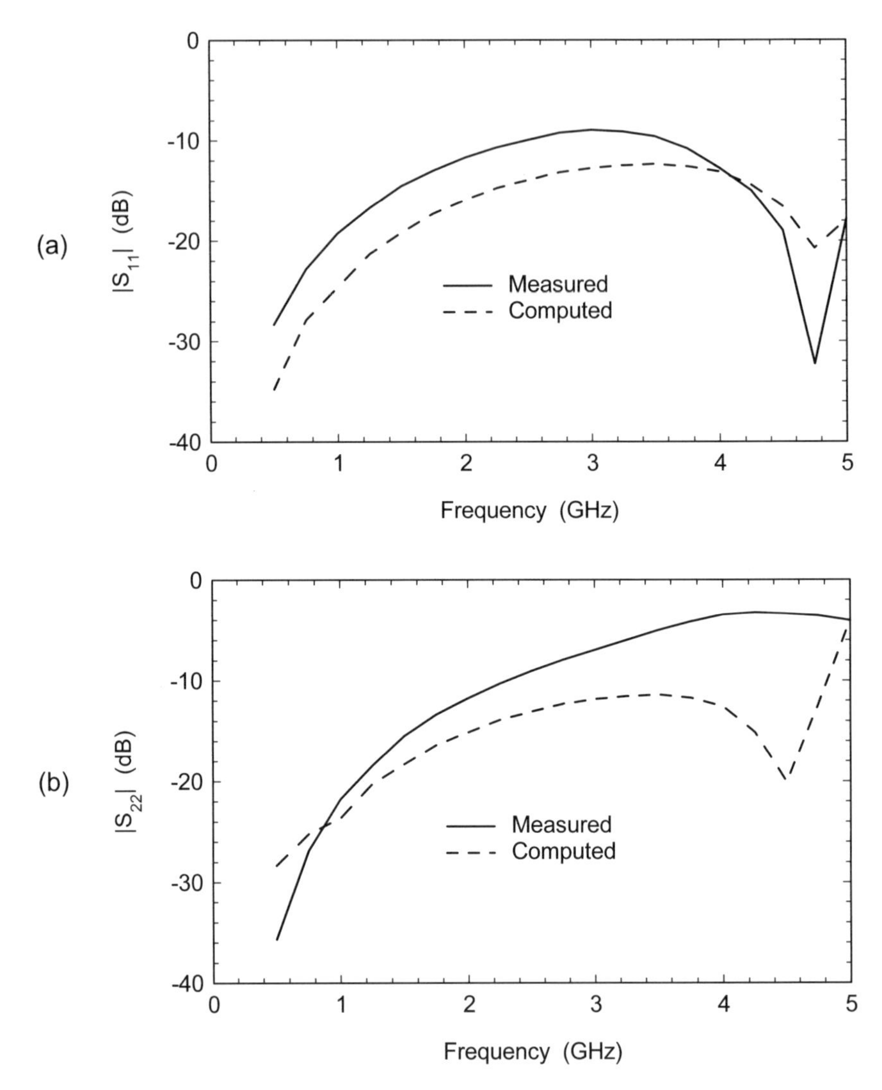

Figure 15.3 First iteration results for the PCS CPW coupler: (a) return loss at Port 1, and (b) return loss at Port 2. © WJ Communications. Reprinted with permission.

After all three problems have been solved, we can combine the multiport *S*-parameters using our favorite linear simulator. The measured versus computed results for Port 1 and Port 2 are shown in Figure 15.3. The return loss prediction at each port is fairly good. The absolute values are close and the field-solver is clearly following the trends. The poor return loss at Port 1 and Port 2 caused us to pursue alternative designs for this coupler. The results for Port 3 and the coupling data are

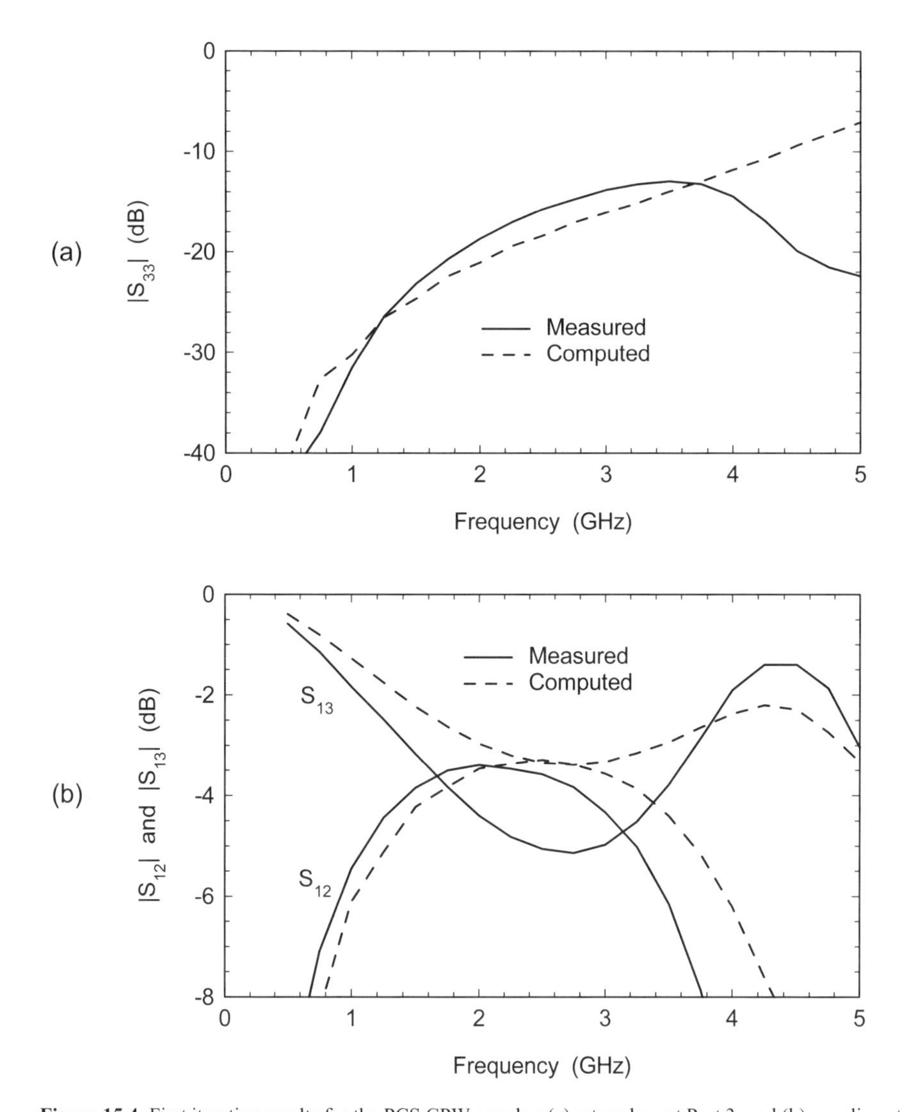

Figure 15.4 First iteration results for the PCS CPW coupler: (a) return loss at Port 3, and (b) coupling at the direct and coupled ports. © WJ Communications. Reprinted with permission.

shown in Figure 15.4. The match at Port 3 is "good enough" as is. When we look at the coupling we see that the computer predicts almost exactly 3 dB while the measured coupler is clearly overcoupled. We have ignored metal thickness in the computer prediction, which may be the major source of the coupling error. But including thickness and loss at this stage of the design would significantly impact the solution time.

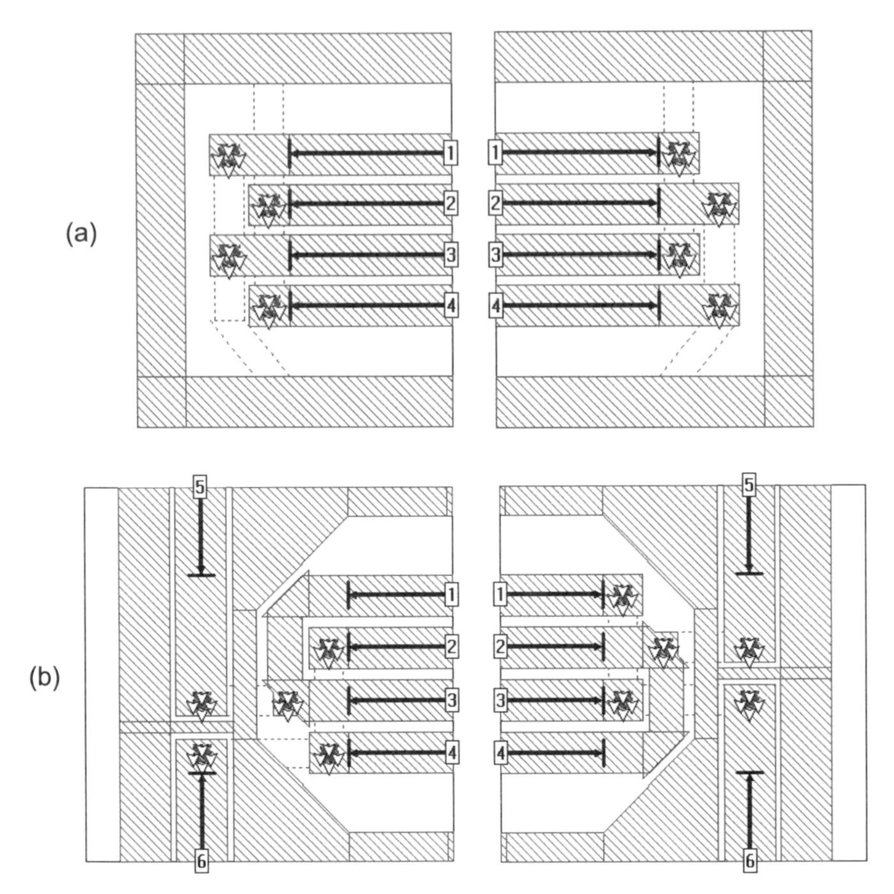

Figure 15.5 End terminations for the PCS CPW coupler: (a) in intermediate iteration that was never built. The missing port connections are one layer down. (b) The final computer iteration. (Sonnet ***em*** Ver. 7.0.) © WJ Communications. Reprinted with permission.

The next iteration of the coupler done on the computer is shown in Figure 15.5(a). Again the goal was to keep the overall length as short as possible. We did several iterations on the field-solver trying different variations on these new end terminations. No significant improvement in return loss was found. The missing port connections are one layer down. This version of the coupler was never built. We saved a lot of fabrication time and experimental time in the lab by doing many experiments on the computer.

Finally, we took a step back and look at the results of all our previous experiments in the lab and on the computer. We abandoned the goal of minimizing the coupler length. This led us to place the I/O lines on the ends of the coupler. The final configuration of the end terminations can be seen in Figure 15.5(b). As before, the terminations are not symmetric.

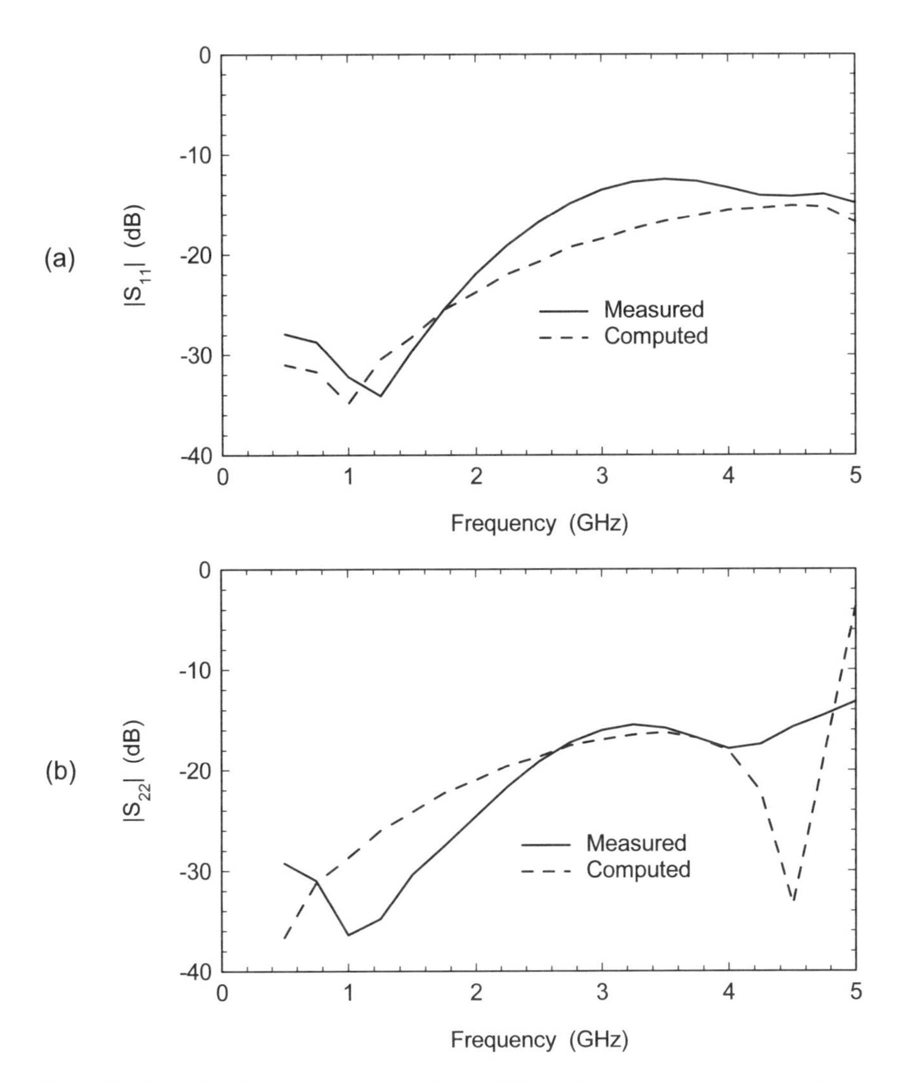

Figure 15.6 Second hardware iteration results for the PCS CPW coupler: (a) return loss at Port 1, and (b) return loss at Port 2. © WJ Communications. Reprinted with permission.

In the final coupler the input port is at the upper right. The coupled port is at the lower right and the direct port is at the lower left. The cross-over connections are one layer down and can be seen as dashed lines. All through the analysis we have been using the same center section computation that we started with. The field-solver simulations are all lossless.

Measured and modeled data for the final iteration of the coupler can be found in Figures 15.6 and 15.7. The measured return loss is now 20 dB or better at all

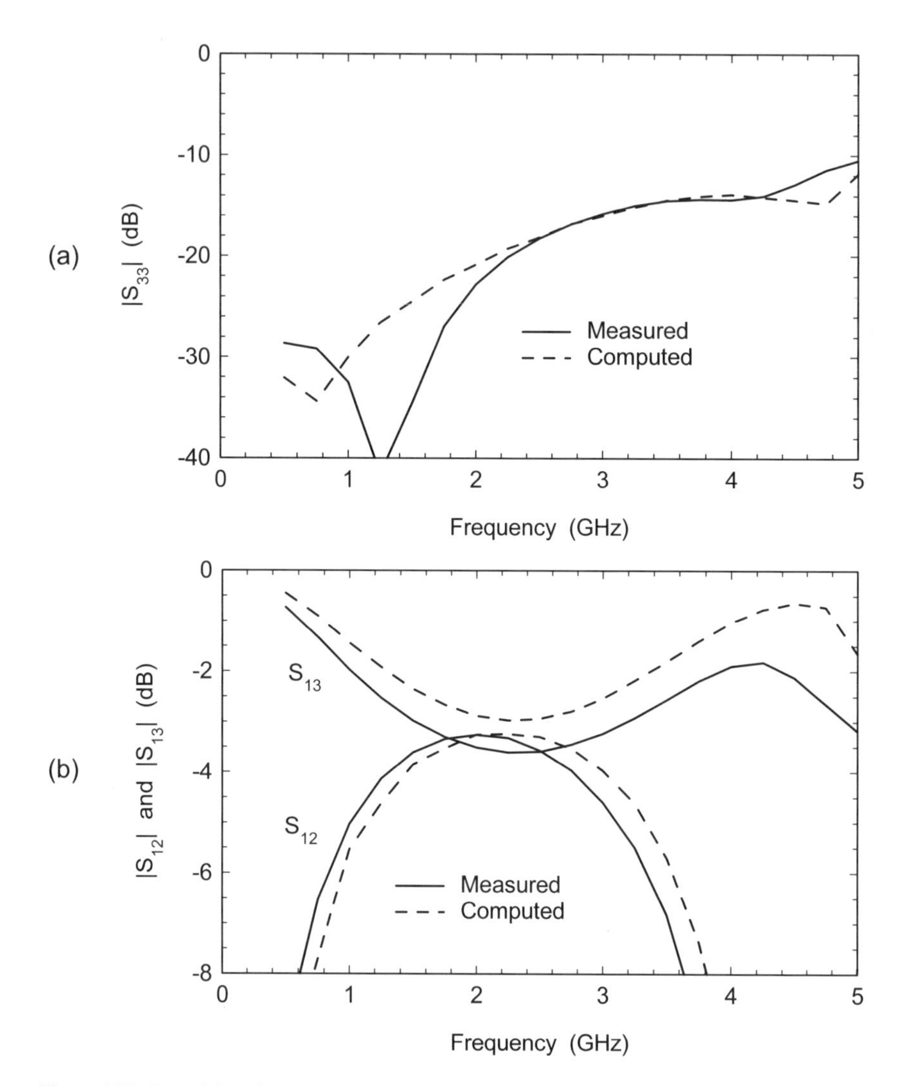

Figure 15.7 Second iteration results for the PCS CPW coupler: (a) return loss at Port 3, and (b) coupling at the direct and coupled ports. © WJ Communications. Reprinted with permission.

three ports of interest. Again, the agreement between measured and modeled is not perfect, but it is probably "good enough" for engineering work. Our measured coupler is slightly overcoupled, which is actually what we wanted. Our prediction of the direct port response (Figure 15.7(b)) has improved but is still not perfect. We are clearly predicting the trends in performance.

This CPW coupler is a good example of how we can use the field-solver to augment the capabilities of standard linear simulators. Our initial coupler design

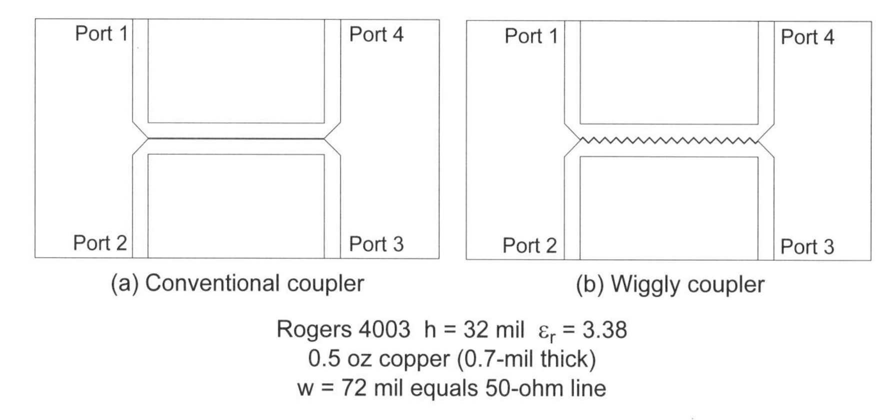

Figure 15.8 10-dB couplers on Rogers 4003: (a) conventional coupler, and (b) wiggly coupler. The "teeth" in the wiggly coupler equalize the even- and odd-mode phase velocities.

used an analytical model to estimate the dimensions of the center coupling section. But, the standard library of models could tell us nothing about the end terminations. Back modeling the first iteration coupler with the field-solver gave us confidence that we could predict what was measured in the lab and hopefully improve on our initial design. Several iterations were made on the computer that were never turned into hardware. This allowed us to optimize fabrication time, fabrication costs, testing time, and testing costs. The third design iteration (second hardware iteration) gave us the performance we desired.

The question remains as to what tool and technique is best for the center section of coupled strips. A 2D cross-section-solver that includes finite strip thickness may be the best answer. If we use a 2.5D planar solver that assumes infinitely thin conductors, we may want to investigate "tricks" that help us approximate finite thickness more accurately. We could also add conductor loss and substrate loss in the final iterations of the design.

15.2 COUPLERS AND METAL THICKNESS

In the previous example the question of metal thickness was left open. This coupler project presented an opportunity to explore the impact of including or ignoring metal thickness in a coupler design. The original goal of this project was to develop a microstrip coupler with improved directivity. In microstrip, the difference between the even- and odd-mode phase velocities causes coupler directivity to degrade. Several techniques have been proposed to improve directivity, including dielectric overlays, lumped capacitive compensation, and the saw-toothed or "wiggly" coupler. In this case we chose to explore the wiggly coupler first proposed by Podel [4–6]. Figure 15.8 shows two layouts for a microstrip 10-dB coupler, a con-

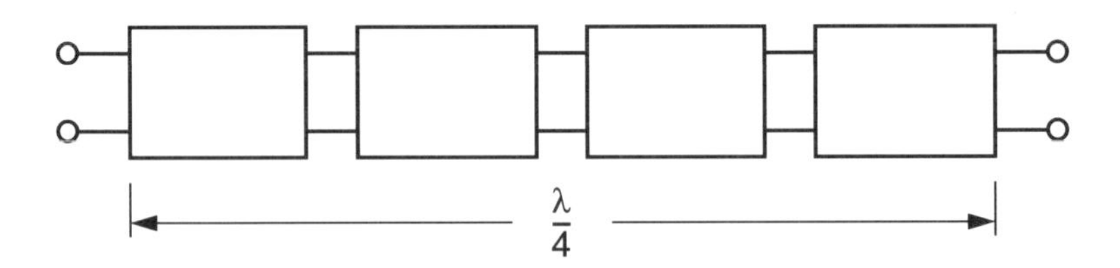

Figure 15.9 The coupled pair is divided into four equal lengths for analysis. The end terminations were not included in the analysis.

ventional design on the left and a wiggly coupler design on the right. The center frequency is roughly 1.8 GHz. A total of four designs were built and tested. Two designs were for wiggly couplers, with and without considerations for metal thickness in the design. And two designs were for conventional couplers, with and without considerations for metal thickness in the design.

We can subdivide our coupler analysis into several smaller components and avoid analyzing the complete structure over and over again on the field-solver. In this case we divided the coupled section into four equal lengths (Figure 15.9). We can analyze just one section and cascade it four times in our circuit simulator. We also ignored the end terminations after performing a numerical experiment that showed they had very little impact.

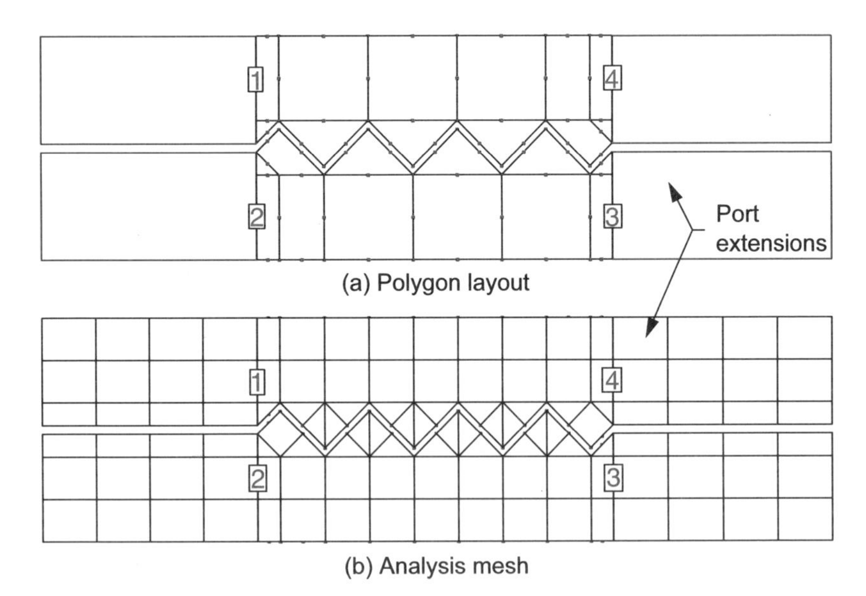

Figure 15.10 Layout of the wiggly coupler problem: (a) the polygon layout, and (b) the resulting analysis mesh. Edge-meshing was not used in this case (Zeland IE3D Ver. 9).

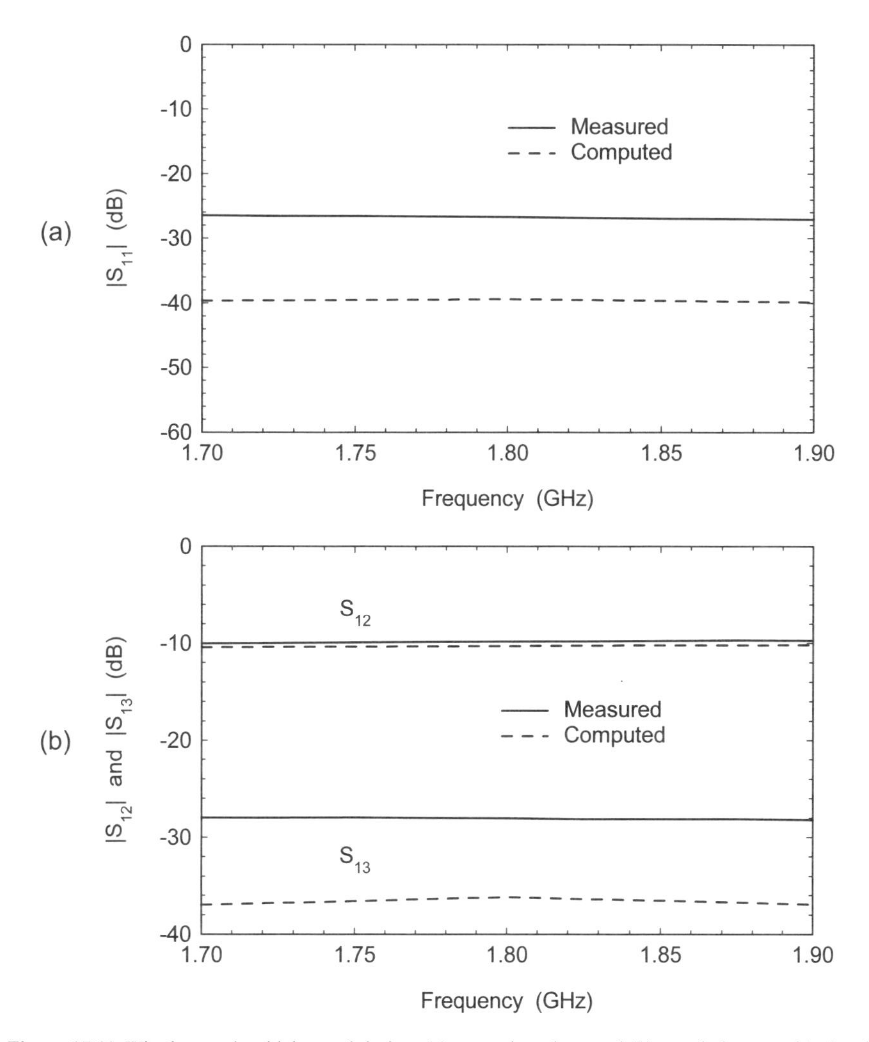

Figure 15.11 Wiggly coupler thick metal design: (a) return loss data, and (b) coupled port and isolated port data. Results are summarized in Table 15.1 (Zeland IE3D Ver. 9).

In order to optimize directivity, we will need to vary the depth and pitch of the saw-tooth pattern. One of the 2.5D laterally open MoM simulators would probably be the best tool, and in this case we chose Zeland IE3D. The layout of the problem is shown in Figure 15.10(a). This particular polygon layout was chosen so we could control the meshing (Section 5.11.2). Edge-meshing would be desirable but we were not able to make it work with a reasonable number of cells. The mesh used for these designs is shown in Figure 15.10(b). The dimensions and results for all four coupler experiments are summarized in Table 15.1.

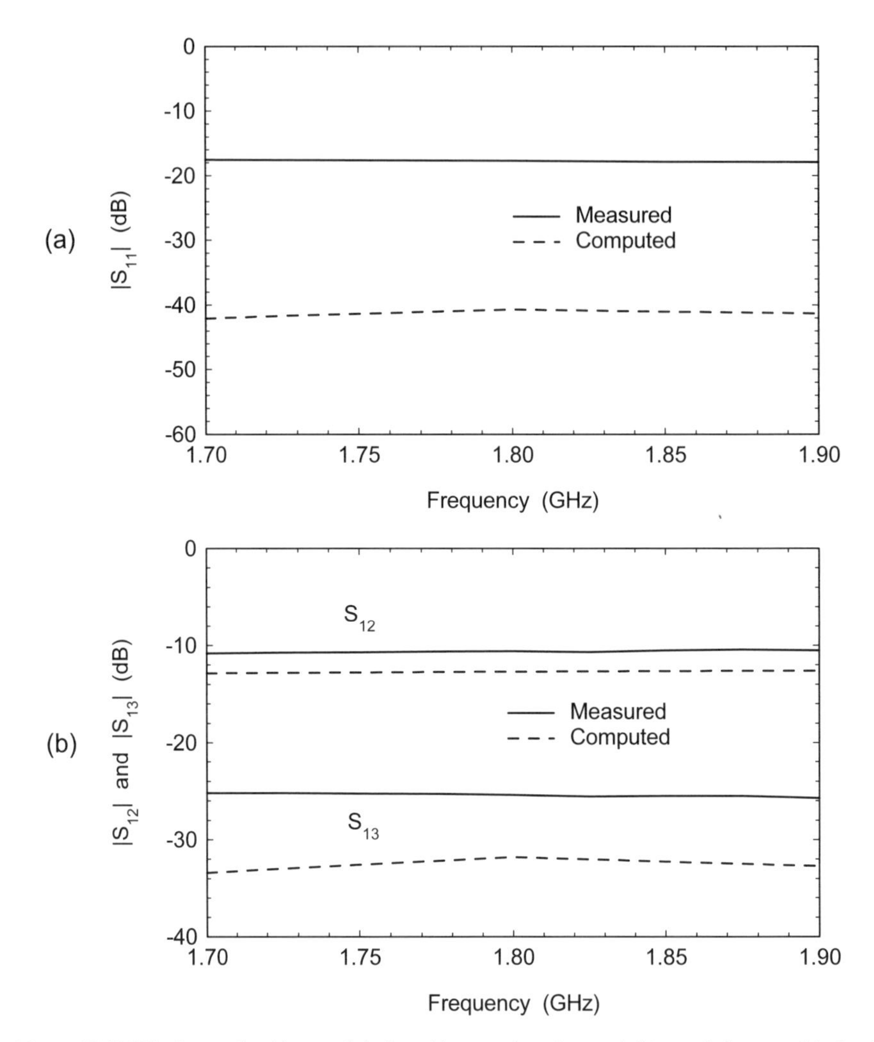

Figure 15.12 Wiggly coupler thin metal design: (a) return loss data, and (b) coupled port and isolated port data (Zeland IE3D Ver. 9).

The first design completed was for a wiggly coupler with thick metal (Figure 15.11). The gap was set to achieve the desired coupling and the coupled line width was adjusted for port match. These two adjustments are nearly orthogonal, so it is easy to optimize a design by hand with only a few iterations. In IE3D we are using two layers of metal, with metal connecting the edges, to model the finite thickness traces. The measured port match ($|S_{11}|$) is not as good as predicted and is limited by the transitions onto the PC board. The measured coupling ($|S_{12}|$) is quite close to the computer prediction. The measured directivity ($|S_{13}|-|S_{12}|$) is probably limited

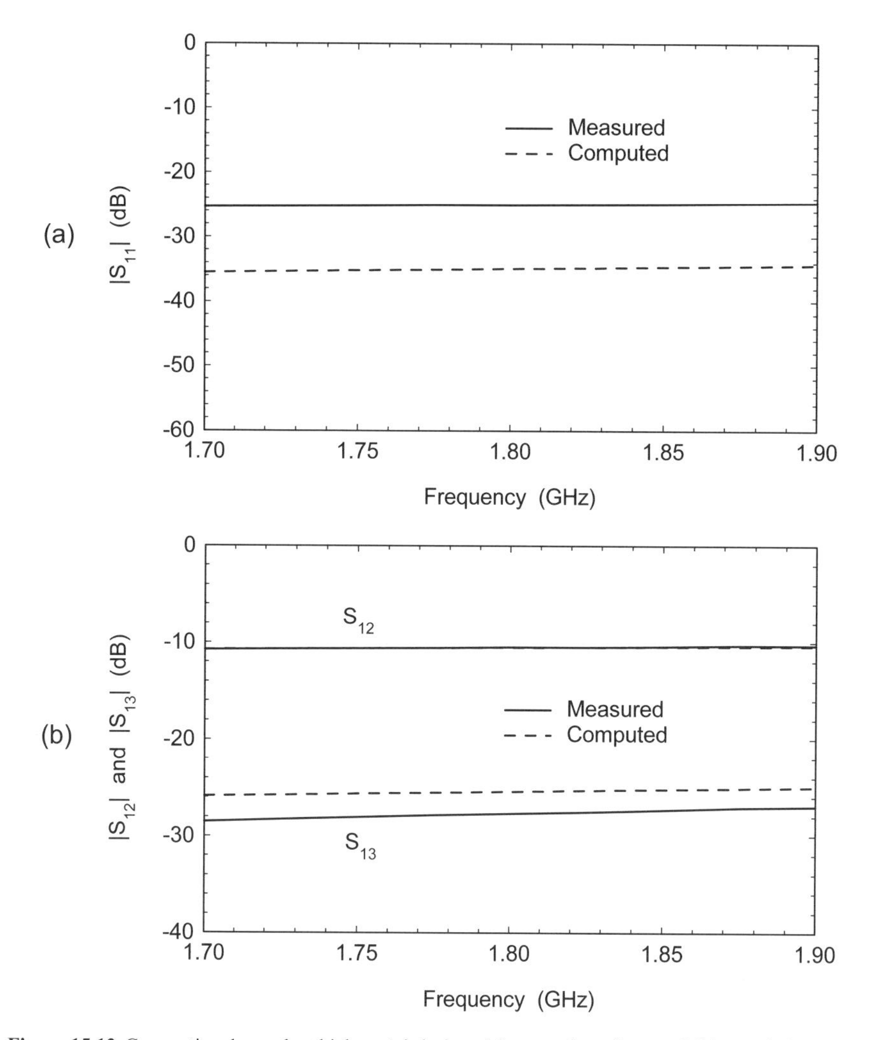

Figure 15.13 Conventional coupler thick metal design: (a) return loss data, and (b) coupled port and isolated port data (Zeland IE3D Ver. 9).

by the port match. After completing the first design, but before any measurements were made, we wondered what differences might be between a design that included metal thickness and one that ignored metal thickness.

Figure 15.12 shows the results for the wiggly coupler design assuming infinitely thin metal. The gap was held constant at the original 5 mil but the line width was adjusted for port match. Note the large change in strip width compared to the thick metal design. The measured port match in this case is about 10 dB worse than the thick metal design. The measured coupling is now off by at least 2 dB. Because

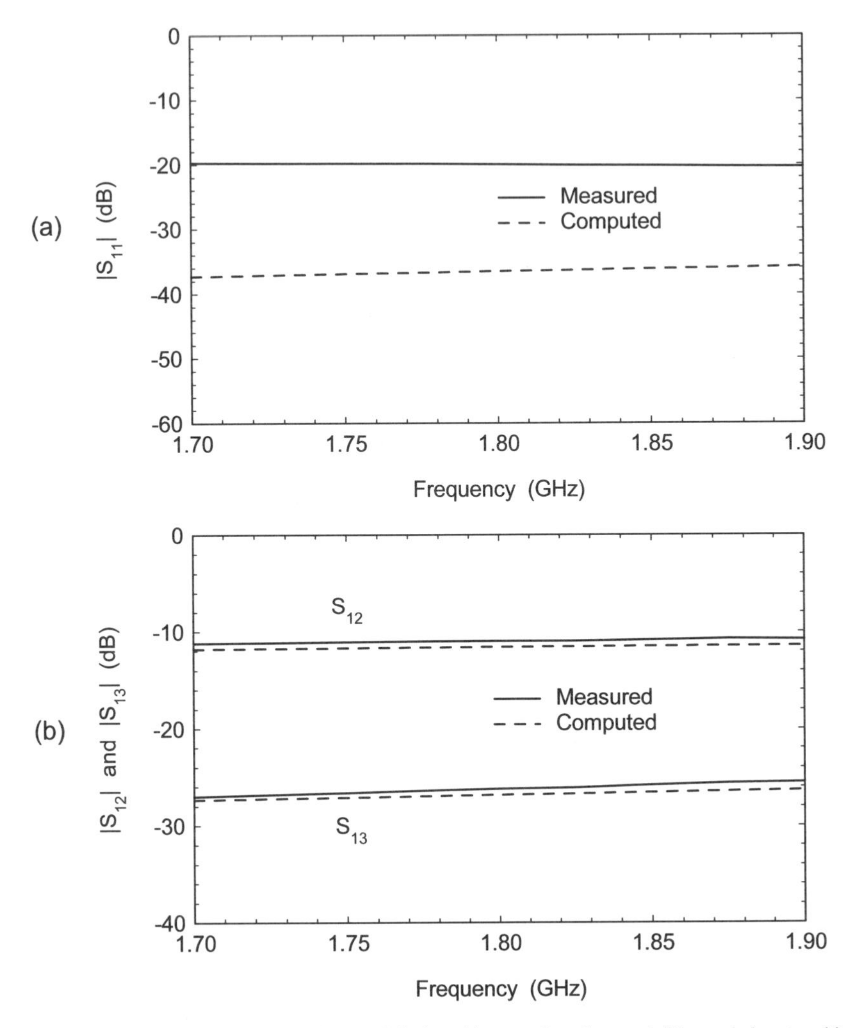

Figure 15.14 Conventional coupler thin metal design: (a) return loss data, and (b) coupled port and isolated port data (Zeland IE3D Ver. 9).

the coupling and port match are worse, the isolation predictions ($|S_{13}|$) are also worse.

The third design was for a conventional coupler with thick metal included in the design (Figure 15.13). The gap was held constant at the original value of 5 mil and the line width was adjusted for port match. The measured port match was comparable to the thick metal wiggly coupler design. The agreement between the measured and modeled coupled port responses was excellent. The measured directivity was actually slightly better than the computer prediction.

Table 15.1

Dimensions and Summary of Results for Coupler Experiments

Coupler type	*Width*	*Gap*	*Length*	*Measured port match*	*Measured directivity*	*Coupling delta*
Wiggly thick	61	5	832	–26.7 dB	18.2 dB	0.4 dB
Wiggly thin	72	5	832	–17.7 dB	14.8 dB	2.1 dB
Standard thick	64	5	832	–25.0 dB	17.2 dB	0.0 dB
Standard thin	69	5	832	–19.9 dB	15.2 dB	0.6 dB

Dimensions are mils, Frequency is 1.8 GHz

Just to complete the cycle a fourth design was done for a conventional coupler assuming thin metal (Figure 15.14). The gap was held constant at 5 mils and the line width was adjusted for port match. Again, note the large change in strip width compared to the thick metal design. The measured port match is about 5 dB worse than the thick metal design. The delta between measured and modeled for the coupled port is not too bad, about 0.6 dB. The isolated port prediction is quite good, but the directivity achieved is slightly worse than the thick metal design.

If we study Table 15.1, we note that we consistently achieved better port match with the thick metal designs. The best measured directivities also correspond to the best port match, which makes sense. The delta between measured and modeled coupling is also smaller for the thick metal designs. However, the improvement in directivity for the thick metal wiggly coupler design was disappointing. Perhaps we did not find the optimum pitch and depth for the teeth of the wiggly coupler.

Building and testing four different coupler designs has given us some feel for the impact of metallization thickness. Most arguments for or against including metal thickness focus on the aspect ratio of the metal thickness to the gap between strips. This assumes that the major impact will be on the capacitive coupling between strips. But we have observed a major impact on the strip width which implies that the inductance per unit length is also changing. Including metal thickness seems to give us better results for both the wiggly couplers and the conventional couplers.

This series of coupler measurements demonstrates how difficult the validation process can be, even for very simple geometries [7, 8]. Our results are based on a sample of one for each coupler type; a statistical sample would clearly be more meaningful. We are also basing our conclusions on a difficult and sensitive measurement, the coupler directivity. Coupler directivity also depends directly on the port match achieved, which depends on the quality of the connector launch onto the board. We should have built a simple through line to verify the return loss of the connector launch onto the board. We also failed to measure the actual dimensions

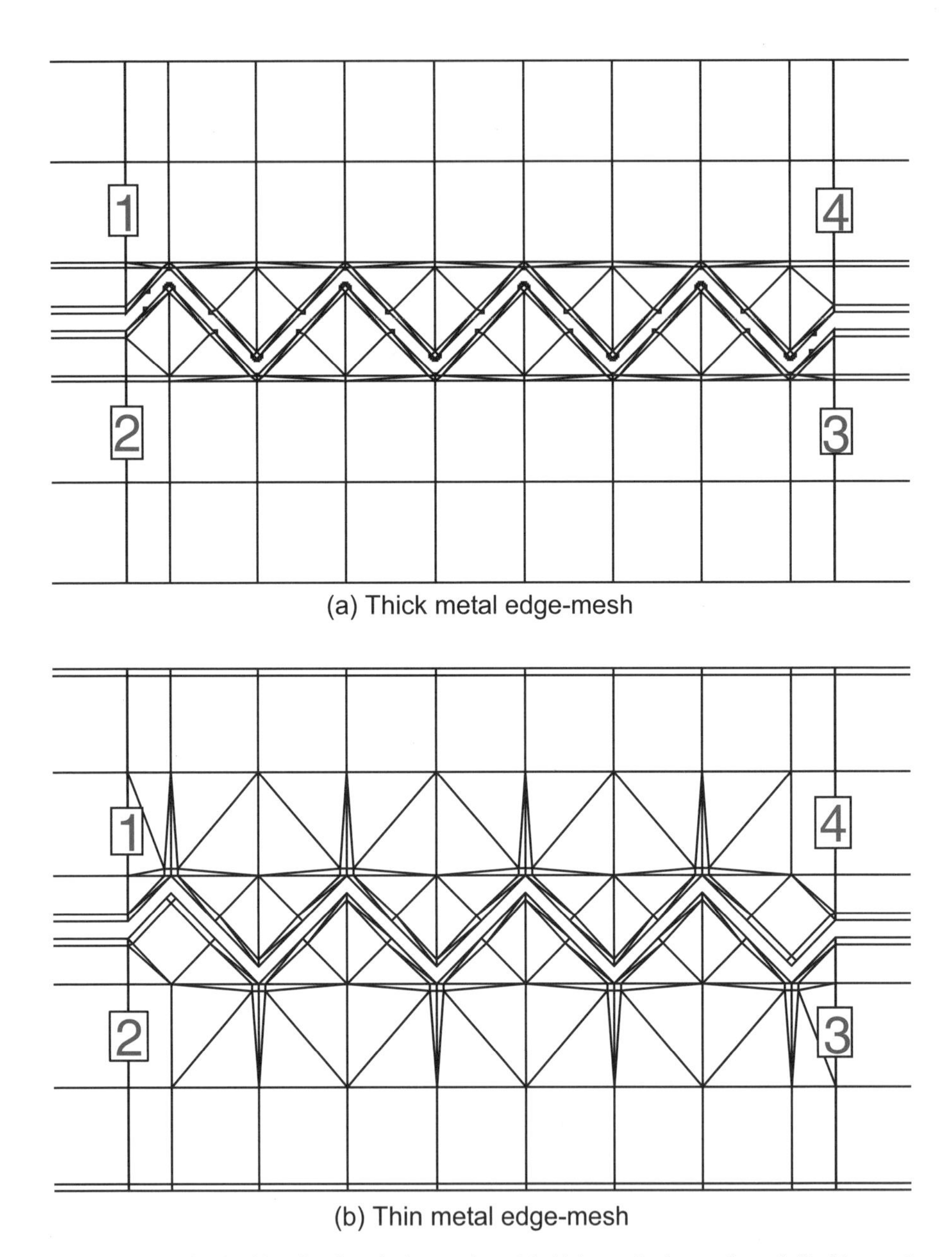

Figure 15.15 Edge-meshing for the wiggly couplers: (a) thick metal edge-mesh, and (b) thin metal edge-mesh (Zeland IE3D Ver. 9).

achieved on the board after etch. We assumed we got what we asked for, which is often not the case.

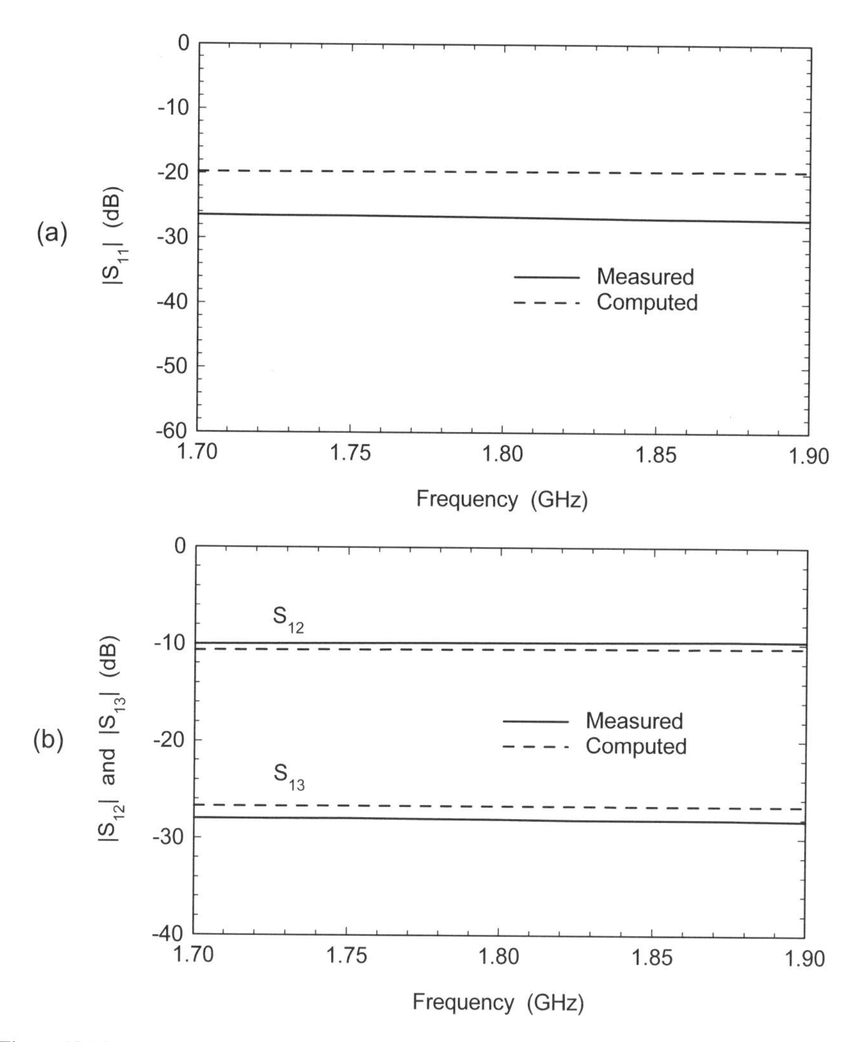

Figure 15.16 Analysis with edge-meshing of wiggly coupler thick metal design: (a) return loss data, and (b) coupled port and isolated port data (Zeland IE3D Ver. 9).

In Chapter 5 we emphasized the importance of edge-meshing, which we have not applied here. IE3D can automatically apply edge-meshing to the thin metal cases fairly easily. Edge-meshing the thick metal wiggly coupler is a rather tedious manual process. After our initial round of experiments was completed, we got some help from the vendor on edge-meshing the two wiggly coupler examples.

The results for the edge-meshed thick metal wiggly coupler can be found in Figure 15.16. Note that we are only back modeling the previous result; the coupler dimensions were not reoptimized for this case. The thick metal mesh (Figure

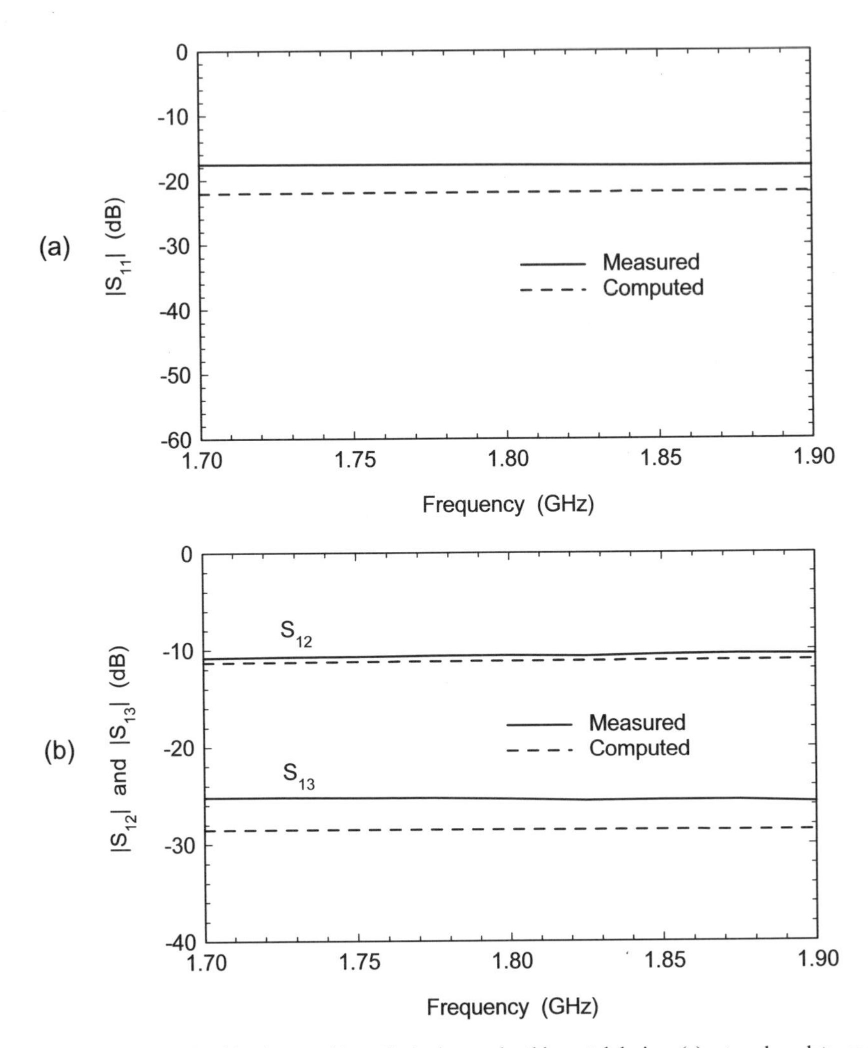

Figure 15.17 Analysis with edge-meshing of wiggly coupler thin metal design: (a) return loss data, and (b) coupled port and isolated port data (Zeland IE3D Ver. 9).

15.15(a)) is actually fairly efficient. The edge cells are 2-mil wide, which may be a little narrow, and there are no edge cells on the outer edges. The port match (Figure 15.16(a)) has changed significantly compared to the thick metal case with no edge-meshing. This change in port match will affect the isolation prediction (Figure 15.16(b)). Overall, including edge-meshing seems to be an improvement. It would be interesting to re-optimize the coupler design using the edge-meshing.

The results for the edge-meshed thin metal wiggly coupler can be found in Figure 15.17. Again, we are only back modeling the previous result. The thin metal

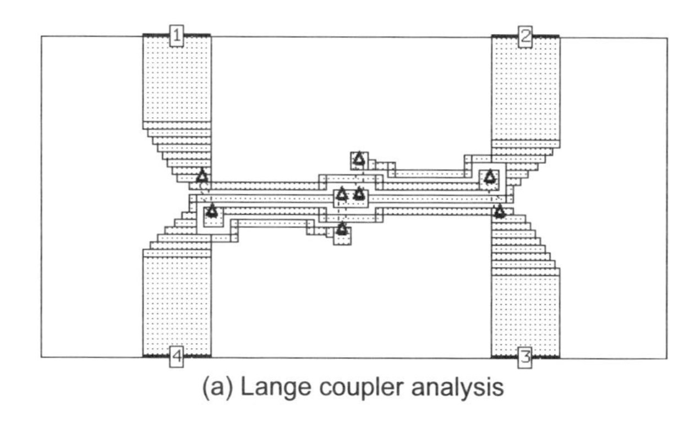

(a) Lange coupler analysis

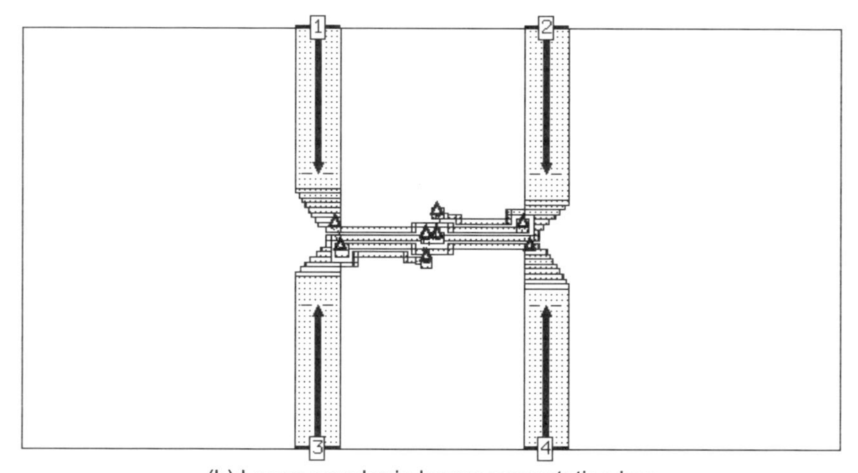

(b) Lange coupler in larger computation box

Figure 15.18 Microstrip Lange coupler analyzed using closed box MoM formulation: (a) initial analysis, and (b) analysis in larger computation box (Sonnet ***em*** Ver. 7.0).

mesh (Figure 15.15(b)) was generated automatically. It could perhaps be more efficient with some manual optimization. The edge cells are again 2-mil wide. The port match (Figure 15.17(a)) has changed significantly compared to the thin metal case with no edge-meshing. The change in port match will also affect thc isolation prediction (Figure 15.17(b)). Including the edge-meshing again seems to be an improvement.

15.3 LANGE COUPLERS

Microstrip Lange couplers are a key component for microwave and millimeter-wave thin-film circuits on ceramic substrates. They are a basic building block for

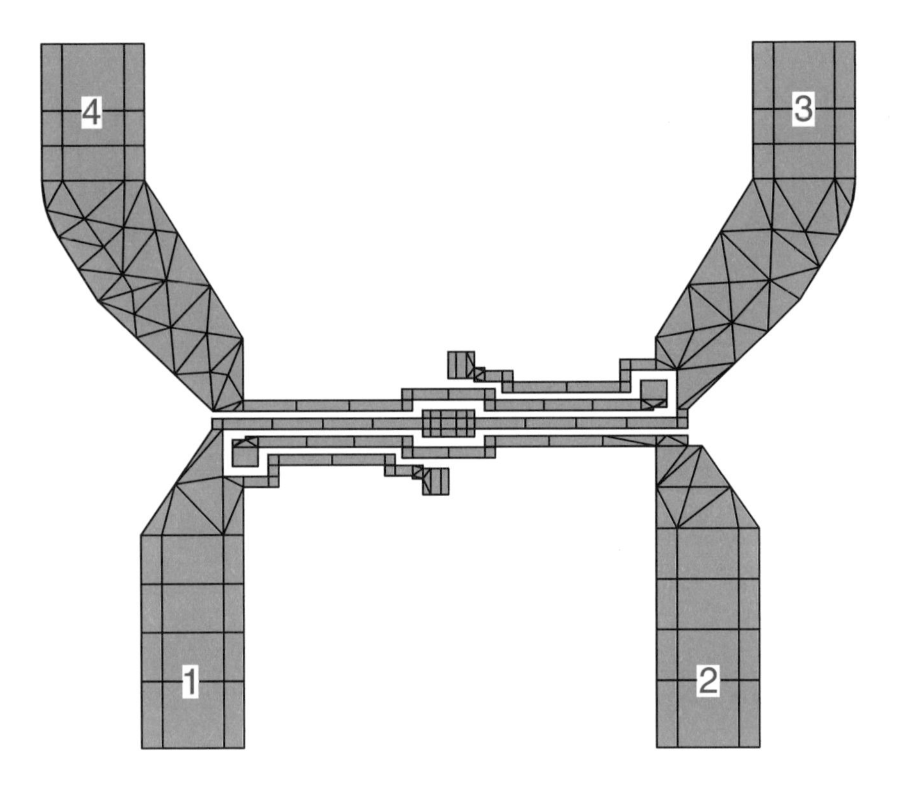

Figure 15.19 Lange coupler analysis using laterally open MoM formulation. This case is for thin metal with edge-meshing, although the edge cells are not shown for clarity. The bondwires are modeled with vias and a second metal layer, not shown (Zeland IE3D Ver. 9).

balanced amplifiers and balanced attenuators. Figure 15.18(a) shows a Ka-band Lange coupler built on 10-mil thick alumina. The critical dimensions are the line widths and gaps in the coupled section, which are both 1 mil in this case.

We only had access to a closed box MoM code when the first analysis of this structure was performed. Luckily, the required dimensions have a common denominator and we could pick a convenient cell size. The layout in Figure 15.18(a) is drawn to scale and the box walls are quite close. In fact the walls are too close and they influence both the computed coupling value and the impedance of the feed lines. We can pull the box walls back (Figure 15.18(b)) but there is a time penalty involved. We have to include all the additional line we have added in the computation. The solution time is now almost double compared to the first simulation. As the box gets bigger there is also the potential for box resonances.

This is a case where the laterally open MoM formulation is probably more desirable. Figure 15.19 shows the same coupler laid out in Zeland IE3D. We can set our ports fairly close to the structure and there are no box walls to influence the

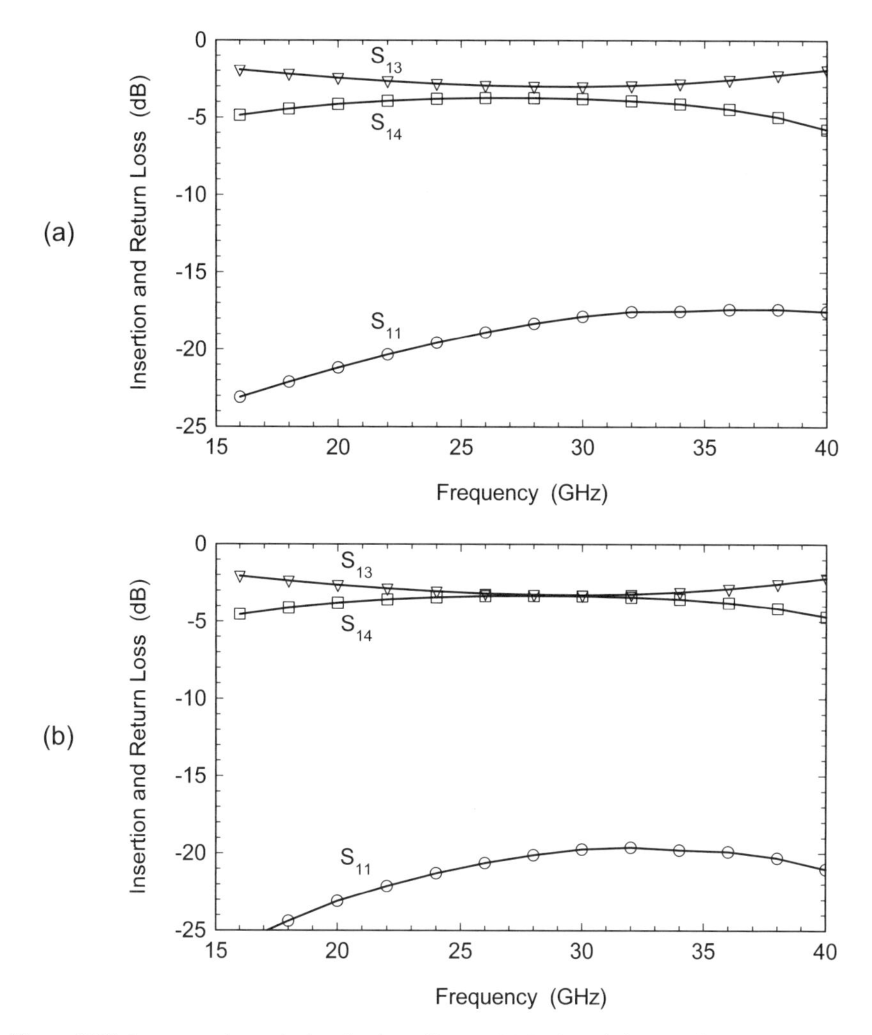

Figure 15.20 Lange coupler analysis using laterally open MoM formulation: (a) thin metal with edge-meshing, and (b) thick metal with no edge-meshing (Zeland IE3D Ver. 9).

solution. Another advantage would be resolution. We got lucky in the closed box analysis—the designer specified 1-mil lines and gaps. If he or she had specified 0.8-mil gaps and 1.4-mil lines, we would have to make some approximations using the closed box formulation. The same specification would not be a problem in a laterally open code.

As in the wiggly coupler example, the questions of how do we treat metal thickness and do we use edge-meshing also arise. The layout in Figure 15.19 is a thin metal solution with 0.1-mil edge cells. Edge-meshing has actually been turned

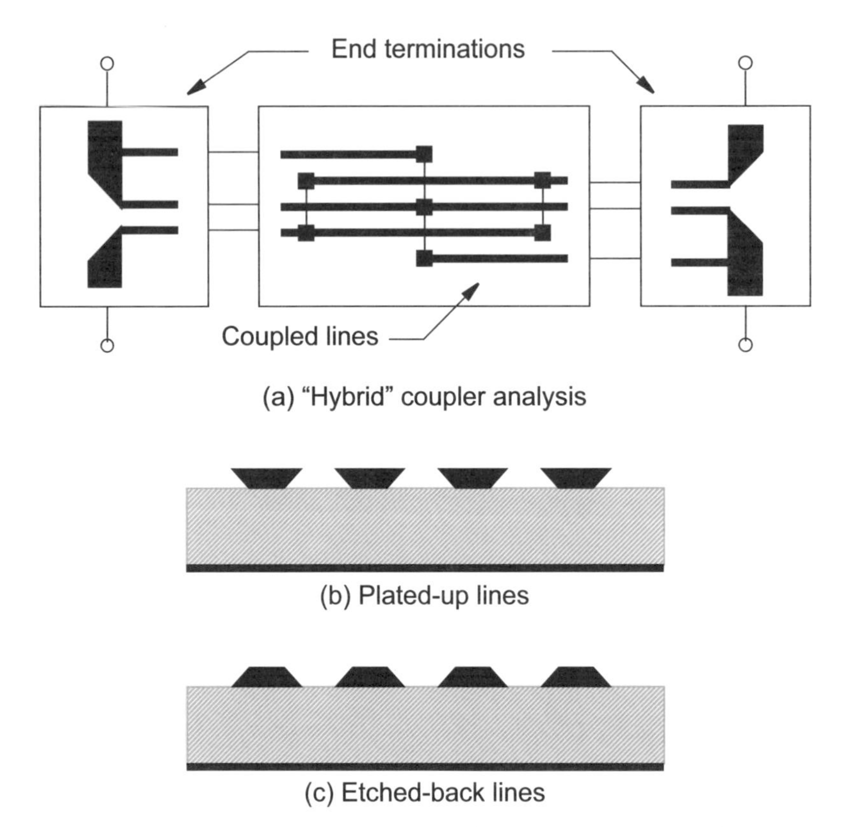

Figure 15.21 Other Lange coupler analysis options: (a) a "hybrid" analysis using a combination of solvers, (b) plate-up lines cross-section, and (c) etched-back lines cross-section.

off in Figure 15.19 just for clarity. The results for the thin metal case can be found in Figure 15.20(a). The simulation indicates that we are slightly undercoupled with fairly good port match. We also analyzed a thick metal case with IE3D. As in the wiggly coupler example, it is more difficult to force edge-meshing when we invoke the thick metal. In this case we did not modify the layout to force edge-meshing on the coupled region. The results for the thick metal case can be found in Figure 15.20(b). The simulation now indicates tighter coupling and slightly better return loss.

The actual coupler may in fact be slightly overcoupled. Depending on the bandwidth required we can intentionally overcouple the coupler to get 3-dB average coupling across the desired band. Unfortunately we do not have an independent measurement of this coupler, but this design was used successfully in a balanced amplifier.

So far our analysis has been very brute force; we put the whole geometry into the simulator and wait for the results. A hybrid approach to the problem (Figure

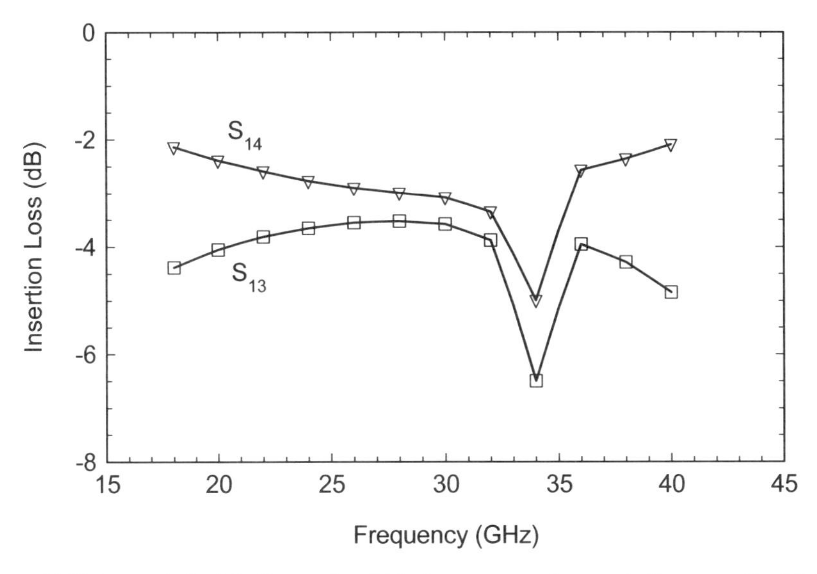

Figure 15.22 One analysis of the Ka-coupler which shows an anomalous result (Sonnet ***em*** Ver. 3.0a).

15.21(a)) may be a better way to go, particularly when we are exploring the boundaries of a new design. Using a 2D cross-section-solver, we can model the narrow strips and gaps very efficiently and with infinite resolution. We can then call on a 2.5D or 3D field-solver to model the end terminations and possibly the central bondwire region [9]. Maas [10] reported an interesting result using LINPAR, a 2D cross-section-solver, for a coupler at 40 GHz. There is also an interesting application note [11] that discusses several strategies for including metal thickness in a Lange coupler EM analysis.

Depending on the numerical method used, the 2D cross-section-solver that we choose may also be optimized for infinitely thin lines; so a correction factor for thickness might be added. A very sophisticated model of the coupled lines would include the actual cross-section shape of the lines. Plated-up lines (Figure 15.21(b)) tend to look like trapezoids with the wide edge up, while etched-back lines (Figure 15.21(c)) look like trapezoids with the wide edge down. The actual cross-section of the line also makes measurement of the realized geometry a tricky proposition.

There is one final issue we can discuss using this coupler as an example: critical analysis of results. When using any CAD tool we should always step back and say, "Do I believe what I am seeing?" This is especially true for field-solvers. There are many parameters to adjust and without some experimental data as benchmarks we may get some misleading results. While analyzing the Ka-band coupler we produced the plot shown in Figure 15.22. Is the sharp discontinuity in the data at 34 GHz real? It could be a box resonance or some other problem. Let's look at a couple of current density plots.

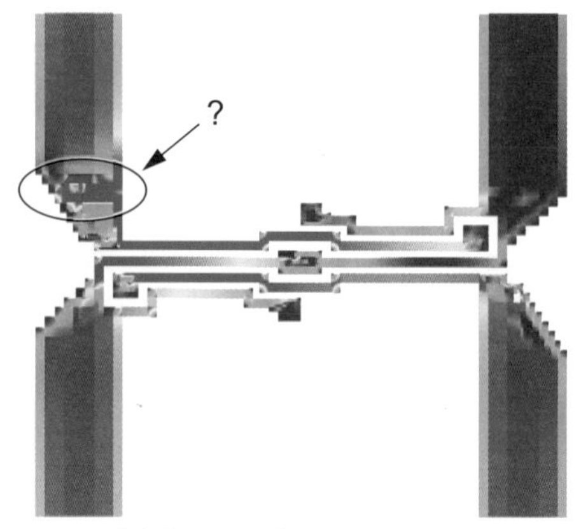

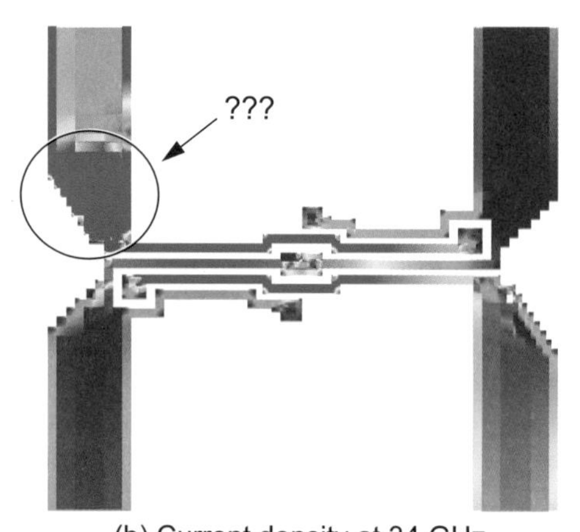

Figure 15.23 Current density plots from the analysis in Figure 15.22: (a) current density at 28GHz with a small anomalous region, and (b) current density at 34 GHz with a much larger anomaly (Sonnet ***em*** Ver. 3.0a).

At 28 GHz where the coupler is still operating correctly we see the normal microstrip current distribution on most of the lines (Figure 15.23(a)). However, there is one small region at the driven port that does not look quite right. At 34 GHz, where we found the discontinuity, the questionable spot has grown to a clear problem (Figure 15.23(b)). There is no logical reason to suddenly have this region of very high current density. We know this based on the current plots we studied earlier for much simpler microstrip structures. It turns out that this was a bug in the meshing algorithm. These codes are large and quite complex; an occasional bug is to be expected. Without looking at the results critically, we may have discarded the design or wasted a lot of time trying to "fix" a spurious solution.

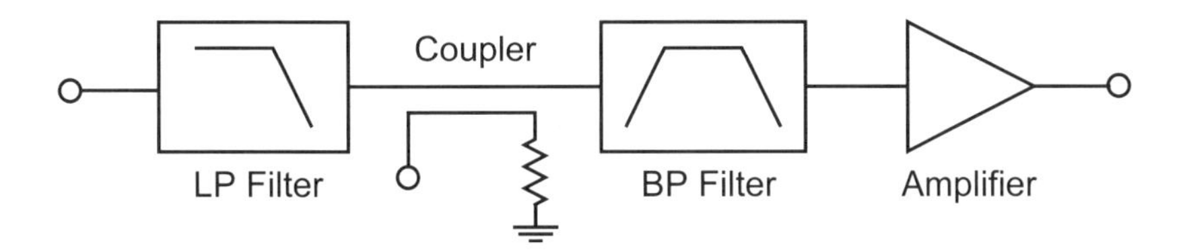

Figure 15.24 Schematic of the "tower top" amplifier. The low noise amplifier mounted at the antenna compensates for cable loss and improves the system noise figure.

We can also safely say that at some point, *all the major commercial codes* have had some bug that would cause an incorrect solution. When a solution does not look right, careful examination of a current plot is a very useful debugging approach for any of the 2.5D or 3D solvers. Small, unintended gaps in the geometry or problems creating a correct mesh will often show up quite clearly as some kind of discontinuity or anomaly in the current density.

15.4 PCS BAND 15-DB COUPLER

Some wireless components for the PCS band are an interesting mix of multilayer board components and low-loss mechanical components. One PCS band application is the "tower top" amplifier (Figure 15.24). This component is a low-noise amplifier and low-loss filter placed near the antenna to overcome cable loss and improve the system noise figure.

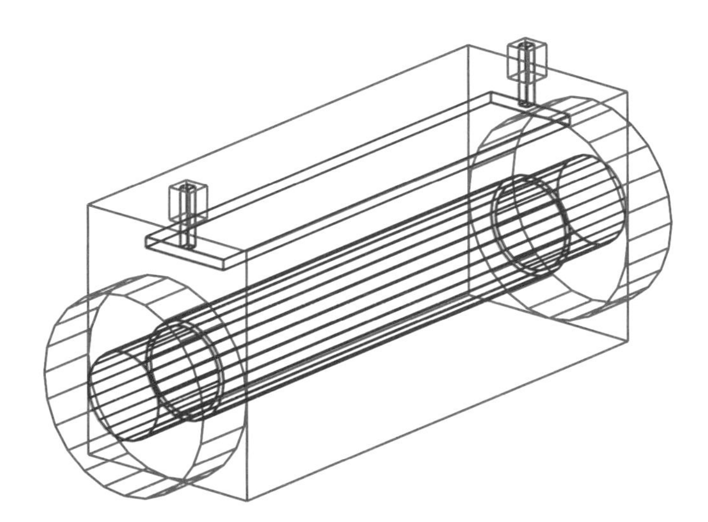

Figure 15.25 Perspective view of the 15dB coupler (Ansoft HFSS Ver. 8.0).

The block diagram includes a low-pass "roofing" filter to guarantee a broad stopband response, a 15-dB coupler to sample the signal at the input, a low-loss bandpass filter and a low-noise amplifier constructed on a multilayer PC board using surface mount technology. The two filters and the coupler are actually mechanical components. Let's take a detailed look at the 15-dB coupler.

A 3D view of the coupler, generated by Ansoft HFSS, is shown in Figure 15.25. The main line through the coupler is an air-filled trough line. The coupled line is a stripline conductor with a rectangular cross-section. The ports at either end of the coupled line were chosen to be square coax just for convenience in the modeling process.

It is tempting to make a guess at the geometry of the coupler, enter that geometry into one of the 3D field-solvers, and begin an analysis or optimization. But using our philosophy of reducing the problem to the lowest order geometry, there is probably a better way. If there are analytical models available we should use them. We can then check or modify the analytical model using a 2D cross-section model. In this case, much of the optimization can be done at the 2D cross-section level. Finally, we can model the full structure using a 3D tool.

In a directional coupler, the desired coupling and the system impedance define the required even- and odd-mode impedances. The electrical length is simply a quarter wavelength in the medium, which is air in this case.

$$C = 20\log\left|\frac{Z_{even} - Z_{odd}}{Z_{even} + Z_{odd}}\right| \tag{15.1}$$

$$Z_0 = \sqrt{Z_{even}Z_{odd}} \tag{15.2}$$

$$Z_{even} = Z_0\sqrt{\frac{1+10^{C/20}}{1-10^{C/20}}} \tag{15.3}$$

$$Z_{odd} = Z_0\sqrt{\frac{1-10^{C/20}}{1+10^{C/20}}} \tag{15.4}$$

If $C = -15$ dB and $Z_0 = 50$ ohm then $Z_{even} = 59.85$ ohm and $Z_{odd} = 41.77$ ohm. If our geometry were microstrip or stripline we could find analytical equations in our favorite linear simulator that would allow us to complete the design of our coupler. In this case we will have to do a little extra work, but it will not be very difficult.

The most efficient way to design this component is to start with a 2D cross-section analysis. Several of the 2D cross-section-solvers that we discussed in Chapter 11 would be suitable. As it happens, there are analytical equations [12, 13] for the cross-section shown in Figure 15.26(a). There are several ambiguities in the refer-

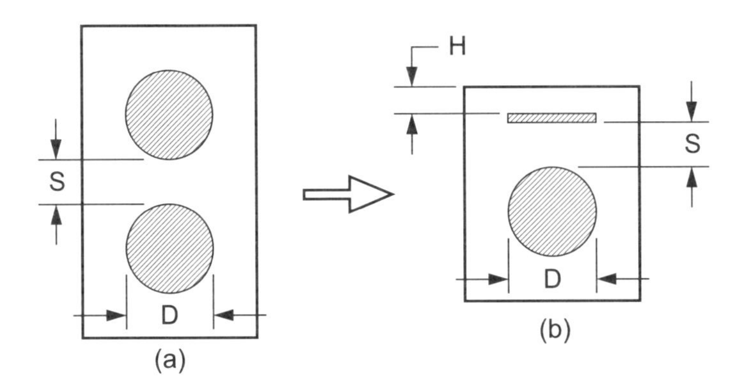

Figure 15.26 A set of analytical equations was found for (a) two coupled rods in a box. This geometry was then transformed to (b) a coupled plate and rod.

ences, but the results were good enough to generate a starting point. This first iteration was then transformed into the configuration in Figure 15.26(b). The spacing between conductors, *S*, was held constant, and the distance to the upper ground plane was set using the desired even-mode impedance and a standard stripline formula.

Now the problem was ready for the 2D cross-section-solver, in this case we used Maxwell SI 2D from Ansoft. The geometry in Figure 15.26(b) was entered and several manual iterations were performed to optimize the coupling and the characteristic impedances of the through line and the coupled line. Only a few minutes were needed to compute each iteration. Only five runs were needed on the 2D

Table 15.2

2D Cross-Section-Solver Results for Figure 15.26(b)

D (in)	*S* (in)	*H* (in)	*K*	*C* (dB)	Z_{input}	Z_{coup}
0.300	0.190	0.300	0.21695	−13.27	51.1	75.1
0.300	0.190	0.250	0.21405	−13.38	51.1	73.8
0.300	0.190	0.100	0.18427	−14.69	50.4	55.1
0.300	0.190	0.080	0.17381	−15.19	50.2	49.2
0.300	**0.190**	**0.083**	**0.17559**	**−15.11**	**50.3**	**50.2**
0.300	0.210	0.083	0.15676	−16.10	50.5	50.2
0.300	0.170	0.083	0.19725	−14.10	50.0	49.9

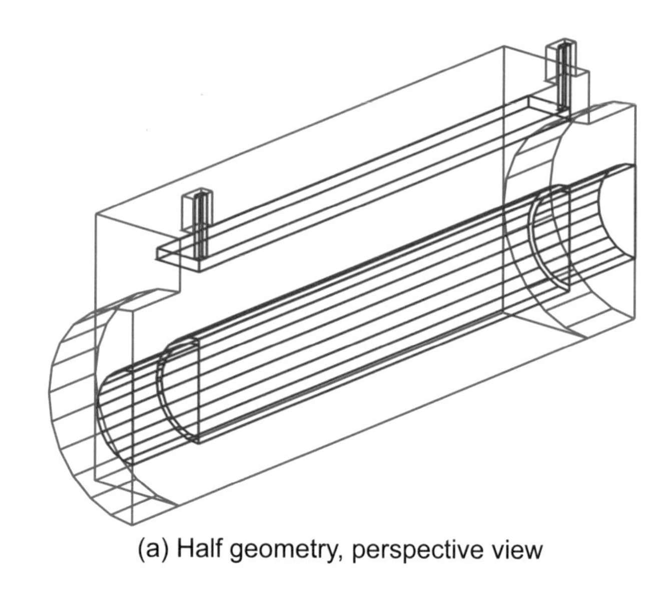

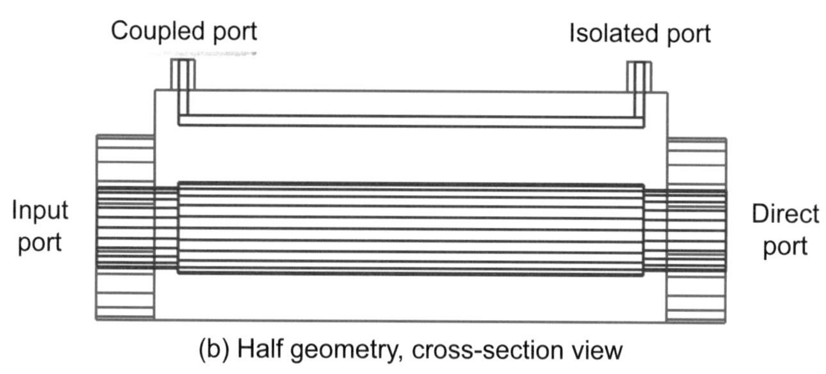

Figure 15.27 Coupler geometry with symmetry plane applied: (a) perspective view, and (b) cross-section view with ports labeled (Ansoft HFSS Ver. 8.5).

solver to generate a solution that is "good enough." Two additional runs were made to test the sensitivity of the coupling parameter to the spacing, S (Table 15.2).

Our 2D analysis indicates we have achieved the desired coupling and that all four ports are well matched to 50 ohms. If we had built several of these structures in the past, we could probably stop right here and construct hardware. If this technology was new to us, then a 3D analysis of the full structure might build our confidence or even uncover a parameter we have overlooked. Earlier we showed the full model of this coupler as entered into Ansoft HFSS. To save time we can take advantage of a symmetry plane and analyze only half of the structure (Figure

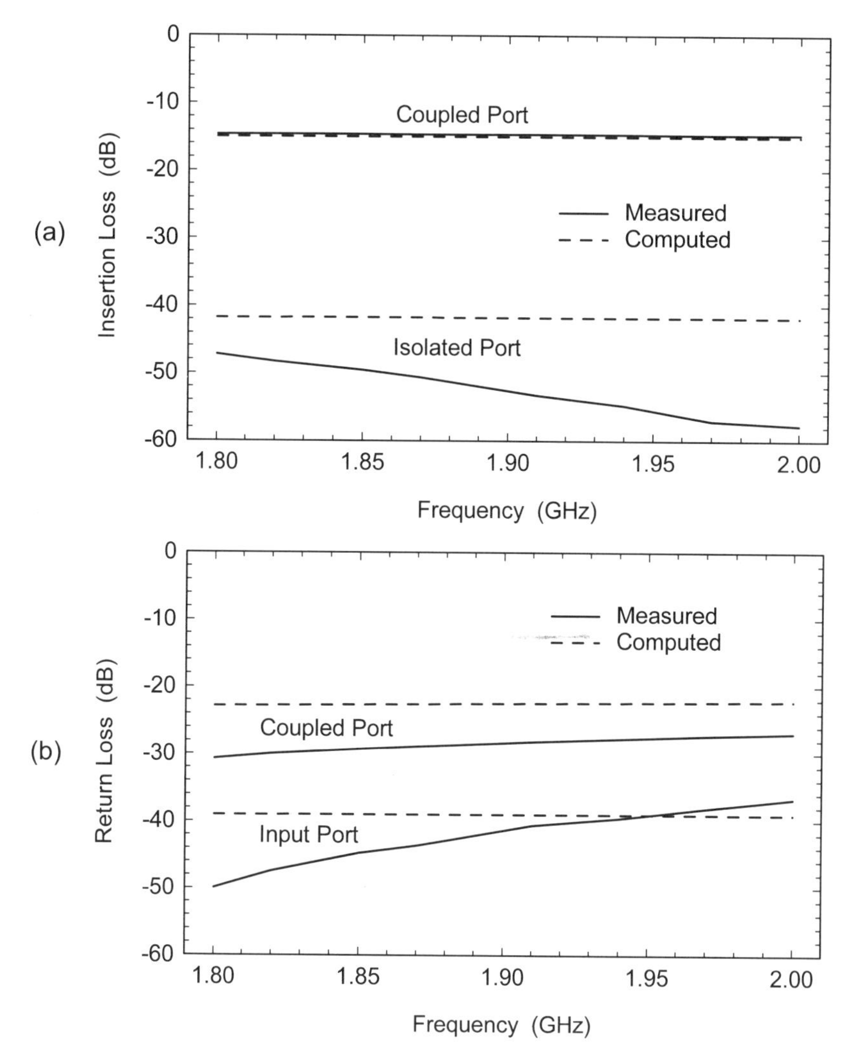

Figure 15.28 Measured versus modeled results for the 15-dB coupler: (a) coupled port and isolated port responses, and (b) return loss at the input and coupled ports (Ansoft HFSS Ver. 8.0).

15.27(a)). A cross-section view might also help visualize the interior of the coupler (Figure 15.27(b)). The ports are labeled to match the response plots that will follow.

We built a prototype coupler and measured it in the lab. The measured and modeled responses at the coupled port and the isolated port are shown in Figure 15.28(a). The coupler directivity is the difference between the isolated and coupled port responses. Figure 15.28(b) shows the measured and modeled port match plots. We would expect the isolated and direct port responses to be similar.

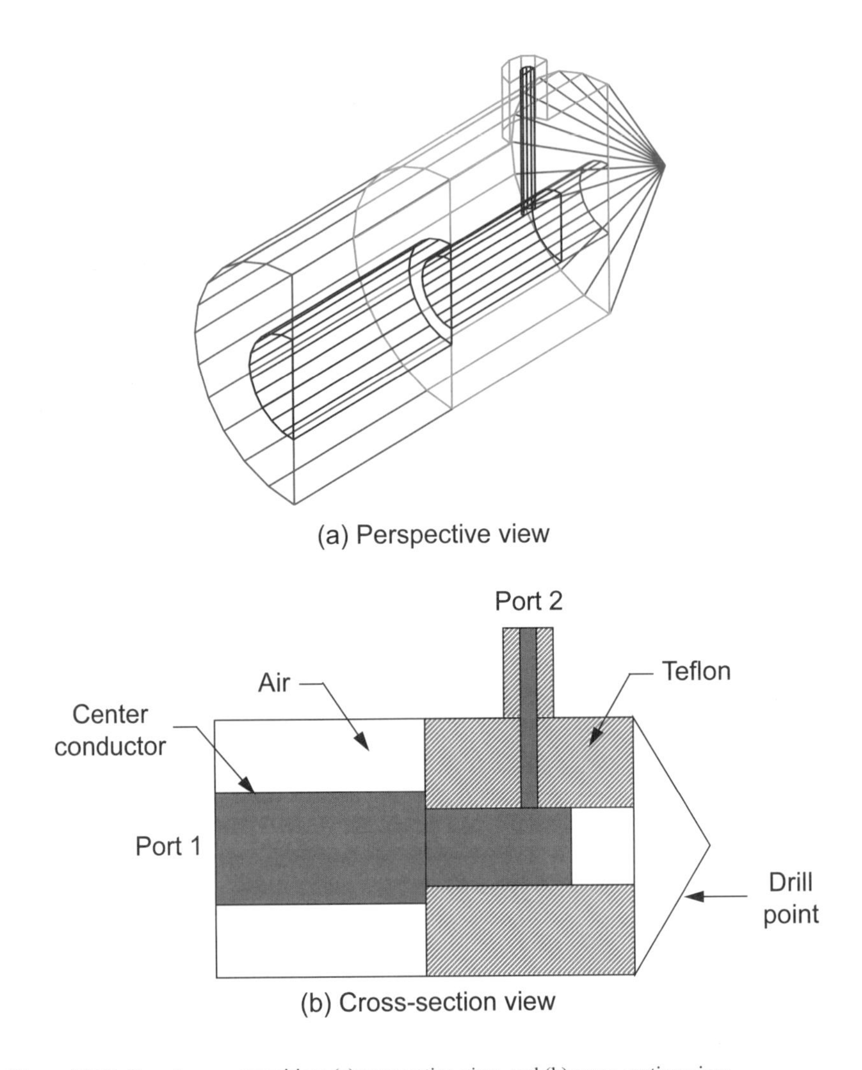

Figure 15.29 Coax-to-coax transition: (a) perspective view, and (b) cross-section view.

This design is a good example of how using the lowest order geometry possible can save design time. We could have started with a full 3D model and tried to optimize the structure. This approach would probably succeed in the end, but would take much more time. Instead, we started with the best analytical approximation we could find and then moved to a 2D cross-section model. All the "optimization" was done in the 2D mode. We built a full 3D model just to check our work. With more experience and confidence, we might have skipped that step.

We could use the full 3D model to improve the transitions at the coupled and isolated ports. Another useful aspect of the 3D model would to be to study manu-

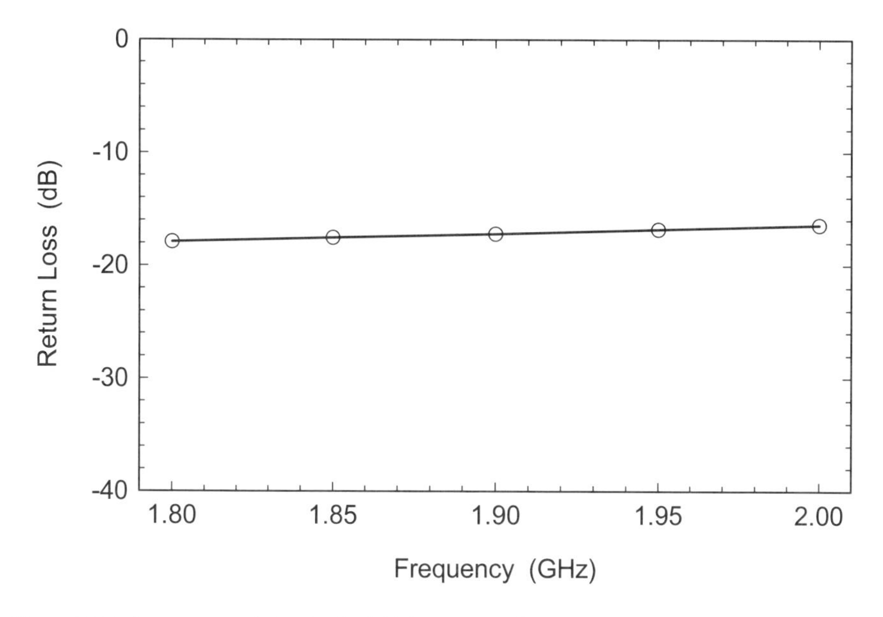

Figure 15.30 Computer prediction for the initial geometry of the coax-to-coax transition (Ansoft HFSS Ver. 8.0).

facturing sensitivities. This could be done using the Optimetrics option in Ansoft HFSS.

15.5 PCS BAND COAX-TO-COAX TRANSITION

In Figure 15.24 the signal has to transition from the main line of the coupler into the bandpass filter. In the actual hardware, the transition includes a right-angle bend and an abrupt change in diameter for the coaxial lines. A 3D view of the right angle coax-to-coax transition is shown in Figure 15.29(a). The larger diameter air-filled coax mates to the 15-dB coupler. The smaller diameter coax is Teflon filled and connects to the bandpass filter. The larger diameter center conductor is supported on its end by a Teflon ring. Although this component is clearly not a coupler, it is covered here in relation to the coupler that it interfaces with.

A cross-section view of the transition (Figure 15.29(b)) will help our understanding of the structure. A Teflon disk supports the open end of the larger structure and the center conductor steps down to maintain a 50-ohm impedance. At the far right is a conical section left behind by the boring operation. This could also be finished flat bottom. Now that we have entered the 3D geometry we can run the first analysis on the field-solver. To save time we will use the obvious symmetry plane and run only half the structure. Our initial guess at the geometry was actually quite

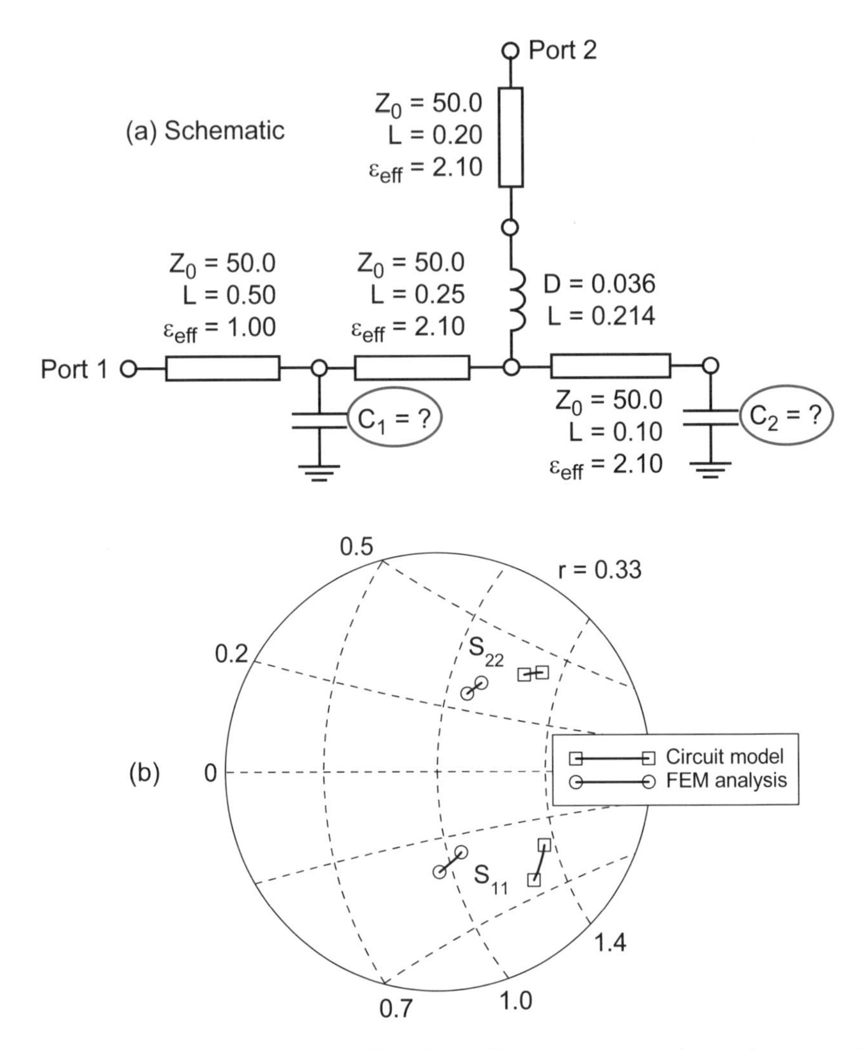

Figure 15.31 Initial circuit model of the right angle transition: (a) schematic with two unknown capacitances; and (b) results plotted on an expanded Smith chart.

good: the return loss is about 17 dB (Figure 15.30). But for the cascade of components in this subsystem we would like to maximize the return loss of each component. Maximizing the return loss will minimize the insertion loss and minimize the interactions between components.

Before we can optimize this structure we need to decide which of the physical dimensions we should vary. If we go by intuition alone, we may be wrong, which is exactly what happened to me on this project. I decided that the step discontinuity region should be optimized, spent a fair amount of time setting up the necessary

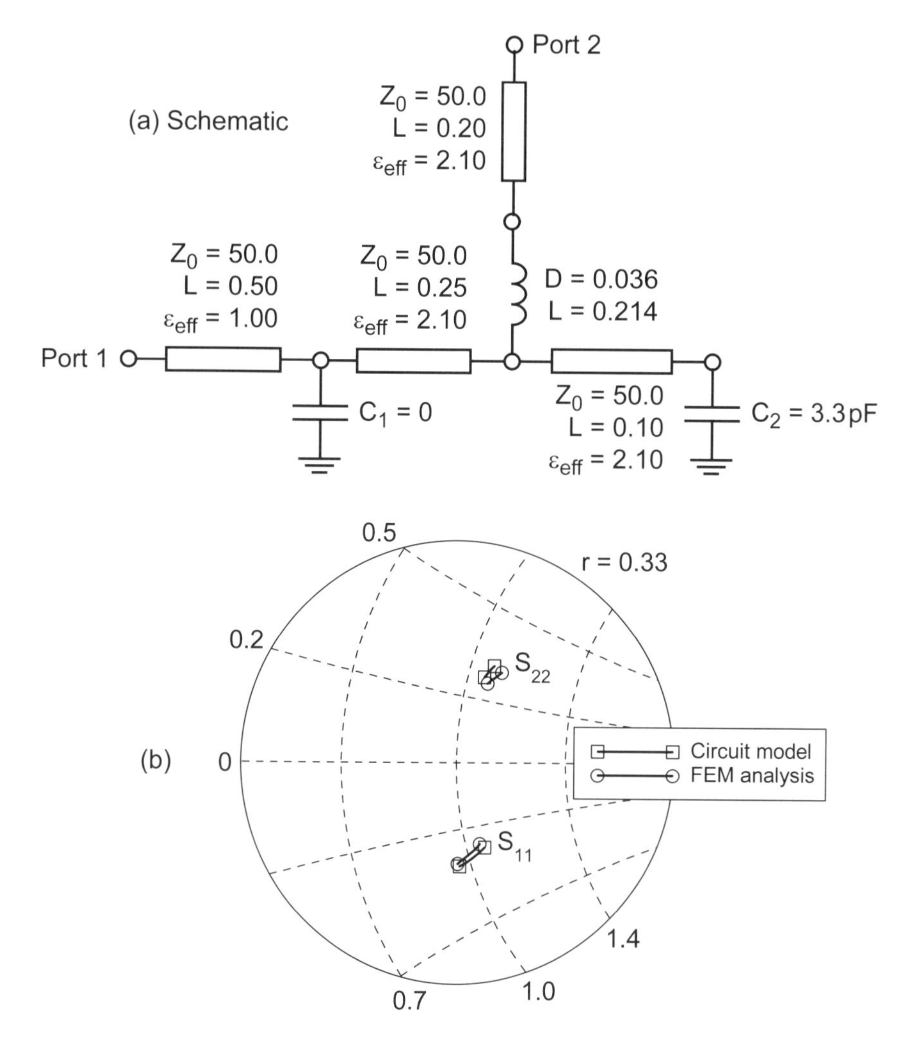

Figure 15.32 Optimized circuit model of the right angle transition: (a) schematic with all elements known; and (b) results plotted on an expanded Smith chart.

models, and wasted an overnight computer optimization that yielded no improvement in performance. But this experience got me thinking about a more efficient way to identify the key optimization variables. Assuming our first 3D analysis is correct, we can match a simple circuit theory model to the field-solver results. Once we have a fast circuit-theory-based model, we can use it to explore the problem and identify the key variables for optimization.

The initial circuit theory model for the right angle transition is shown in Figure 15.31(a). The parameters of the basic coaxial elements can be found from the

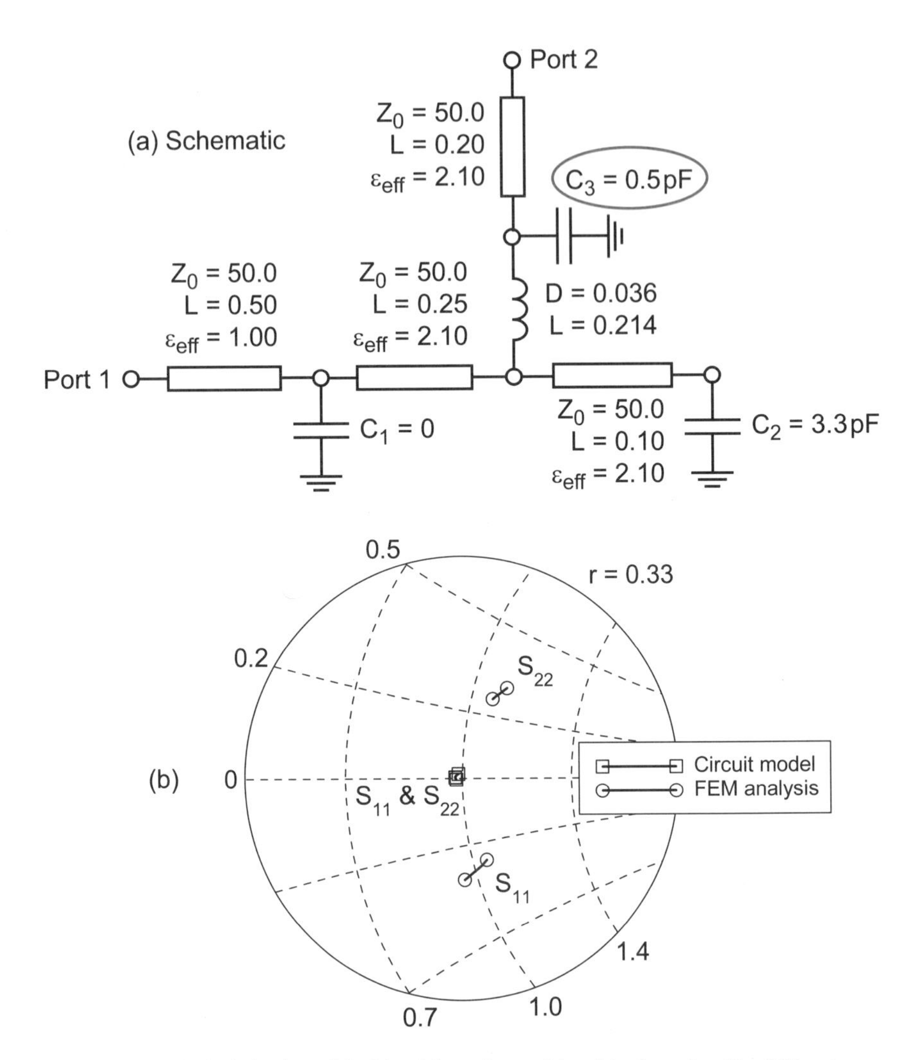

Figure 15.33 Optimized circuit model of the right angle transition: (a) schematic with additional capacitor to compensate the transition; and (b) results plotted on an expanded Smith chart.

known dimensions. We have modeled the connection between the coax lines as a wire. There are a couple of capacitors in the circuit theory model that we are initially unsure of. Capacitor C_1 represents the excess capacitance of the step junction. Capacitor C_2 represents the unknown open-end capacitance of the center conductor at the far right. The initial fit of the circuit model to the field-solver analysis is shown in Figure 15.31(b).

A quick optimization tells us that C_1 is zero and C_2 is about 0.33 pF. The fit of the circuit model to the field-solver analysis is shown in Figure 15.32(b). The

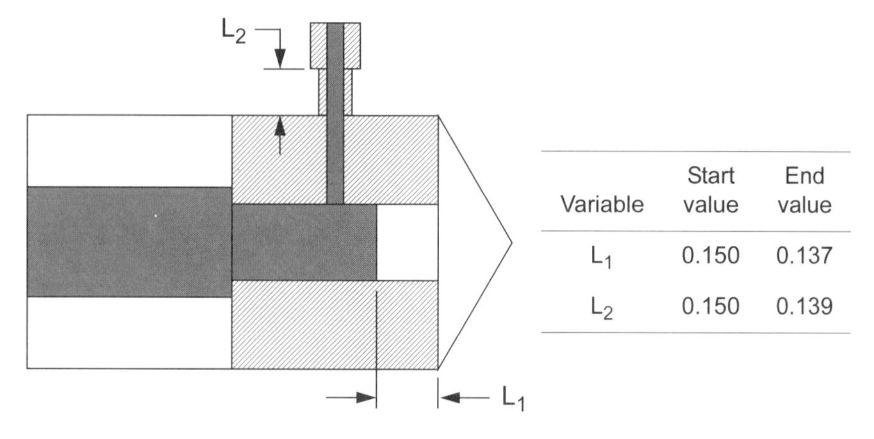

Variable	Start value	End value
L_1	0.150	0.137
L_2	0.150	0.139

Figure 15.34 Variables chosen for the optimization of the right angle transition based on the circuit theory analysis and values at start and end of the optimization.

agreement between the two models is quite acceptable. Now we can use the circuit model to explore optimization strategies for the 3D field-solver model.

Its quite fast and easy to experiment with our circuit model and decide how to approach the compensation problem. If the connection between the two 50-ohm coax lines is inductive, then perhaps some shunt capacitance at both ends of the inductor will compensate the discontinuity. We already have some excess capacitance at one end of the inductor due to the open-ended stub so we added a lumped capacitor at the other end of the wire model. A quick tuning of the circuit tells us that this new capacitor needs to be about 0.5 pF. With the additional compensation, the circuit model is now very well matched to 50 ohms.

One way to create the additional capacitance we need near Port 2 is to reduce the diameter of the output coaxial line for a short distance. Just for good measure we will also vary the length of the open stub. The first optimization of this structure used Empipe3D [14] in conjunction with Ansoft HFSS. The Empipe3D technology was later acquired by Hewlett-Packard (now Agilent). Today, we can do the same optimization with the Optimetrics module in Ansoft HFSS. To set up the problem we first draw the base or reference structure. Then we draw one new structure for

Table 15.3

Trajectory of Field-Solver Solutions

Solution no.	1	2	3	4	5	6	7	8	9	10
L_1	0.15	0.16	0.15	0.14	0.15	0.14	0.13	0.12	0.13	0.14
L_2	0.15	0.15	0.16	0.14	0.14	0.15	0.13	0.13	0.14	0.13

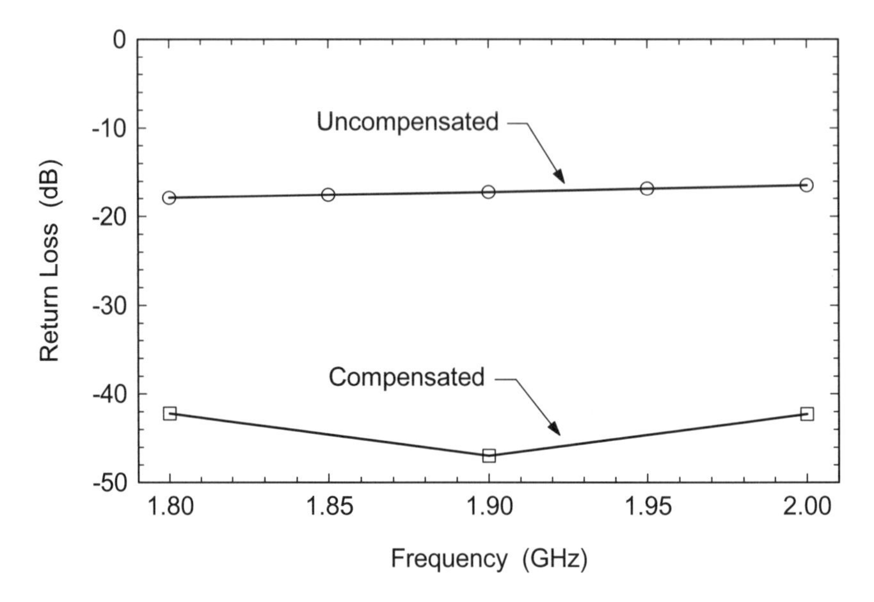

Figure 15.35 Computed return loss of the original transition and computed return loss of the compensated transition (Ansoft HFSS Ver. 8.0).

each of the variables that have been defined. In the new models, small changes are made to the dimensions that correspond to the variables. We are not allowed to change the number or names of the objects in the model. With a very good starting point from our circuit theory model, the optimization proceeds quite quickly. The starting values and optimized values for both variables are shown in Figure 15.34.

A total of 10 field-solver solutions were computed. It is interesting to look at the "trajectory" of the field-solver solutions (Table 15.3). We specified 10-mil steps in both variables for the field-solver solutions. Notice that the final solution falls off this "grid" of known solutions because the software can interpolate between known solutions. In Figure 15.35 we have plotted the return loss of the original transition and the return loss of the compensated transition.

A few weeks after this design was completed, the machinist and the mechanical engineer decided to change the manufacturing process, which resulted in a flat bottom bore rather than the drill point we analyzed earlier. The flat bottom version would be fabricated with a large end mill or a flat bottom boring tool. The two structures are shown in Figure 15.36 for comparison. The optimized dimensions for both versions are also shown in Figure 15.36. There is a significant difference between the two. By "significant" we mean the difference is greater than the tolerance we would expect to hold on these dimensions. An independent measurement of this transition was never made. However, this design was used successfully in the fully integrated tower top amplifier product.

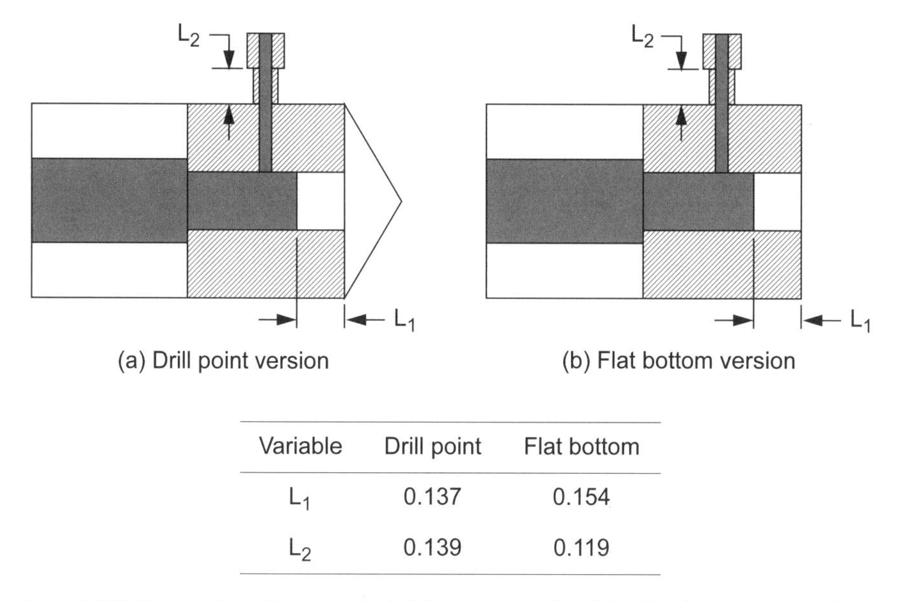

Variable	Drill point	Flat bottom
L_1	0.137	0.154
L_2	0.139	0.119

Figure 15.36 Two versions of the optimized right angle transition: (a) drill point version, and (b) flat bottom version. The optimized dimensions for both versions are also shown.

Our initial guess at the dimensions for this transition was actually quite good. But it was an interesting exercise to see if the performance could be improved. After our first field-solver analysis we were able to build a simple circuit theory model that helped us understand our problem better. There are no analytical models in any linear simulator that will tell us the capacitance of the open-ended stub. A very quick fit to the field-solver results gave us that capacitance and told us the step capacitance was essentially zero.

We then used our circuit theory model to explore various optimization strategies. This is much more efficient than trying educated guesses on the field-solver. Finally it was time to set up the optimization problem. Using the available software tools, we only have to generate one new geometry file for each variable. Because our starting point was so good, the optimization proceeded quite quickly. If time allowed we could continue this study with a tolerance analysis or even a yield optimization.

References

[1] Oliver, B. M., "Directional Electromagnetic Couplers," *Proc. IRE*, Vol. 42, No. 11, 1954, pp. 1686–1692.

[2] Lange, J., "Interdigitated Stripline Quadrature Hybrid (Correspondence)," *IEEE Trans. Microwave Theory Tech.*, Vol. 17, No. 12, 1969, pp. 1150–1151.

[3] Private communication, Eric Frick, WJ Communications, Inc.

[4] Podel, A., "A High Directivity Microstrip Coupler Technique," *IEEE MTT-S Int. Microwave Symposium Digest*, Newport Beach, CA, May 11–14, 1970, pp. 33–36.

[5] Taylor, J., and D. Prigel, "Wiggly Phase Shifters and Directional Couplers for Radio-Frequency Hybrid-Microcircuit Applications," *IEEE Trans. on Parts, Hybrids, and Packaging*, Vol. 12, No. 4, 1976, pp. 317–323.

[6] de Ronde, F. C., "Wide-Band High Directivity in MIC Proximity Couplers by Planar Means," *IEEE MTT-S Int. Microwave Symposium Digest*, Washington, DC, May 28–30, 1980, pp. 480–482.

[7] Rautio, J. C., "Experimental Validation of Electromagnetic Software," *Int. J. MIMCAE*, Vol. 1, No. 4, 1991, pp. 379–385.

[8] Swanson, Jr., D. G., "Experimental Validation: Measuring a Simple Circuit," *IEEE Int. Microwave Symposium Workshop WSMK Digest*, Atlanta, GA, June 14–18, 1993.

[9] Gentili, G. G., et al., "Accurate Modeling of Lange Couplers for CAD Applications," *21st European Microwave Conference Proceedings*, Stuttgart, Germany, September 1991, pp. 1556–1561.

[10] Maas, S. A., "Accurate Design of Lange Couplers on GaAs," *Microwave Journal*, Vol. 39, No. 8, 1996, pp. 90–98.

[11] "Precise Electromagnetic Analysis of Lange Couplers Using ***em***™," Sonnet Application Note 40-01, Sonnet Software, Liverpool, NY, 1997.

[12] Stracca, G. B., et al., "Numerical Analysis of Various Configurations of Slab Lines," *IEEE Trans. Microwave Theory and Tech.*, Vol. 34, No. 3, 1986, pp. 359–363.

[13] Agarwal, A., et al., "Coupled Bars in Rectangular Coaxial," *Electronics Letters*, Vol. 25, No. 1, 1989, pp. 66–67.

[14] Empipe3D, Optimization Systems Associates, Dundas, Ont., Canada.

Chapter 16

Microstrip Filters

A system designer has many different filter technologies to choose from with a wide range of unloaded quality factors (Q_u). The required filter bandwidth and the Q_u of the filter technology determine the resulting insertion loss of the filter. The achievable Q_u is generally proportional to the volume available. Waveguide and dielectric resonator filters have Q_us of 10,000 or more. Filters with less than 1% bandwidth and insertion loss less than 1 dB can be realized. Combline and other coaxial TEM filters have Q_us in the 2,000 to 5,000 range. Suspended substrate and ceramic coaxial resonators achieve Q_us in the 300 to 500 range. By the time we get to microstrip, where the fields are confined to a fairly small volume, the maximum Q_u has dropped to about 150. In microstrip, filters with 20% bandwidth or more can have insertion losses less than 2 dB. Below 10% bandwidth the insertion loss rises rapidly.

Although microstrip is not the highest performance filter technology, it is the preferred choice in many thin-film on ceramic and printed circuit board applications. RF preselector filters, image rejection filters, local oscillator (LO) filters and intermediate frequency (IF) filters can all be realized in microstrip. The goal is to create an exact design, fabricate the filter to very tight tolerances, and do no tuning on the filter in production. Of course, the real design and production environments are slightly less than ideal. However, the EM field-solver can have a large impact on the design process. Every improvement in accuracy at the design level raises the chances of success that much higher.

Planar filters may be another class of components that we can use to benchmark the achievable accuracy of EM field-solvers. In a filter there is an exact center frequency, bandwidth, and return loss level that we are trying to achieve. We can easily measure the differences between measured and modeled results and report the errors. There are also no active devices to add uncertainty to the fabricated results. However, we often do not know the exact electrical parameters of the substrate used and we may not even know the exact physical dimensions of the substrate. We also often fail to carefully measure the dimensions actually delivered and compare them to what was specified.

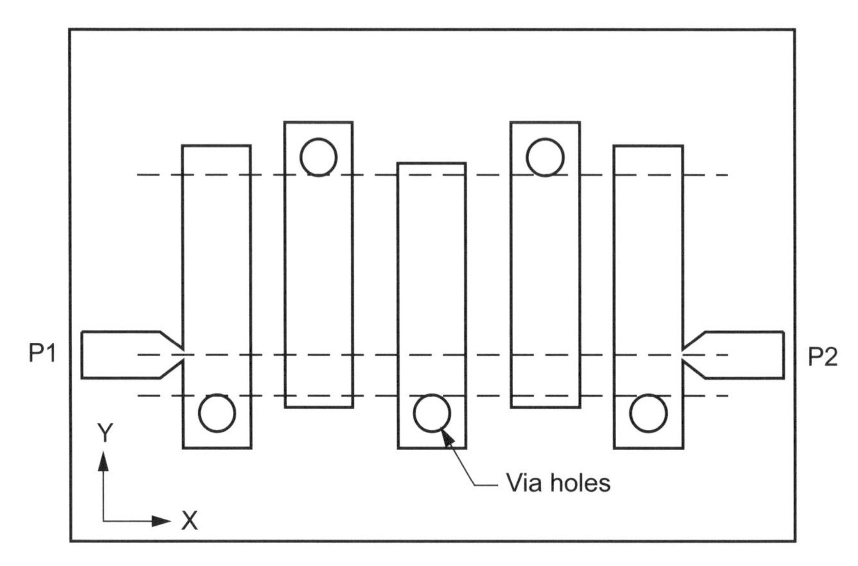

Figure 16.1 A fifth-order microstrip interdigital filter with tapped input and output resonators. The dashed lines indicate the planes used to subdivide the geometry for analysis. © 1995 Nexus Media Ltd. [1].

Despite these caveats, planar filters have provided many opportunities to test the capabilities of EM field-solvers and challenge the ingenuity of filter designers. In this chapter we offer a small sample of the many designs attempted by the author over the last decade.

16.1 INTERDIGITAL FILTERS

The microstrip interdigital filter, Figure 16.1, consists of quarter wavelength resonators shorted at one end. The shorted ends alternate, forming the "interdigitated" set of resonators. We can tap into the first and last resonators or couple into the filter with redundant resonators. In the example shown, all the resonators have the same width and roughly the same impedance. You may also see designs where the resonators have different widths. Either approach is valid because in fact there is no unique solution; there are an infinite number of width and gap combinations that will realize the same filter. For design and optimization purposes, a design with equal width strips has fewer variables to deal with.

The interdigital filter continues to be popular due to its compact layout. An $N = 5$ filter occupies a rectangular area roughly $\lambda/4$ on a side. One disadvantage of the microstrip interdigital filter is its sensitivity to misalignment between the metal pattern and the via holes. The vias are generally drilled first with some finite tolerance on their location. Then the metal pattern is aligned to the hole pattern with

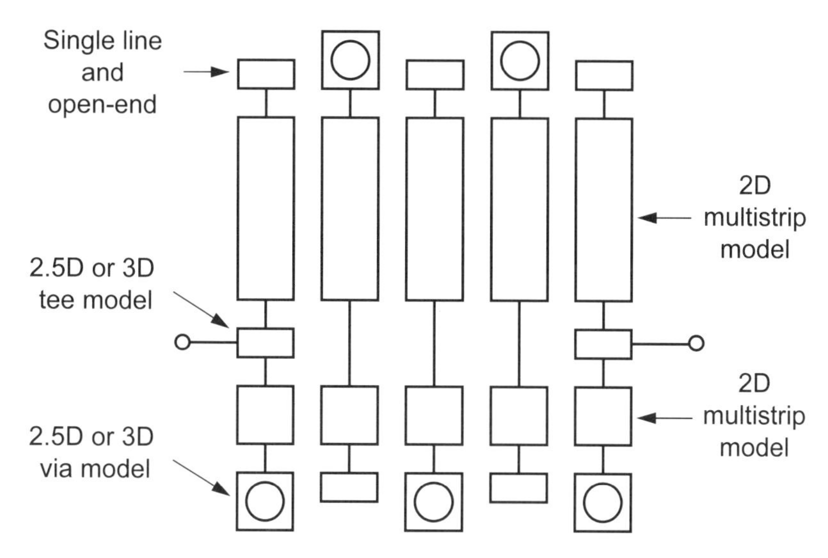

Figure 16.2 Schematic of the filter analysis. The "hybrid" solution combines analytical models, multistrip models (2D cross-section engine), and results from a 2.5D or 3D solver for the tee-junctions and the vias. © 1995 Nexus Media Ltd. [1].

some finite tolerance in the photolithography. If the metal pattern is offset in the positive or negative Y-direction (Figure 16.1), one set of resonators is too long while the other is too short. Or if one hole is out of position in the Y-direction, that resonator will be too long or too short. Misalignment in the X-direction is less of a problem. There may be grounding systems that reduce the alignment sensitivity somewhat.

There are several commercially available programs that will design this type of filter—one of the better ones is IDM [2]. If we wish to do our own analysis and optimization, we need a strategy for subdividing the problem into manageable pieces. The dashed lines in Figure 16.1 indicate the planes we might use to subdivide the geometry into available model types. Figure 16.2 is a schematic view of the resulting model. We have two sets of multiple coupled lines, the vias, two tee-junctions, uncoupled single lines, and some open-ends.

But if we think about the current distributions we looked at for the tee-junctions and the via we might question the validity of this approach. If we imagine current coming in on the feed line from the left, we know that the current coming around the corners of the tee-junction takes a finite amount of time and distance to reestablish the normal microstrip current distribution. On the via side of the tee, before the normal distribution is reestablished, the current hits the via and gets pulled into the center of the strip and down to ground. Meanwhile, the multiple coupled strip models on both sides of the tee-junction assume a standard, undisturbed microstrip current distribution along their whole length. At the open ends of the res-

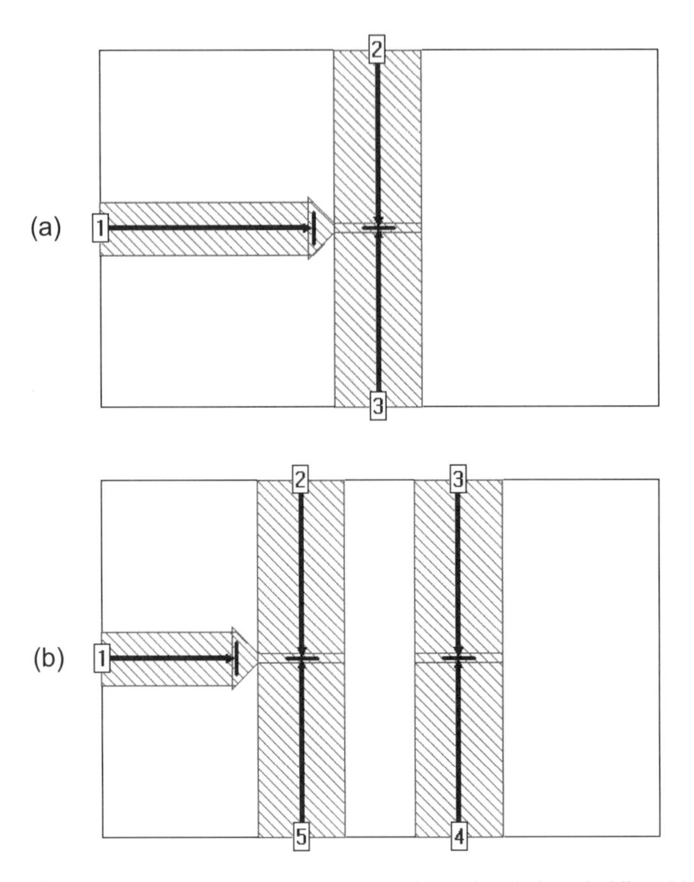

Figure 16.3 Tee-junction with tapered common arm and zero length de-embedding: (a) tee-junction alone, and (b) tee-junction with next line over (Sonnet ***em*** Ver. 8.0).

onators, where the multistrip model meets the open-ends, the multistrip line magically becomes a single, isolated line in zero distance. On the next line over, the multistrip model meets a via. In fact, this boundary that we imposed arbitrarily, based only on visual clues, marks a region on each strip where the actual current distribution may be fairly complex.

While standard, individual discontinuity models may not capture the complexity of the true current distributions, field-solver models can capture the true behavior. If we think about the tee-junction model (Figure 16.3(a)), we first do an analysis of the whole geometry, then we de-embed using lengths of uniform, ideal line. When we join the models together, we add some of that ideal line back into the global model. But the complexity of the junction region is still there, stored in the *S*-parameters of the model. This "memory" of complex behavior is independent of how much uniform line we remove during de-embedding. De-embedding to zero

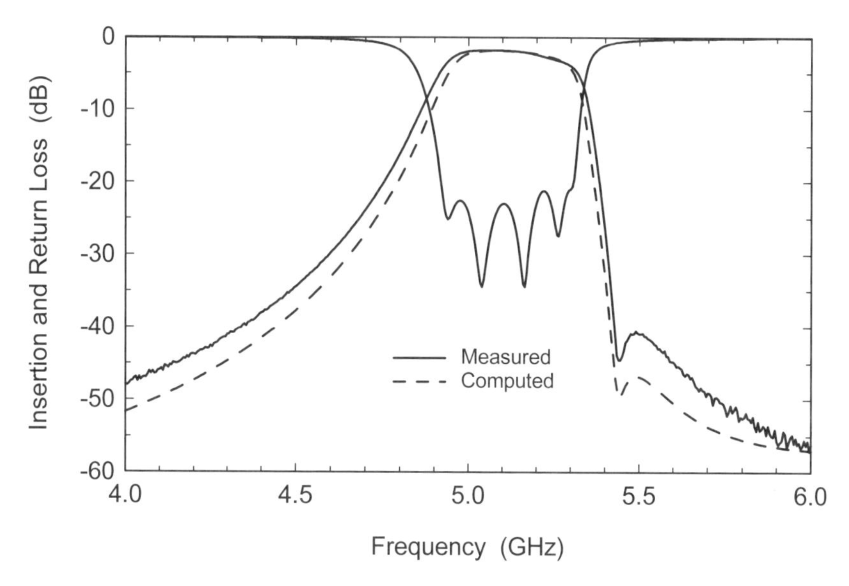

Figure 16.4 Tuned result for a fifth-order microstrip interdigital filter. The transmission zero on the high side of the passband is due to the nonadjacent couplings between the strips. © 1995 Nexus Media Ltd. [1].

length on Port 2 and Port 3 of the tee-junction also makes the global model more efficient. If we did not de-embed to zero length on those arms, we would have to add a third multistrip model to the schematic to account for the physical width of the tee-junction model.

The tee-junction in this filter is not really isolated; there is a second line nearby. The gap between lines is typically one substrate thickness or less and it is also one line width or less. Solving for the tee-junction in the presence of the second line (Figure 16.3(b)) might give us better results. This five-port model could be updated when significant changes in the gap dimension occurred. We should also note the taper in the feed line at the common junction in the tee models. Experimentally this was found to work better than a full width common arm. My intuition and the current plots tell me that this forces the non-uniform current region to be as small as possible.

The tee-junction and via models needed for the filter analysis are "static" in the sense that their dimensions do not change in the optimization process. The widths of the resonators are all equal and do not vary during optimization. So these models can easily be *S*-parameter files that are computed once and used over and over again.

In the mid-1980s when we first started to look at this type of filter, there were no multiple coupled line models in the commercial circuit simulators. There were several approximate techniques [3], and we came up with our own approximate

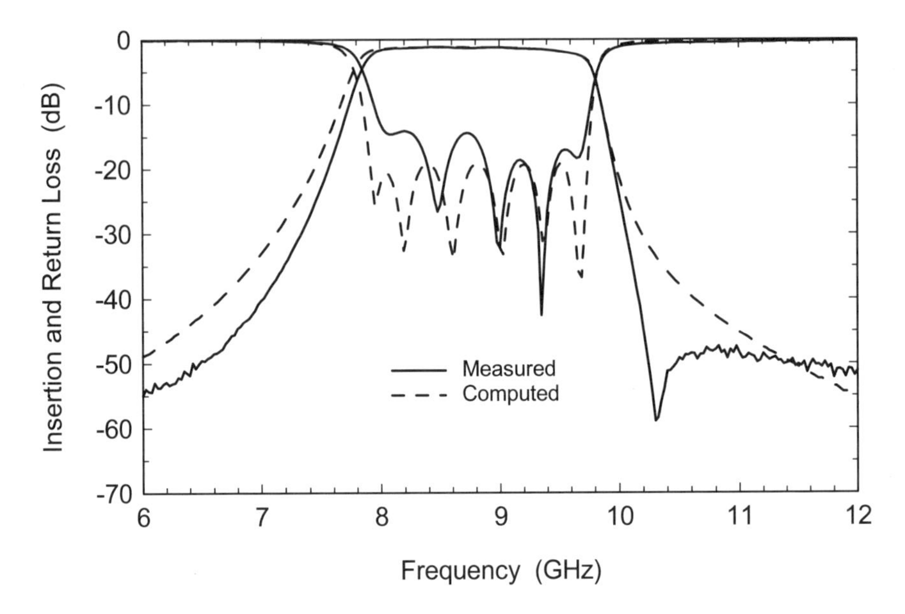

Figure 16.5 Turn-on performance of a seventh-order microstrip interdigital filter. The transmission zero prediction has moved off the plot to the right due to a problem in the 2D cross-section-solver.

technique [4] that is occasionally still used today. Later, full-wave, 2D cross-section-solvers became available in Super-Compact (now Designer from Ansoft) and LINMIC+/N from AC Microwave. When 2.5D and 3D field-solvers became available in the late 1980s and early 1990s, we had all the tools needed for fairly efficient analysis and optimization of interdigital filters.

Figure 16.4 shows the results for an interdigital filter designed using this "hybrid" combination of 2D cross-section and 2.5D planar simulators. It is a fifth-order interdigital filter in C-band. The transmission zero on the high side of the passband is caused by the nonadjacent couplings between the strips. The multistrip model is able to predict this zero fairly accurately. The center frequency error is less than 1% and the bandwidth error is 50 MHz, or about 9% of the desired bandwidth.

A second filter example is shown in Figure 16.5. This is the turn-on performance of a seventh order interdigital filter centered near 9 GHz. The substrate is 15-mil thick alumina, $\varepsilon_r = 9.8$. The bandwidth error is about 70 MHz, or 3% of the desired bandwidth. Notice that transmission zero has disappeared from the computer prediction. In fact, it is there but it has shifted to the right and off the plot. This is the result of a fairly subtle problem in the full-wave 2D cross-section-solver [5]. If we use a quasi-static solution of the multiple strips, there is an unambiguous conversion from the [L] and [C] matrices we compute to the Y-parameters we need for the circuit simulator. In the 2D full-wave solver, there are hybrid, non-TEM

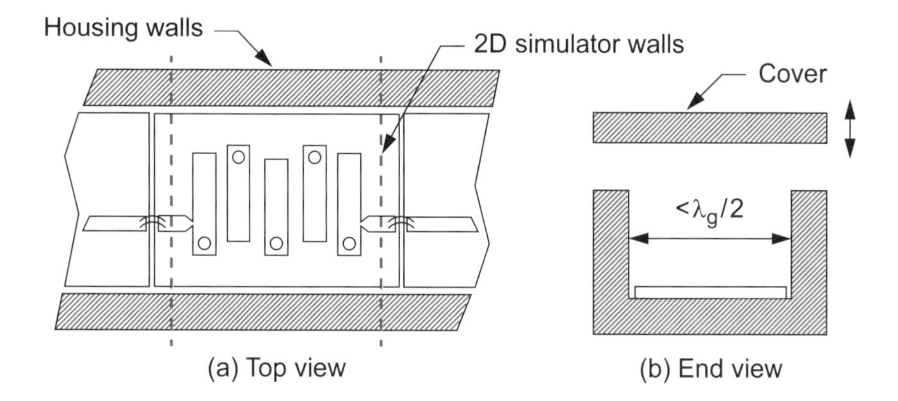

Figure 16.6 Interdigital filter installed in a channelized housing: (a) top view, and (b) end view. The width of the channel must be less than $\lambda_g/2$ to form a waveguide below cutoff. The 2D simulator walls (red dashed lines) have the wrong orientation to model the channel effects.

field components in the solution. These non-TEM components get stronger as frequency increases and it is not clear how to handle them in the conversion to *Y*-parameters. So a quasi-static solver predicts the relative position of the transmission zero at all frequencies, but as frequency increases its center frequency prediction get less accurate. The full-wave solver gives a more accurate center frequency prediction and can find the transmission zero at low frequencies, but loses it at higher frequencies.

There is another second-order effect observed in these filters that has recently been solved. These filters are normally mounted in a package or channelized housing whose cross-section forms a cutoff waveguide (Figure 16.6). If you take the cover off the package while measuring one of these filters, the lower band edge suddenly expands down in frequency. The shift is significant and easily observed by eye. For years we tried to model this using a change in the electrical parameters of the strips due to the loading of the cover but the numbers never made sense. One day we happened to lay a paper clip across the package, instead of the cover, and the lower band edge pulled in. So it was clearly not the loading of the cover, but coupling to evanescent waveguide modes in the channel. The paper clip forced the two sides of the channel to the same potential, just like the cover would. Again, any 2D closed box, cross-section analysis of the strips cannot pick this up because its solution box is oriented 90 degrees from the true package walls. Only a full 2.5D or 3D analysis with the package walls in their true position will fully capture this effect. Recently, Matthai and Rautio [6, 7] documented the effect of the package on multiple coupled strips and explained the bandwidth contraction in the filter.

Why not just do a complete 2.5D or 3D analysis of the filter and be done with it? Because, even with today's computational resources, that is not very efficient.

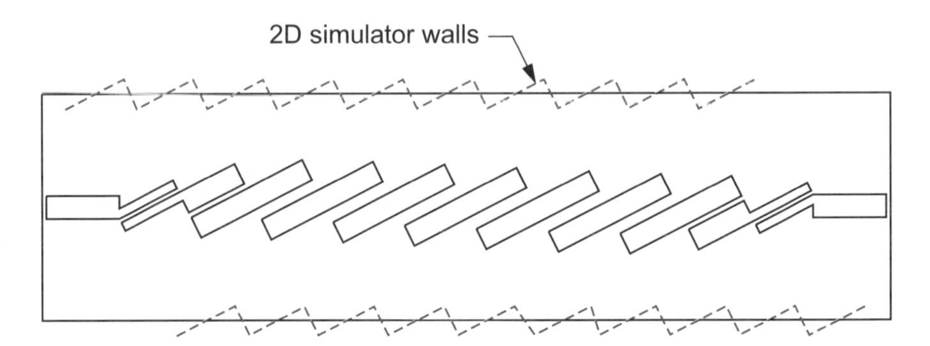

Figure 16.7 A ninth-order millimeter-wave edge-coupled filter. The substrate is 10-mil thick alumina, 385-mil long by 100-mil wide. The filter is modeled using a cascade of 2D cross-section solutions. The red dashed lines indicate the location of the 2D simulator walls (Ansoft Serenade). © 1995 Nexus Media Ltd. [1].

It's hard to do a complete filter with the closed box MoM codes because of grid resolution issues. The laterally open MoM codes overcome the resolution problem but, until fairly recently, could not include the box walls in the simulation. We can do the problem with a 3D FEM code, but again the solution time will be too long for efficient optimization. What we can do is optimize a design using the hybrid method, run a full 3D simulation of that geometry to find the bandwidth error due to the package effects, then go back to the fast, hybrid method and add the bandwidth correction to the original specification.

16.2 EDGE-COUPLED FILTERS

The microstrip edge-coupled filter (Figure 16.7) consists of half wavelength resonators open circuited at both ends. The resonator is coupled to its adjacent resonators over a distance of one quarter wavelength. The input and output couplings can be quarter wavelength coupled sections, quarter wavelength transformers [8], or the first and last resonators can be tapped. In the example shown, all the interior resonators have the same strip width. In some designs each quarter wavelength coupled section, up to the midpoint of the filter, has a different strip width. This is again unnecessary and introduces additional step discontinuities in the interior of the filter. As in the interdigital filter, there is no unique combination of strip widths and gaps for a given filter design. We typically make the strip widths as wide as possible to maximize the Q_u. However, as we increase the strip widths the first gap gets smaller, which puts a practical limit on the width we can realize.

At higher microwave and millimeter-wave frequencies the microstrip edge-coupled filter is a popular topology. Its long, narrow aspect ratio makes it easy to pack several filters side-by-side in a module and maintain a high waveguide cutoff

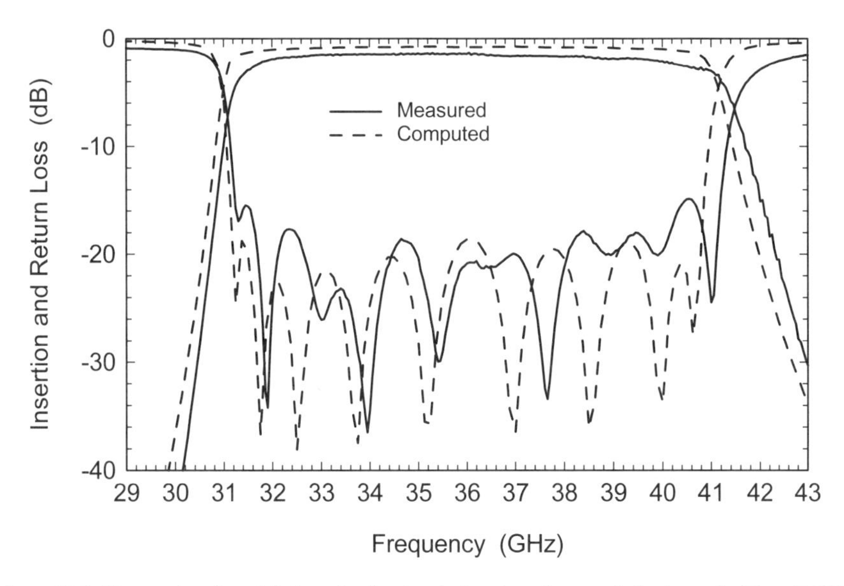

Figure 16.8 Measured and modeled results for the ninth-order edge-coupled microstrip filter. © 1995 Nexus Media Ltd. [1].

frequency in the filter channel. In the interests of circuit density, we have started to bring the sidewalls as close as two or three substrate thicknesses away from the filter. We have found that the sidewall location has a bigger influence than the cover height, which is typically eight to 10 substrate thicknesses. Like the interdigital, the edge-coupled filter couples to evanescent modes in the channel and the bandwidth contracts when we install the cover. For applications below 6 GHz an edge-coupled filter becomes physically very long and we often switch to other topologies. Other options include the hairpin and a pseudo-lumped element topology that we will discuss in later sections. There are several commercially available programs that will design edge-coupled filters; one of the better ones is ECM [9].

We should also mention that all the distributed filter topologies tend to launch quite a bit of energy into the substrate. The air/dielectric interface at the substrate surface tends to trap this energy and allows it to propagate away from the filter in the substrate. This effect is, of course, worse for the higher dielectric constants. Putting filters on individual substrates, as in Figure 16.6, helps to isolate them from other circuits. We have seen cases where filters fabricated on the same ceramic substrate could not be made to perform correctly.

The breakthrough in CAD for this filter has again been the 2D cross-section-solver, which allows us to include the sidewall position, at least in an average sense. Figure 16.7 is a typical layout for an $N = 9$ millimeter-wave edge-coupled filter [1]. The filter is analyzed using a cascade of 2D cross-section solutions. The dashed lines indicate the location of the sidewalls in the 2D cross-section-solver. The sub-

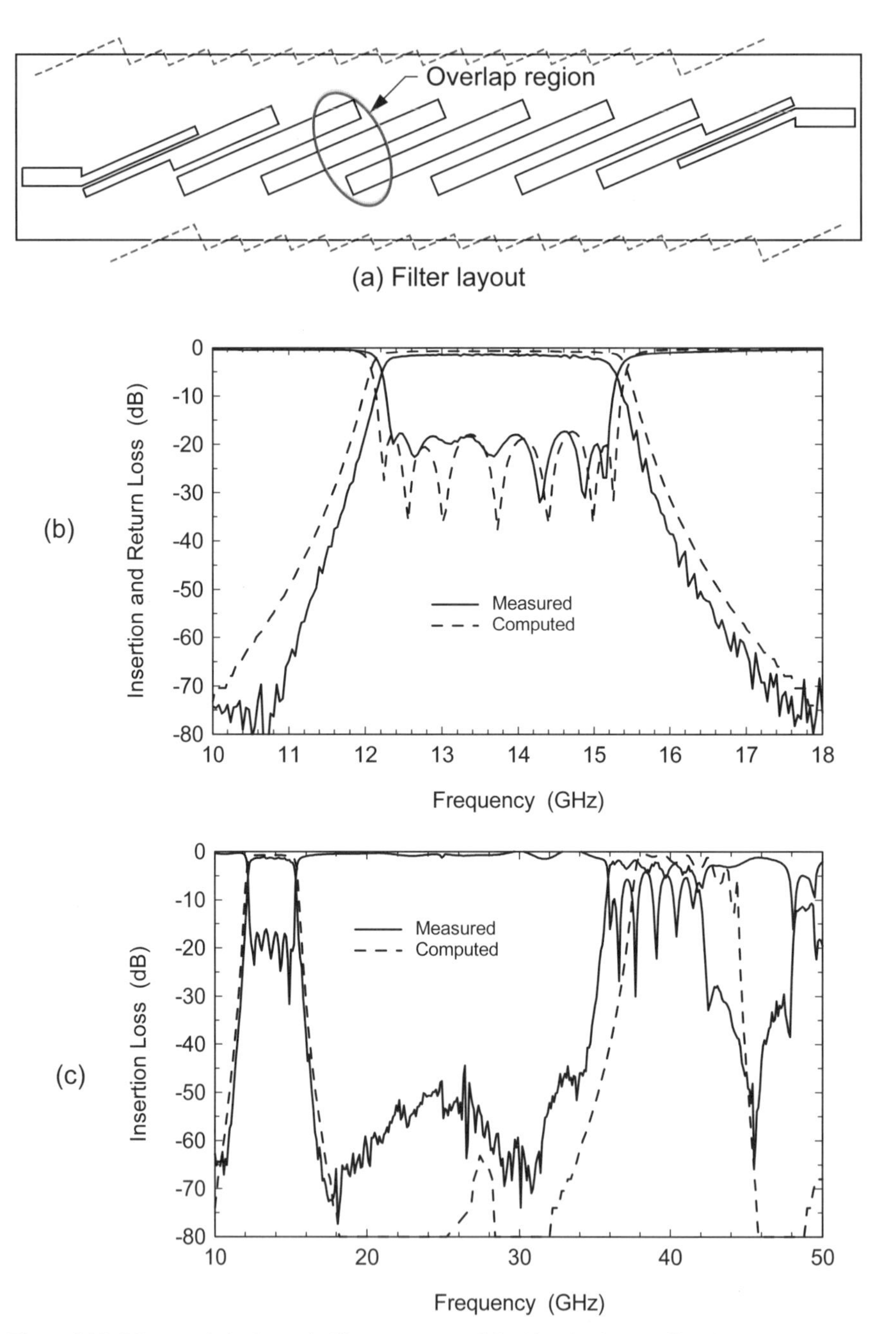

Figure 16.9 Edge-coupled microstrip filter compensated for $2f_0$ rejection: (a) filter layout, (b) narrowband response, and (c) broadband response (LINMIC+/N Ver. 6). © 1995 Nexus Media Ltd. [1].

strate is 10-mil alumina, 385-mil long by 100-mil wide. All the interior resonators have been set to the same width to minimize the number of step discontinuities. The remaining step discontinuities are either analytical models or 2.5D field-solver models, depending on the aspect ratio of the step. The open-end models are presently analytical, although we have experimented with models from the field-solver.

Figure 16.8 shows the measured and modeled responses for this filter; the computed response is from Ansoft Serenade. The measured curve is turn-on data and is typical of all the responses from a 2 by 2 inch substrate. The center frequency error is about 140 MHz or 0.4% and the bandwidth error is 200 MHz or 2% of the desired bandwidth.

One disadvantage of the edge-coupled filter is a $2f_0$ response due to the difference between the even- and odd-mode phase velocities. Several techniques to compensate the phase velocities have been published and we often use the technique by Riddle [10]. His approach extends the resonator lengths at the quarter-wave section junctions, which creates a three-coupled strip region (Figure 16.9(a)). Again we use the 2D cross-section-solver to model the two-strip and three-strip regions with the average sidewall positions included. The length of the overlap region is adjusted to maximize the rejection at the $2f_0$ frequency.

Figure 16.9 shows narrowband and broadband plots for this X-band filter; the computed responses are from LINMIC+/N by AC Microwave. The center frequency error is essentially zero and the bandwidth error is 250 MHz or 8%. In the broadband plot we can see nearly 50 dB of rejection at the $2f_0$ frequency. Without compensation, the rejection at $2f_0$ might only be 5 to 15 dB. The computer prediction above 30 GHz can be improved by adjusting the parameters of the 2D solver in LINMIC+/N.

16.3 22.5-GHZ BANDPASS FILTER

There is another potential solution to the high-frequency spurious responses found in distributed filters. The bandpass filter shown in Figure 16.10(a) is a pseudo-lumped topology fabricated on a thin, low dielectric constant substrate [11]. High impedance transmission lines form series inductors, while pairs of rectangular patches, separated by narrow gaps, form capacitor pi-networks. We call this topology "pseudo-lumped" because all of the printed inductive and capacitive elements are small in terms of wavelengths at band center. A fragment of the lumped element prototype for this filter is shown in Figure 16.10(b). Parallel plate chip capacitors are used across the outermost gaps to increase the coupling. The parasitic shunt capacitance of the printed inductors is absorbed in the adjacent shunt capacitors.

This topology has been used very successfully at lower microwave frequencies [12, 13]; its principal advantage is spurious-free performance out to $4f_0$ or even $5f_0$. Unlike distributed filters, these filters do not launch so much energy into the substrate and they couple much less to evanescent modes in the channel formed by the packaging. In this case the substrate material, Trans-Tech D450, is 10-mil thick

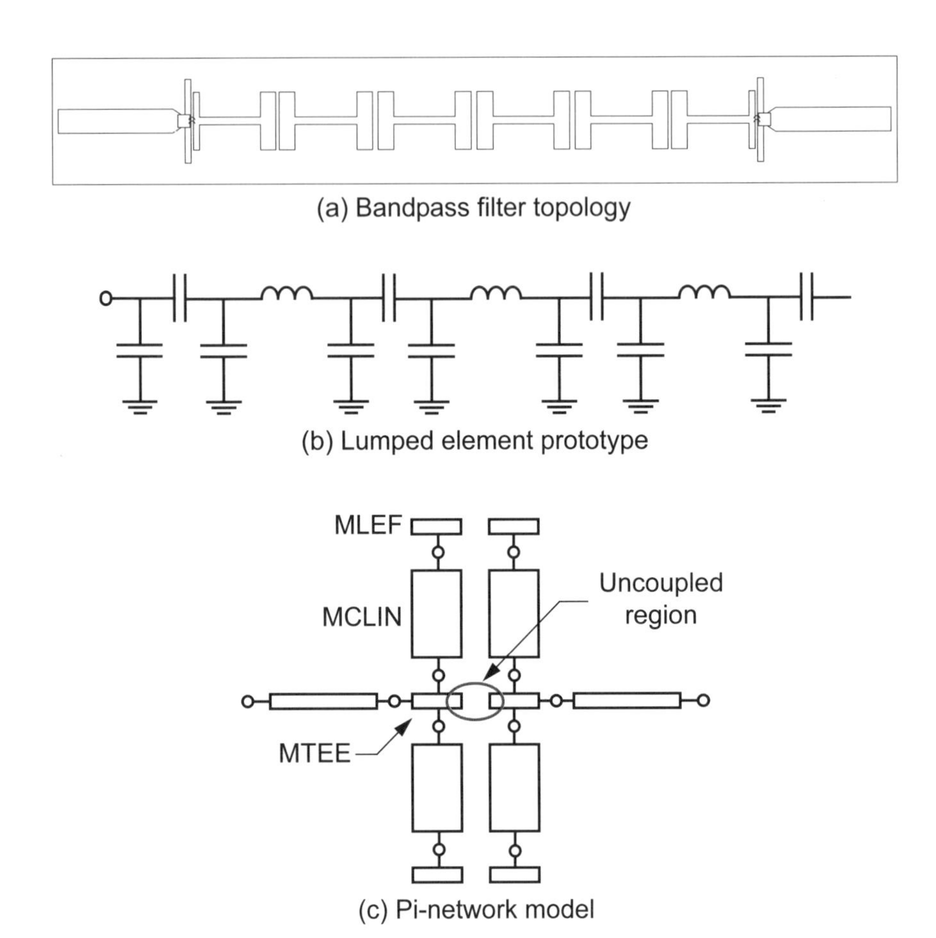

Figure 16.10 22.5-GHz bandpass filter: (a) top view of filter topology, (b) partial schematic of the lumped element prototype, and (c) schematic of the circuit theory model for one of the capacitive pi-networks. © 1995 John Wiley & Sons, Inc. [11].

with a relative dielectric constant of 4.5. The substrate size is 690 by 100 mil. The low dielectric constant makes small shunt capacitors realizable and helps prevent higher order modes in the substrate. We also attempted to fabricate this type of filter on a low dielectric constant, soft substrate with copper metallization, but it was impossible to get enough resolution in the etching process.

Figure 16.10(c) is a schematic showing how one of the pi-networks might be modeled using the standard library of elements in any linear circuit simulator. The geometry is so simple we assumed we could use analytical models and avoid a field-solver analysis. A width of 6 mil was chosen for the series transmission lines. With a 6-mil line width, the MTEE model introduces an uncoupled region in the center of the pi-network. We expected this uncoupled region to introduce a small,

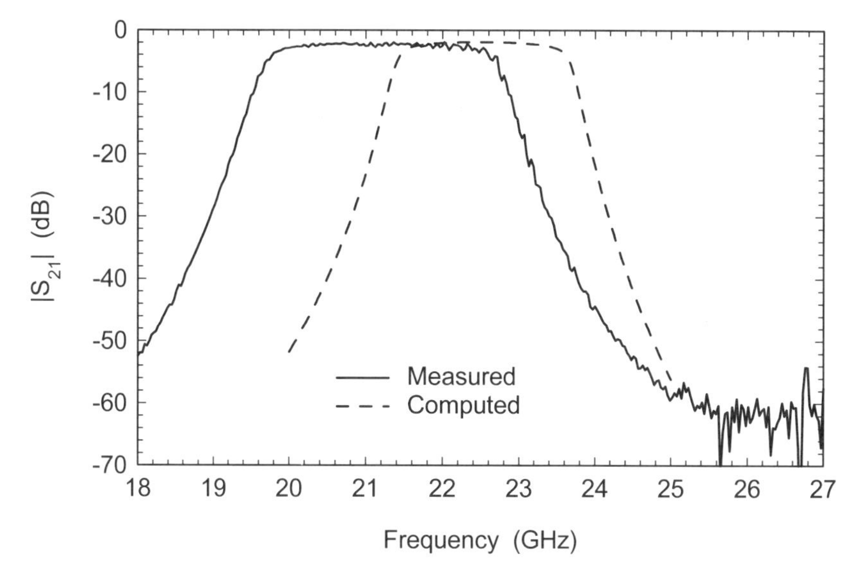

Figure 16.11 The measured versus modeled results for the first iteration of the 22.5-GHz bandpass filter. The center frequency error is about 5.8% and the bandwidth error is about 33%. © 1995 John Wiley & Sons, Inc. [11].

but acceptable, error into the final design. In fact, there were other errors that were far more significant.

The measured versus modeled results for the first iteration filter are shown in Figure 16.11. The correlation is quite poor between the measured results and the computer model. The center frequency error is 1.31 GHz or 5.8%, and the bandwidth error is 830 MHz or 33%. After carefully checking the design file and the as-built dimensions we finally back modeled the first iteration filter using an EM field-solver. The agreement between the field-solver-based analysis and the measured data was fairly good. But the question remained, how could a circuit-theory-based analysis of such a simple geometry be so far off?

A qualitative interpretation for the behavior of this filter can be found by examining the current distribution on the pi-networks. First let's look at the conventional coupled line case. Figure 16.12(a, b) show the even- and odd-mode current distributions on a pair of coupled lines. Port 1 and Port 2 are the driven ports. In the even-mode, the current is nearly twice as large on the outer edges of the strips compared to the inner edges. The odd-mode case is just the reverse: the current is nearly twice as high on the inner edges of the strips. These current distributions are consistent with the conventional theory for coupled lines.

The equivalent even- and odd-mode current distributions on the pi-networks can be found in Figure 16.12(c, d). In the even-mode there is more current on the outer edges of the strips and very little current on the inner edges. The odd-mode

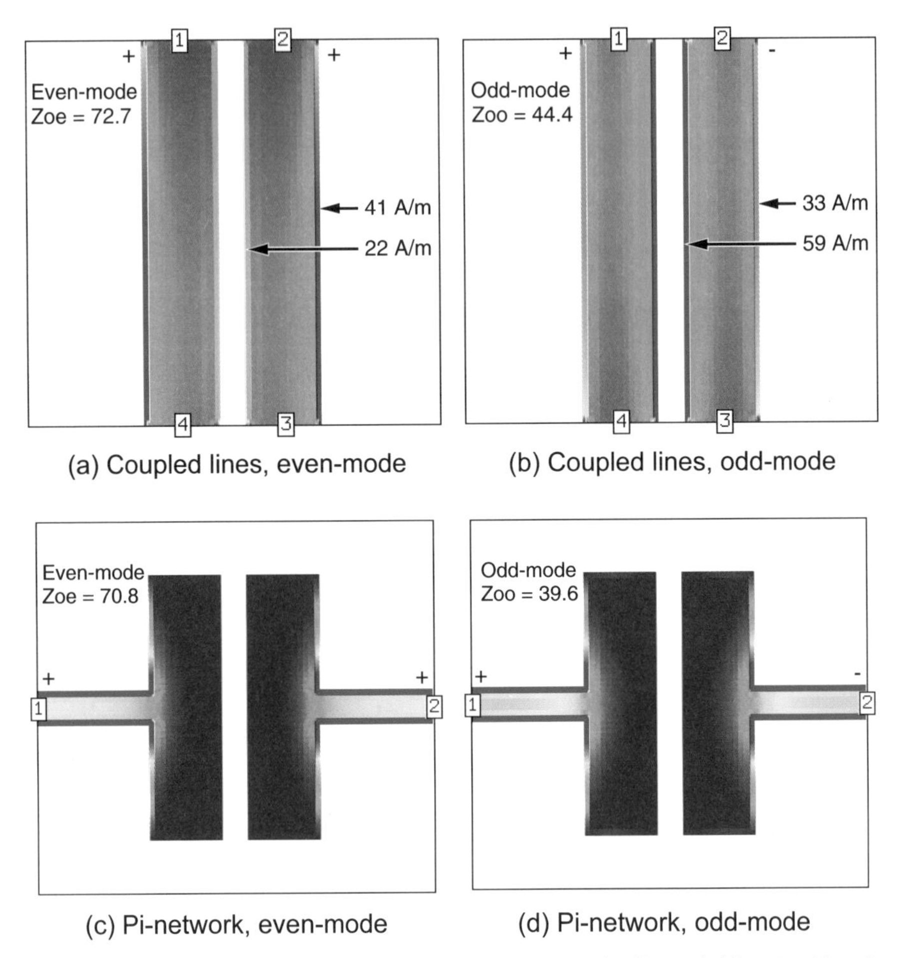

Figure 16.12 Current distributions for: (a) coupled lines in even-mode; (b) coupled lines in odd-mode; (c) pi-network in even-mode; and (d) pi-network in odd-mode. The scale in (a) and (b) is 0 to 40 A/m; and in (c) and (d) it is 0 to 60 A/m. © 1995 John Wiley & Sons, Inc. [11].

current distribution is very similar to the even-mode; the current tends to maximize on the outer edges of the strips.

The current distributions on the pi-networks are the key to understanding the large error in the circuit theory model. Using the conventional model library, we assumed that the pi-network patches could be described by coupled lines with normal even- and odd-mode current distributions. By feeding the pi-networks in the center of the strips, we have forced a current distribution that is quite different from the conventional coupled line case. In [11] an even- and odd-mode analysis on the pi-networks concludes that the even-mode impedance is virtually unchanged while the odd-mode impedance is 13% lower than the conventional coupled line case. So

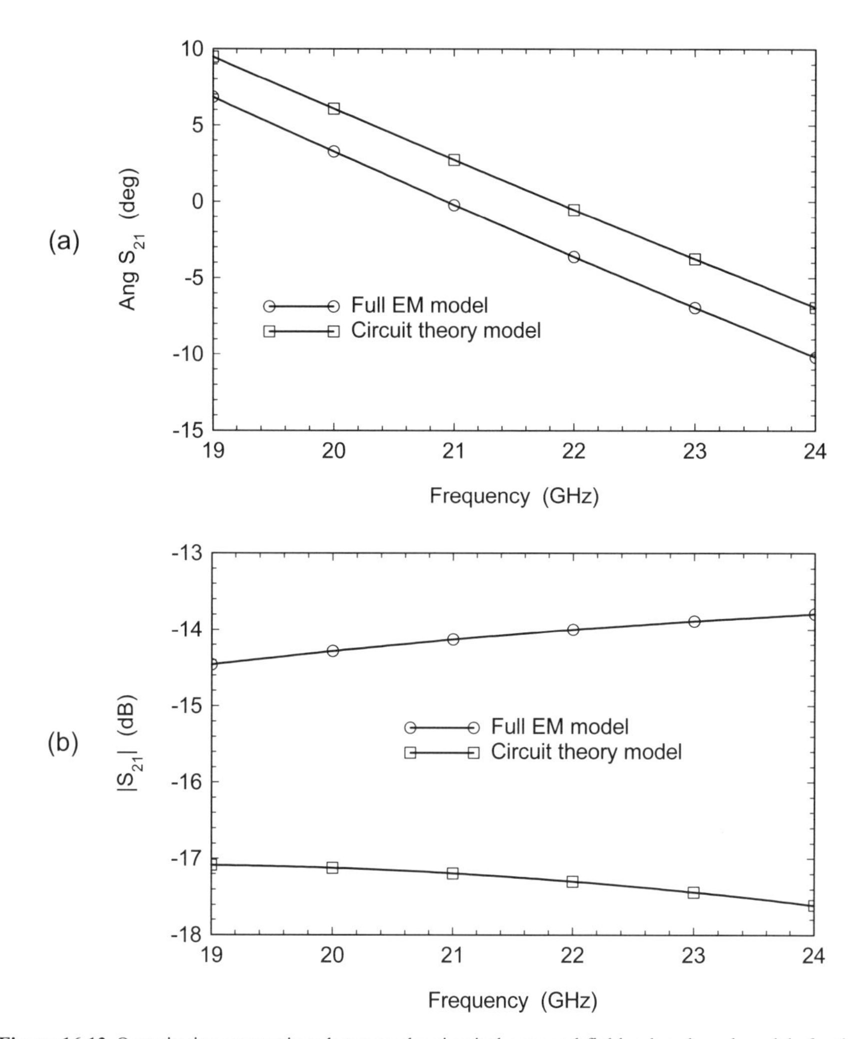

Figure 16.13 Quantitative comparison between the circuit theory and field-solver-based models for the pi-networks: (a) S_{21} phase, and (b) S_{21} magnitude. These errors account for the center frequency and bandwidth errors in the first iteration filter. © 1995 John Wiley & Sons, Inc. [11].

stated very simply, the actual current distribution on the pi-networks does not match the assumed current distribution in the coupled line circuit theory model, which causes the analysis to fail.

Visualization gives us a qualitative understanding of what went wrong in this design. We can go back to the field-solver and put some quantitative numbers on the error in this design. The two graphs in Figure 16.13 compare the original circuit-theory-based model to the full pi-network model computed on the field-solver.

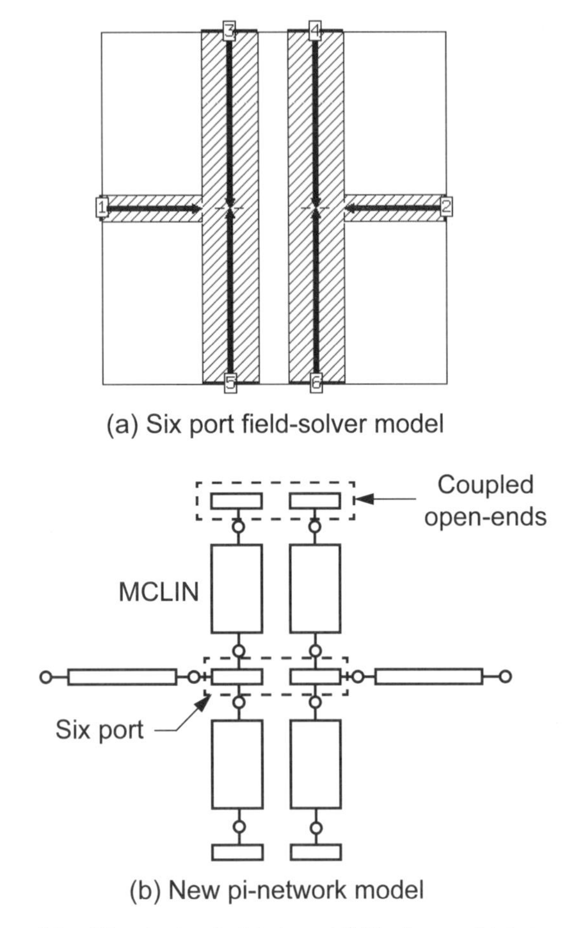

Figure 16.14 New models of the pi-network: (a) six-port field-solver model that captures the correct current distribution; and (b) schematic of new pi-network model with coupled open-end model and MCLIN model for length adjustment. © 1995 John Wiley & Sons, Inc. [11].

The S_{21} phase angle shows a large difference between the two analysis techniques. The phase angle error accounts for the center frequency shift in the first iteration filter. Perhaps the most startling error is in the S_{21} magnitude. This is, of course, directly related to the change in odd-mode impedance that we noted earlier. The error between the two analysis methods is 3.4 dB or 24% at 22.5 GHz. This accounts for the large bandwidth error in the first iteration filter.

A hybrid approach was again used to model the pi-networks in the second iteration filter. The field-solver was used to generate a six-port model (Figure 16.14(a)) that captures the unconventional current distribution on the pi-networks. Note that the indicated de-embedding implies that we are exporting a model with almost no physical size to the circuit simulator. The field-solver was also used to generate a

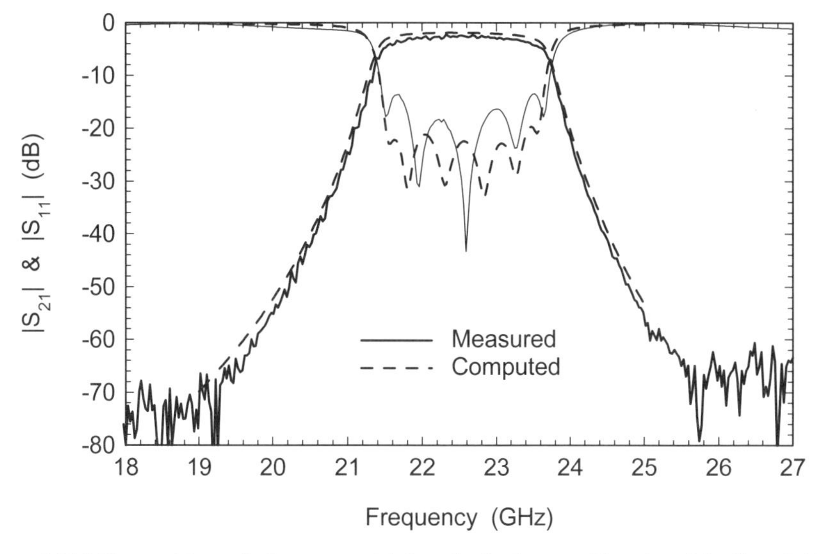

Figure 16.15 Measured (no tuning) versus modeled results for the second iteration of the filter. © 1995 John Wiley & Sons, Inc. [11].

coupled, open-end model. The schematic for the new pi-network model is shown in Figure 16.14(b). Note that we are again using the circuit theory coupled line model, after spending several pages proving that it is wrong for this case. The circuit theory coupled line model in Figure 16.14(b) is basically a mathematical convenience at this point. We are using it to adjust the length of pi-network during optimization. This works because the field-solver and the circuit theory models do agree for a simple coupled line pair (Figure 16.12(a, b)). The field-solver uses the simple coupled line pair to de-embed and we use the coupled line circuit theory model to add length back in the circuit simulator. But the complex behavior of the pi-network is still captured in the six-port field-solver model, no matter how much coupled line length we de-embed.

The turn-on results (no tuning) for the second iteration filter are shown in Figure 16.15. The pi-network gaps increased by 60% to 80% and the series line lengths decreased by 8% compared to the first iteration filter.

As we use the field-solver more, we will occasionally find cases like this one where circuit theory fails. The assumed current distributions on the ideal library models do not match the current distributions on our actual circuit. Does this mean we must eventually abandon circuit theory? Not at all. It only means we must carefully consider the assumptions made in our analytic modeling approach and apply the field-solver judiciously to those key areas where conventional models may break down. Visualization of currents can help us understand the failure mechanism in the model. Unfortunately, we often only discover these problems after the first set of hardware is built.

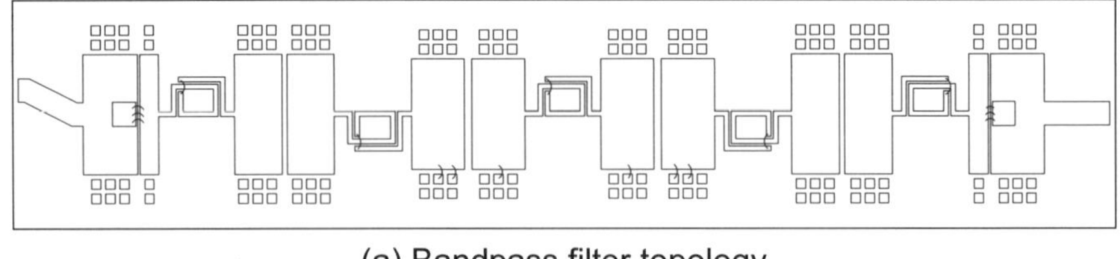

(a) Bandpass filter topology

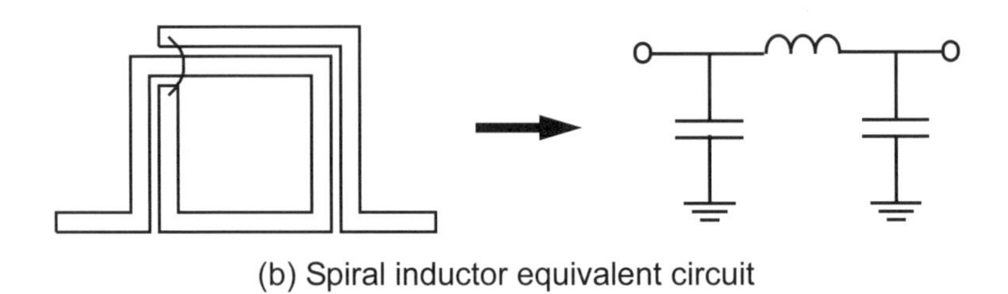

(b) Spiral inductor equivalent circuit

Figure 16.16 3.7-GHz bandpass filter: (a) top view of filter topology; and (b) printed spiral inductor equivalent circuit. © 1995 John Wiley & Sons, Inc. [14].

16.4 3.7-GHZ BANDPASS FILTER

Below about 6 GHz many of the popular distributed filter topologies become physically quite large. One option is to revert to truly lumped element topologies. Chip capacitors and very small, wire-wound inductors can be used to realize many filter topologies. Manufacturing so-called "chip and wire" filters in large quantities can be a challenge. Another alternative is a printed, pseudo-lumped element filter, like the one in the previous section. We can print various types of capacitors and inductors on a thin-film substrate with very high resolution. In the frequency range where the printed elements are small in terms of wavelengths they behave as lumped elements. However, at some higher frequency they will start to show distributed behavior and eventually resonate. The achievable Q for printed spirals is typically lower than a quarter- or half-wavelength resonator. However, if the bandwidth is not too narrow we can still realize usable filters.

Another example of this pseudo-lumped element approach is shown in Figure 16.16(a). It is a bandpass filter centered at 3.7 GHz [14]. The large metal patches form capacitors to ground. The gaps between the metal patches form series capacitors. On the ends of the filter we need more series capacitance than we can realize with the gap alone. Chip capacitors are bonded across the outermost gaps to increase the series capacitance. The printed spiral inductors are in series with the capacitor pi-networks (PINETs). This filter uses the same lumped element prototype as the previous example, Figure 16.10(b). This is the so-called "tubular" or "dumb-bell" prototype. This form of filter can also be integrated into a coaxial line. The prototype can synthesized with S/FILSYN [15].

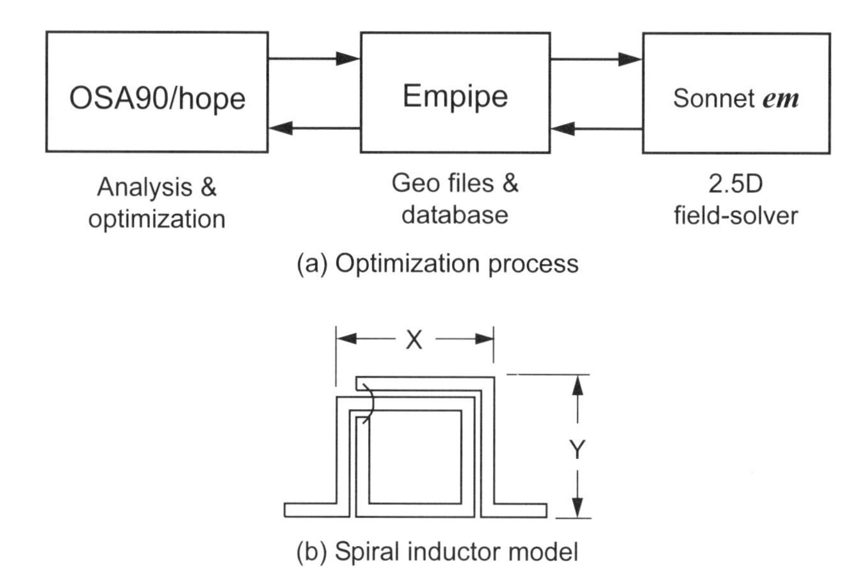

Figure 16.17 Optimization of planar circuits using OSA90/hope and Empipe. (a) Empipe generates new geometry files for the field-solver and stores the results in a database that can be interpolated. (b) Typical spiral inductor where X or Y is allowed to vary during optimization. © 1995 John Wiley & Sons, Inc. [14].

Once we have synthesized the prototype we begin the conversion to printed elements. First we realize the spiral inductors. The one and three-quarter turn geometry used here was found to be a good compromise between useful inductance and maximum self-resonant frequency. The simplest equivalent circuit for the printed spiral is a series inductance with shunt capacitors on each end (Figure 16.16(b)). In this case the shunt capacitors can easily be absorbed into the shunt capacitors of the PINETs. So, as we introduce the printed spiral inductors the values of all but the outermost shunt capacitors in the prototype are reduced. After all the printed inductors have been introduced, we can convert the prototype PINETs to printed form. Based on our experience with the previous example, we decided to model the complete PINET structure in the field-solver.

This example also demonstrates how the field-solver can be used to optimize planar circuits (Figure 16.17(a)). The filter was subdivided into three unique pi-network elements and one spiral inductor element. The analysis and optimization of these circuit elements were controlled by a linear simulator, OSA90/hope [16], with an auxiliary interface to the field-solver, Empipe [17]. The analysis and optimization proceed by first building a database of field-solver solutions around the starting point and then interpolating in the existing database or adding new solutions to the database. One side benefit of this approach is that it frees the user from the fixed grid in closed box MoM simulators. That is, solutions can be found with dimen-

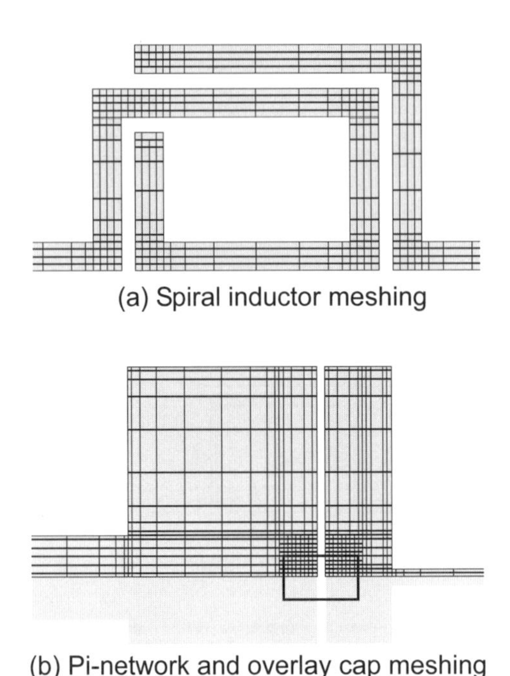

Figure 16.18 Meshing used in the filter case study: (a) spiral meshing using 1-mil grid; and (b) PINET meshing including overlay capacitor (Sonnet ***em*** Ver. 7.0). © 1995 John Wiley & Sons, Inc. [14].

sions that do not fall on the analysis grid. More details on direct driven electromagnetic optimization can be found in [18, 19].

In this filter the insertion loss and stopband rejection requirements forced a compromise to be made in the spiral inductor design. Wide traces minimize the insertion loss, but a narrow trace maximizes the self-resonant frequency. In the field-solver modeling there is also a question of convergence. We covered spiral inductor meshing (Figure 16.18(a)) in Section 5.11.3 using this example. We should also review the meshing of printed capacitors in Section 5.11.4 and meshing overlay capacitors in Section 5.11.5. The chip capacitors in this design are approximated by a small patch in a second metallization layer with a very thin (0.02 mil) air-dielectric layer between the metal layers (Figure 16.18(b)). Using air guarantees that the rest of the circuit will not be modified by the extra dielectric layer. Remember that MoM solvers require exact mesh alignment between layers for this case.

Four filter designs were fabricated and tested for this case study. In three of the designs, different widths and gaps were used for the spiral inductor to explore the trade-off between Q and self-resonant frequency. The first design Figure 16.19(a) used spiral inductors with 2-mil wide traces and 1-mil gaps. The turn-on insertion loss and return loss are compared to the computer prediction. The goal for the filter

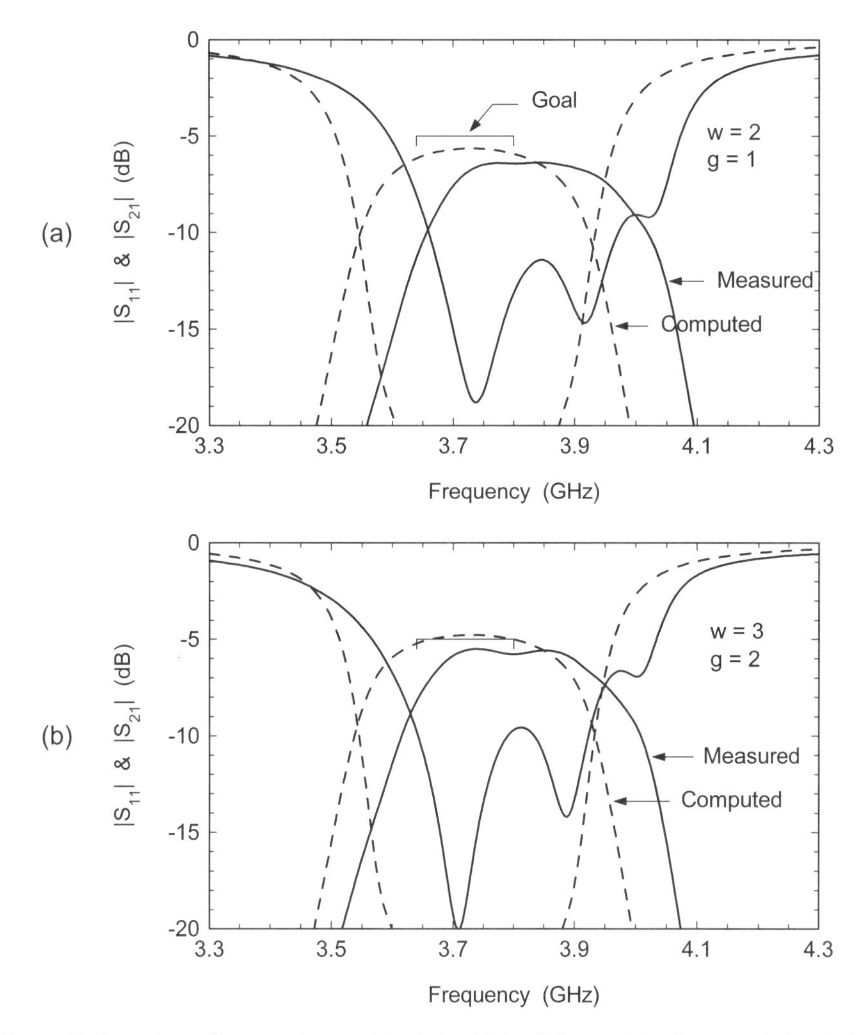

Figure 16.19 Bandpass filter experiments: (a) spirals with 2-mil lines and 1-mil gaps; and (b) spirals with 3-mil lines and 2-mil gaps. © 1995 John Wiley & Sons, Inc. [14].

was less than 5 dB insertion loss from 3.64 to 3.80 GHz. The measured inductor Q for this sample was 55, and the center frequency error was 100 MHz, or 2.7%. In the second design, the spirals had 3-mil wide traces and 2-mil gaps. The turn-on data is compared to the measured results in Figure 16.19(b). The inductor Q has increased to 65 and the frequency error has decreased to 76 MHz, or 2%.

The third design (Figure 16.20(a)), had spirals with 4-mil traces and 2-mil gaps. The inductor Q is now close to 75, the insertion loss has decreased almost 2 dB, and the frequency error is down to 50 MHz, or 1.3%. These first three filters

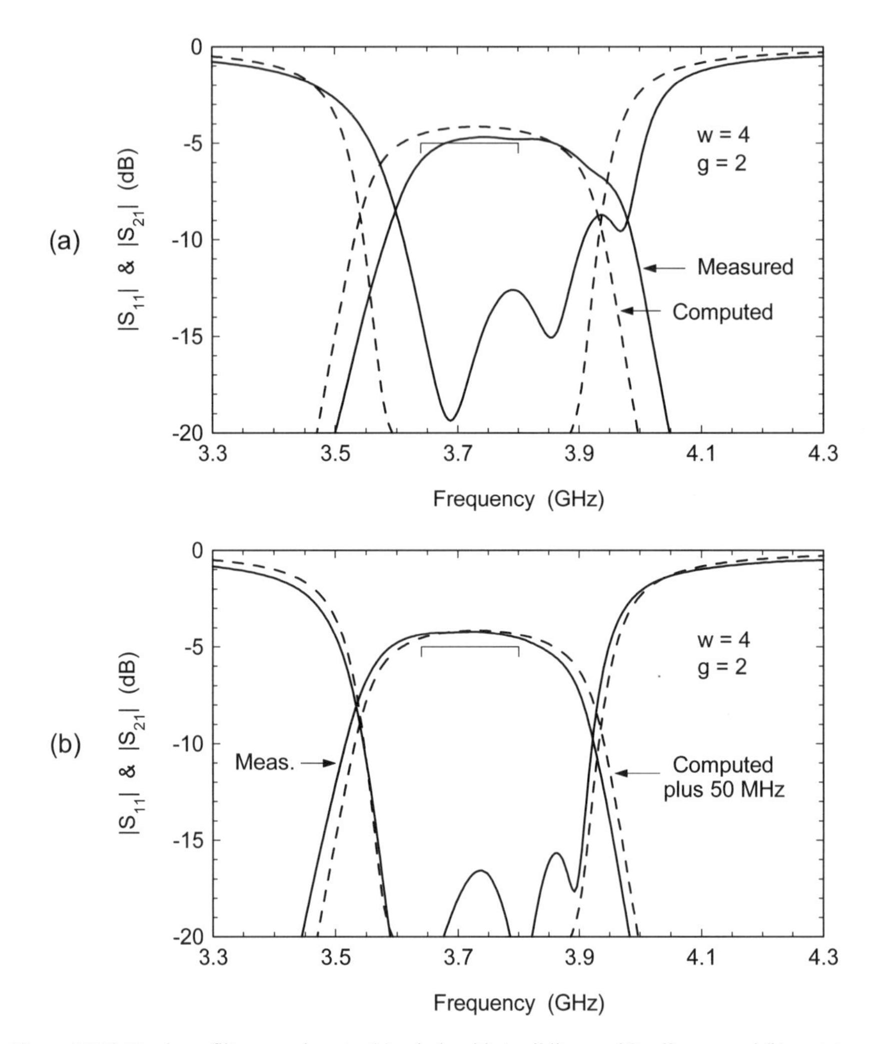

Figure 16.20 Bandpass filter experiments: (a) spirals with 4-mil lines and 2-mil gaps; and (b) prototype shifted by 50 MHz to account for center frequency error. [14].

were fabricated on the same 2 by 2-inch substrate. Why did the frequency error decrease as the trace width got larger? A 1-mil grid was used for the spiral analysis in all three cases; the number of subsections across the width increased for each design.

When this filter was transferred to production, it was desirable to more carefully center the turn-on frequency. To account for the remaining center frequency error, the ideal prototype was shifted 50 MHz low, and the optimization was run

Table 16.1

Printed Spiral Inductor Meshing Experiments

Trace width (mil)	*Pattern of subsection widths* (mil)	*Grid size* (mil)	*Number of subsections*	*Solution time** (min:sec)	*Filter f_0 error*
2	1-1	1.0	298	1:25	2.7%
3	1-1-1	1.0	556	1:40	2.0%
4	1-1-1-1	1.0	844	2:38	1.3%
4	1-2-1	1.0	599	7:34	1.3%
4	0.5-1-1-1-0.5	0.5	1416	10:53	0.8%
4	0.5-3-0.5	0.5	705	4:18	0.8%

**50-MHz Sparc-10 with 64-MB RAM, circa 1994*

one more time. The Y-dimension of the spirals (Figure 16.17(b)), changed 0.9 mil in this final optimization. The tuned filter results are compared to the original specification in Figure 16.20(b). If you look carefully at Figure 16.16(a) you can see the tuning; six small tuning pads are bonded in on the center-most PINETs. The final center frequency error was about 0.5%. Because this fourth filter was fabricated on a different substrate and on a different day, it seems to indicate that the center frequency error in the third design is at least repeatable.

We can hypothesize that the remaining filter center frequency error lies in the spiral inductor modeling. The obvious way to improve accuracy is to increase the number of subsections used to describe the metallization pattern. However, solution time will increase rapidly with the number of subsections, so we would like to apply any additional subsections in an intelligent way. The results of the meshing experiments for the spiral inductor are repeated here in Table 16.1. With edge-meshing we can reduce the center frequency error to about 0.8%. In later, unpublished experiments we determined that some of the remaining error may be due to interactions between the spiral inductors and the pi-networks. Unfortunately, this interaction is ignored when we cascade individual field-solver solutions for each component.

16.5 1.5 TO 5.5-GHZ BANDPASS FILTER

This example is another pseudo-lumped bandpass filter which covers 1.5 to 5.5 GHz. This topology includes both series and shunt resonators (Figure 16.21). The four large metal patches in the center of the layout are shunt capacitors to ground. Each of these four patches has an inductor to ground which forms a shunt resonator. The four shunt resonators are top coupled with series capacitors. Part of coupling capacitance comes from the gap between the patches and the rest from the

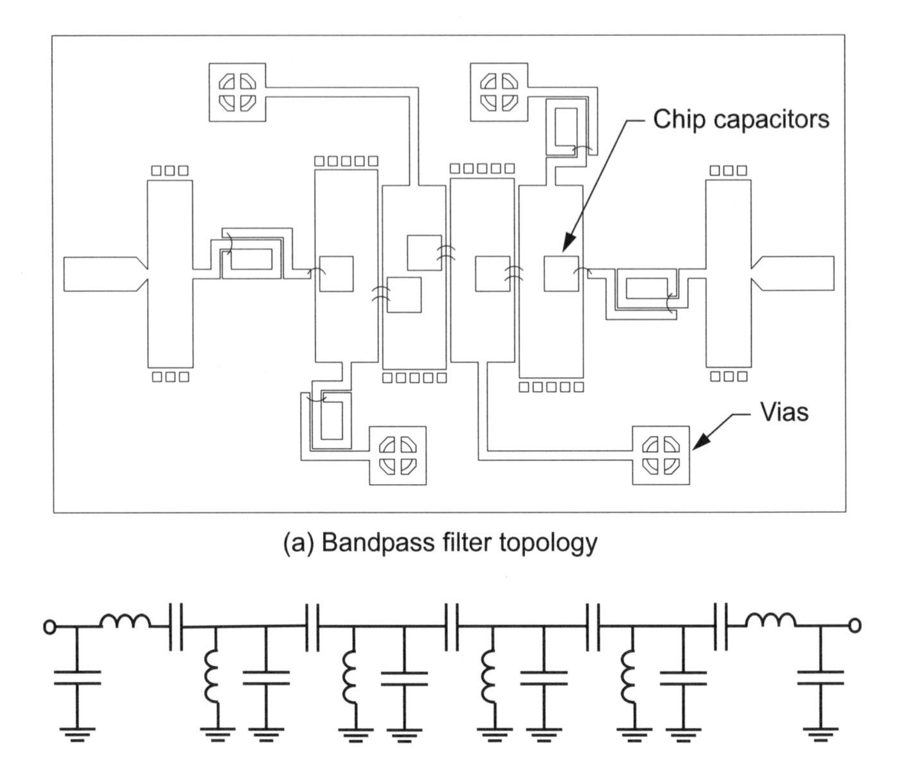

Figure 16.21 Printed, broadband bandpass filter. (a) Top view of the layout. There are four shunt resonators in the center of the layout and a series resonator on each end. (b) The lumped element prototype for this topology. © 1995 Nexus Media Ltd. [1].

chip capacitors. At each end of the filter there is a series *LC* resonator formed by the printed spiral inductor and the chip capacitor it is connected to. The substrate is 15-mil alumina, 350-mil long by 205-mil wide. Although it is possible to describe this geometry using a combination of analytical and 2D cross-section models, the results will be disappointing. The multiple feed points on the larger patches can set up current distributions that are not accurately described by a combination of standard analytical models.

This design also demonstrates how a 2.5D field-solver can be used inside an optimization loop. The filter is divided into four subnetworks: the series inductor, the shunt inductor, the single shunt patch, and the group of four shunt patches in the center of the filter. The later subnetwork is a 12-port problem. Six of those ports connect this subnetwork to the other subnetworks. Three additional pairs of ports allow ideal capacitors to be connected across the gaps in the linear simulator. At the time this work was done, it was not possible for the field-solver to automatically

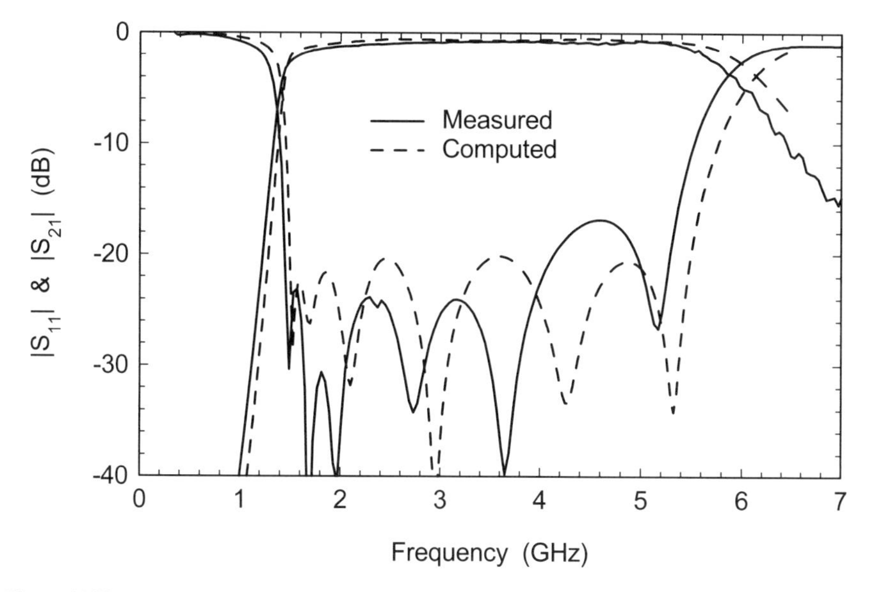

Figure 16.22 Measured versus modeled results for the $N = 6$ pseudo-lumped bandpass filter. The capacitive couplings between resonators steepen the rejection rate on the low side of the passband. © 1995 Nexus Media Ltd. [1].

calibrate the internal ports needed to connect the capacitors. This calibration was done manually by deriving the port discontinuities with some independent field-solver problems.

In this case the linear simulator was OSA90/hope and the field-solver was Sonnet ***em***. An interface program called Empipe managed the database of field-solver solutions and interpolates between existing solutions. Each subnetwork is an independent field-solver problem. Because each subnetwork is not resonant, we only need to compute three or four frequency points across the passband. The linear simulator can then generate a fine frequency sweep by interpolating between the points computed by the field-solver. Again, the danger in this subdivision approach is that some important interaction between subnetworks will be discarded.

The measured versus modeled data for this example are shown in Figure 16.22. The center frequency error is about 50 MHz or 1.4% and the bandwidth error is 130 MHz or 3%. The steeper selectivity on the low side of filter is due to the capacitive couplings between resonators, which are all highpass type elements.

16.6 22.5-GHZ BANDSTOP FILTER

The bandstop filter shown in Figure 16.23(a) is a microstrip version of a topology proposed by Schiffman and Matthaei [20]. It is a three-resonator filter designed to

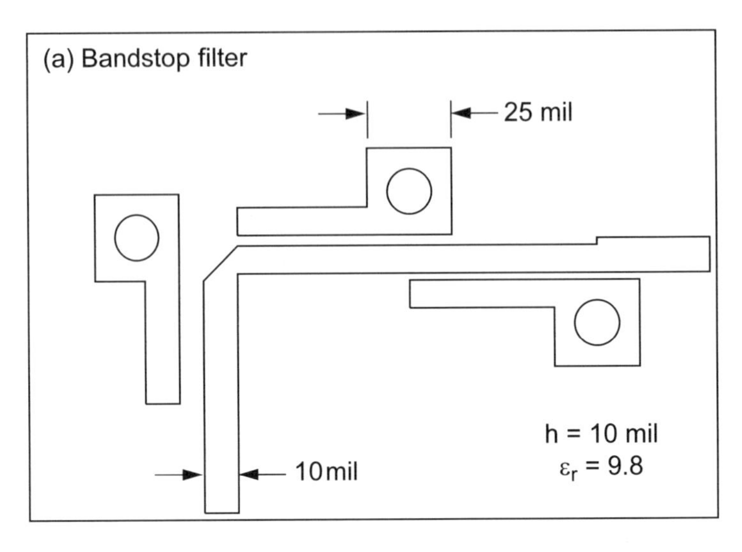

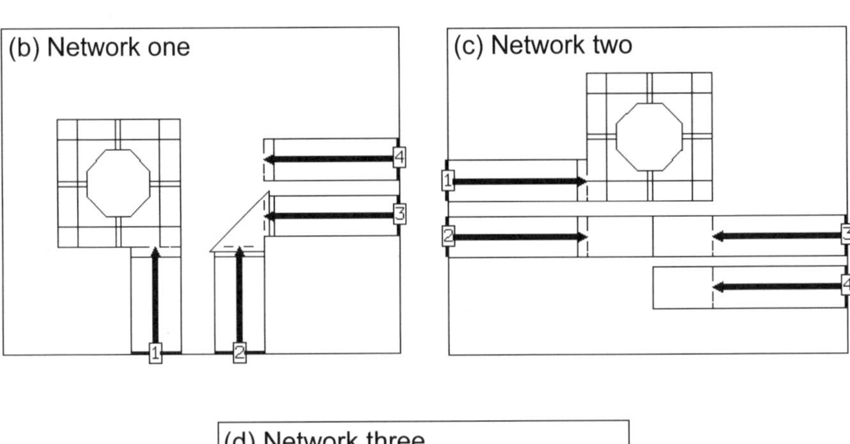

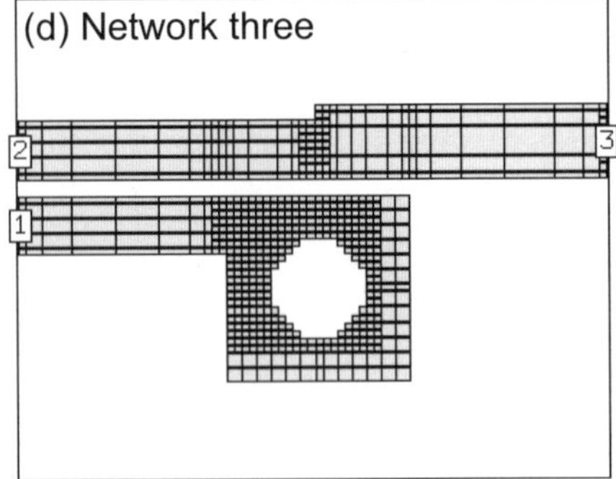

Figure 16.23 Microstrip bandstop filter: (a) top view of filter topology; (b) grouped discontinuities in network one; (c) discontinuity group in network two; and (d) network three group with actual analysis mesh (Sonnet ***em*** Ver. 7.0).

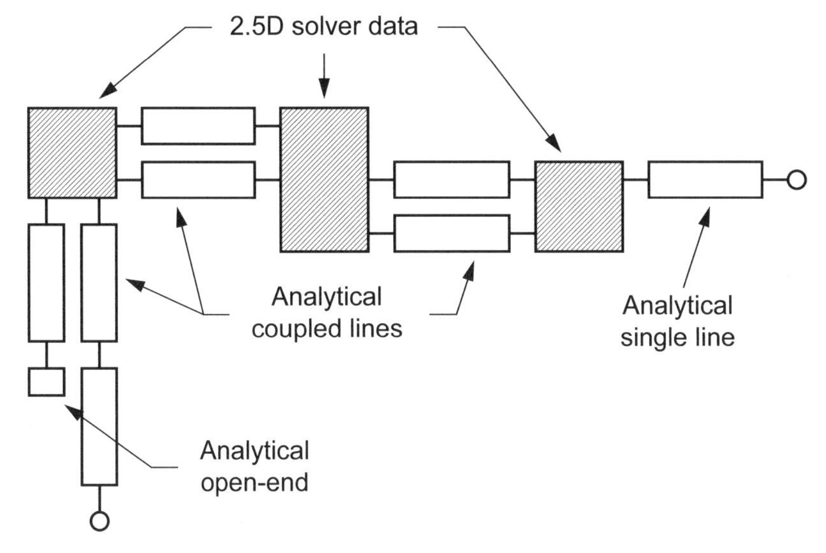

Figure 16.24 Analysis of the bandstop filter using "black box" data from the field-solver connected by analytical coupled line and single line models.

reject signals from 21.5 to 23.5 GHz and pass signals from 25 to 40 GHz. The layout shown is on 10-mil thick alumina and was designed to fit the I/O locations of an existing substrate. We might suspect that we need to use the field-solver for at least the via holes.

After doing a preliminary layout it became obvious that there are some regions not easily described using a cascade of circuit theory models. The design was subdivided into three multiport field-solver problems connected by analytical coupled line models in a linear simulator. The first field-solver network (Figure 16.23(b)) is centered on the via at the left. Note there are several discontinuities in close proximity: the via and surrounding pad, an asymmetrical step into the via, the mitered bend, and the open-end. At the time this work was done we were experimenting with very aggressive control of the meshing process. The figure shows the polygon layout used to control the meshing process.

The second field-solver problem (Figure 16.23(c)) tackles the area around the central via. There is an ambiguous region between the second and third resonators due to the size of the via pad. Is the connecting line between the second and third resonators a single line or a coupled line? Where does the second resonator end, at the edge of the pad or the edge of the hole? The outlines show how each network was described to the field-solver to guarantee a good mesh.

The final field-solver problem concentrates on the via at the far right (Figure 16.23(d)). Again there is some ambiguity regarding the length of the third resonator depending on how the current terminates on the via hole. Here we have shown the subsectioning of this problem. Using the field-solver on these three networks takes

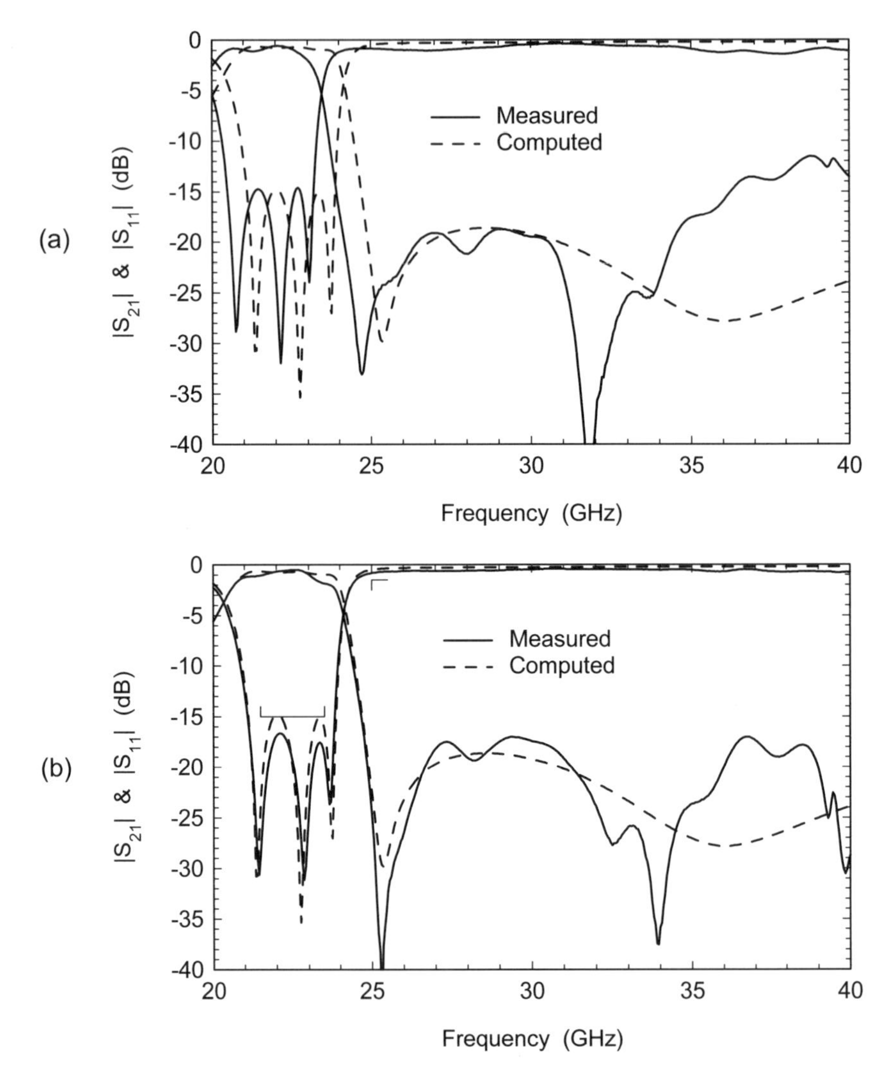

Figure 16.25 Measured versus modeled results for the bandstop filter: (a) turn-on results, and (b) tuned result.

much of the uncertainty out of this design. We can get good analytical data on each network without fully understanding the details of how each one behaves.

The schematic in Figure 16.24 shows how the hybrid solution is built using the field-solver data. An initial estimate of the circuit dimensions is made and a set of field-solver data is computed. The circuit is then optimized; coupled line gaps and line widths are allowed to change. A second set of field-solver data is then computed using the new dimensions and the circuit is optimized again. It is seldom necessary to iterate in this way more than two or three times. Today this circuit could also be optimized automatically using one of the 2.5D MoM simulators.

Finally, Figure 16.25(a) shows the measured versus modeled results for the bandstop filter at turn-on. The stopband is shifted about 700 MHz low, almost 3% error at 25 GHz. However, it looks like the bandwidth is correct. The filter was tuned by scribing off a corner of each resonator at the open-end (Figure 16.25(b)). The bandwidth is indeed correct, which indicates that the gaps are correct. Still, the turn-on response was somewhat disappointing and we were not able to determine the exact source of the initial error.

It is fairly easy to dream up geometries that cannot be easily described as a cascade of standard analytical models. We can either accept the limitations of the standard library, or use the field-solver to model a group of discontinuities in close proximity. This approach allows us to disregard any preconceived notions about what our circuit must look like. The analytical data we get from the field-solver analysis will be accurate even if we do not fully understand all the interactions going on in the circuit. We can also combine the "black box" data from the field-solver with analytical models to perform analysis or optimization.

References

[1] Swanson, Jr., D. G., "First Pass CAD of Microstrip Filters Cuts Development Time," *Microwaves and RF 95*, London, UK, October 10–12, 1995, pp. 8–12.

[2] Interdigital Microstrip (IDM), Forem USA, Amesbury, MA, USA.

[3] Denig, C., "Using Microwave CAD Programs to Analyze Microstrip Interdigital Filters," *Microwave Journal*, Vol. 30, No. 3, 1989, pp. 147–152.

[4] Swanson, Jr., D. G., "A Novel Method for Modeling Coupling Between Several Microstrip Lines in MICs and MMICs," *IEEE Trans. Microwave Theory and Tech.*, Vol. 39, No. 6, 1991, pp. 917–923.

[5] Jansen, R. H., "Some Notes on Hybrid-mode versus Quasi-static Characterization of High Frequency Multistrip Interconnects," *23rd European Microwave Conference Proceedings*, Madrid, Spain, September 1993, pp. 220–222.

[6] Rautio, J. C., and G. L. Matthaei, "Tracking Error Sources in HTS Filter Simulations," *Microwaves and RF*, Vol. 37, No. 12, 1998, pp. 119–130.

[7] Matthaei, G. L., J. C. Rautio, and B. A. Willemsen, "Concerning the Influence of Housing Dimensions on the Response and Design of Microstrip Filters with Parallel-Line Couplings," *IEEE Trans. Microwave Theory and Tech.*, Vol. 48, No. 8, 2000, pp. 1361–1368.

[8] Kirton, P. A., and K. K. Pang, "Extending the Realizable Bandwidth of Edge-Coupled Stripline Filters," *IEEE Trans. Microwave Theory and Tech.*, Vol. 25, No. 8, 1977, pp. 672–676.

[9] Edge-Coupled Microstrip (ECM), Forem USA, Amesbury, MA, USA.

[10] Riddle, A., "High Performance Parallel Coupled Microstrip Filters," *IEEE MTT-S Int. Microwave Symposium Digest*, New York, NY, May 25–27, 1988, pp. 427–430.

[11] Swanson, Jr., D. G., "Using a Microstrip Bandpass Filter to Compare Different Circuit Analysis Techniques," *Int. J. MIMICAE*, Vol. 5, No. 1, 1995, pp. 4–12.

[12] Swanson, Jr., D. G., "Thin-Film Lumped-Element Microwave Filters," *IEEE MTT-S Int. Microwave Symposium Digest*, Long Beach, CA, June 13–15, 1989, pp. 671–674.

[13] Swanson, Jr., D. G., R. Forse, and B. Nilsson, "A 10GHz Thin Film Lumped Element High Temperature Superconductor Filter," *IEEE MTT-S Int. Microwave Symposium Digest*, Albuquerque, NM, June 2–4, 1992, pp. 1191–1193.

[14] Swanson, Jr., D. G., "Optimizing a Microstrip Bandpass Filter Using Electromagnetics," *Int. J. MIMCAE*, Vol. 5, No. 5, 1995, pp. 344–351.

[15] S/FILSYN, ALK Engineering, Salisbury, MD.

[16] OSA90/hope, Optimization Systems Associates, Dundas, Ont., Canada.

[17] Empipe, Optimization Systems Associates, Dundas, Ont., Canada.

[18] Bandler, J. W., et al., "Minimax Microstrip Filter Design Using Direct EM Field Simulation," *IEEE MTT-S Int. Microwave Symposium Digest*, Atlanta, GA, June 14–18, 1993, pp. 889–892.

[19] Bandler, J. W., et al., "Microstrip Filter Design Using Direct EM Field Simulation," *IEEE Trans. Microwave Theory and Tech.*, Vol. 42, No. 7, 1994, pp. 1353–1359.

[20] Schiffman, B. M., and G. L. Matthaei, "Exact Design of Bandstop Microwave Filters," *IEEE Trans. Microwave Theory and Tech.*, Vol. 12, No. 1, 1964, pp. 6–15.

Chapter 17

Other Microwave Filters

In the previous chapter we studied several planar filter examples, but there are many other filter technologies that are equally interesting. Waveguide filters are often used in high-performance satellite filters and multiplexers. Various types of combline and interdigital filters can be found in many high-performance military and commercial systems. These filters typically use metal rods as the resonators in an air-filled cavity. Dielectric resonator filters are also found in space applications and wireless basestations. The system specifications on these filters often require spurious-free performance out to three or four times the filter center frequency. We often add an additional lowpass cleanup or roofing filter to meet this requirement.

All of these filter types present opportunities to apply a field-solver to improve the design process. In this chapter we will pick a few examples that emphasize the fundamental concepts we are developing in this book.

17.1 COAXIAL LOWPASS FILTERS

Earlier in our discussion of FEM meshing techniques we took a detailed look at coaxial step discontinuities (Section 6.4.3). The motivation for that work was a study of several coaxial lowpass filter designs. In these filters the primary feature is the step discontinuity between the high impedance and low impedance lines in the filter. These filters are used as cleanup or roofing filters in wireless basestations.

Filters present a special problem for the automatic mesh refinement process used in FEM solvers. Our basic guideline for meshing requires a cell size of $\lambda/10$ to $\lambda/30$ at the highest frequency of interest. However, for lowpass and bandpass filters the highest frequency of interest is probably in the stopband of the filter. Most filters work by reflection in the stopband. Energy is reflected at the ports and relatively little energy penetrates into the interior of the filter. If we mesh at a frequency deep in the stopband, the automatic meshing algorithm tends to focus on the ports and ignores the interior of the filter. For filters, it is generally more productive to mesh in the passband or at least near one of the band edges. We may also have to

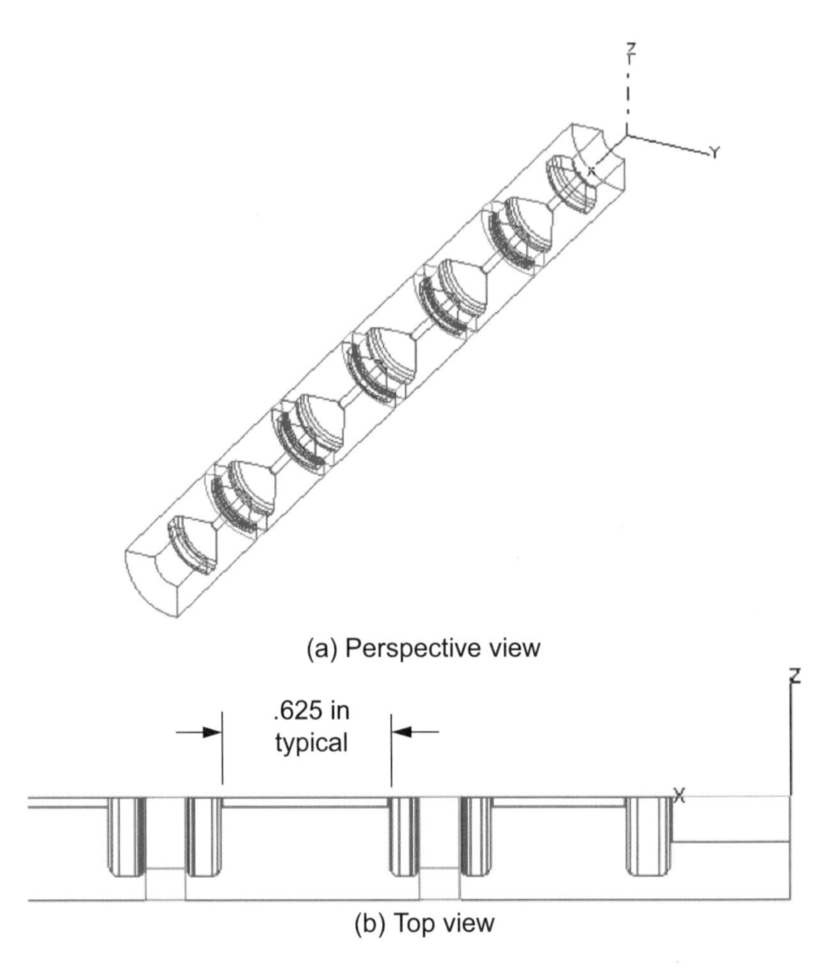

Figure 17.1 $N = 13$ coaxial lowpass filter: (a) perspective view of quarter symmetry model; and (b) top view of quarter symmetry model (Ansoft HFSS Ver. 8.0).

specify some "seeding" of the mesh in the interior of the filter. Seeding directs the software to put elements of some minimum size in a region that we specify. The seeding process helps the software establish a connection between the ports, even if we are meshing in the stopband.

The first lowpass filter example [1] is a fairly conventional cascade of high impedance and low impedance transmission lines. One quarter of the filter geometry is shown in Figure 17.1(a). A partial cross-section view is shown in Figure 17.1(b). In this particular topology, high impedance lines approximate series inductors and low impedance lines approximate shunt capacitors. Both element types are small in terms of wavelengths. Dielectric rings support the filter structure and keep

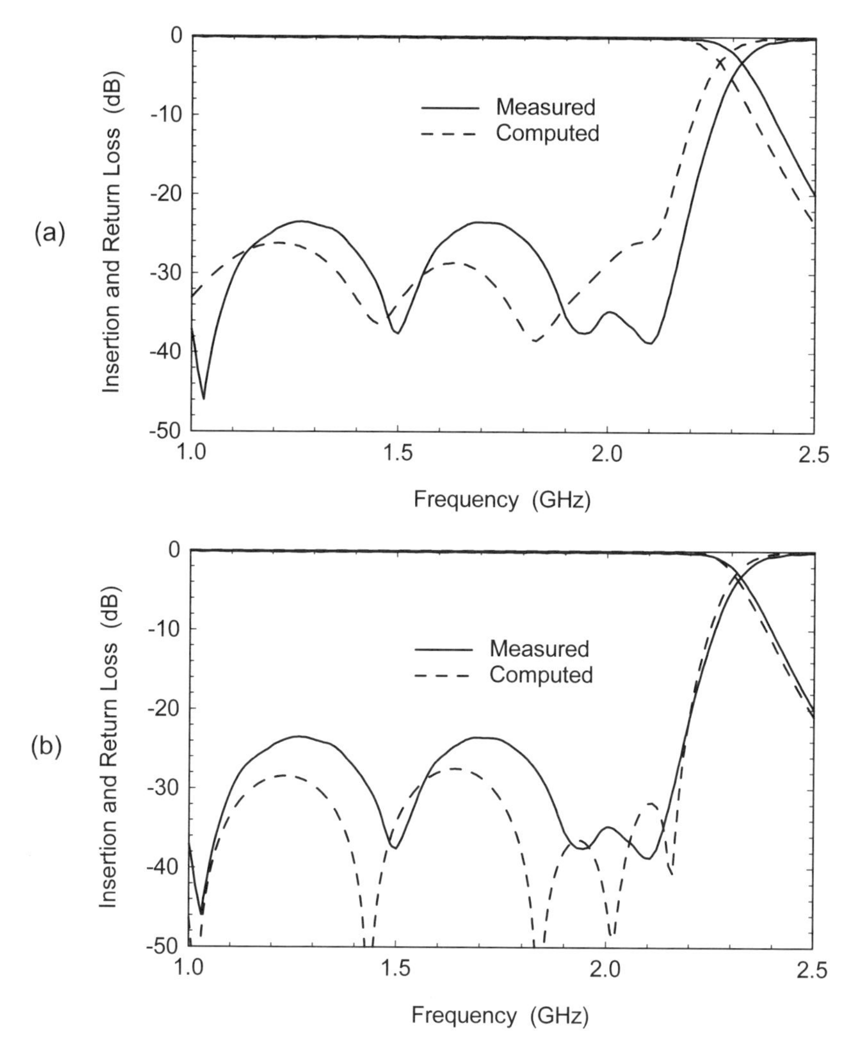

Figure 17.2 Results for the lowpass filter in Figure 17.1: (a) default meshing parameters and mesh frequency of 1.8 GHz; and (b) 40% refinement and 0.1-in seed by length near the steps (Ansoft HFSS Ver. 8.0).

it centered in the outer conductor. In this example, the edges of the low impedance sections have been radiused to enhance the power handling capability.

The plot in Figure 17.2(a) shows the initial analysis of this structure using the default meshing parameters and a mesh frequency of 1.8 GHz (inside the passband). The error in cutoff frequency is about 50 MHz but the return loss curve looks fairly reasonable. The troubling aspect of this result is that without the measured data to compare to, we might assume that this result is perfectly correct. The

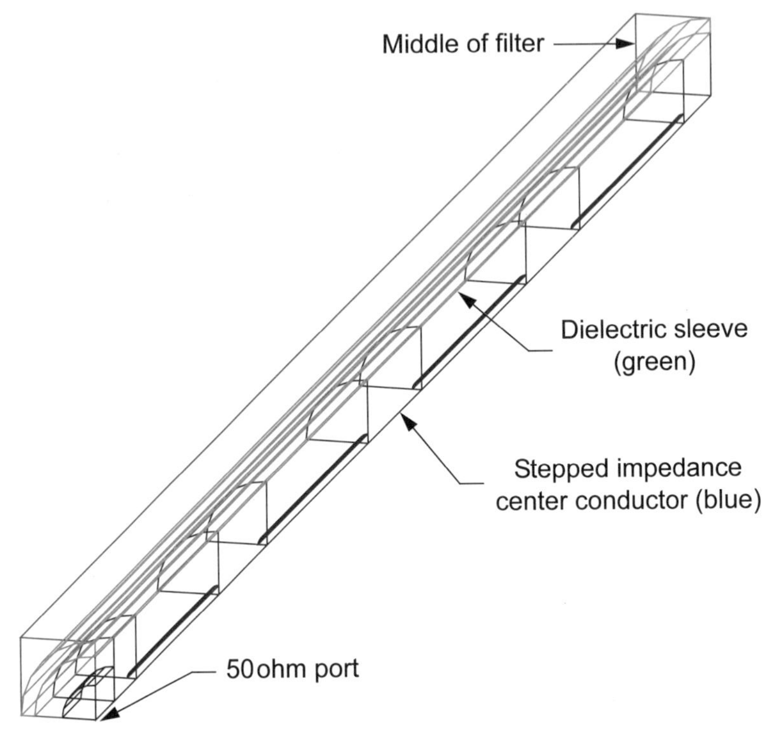

Figure 17.3 Model of the $N = 17$ lowpass filter with quarter symmetry in the cross-section and half symmetry along its length (Ansoft HFSS Ver. 8).

error in cutoff frequency is small as a percentage of the cutoff frequency. But in the stopband a key rejection point may have shifted 5 to 10 dB, which was probably most of the margin in the design.

The result in Figure 17.2(a) was actually the data that launched the convergence study in Section 6.4.3. The plot in Figure 17.2(b) shows the results with 40% refinement and 0.1-in seeding by length in a dummy region around each step. Although the agreement is not perfect, it is greatly improved. The field-solver model does not include transition details at both ends of the filter, which results in a more optimistic return loss prediction.

Figure 17.3 shows a second, slightly different $N = 17$ lowpass filter implementation that was also studied [2]. This filter again uses a cascade of high impedance and low impedance lines. However, in this case the outer conductor is a square, 0.250 in on a side. The center conductor assembly is slid inside a PTFE sleeve, which is tangent to the outer conductor walls. To save time we again analyze one quarter of the cross-section geometry. Because the filter is also symmetrical along its length, we can analyze one half of the filter lengthwise and cascade the results back-to-back in our circuit simulator.

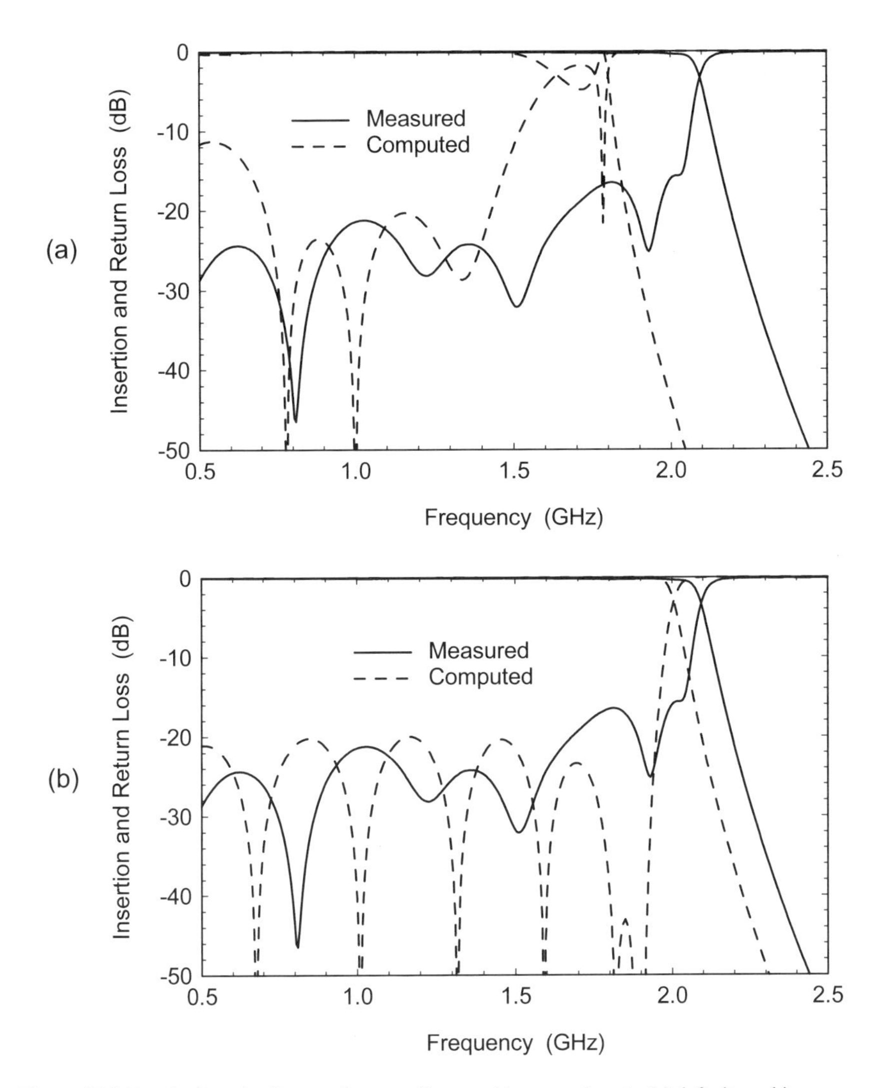

Figure 17.4 Results from the first two lowpass filter meshing experiments: (a) default meshing parameters and a mesh frequency of 3.25 GHz; and (b) 40% refinement, seeding of 0.05 in near the steps and a mesh frequency of 1.8 GHz (Ansoft HFSS Ver. 8).

We performed 10 different meshing experiments on this geometry. The first experiment, Figure 17.4(a), used the default meshing parameters and a mesh frequency of 3.25 GHz. It is obvious that this naive approach to meshing does not give useful results. A second experiment, Figure 17.4(b) was run using 40% refinement, a seed value of 0.05 in near the steps, and a mesh frequency of 1.8 GHz. We are getting closer to the correct solution, but the cutoff frequency error is still 90 MHz.

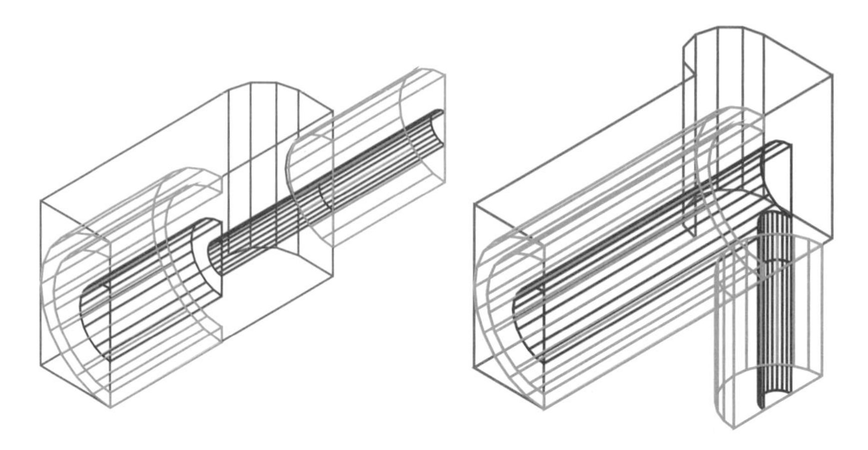

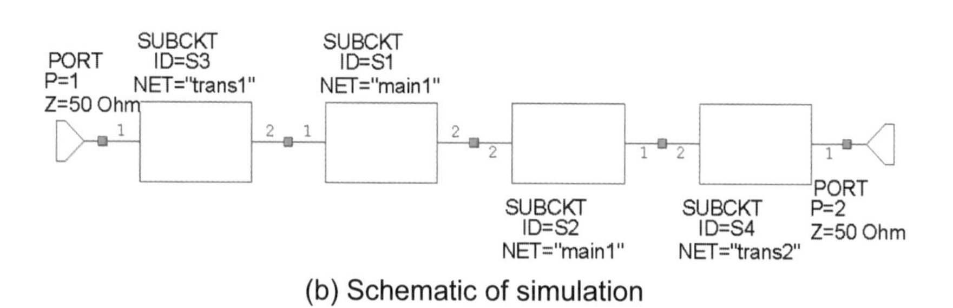

Figure 17.5 (a) Transitions to SMA connectors at each end of the filter (Ansoft HFSS Ver. 8). (b) The schematic of circuit level simulation (Microwave Office 2001).

For the next round of experiments we examined the experimental hardware and decided to include the transitions to SMA connectors at each end of the filter (Figure 17.5(a)). Neither transition was optimized for return loss performance in the band of interest. Now our circuit simulation file includes two copies of the half filter simulation and one file each for the transitions at the ends (Figure 17.5(b)).

The analysis results for the latest iteration are shown in Figure 17.6(a). At this point we also tried a different meshing scheme. If we mesh in the passband near the cutoff frequency, but don't do any seeding, we notice that the mesh is denser near the ports and less dense near the step discontinuities. If we mesh at much lower frequency, say 0.1 GHz, the reactance of the discontinuities is much lower. Or in other words, the network is even less distributed and more lumped [3]. If we adjust the LAMBDA_REFINE_TARGET variable to 0.00084, we will still get enough mesh per wavelength near the cutoff frequency. We compared these results to a mesh seeded at 0.05 in and the results were nearly identical.

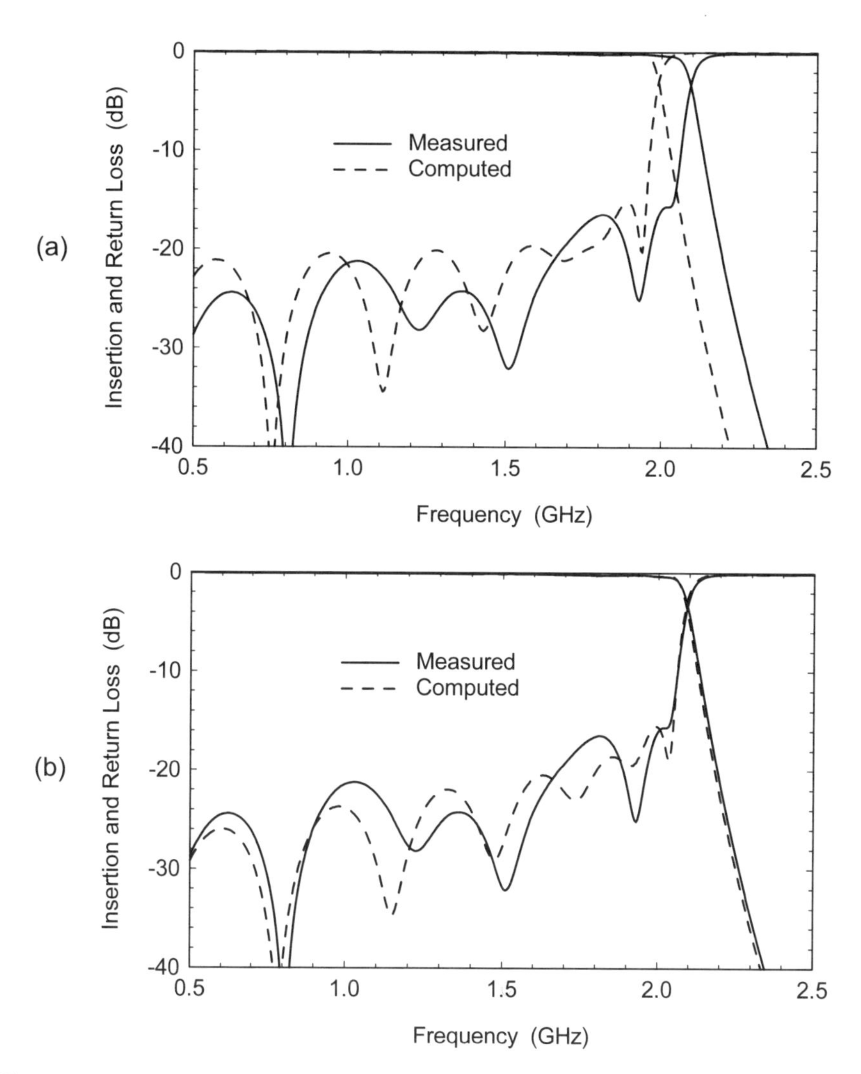

Figure 17.6 Lowpass filter results: (a) advanced meshing and transitions models; and (b) transition models and tolerances added for air gaps.

Although the slope and shape of the computed return loss plot now matches the measured results more closely, we still have about 100 MHz error in the cutoff frequency. Again, we went back to the experimental hardware and the fabrication drawings. Up to this point our field-solver models have assumed that the PTFE sleeve makes perfect contact with the low-impedance transmission line sections. The model also assumes that the PTFE sleeve is perfectly tangent to the square outer conductor on each side. After examining the tolerances on the prints and the

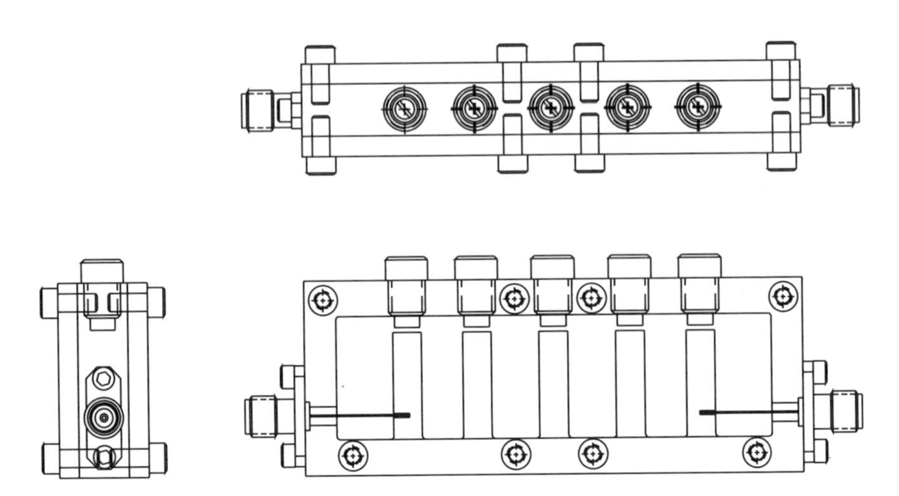

Figure 17.7 Experimental hardware for $N = 5$ combline filter.

actual hardware, we determined that the outer radius of the PTFE sleeve could be 0.003 to 0.005 in less than the full distance to the outer wall. We chose 0.004 in for the next model and added loss as well.

The agreement between model and measurement is now quite good (Figure 17.6(b)). From the last two computer runs we can determine that the sensitivity of cutoff frequency to air gaps is about 25 MHz per mil of radius. As always, a larger sample of measured components is needed make our results statistically significant.

These two lowpass filter examples were the original justification for the step discontinuity convergence study we presented earlier. Once we understand the convergence behavior for the steps, we can apply that knowledge to the complete filter. Filters present additional challenges to the automatic mesh refinement process. If we mesh deep in the stopband, the meshing algorithm tends to focus on the high reflection near the ports and ignores the interior of the filter. One strategy is to seed the step regions and mesh in or very near the passband. For the pseudo-lumped lowpass another strategy is to mesh at a very low frequency and adjust the lambda refine variable to give us enough cells per wavelength near the cutoff frequency. Finally, as with any CAD tool, our results are only as good as the information we feed in. Good agreement between measured and modeled in our second example required a more careful tolerance analysis.

17.2 3.5-GHZ COMBLINE FILTER

The combline filter is one of the more popular filter topologies. The resonators are metal cylinders or bars with one end shorted to ground. The resonators are less than a quarter wavelength long, typically 30 to 70 electrical degrees. They are brought to

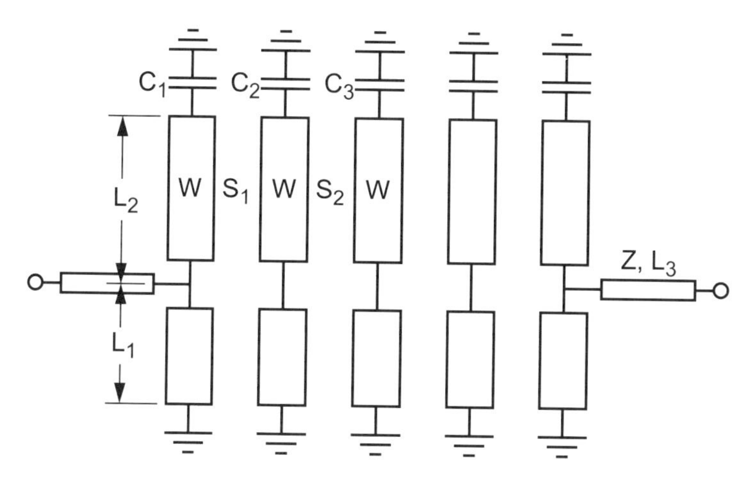

Figure 17.8 Schematic for the combline filter analysis with two groups of multiple coupled lines and tuning capacitors.

resonance with various types of capacitive loading at the open circuit end. The resonators can be arranged in a linear array between parallel ground planes or they might be located in individual cavities that are coupled through irises. A wide range of center frequencies and bandwidths can be realized. The main driver is the bandwidth, which determines the spacing between resonators and the type of coupling structure needed at the input and output. Bandwidths below 1% and greater than an octave have routinely been achieved depending on the application.

Figure 17.7 shows an $N = 5$ combline bandpass filter. The desired bandwidth is 2% at 3.5 GHz. The main body of the filter is numerically controlled (NC) or electron discharge (ED) machined from a solid block. The resonators have a rectangular cross-section. Covers are then bolted to both sides of the main body to form the ground planes [4]. This filter uses tapped input and output resonators. The vertical position of the tap line on the resonator determines the external Q (Q_{ex}). Tuning elements at the open-end of each resonator adjust their frequency. In this case there are no adjustments for the couplings between resonators.

Several filters like this one were designed and built to test a new model for an array of coupled slab lines [5]. As always, our first choice is to use the simplest, lowest order analysis scheme to do the design. So our first attempt was a circuit-theory-based model based on an analytical description of the coupled array. The schematic for this approach is shown in Figure 17.8. It is very similar to the schematic we used earlier for the microstrip interdigital filter. We can force all the bars to have the same width, which reduces the number of variables. And of course we should take advantage of symmetry. In the end, there are seven variables needed to optimize the geometry. The diameter and length of the tap lines are somewhat arbitrary and are fixed at an early stage in the design process.

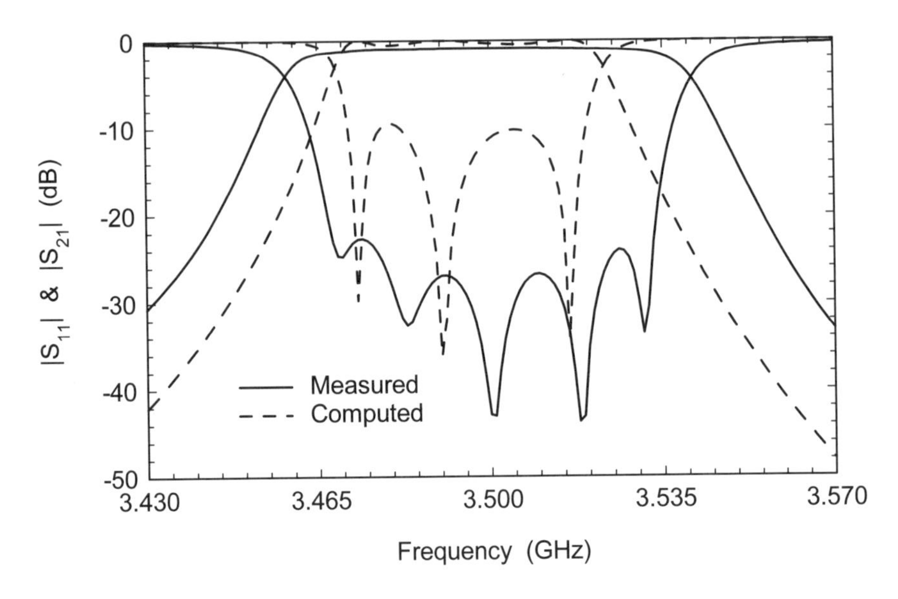

Figure 17.9 Measured versus modeled results using the original analytical circuit model. The measured bandwidth is much greater than what is predicted. © 1999 IEEE [13].

There are analytical equations [6] and design curves [7] for coupled slabs in the literature. We can then use approximate techniques [8, 9] to generate the *Y*-matrix for an array of bars. Once the details were worked out it was all programmed and turned into a compiled, user-defined model in Touchstone Sr. [10]. So we had a fast, geometry-driven model we could use to analyze and optimize the combline filter. The only problem was, it did not give the correct answer.

Figure 17.9 shows measured data from a filter like the one in Figure 17.7 and the computer prediction from the analytical model we just described. There is clearly a huge error in the predicted bandwidth. This problem is actually well known in the filter community. It is understood that any TEM model of the combline filter does not give the correct bandwidth. The errors in bandwidth can be anywhere from 10% to 40% depending on the ground plane spacing. Combline filter designers have clever ways to account for the bandwidth error and correction factors are built into design programs like CLD [11].

There is some debate as to the exact mechanism for the bandwidth error. Between the ground planes the filter cavity is a waveguide below cutoff. The resonators tend to excite and couple to evanescent modes in the cavity. Our TEM model and any 2D cross-section solution of the coupled bar array does not include the coupling to the waveguide channel. At this point it may be tempting to abandon our first approach and go straight to a 3D field-solver. The 3D field-solver should get the physics exactly right and capture all the first and second order effects in the fil-

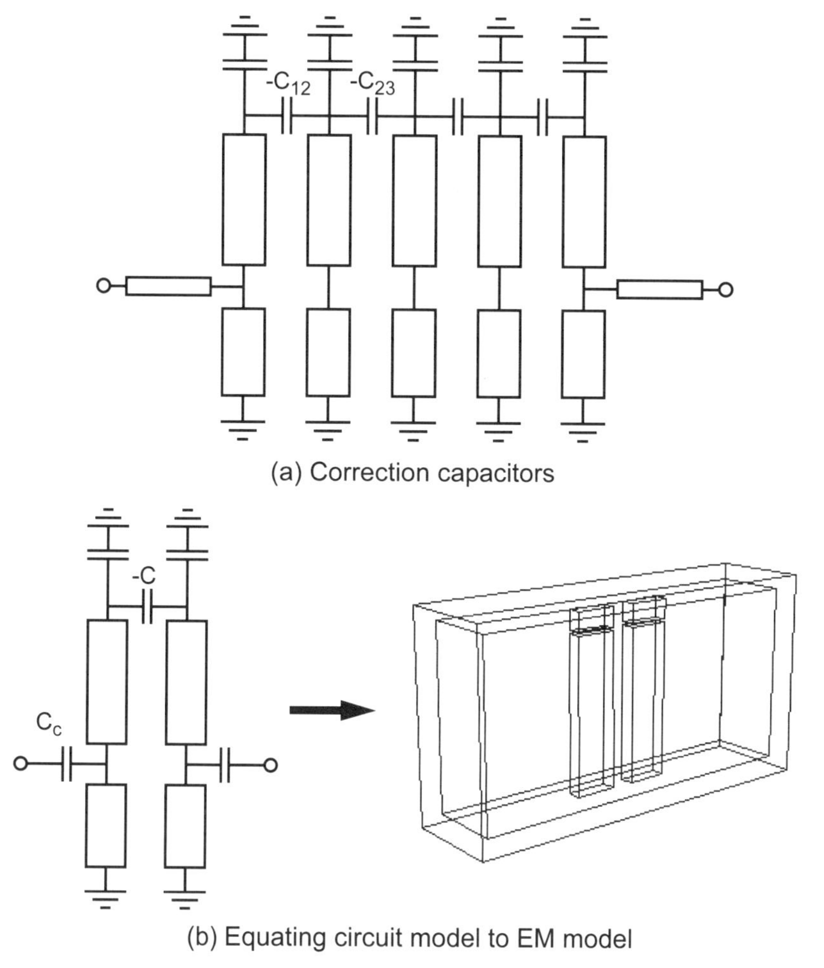

Figure 17.10 Combline filter modeling: (a) adding coupling correction capacitors to the model; and (b) extracting the correction capacitance by equating the circuit model to a field-solver based model. © 1999 IEEE [13].

ter cavity. Our $N = 5$ filter only has seven variables; with today's computer resources we might be able to optimize the complete 3D geometry. But what if we wanted to do a larger structure with more resonators and more variables? At some point the brute force analysis of the complete geometry will become unwieldy.

But perhaps there is a way to correct the circuit theory model using data from the field-solver. In fact, Shapir and Sharir [12, 13] have published a simple correction method that is quite easy to understand and implement. The fundamental error between the model and the measured hardware is in the bandwidth, which implies that the couplings between the resonators are wrong. So, if we can correct the cou-

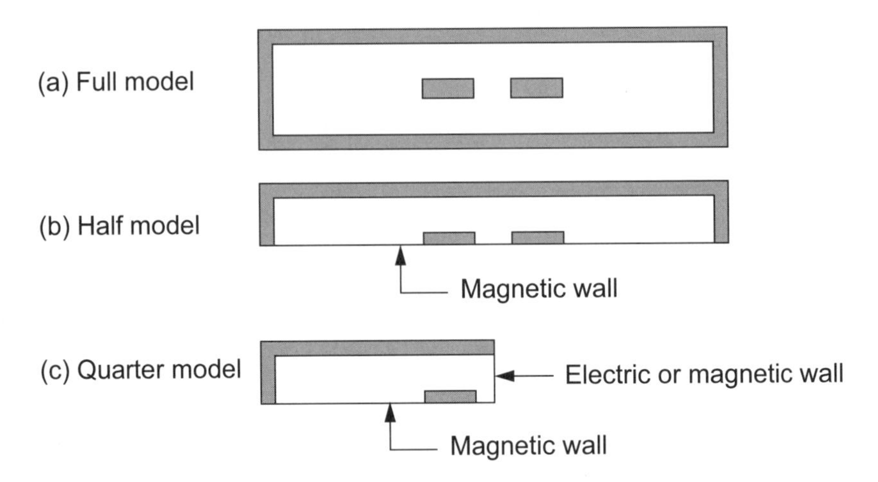

Figure 17.11 Symmetry considerations for the 3D TLM model: (a) top view of the full coupled resonator model; (b) half model with symmetry plane on the long axis; and (c) quarter model with second symmetry plane between the coupled strips.

pling in the model it might become usable. It turns out that connecting a negative capacitor across the tops of a resonator pair drives the coupling, and hence the bandwidth, in the correct direction (Figure 17.10(a)). Like a magnetic wall, the negative capacitor is a fictitious element, but we are free to use it if it is useful.

To find the correction factor we can compare our circuit model to a 3D field-solver model of a pair of resonators (Figure 17.10(b)). The value of the correction capacitance will be a function of the spacing between the resonators. We have several 3D numerical methods to choose from, both time domain and frequency domain methods. In the frequency domain we may have to analyze a fairly large number of frequencies to find the resonant frequencies of the coupled pair. An FEM eigenmode-solver would be a good choice. With a time domain tool it would be easy to find the resonances after we FFT the data. Symmetry considerations will apply to any method that we choose. Convergence issues will be important no matter what method we choose.

At the time this work was done we had access to the 3D TLM tool, Micro-Stripes. So we built a 3D model for a pair of coupled bars (Figure 17.10(b)). In the time domain we can pulse the structure across one of the gaps between the resonator and its tuning element. We can sense the resulting field at any point in the cavity. With this approach we don't have to model any coupling structures into the cavity. A top view of the complete geometry is shown in Figure 17.11(a). It would clearly be inefficient to model the complete geometry when we have two obvious symmetry planes. We can place a magnetic wall down the center line of the longer dimension and analyze only half the geometry (Figure 17.11(b)). Typical results for this analysis are shown in Figure 17.12(a). The coupling coefficient between the

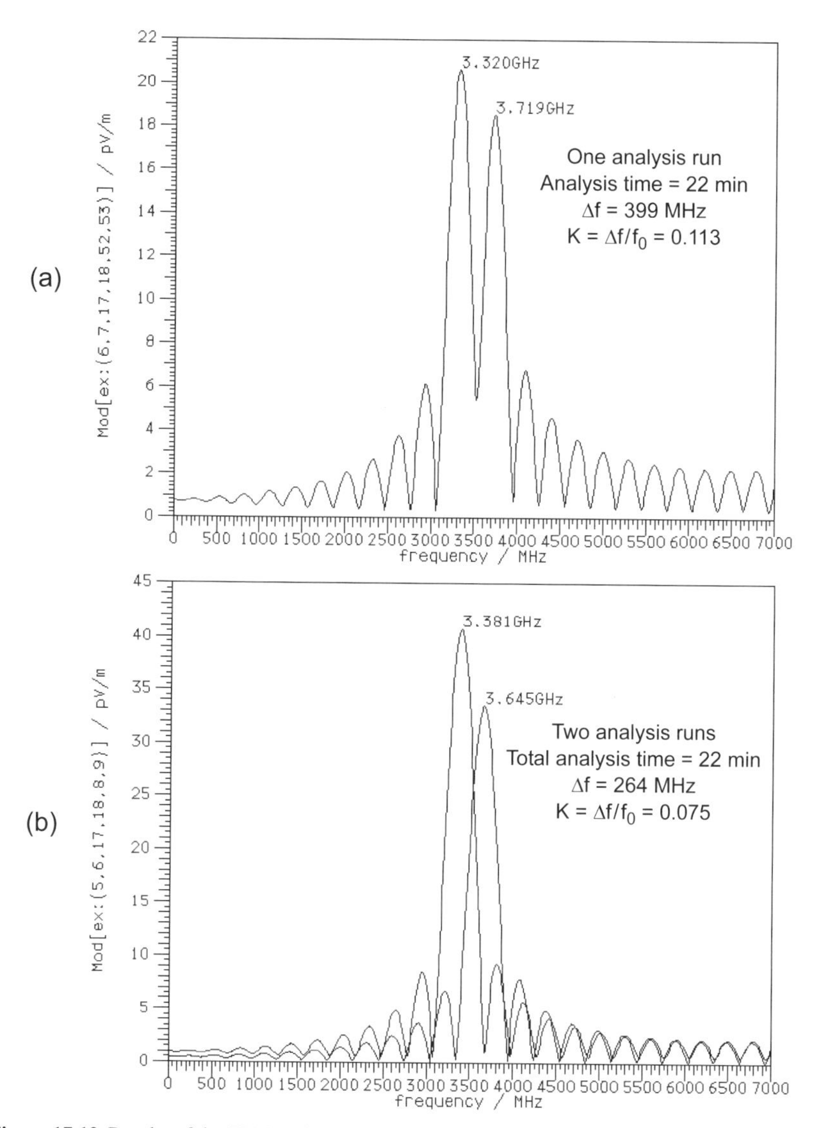

Figure 17.12 Results of the TLM analysis for (a) the half model, and (b) two runs of the quarter model. The coupling coefficient is derived from the frequency peak data (Flomerics Micro-Stripes).

resonators, K, is defined as the delta in frequency between the two peaks divided by the center frequency [14, 15].

We can apply a second symmetry plane between the two resonators (Figure 17.11(c)). But now we have to do two analysis runs, one with electric wall and one with a magnetic wall at the second symmetry plane. The results for this analysis are

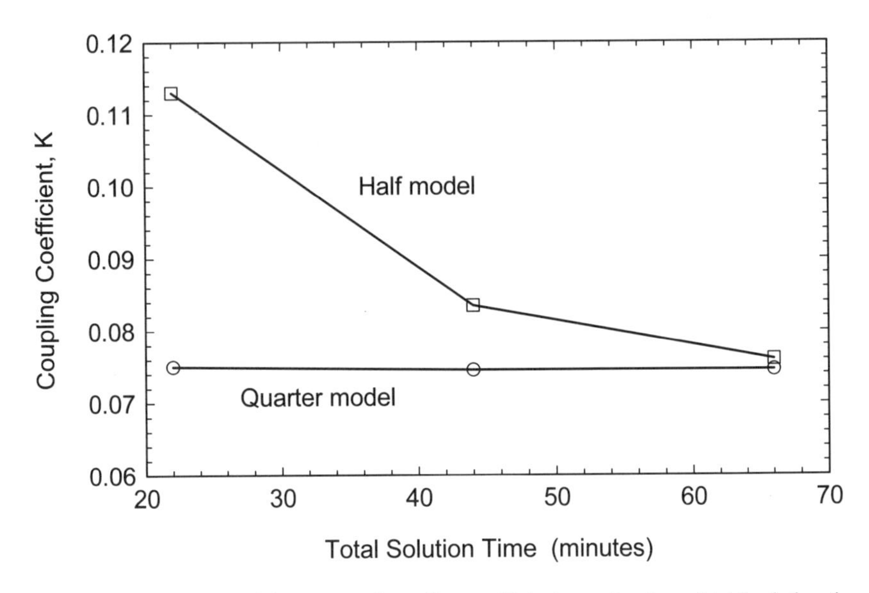

Figure 17.13 Convergence of the computed coupling coefficient as a function of total solution time, circa 1997. The quarter model (two analysis runs) converges much more quickly than the half model (Flomerics Micro-Stripes).

shown in Figure 17.12(b). The two analysis runs are overlaid on the same plot to save space. The total analysis time is the same as Figure 17.12(a) but we are getting a quite different value for *K*. In fact, the second solution is more accurate. This makes sense if we review our basic FFT theory. In Figure 17.12(a) we are asking the FFT process to resolve two closely spaced resonances. To get an accurate solution we need a lot of time steps. In Figure 17.12(b) the FFT process only has to identify a single resonant peak in each analysis. It can achieve the same or better accuracy with far fewer time steps. And the total solution time will be lower, even though we are running two problems instead of one. In Figure 17.13 we show the convergence in computed coupling coefficient for the half model case and the quarter model case as a function of total solution time. The solution times are typical of a 50-MHz SUN SPARC-10, circa 1997. We can achieve the same level of accuracy with both approaches, but the quarter model is clearly far more efficient. This is another very simple example of using our understanding of how the software works to improve the efficiency of the design cycle.

We also need to look at the convergence of the coupling coefficient computation as a function of meshing. Figure 17.14 shows three different meshing experiments. This rectangular geometry is a simple one for the cubic meshing that is fundamental to time domain solvers. In this example the spacing between resonators is 0.1 in. In Figure 17.14(a) we apply a uniform mesh to the entire computation region. The cells inside the metal are ignored by the solver and not computed. In

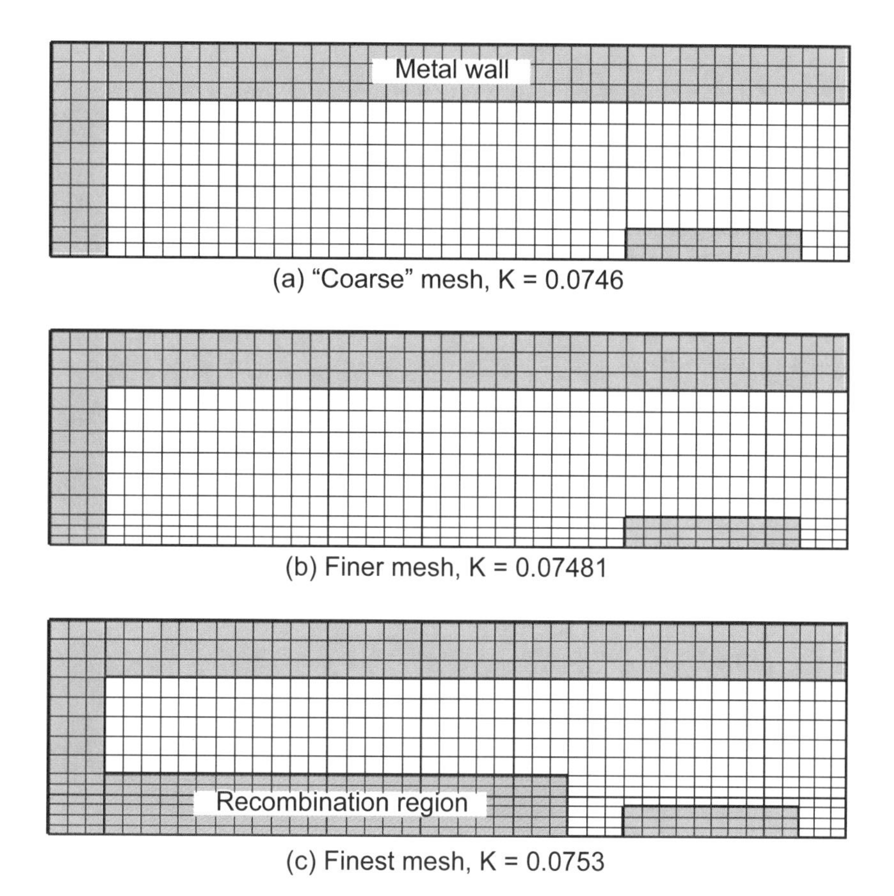

Figure 17.14 Three different meshings for the coupled rectangular bars: (a) uniform mesh; (b) finer mesh in the gap; and (c) finer mesh on all sides of the bar (Flomerics Micro-Stripes).

Figure 17.14(b) a finer mesh has been applied in the gap. Due to the nature of the meshing process, this finer mesh also propagates to the outer boundary of the cavity where it is not really needed. In Figure 17.14(c) we applied a finer mesh on all sides of the resonator. We were also able to "recombine" some of the smaller cells into larger cells in the region to the left of the resonator. The shift in computed coupling coefficient between the starting mesh and the finest mesh is almost 1%.

After we are convinced that the field-solver is being applied correctly, we are ready to find the correction factors needed for our analytical coupled bar model. Figure 17.15(a) shows the coupling between bars computed by Micro-Stripes. The fact that the data fits a simple parabolic curve also adds to our confidence that it is correct. Next we compare the 3D TLM coupling value to our circuit model value and find the value of correction capacitance that makes the two match (Figure 17.10(b)). Figure 17.15(b) shows the derived correction capacitance as a function

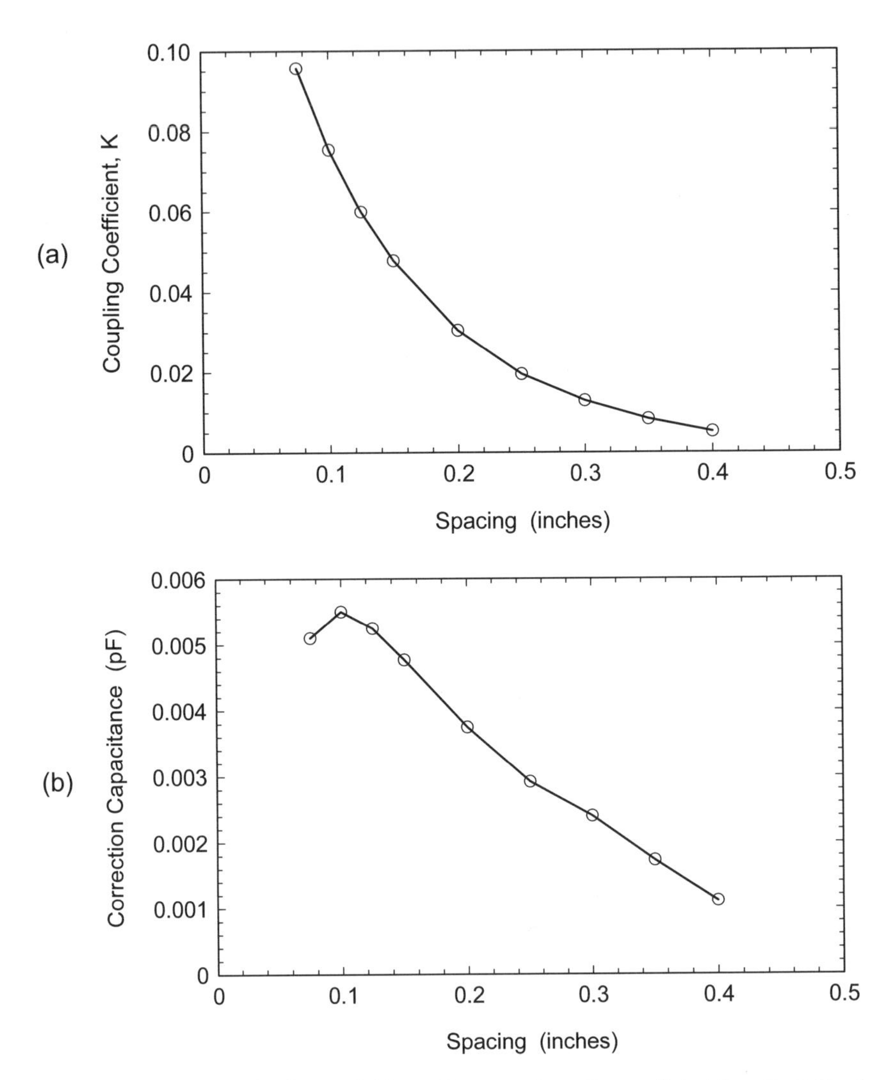

Figure 17.15 Design data for the coupled bars: (a) computed coupling coefficient from the field-solver; and (b) capacitance needed to correct the TEM coupled bar model. © 1999 IEEE [13].

of the spacing between the bars. The absolute values of the corrections are incredibly small.

Of course, the final step is to use the corrected analytical model to design a filter. The goal was again a 2% bandwidth filter at 3.5 GHz. The ground plane spacing was 0.318 in, the width of each resonator was 0.180 in and the thickness of each resonator was 0.062 in. Figure 17.16 shows the measured versus modeled results for this experiment. The actual dimensions of the experimental hardware were care-

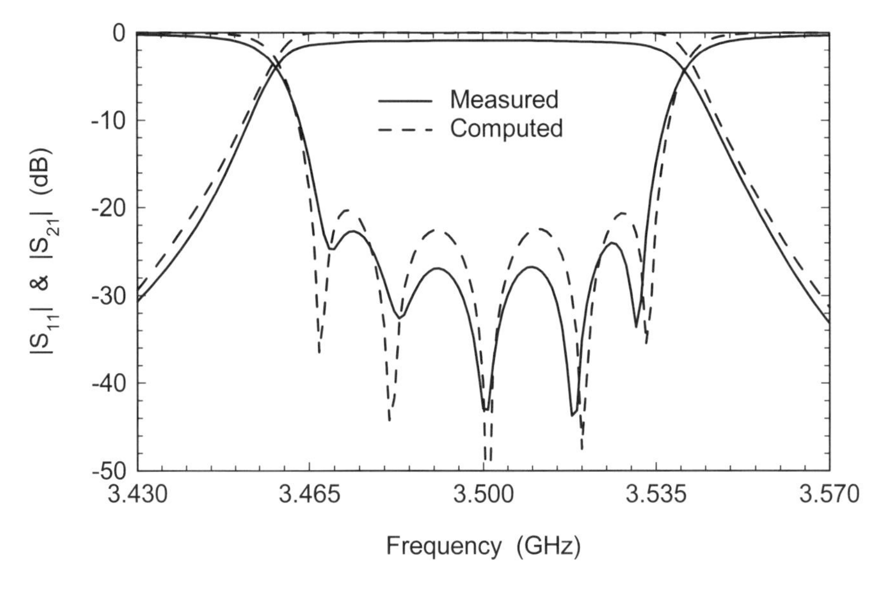

Figure 17.16 Results for the second experimental filter designed with the corrected analytical model. The ground plane spacing is 0.318 in. The resonators are 0.180 in wide and 0.062 in thick. © 1999 IEEE [13].

fully measured and these dimensions are used for the computer prediction. Again we should emphasize that the experimental hardware had no adjustments for the couplings between resonators, so the bandwidth could not be adjusted. Table 17.1 is a summary of the computed and measured dimensions for this filter. We have included the dimensions computed with the new model, measured dimensions of the hardware, the dimensions computed by the design program CLD, the dimensions extracted directly from the TLM solutions, and the prototype coupling coeffi-

Table 17.1

Dimensions for Second Experimental Filter

Key param	*Circuit theory*[1]	*Measured dimen*[2]	*CLD dimen*	*3D TLM dimen*	*K's and Q's*
Tap	0.157	0.156	0.159	0.146	47.88
S1	0.261	0.257	0.260	0.264	0.01746
S2	0.299	0.295	0.296	0.300	0.01278
S3	0.299	0.296	0.296	0.300	0.01278
S4	0.261	0.258	0.260	0.264	0.01746
Tap	0.157	0.157	0.159	0.146	47.88

1. Tolerance of ±0.002 on print. 2. Bars average 0.182 wide and 0.061 thick.

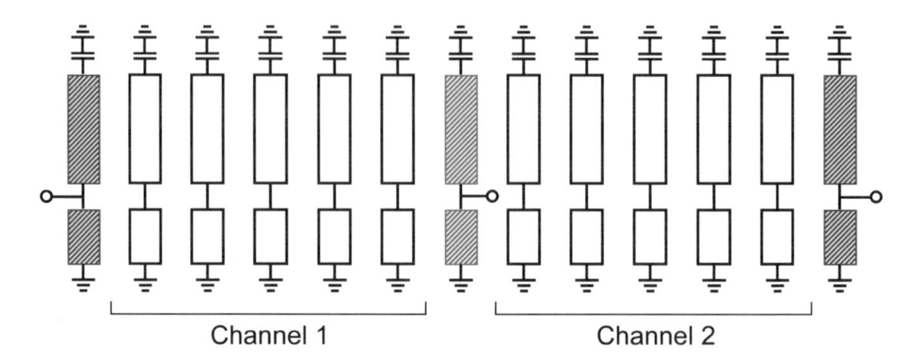

Figure 17.17 Circuit model for the diplexer. The redundant resonator at the common junction is shown in red. The outer redundant resonators (blue) introduce an extra degree of freedom and allow all three taps points to be same height above ground.

cients and external *Q*s. The data from the TLM solutions indicates the tap points should be shifted slightly, which was confirmed in later experiments.

The original motivation for the analytical, coupled slabline model was a requirement for a diplexer at 3.5 GHz. Diplexers can be various combinations of lowpass, highpass, and bandpass filters. In this case the diplexer is two bandpass filters joined at a common junction forming a three-port network. This also happened to be a noncontiguous diplexer; there is a small guard band between the two channels. Probably the easiest design method for this case is the phasing method. We start with two doubly terminated narrowband filters. Then we add a length of transmission line from the input of each filter to the common junction. The length of each line is adjusted to present an open circuit in passband of the opposite channel. We then retune both channels for equal ripple performance.

A more compact diplexer geometry uses a redundant resonator at the common junction which couples to both filter channels. The redundant resonator is sometimes called a susceptance-annulling network [16]. In this diplexer design we wanted to use a geometry very similar to Figure 17.7 but with simplifications that would make it more suitable for high volume manufacturing. The circuit theory model for the diplexer is shown in Figure 17.17. Each channel of the diplexer is an $N = 5$ bandpass filter. The center resonator (red) is the susceptance-annulling network between the two filters. Redundant resonators (blue) at each end give us an extra degree of freedom and allow us to make all three tap points come out at the same height above ground. This greatly simplifies and speeds up the circuit model. In all there are 20 variables to optimize so a direct analysis and optimization of the complete structure using a field-solver was out of the question.

Figure 17.18 shows the measured versus modeled results for the first diplexer prototype. Again, there are no coupling screws in the hardware that would allow us

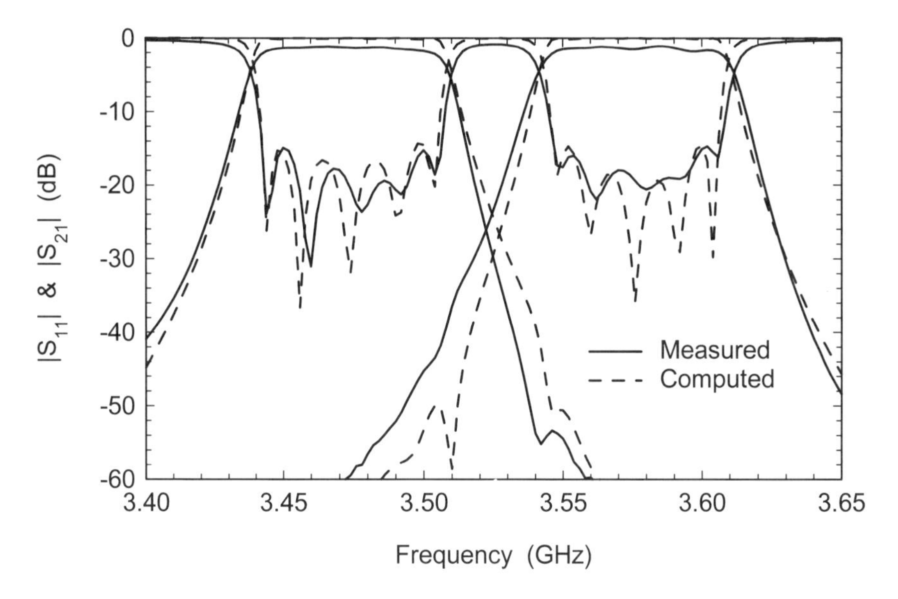

Figure 17.18 Measured versus modeled results for the diplexer prototype. No coupling screws were used in this tuning.

to fudge the bandwidth results. Adding just two coupling screws allowed us to retune the filter and achieve 20 dB return loss in each channel.

17.3 2.14-GHZ COMBLINE FILTER

In the previous example we optimized a combline filter and a combline diplexer with the 2D cross-section engine inside the optimization loop. That particular 2D engine was based on analytical equations with some field-solver-based corrections, so it was fairly fast. Using a 3D field-solver inside the optimization loop may not always be practical. In the case of combline filters, we need a resolution of less than 0.001 in on the tuning screw depths to reach equal ripple tuning. The spacing between resonators or the iris openings between cavities must also be determined to high accuracy.

But the tuning mechanism in a combline filter is actually quite well understood. As the tuning screw approaches the end of the resonator, capacitance increases. The screw might also enter a pocket in the top of the resonator, or the end of the resonator may also enter a pocket in the cover of the housing. All three styles of tuning can easily be modeled with transmission lines and capacitors.

So the question becomes, why tune the resonators in the EM domain at all? Why spend a lot of field-solver time making fractional adjustments to tuning

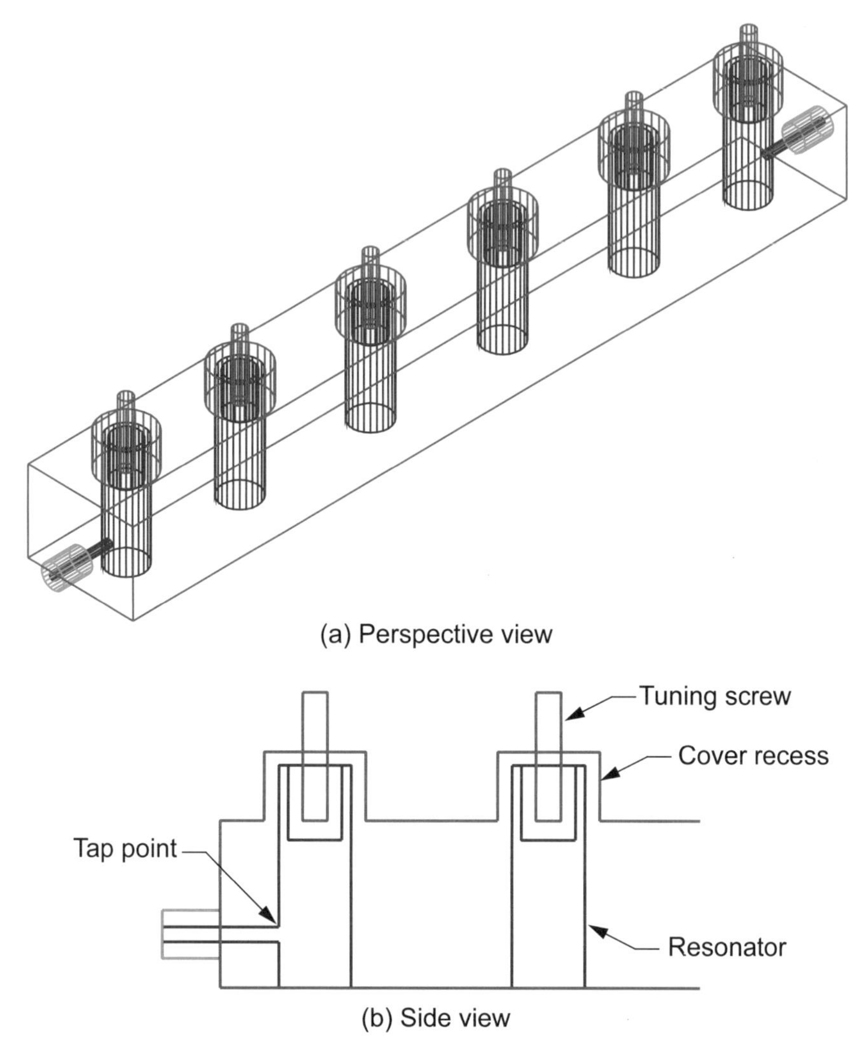

Figure 17.19 Field-solver model for the $N = 6$ combline filter: (a) perspective view of the complete filter; and (b) side view of the half model. The resonators fit in pockets in the cover and the tuning screws fit in pockets in the resonators (Agilent HFSS Ver. 5.6).

screws and resonator spacings? It turns out there is a very simple way to tune the filter in the circuit simulator and do only one or two field-solver simulations of the complete filter [17, 18].

Figure 17.19(a) shows an $N = 6$ combline filter with cover loading. The desired passband is 2.083 to 2.180 GHz and the desired return loss is 26 dB. The ends of the resonators extend into pockets in the cover and the tuning screws extend into

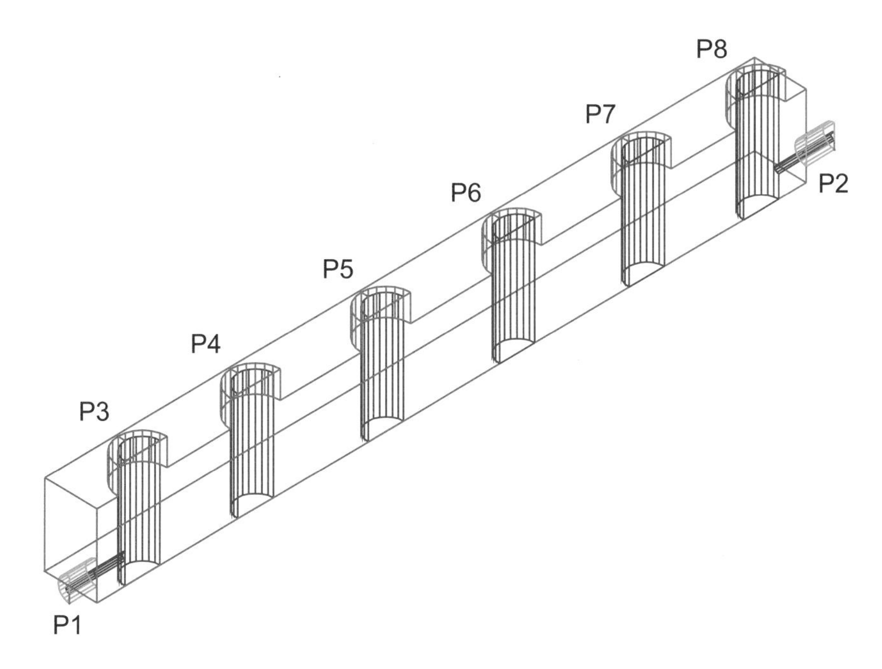

Figure 17.20 The modified field-solver model for the combline filter that simplifies the tuning geometry and places a port at each tuning screw location (Agilent HFSS Ver. 5.6).

pockets in the resonators. Figure 17.19(b) shows a cross-section of the resonator detail. We can simplify the field-solver problem by eliminating a lot of the tuning screw detail and placing ports at the tuning screw locations (Figure 17.20). Now we are solving for an eight-port network in the field-solver, rather than a two-port with movable screws. We can take the eight-port network to our favorite circuit simulator and tune the filter by connecting positive or negative capacitors to Ports 3 through 8.

The initial design for this filter can be obtained with commercial software or by the K and Q_{ex} method [19, 20] on the field-solver. To get the spacings between the resonators we need a look up table for coupling coefficient as a function of resonator spacing. The geometry in Figure 17.21(a) is one way to get this information. We connect capacitors to Port 1 and Port 2 to tune the resonators to band center. We then use the probes at Port 3 and Port 4 to measure the S_{21} through the structure. This is exactly how we used to measure test structures in the lab. We could also capacitively couple directly to Port 1 and Port 2. The two outer resonators represent adjacent resonators in the final filter and have some influence on the computed coupling coefficient [21]. The outer resonators should be short-circuited for this measurement. A second field-solver model, Figure 17.21(b), is used to find the Q_{ex} (the tap height) and the K_{12} coupling coefficient. The K_{12} coupling as a function of res-

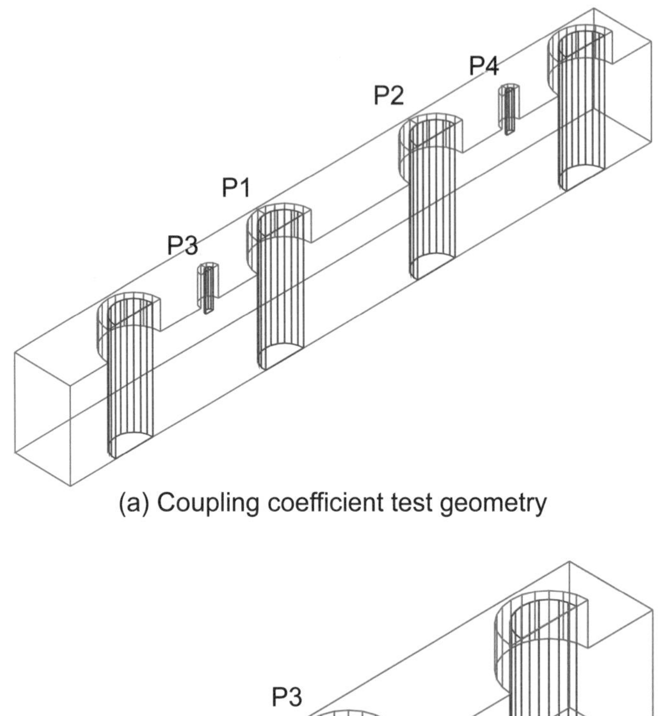

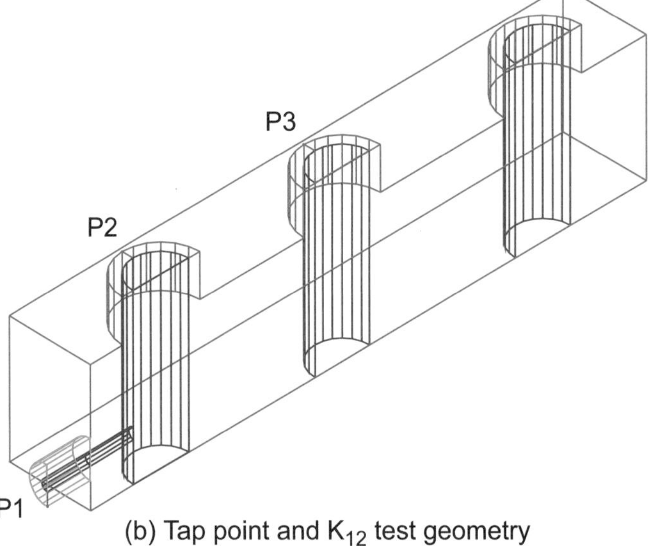

Figure 17.21 Field-solver geometries used to generate (a) coupling coefficient data for the interior of the filter, and (b) tap point location and K_{12} coupling data (Agilent HFSS Ver. 5.6).

onator spacing is slightly different than the interior resonators due to the presence of the tap.

With our design data in hand we produce a prototype of the complete filter in the field-solver and tune the prototype in our circuit simulator. Couplings between

Table 17.2

Dimensions for the Combline Filter in Figure 17.19(a)

Feature	*Dimension (in)*
Ground plane spacing	0.750
Resonator diameter	0.250
Resonator length	0.560
Resonator inner diameter	0.186
Cover recess diameter	0.350
Cover recess depth	0.230
Tuning screw diameter	0.086
Tap line diameter	0.050
Tap line length	0.200
Tap position (from bottom)	0.178
Spacing Reso 1 to Reso 2*	0.823
Spacing Reso 2 to Reso 3*	0.919
Spacing Reso 3 to Reso 4*	0.934

*Center to center

resonators can be trimmed by added elements between the ports. In the case of a combline filter, a short-circuited transmission line that is the same electrical length as the combline resonator is the same type of element we would use in a circuit-theory-based model of the filter. The variable element is then the impedance of the transmission line, which can be positive or negative for tuning purposes. Once the prototype is tuned, we can measure the realized coupling coefficients and make corrections to our geometry. With a good starting point, typically only one round of corrections is needed.

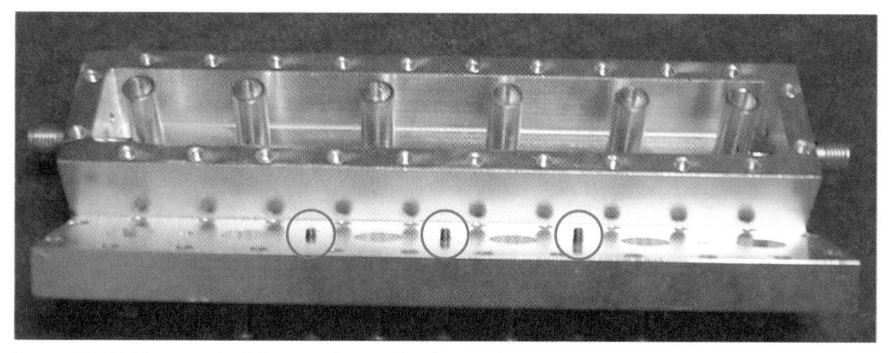

Figure 17.22 Photograph of the experimental hardware with the cover removed after tuning. The penetration of three coupling screws can be seen (red circles). Photo courtesy of Forem USA.

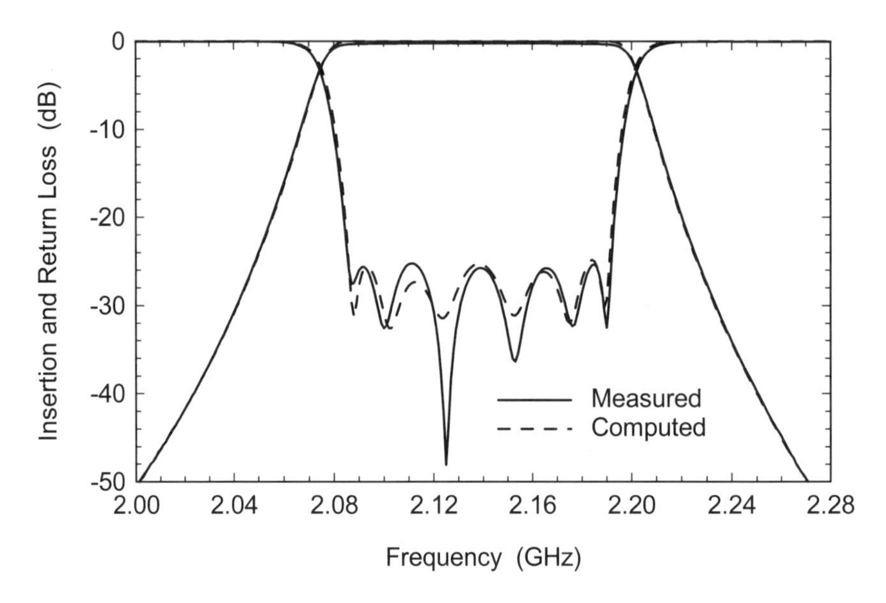

Figure 17.23 Measured versus modeled results for the combline filter. No corrections to the couplings between resonators were made in the tuned field-solver model.

The filter prototype dimensions are listed in Table 17.2. A photograph of the prototype hardware can be found in Figure 17.22. The cover was removed after tuning to show the penetration of the coupling screws. Notice that only three coupling screws have any appreciable depth. The measured versus modeled results are shown in Figure 17.23. When the computer model was tuned, no coupling corrections were used.

The "port tuning" method has some interesting aspects. In the field-solver model, the ports completely de-Q and detune the resonators. We have taken a very sensitive, high Q, resonant structure and turned it into a very benign multiport network. The variation in the resulting S-parameters across the passband is quite low. We are essentially just measuring coupling coefficients on the field-solver. A similar concept has been used in the analysis of some planar filters. It that case the desired geometry is split into two or more nonresonant pieces for EM analysis then recombined using circuit theory [22]. At first this all seems a little counter-intuitive but I believe there is a network theory theorem that states "the response of a passive network is invariant to its terminations." We are simply computing the response of the model with one set of terminations and tuning it up with a different set of terminations.

Because the network response is so benign, we only need a few points in the passband and a few points in the stopband to find the exact filter response, even below −100 dB in the stopband [23]. In the case of the frequency domain solvers this has a huge impact on total solution time. This technique can easily be extended to address larger and more complex filters.

References

[1] Private communication, Raja Reddy, Forem USA, September 2001.

[2] Private communication, Eric Wiehler, Forem USA, September 2001.

[3] Private communication, Brad Brim, Ansoft Corp., September 2001.

[4] Edelman, J., and G. Greer, "CAE Tools Shape Square Rods for Slabline Filters," *Microwaves & RF*, Vol. 26, No. 11, 1988, pp. 119–126.

[5] Swanson, Jr., D. G., "Optimizing Combline Filter Designs Using 3D Field-Solvers," *IEEE MTT-S Int. Microwave Symposium Workshop Notes*, Denver, CO, June 8–13, 1997.

[6] Perlow, S., "Analysis of Edge-Coupled Shielded Strip and Slabline Structures," *IEEE Trans. Microwave Theory Tech.*, Vol. 35, No. 5, 1987, pp. 522–529.

[7] Getsinger, W., "Coupled Rectangular Bars Between Parallel Plates," *IRE Trans. Microwave Theory Tech.*, Vol. 10, No. 1, 1962, pp. 65–72.

[8] Chen, Z., "Fast Computation of Multiport Parameters of Multiconductor Coupled Microstrip Lines," *IEEE Trans. Microwave Theory Tech.*, Vol. 43, No. 6, 1995, pp. 1393–1395.

[9] Swanson, Jr., D. G., "A Novel Method for Modeling Coupling Between Several Microstrip Lines in MICs and MMICs," *IEEE Trans. Microwave Theory Tech.*, Vol. 39, No. 6, 1991, pp. 917–923.

[10] Touchstone Sr.™, Agilent EEsof EDA, Santa Rosa, CA.

[11] Combline Design (CLD), Forem USA, Amesbury, MA.

[12] Shapir, I., and V. A. Sharir, "Modeling Structure Parasitics in Comb-Line Filters," *IEEE MTT-S Int. Microwave Symposium Digest*, San Francisco, CA, June 17–21, 1996, pp. 477–480.

[13] Shapir, I., V. A. Sharir, and D. G. Swanson, Jr., "TEM Modeling of Parasitic Bandwidth Expansion in Combline Filters," *IEEE Trans. Microwave Theory Tech.*, Vol. 47, No. 9, 1999, pp.1664–1669.

[14] Zaki, K. A., and C. Chen, "Coupling Between Hybrid Mode Dielectric Resonators," *IEEE MTT-S Int. Microwave Symposium Digest*, Las Vegas, NV, June 9–11, 1987, pp. 617–620.

[15] Hong, J. S., and M. J. Lancaster, *Microstrip Filters for RF/Microwave Applications*, New York: John Wiley & Sons, Inc., 2001, Chapter 8.

[16] Matthaei, G. L., and E. G. Cristal, "Theory and Design of Diplexers and Multiplexers," in *Advances in Microwaves, Vol. 2*, New York: Academic Press, 1966.

[17] Swanson, Jr., D. G., and R. J. Wenzel, "Fast Analysis and Optimization of Combline Filters Using FEM," *IEEE MTT-S In. Microwave Symposium Digest*, Phoenix, AZ, May 20–25, 2001, pp. 1159–1162.

[18] Swanson, Jr., D. G., "Fast Analysis and Optimization of a Compline Filter Using FEM," *IEEE MTT-S Int. Microwave Symposium Workshop WSA Notes*, Phoenix, AZ, May 20–25, 2001.

[19] Dishal, M., "A Simple Design Procedure for Small Percentage Bandwidth Round-Rod Interdigital Filters," *IEEE Trans. Microwave Theory Tech.*, Vol. 13, No. 9, 1965, pp. 696–698.

[20] Wong, J. S., "Microstrip Tapped-Line Filter Design," *IEEE Trans. Microwave Theory Tech.*, Vol. 27, No. 1, 1979, pp. 44–50.

[21] El Sabbagh, M., K. Zaki, and M. Yu, "Non-Adjacent Resonators Effects on Coupling and Resonant Frequency in Combline Filters," *IEEE MTT-S Int. Microwave Symposium Digest*, Phoenix, AZ, May 20–25, 2001, pp. 1313–1316.

[22] ***Em*** User's Manual, Chapter 9, Circuit Subdivision - A Filter Example, Sonnet Software, Inc., Liverpool, NY.

[23] Wenzel, R. J., and D. G. Swanson, Jr., "Observations on the Stopband Performance of Tapped Line Filters," *IEEE MTT-S Int. Microwave Symposium Digest*, Phoenix, AZ, May 20–25, 2001, pp. 1619–1621.

Chapter 18

Choosing the Right Software

When teaching this material to other engineers, I (D.S.) am often asked, "Which software tool is best?" The most honest answer, but not the one hoped for is, "it depends." It depends very much on the particular problem you are trying to solve. We can generalize and suggest a particular code for a certain class of problems, but there will always be exceptions. Hopefully some of the examples presented in this book have made some of the trade-offs and decisions to be made more clear. If you are faced with the prospect of buying the first field-solver in your organization, hopefully you can find one that meets 60 to 80 percent of your present needs. A second package might then be purchased in a year or two to fill in some of the gaps. At some point it is nice to have access to a 2D solver, a 2.5D solver, and one of the many 3D solvers. Choosing software also has a subjective component that has to do with ease of use and how the interface is organized. I would rather leave you with a list of comprehensive questions to ask yourself and the software vendor rather than impose my own list of favorite tools.

18.1 THE SOLUTION PROCESS FROM START TO FINISH

Any useful numerical tool must have some basic characteristics no matter what the underlying method. First we have to capture the geometry of the physical object or objects we are trying to model. For 2D or 2.5D tools this is not too difficult. Capturing 3D geometry descriptions is much more challenging. Once the geometry is entered we can assign material properties and boundary conditions. Some of the boundary conditions are most naturally satisfied by the numerical method itself. For example, many methods assume the outermost boundary is a perfect conductor. Sometimes we introduce nonphysical boundaries in order to take advantage of symmetry planes. Next, we must create or mesh or discretize the problem. Generally we are creating a surface mesh or a volume mesh depending on the numerical method. After meshing we can finally solve the problem. In the case of MoM and FEM this involves a matrix inversion. For FDTD and TLM we generally apply some kind of pulse excitation and time step until convergence is reached. The final step is postprocessing of the results. This may include generating *S*-parameters as well as various field quantity plots. To summarize, the solution process from start to finish is:

- Capture the geometry of the physical object(s);
- Assign material properties and boundary conditions;
- Create an analytical mesh;
- Solve (explicit or implicit);
- Postprocess.

18.2 FEATURES ALL TOOLS MUST HAVE

Many of the problems we are concerned with have ports. Ports allow us to excite the circuit or antenna and measure the results. In some cases we might use source and sense points somewhere in the solution space rather than a physical port. To be really useful, ports must be calibrated. Field-solvers have numerical port discontinuities just like network analyzers or test fixtures have physical port discontinuities. The easier type of port to implement is on the boundary of the problem space. Some solvers also allow access to ports that are internal to the problem geometry. Internal ports are generally more difficult to implement and calibrate. Sometimes when we measure an active or passive device in a fixture we would like to remove the effects of the fixture; this process is called de-embedding. We also use de-embedding in field-solvers to separate numerical port and fixture effects from our device under test. De-embedding is actually easier and more flexible in a field-solver than in the laboratory. Multiport de-embedding that would be quite difficult in the lab is actually quite easy in the field-solver. To summarize the most important features:

- Ports or source/sense points;
- Port calibration:
  - Numerical port discontinuity;
  - Ports on boundaries versus internal ports.
- De-embedding:
  - Single ports;
  - Multiple ports;
  - Internal ports.

18.3 FEATURES THAT ARE NICE TO HAVE

While not absolutely necessary, there are some features that help the user work more effectively. These features mostly apply to the solution process. Some kind of intelligent display of the solution progress is most welcome. Port and or de-embedding impedances help the user catch errors in the problem setup. It is always a good idea to check the port or de-embedding impedances early in the solution process. Matrix fill and invert times or the number of time steps give the user an estimate of total solution time. Displaying *S*-parameters as they are computed lets the user check the first frequency for the expected results. The user can also check the solution time for the first frequency and estimate the total run time for the job. All of this information should be readily available in one window with automatic updates. As users we are looking for:

- Intelligent display of solution progress;
- Port and/or de-embedding impedances:
  - Quickly find errors in problem setup;
  - Good first step.
- Matrix fill and invert times or number of time steps;
- *S*-parameters as they are computed:
  - Check first frequency for expected results;
  - Check solution time of first frequency.
- All information is one place with automatic updates.

18.4 VISUALIZATION

Visualization is one of the more compelling aspects of using a field-solver. The fact that we can now see many of the effects that we could only describe mathematically in the past is quite incredible. However, the options and available features for visualization will vary somewhat between tools. In general, the 3D tools have the most flexibility for displaying various field quantities. The MoM tools compute conduction currents and display them quite well. Some of them also have an E-field display. The available features for displaying far field parameters will vary from tool to tool. Symmetry planes in the problem description may limit the display of far field parameters. The user must examine the features of a potential tool carefully to be sure a desired capability is present. Questions to ask are:

- What does each tool do best?

- What parameters do we want to look at?
  - Surface conduction currents;
  - Mag E-field or vector E-field;
  - H-field quantities;
  - Radiation patterns.
- Do we want to sweep frequency or phase?
- Would time stepping tell us more?

18.5 EASE OF USE AND TOTAL SOLUTION TIME

The total solution time for a field-solver problem depends on more than just the clock speed of the computer. First there is the time required to set up the model. Ease of use enters into the equation but it is very hard to quantify. The newer tools that have focused on the PC platform only and have invested in the ACIS tool box seem to have an advantage. The experience level of the user will also influence setup time greatly. In general, setup time for 2D will be shorter than 2.5D which will be much shorter than 3D. If automatic or manual optimization of the model is required, it takes time to determine the correct variables to use. Again, an experienced engineer will have more intuition when choosing variables. We have seen that having a circuit model that fits the first analysis run is a very valuable tool for choosing variables. Finally, the actual compute time can be greatly impacted by approximations to the geometry and the quality of the mesh. The application of symmetry, where possible, will also have a large impact on compute time. So the total solution time issues are:

- Time to set up the model:
  - Ease of use;
  - Experience level of the user;
  - 3D takes longer than 2.5D, which takes longer than 2D.
- Time to determine the correct optimization variables:
  - Experience level of the user and intuition;
  - Ability to make a circuit model.
- Actual compute time:
  - Intelligent approximations to the geometry;
  - Quality of the mesh and symmetry.

18.6 THE RIGHT TOOL FOR THE JOB

Answering the following questions will help you choose the first field-solver you will purchase to augment your existing CAD tools. If more than one tool is available to you, the same list will help you pick the right tool for a particular design problem. Many of these questions originally appeared in an excellent review article by Veidt [1]. They are repeated here with additions and modifications by this author.

- *What do the majority of my problems look like?*
  Hopefully you can identify one tool that will solve the bulk of your problems efficiently. If not, you may need to add a second or third tool in the future to address some additional need.
- *Are material losses significant in the simulation problems?*
  With no losses, most solvers revert to real math to speed up the solution process. When the first lossy element is added to the model, all math becomes complex and computation time increases significantly.
- *Is the ratio between the largest and smallest feature large?*
  If this is the case, the gridding method must be examined closely. In 3D, FEM codes will do the best job adapting to this situation. For planar problems, closed-box MoM codes have trouble with high-resolution structures like Lange couplers.
- *Do the structures have curved surfaces or are they orthogonal?*
  Curved boundaries in 3D require FEM or a time domain code with advanced meshing. Curved boundaries in 2.5D can be best approximated with a laterally open MoM code.
- *Do the problems involve a plane of symmetry?*
  Symmetry can greatly reduce computation time, but not all codes support it. Also check to see if far-field patterns can be calculated if a symmetry plane is present.
- *What types of port structures do your problems require?*
  Most solvers now handle the typical cases of waveguide, coax, and microstrip ports quite well. CPW and slotline ports are still a problem for some simulators. Ports involving multiple coupled strips on a single port face are still a problem for many 3D simulators. The options for calibrated internal ports are severely limited for all simulators.
- *What information is to be obtained from the computer simulation?*
  Typically, a combination of *S*-parameters, surface-currents, near-field quantities, and far-field radiation patterns are required. MoM codes display surface currents well, but not arbitrary field quantities. FEM codes display field quantities very well but do a poor job with surface currents.

- *Is broadband frequency data needed?*
  Time domain simulators have traditionally been the first choice when very broadband data is needed. However, many frequency domain simulators now have a fast sweep option that may give equally good results.
- *Is an eigenmode-analysis more helpful than swept data?*
  Some field-solvers offer an eigenmode-solver as an option. For a structure without any ports, the eigenmode-solver finds any number of resonant modes specified by the user. No prior knowledge of mode frequencies is required.
- *Will the field-solver be used to compute multiport* S*-parameter data?*
  Frequency domain methods can solve for all ports in a single simulation. Time domain solvers must do additional simulation runs for lossy multiports.
- *Will radiation problems need to be simulated?*
  A buffer region is required between the structure and the radiation boundary, which increases the problem size. ABCs and PMLs also require significant computation resources.
- *Are models of biological structures needed?*
  Biological structures are not composed of geometrical shapes. A model entry scheme that applies properties to cells or groups of cells may be useful. If the software vendor can provide biological models, that may be very valuable despite other limitations.
- *Do the structures use wires, narrow slots, or thin-films?*
  Some of the time domain simulators have special elements for those cases that do not require a fine mesh. There are also specialized MoM codes for thin wires combined with conducting plates [2].

References

[1] Veidt, B., "Selecting 3-D Electromagnetic Software," *Microwave Journal*, Vol. 41, No. 9, 1998, pp. 126–137.

[2] Burke, G. J., and A. J. Poggio, "Numerical Electromagnetics Code (NEC-2), Lawrence Livermore Laboratory, January 1981.

Appendix A

Survey of Field-Solver Software

We will attempt to list the most widely available software packages of interest to the RF/microwave engineer. There is another group of tools used in the high-speed digital community that we have specifically not listed here because we do not have first-hand experience using them. Inclusion in this list does not imply an endorsement by the authors. Likewise, omission from this list does not imply rejection by the authors. We have adopted our earlier system of classification by geometrical complexity rather than the specific numerical method used.

Also, this appendix is not intended to be a comprehensive list of features for each tool. Unfortunately, this type of information becomes dated rather quickly. Rather, it is again intended as a guide to what is possible and what is available from the various vendors. The software developers are constantly updating their products and the reader is encouraged to contact them directly for the most current product information. Contact information, in alphabetical order by vendor name, can be found in Appendix B.

A.1 2D CROSS-SECTION-SOLVERS

We can get access to this software in three ways. Some tools are stand-alone, some are integrated into a linear or nonlinear circuit simulator, and some 2.5D and 3D tools will give you the impedance and phase velocity of single strip at a port.

A.1.1 Stand-Alone Software–PDE Solvers

For the case of single or coupled strips/slots, a simple stand-alone tool can often compute the impedance and phase velocity information that you need very efficiently. Multistrip problems are more difficult and we will discuss a better approach for those later. These three tools are actually general purpose partial differential equation solvers and can be used for other applications besides 2D electrostatics.

FEMLAB - COMSOL

Numerical Method: FEM

Platforms: Windows

Features: Handles arbitrary cross-sections and rotationally symmetric problems.

Works in conjunction with MATLAB.

Automatic mesh refinement.

Eigenmode-solver for modal solutions.

Various solution types can be linked.

Comments: Wide range of options available with MATLAB.

FlexPDE - PDE Solutions

Numerical Method: FEM

Platforms: Windows

Features: Handles arbitrary cross-sections and rotationally symmetric problems.

Automatic mesh refinement.

Eigenmode-solver for modal solutions.

3D version is available as well

Comments: Text file input rather than GUI.

Flexible options for graphics.

QuickField - Tera Analysis

Numerical Method: FEM

Platforms: Windows

Features: Handles arbitrary cross-sections and rotationally symmetric problems.

Also handles magnetics, currents, thermal, and stress.

Various solution types can be linked (e.g. thermal with stress).

Manual meshing, no automatic refinement.

Comments: Student and professional versions available.

A.1.2 Stand-Alone 2D Electrostatic Solvers

In addition to the general purpose PDE solvers there are dedicated electrostatic solvers. These tools are generally set up to handle multiple strips in an efficient manner.

Maxwell 2D - Ansoft

Numerical Method: FEM

Platforms: Windows and UNIX

Features: Handles arbitrary cross-sections.

Computes [L], [C], [R], and [G] matrices for multiconductor systems.

Also computes impedance matrix.

Comments: Common GUI with other Ansoft products.

LINPAR and MULTLIN - Artech House Publishers

Numerical Method: MoM

Platforms: Windows

Features: Handles a large number of multiconductor transmission line cross-sections.

Microstrip, stripline, coplanar waveguide, coupled rectangular bars, multilayer planar structures, and user-defined structures can be analyzed.

LINPAR computes [L], [C], [R], and [G] matrices for multiconductor systems. It will also present impedance and scattering parameters for single and coupled lines.

MATPAR accepts [L], [C], [R], and [G] matrices and computes multiport *S*-parameter files or SPICE models.

C_NL2 from Artech has a multiple coupled line model that reads the output file from LINPAR.

Comments: Used in both the RF and high-speed digital communities.

Relatively inexpensive.

ElecNet - Infolytica

Numerical Method: FEM

Platforms: Windows

Features: Handles arbitrary cross-sections and rotationally symmetric problems.

3D option also available.

Active X scripting with Visual Basic, Java, or Pearl.

Use MATLAB or Excell to drive optimization.

Comments: No direct computation of capacitance or impedance.

ELECTRO - Integrated Engineering Software

Numerical Method: BEM

Platforms: Windows

Features: Handles arbitrary cross-sections and rotationally symmetric problems.

Automatic mesh refinement.

Automatic calculation of [L], [C], and [Z].

Parametric and batch mode analysis.

A.1.3 Summary for Stand-Alone 2D Solvers

The most flexible tools in this group have to be the general purpose PDE solvers. They can do many things, if you can figure out how to formulate the problem. In Chapter 11 you will find examples of how to use these tools for transmission line problems. The second group are the dedicated electrostatic solvers. Their capabilities vary; some output L and C matrices directly, others require the user to do some additional post-processing.

A.1.4 Integrated 2D Field-Solvers

Probably the most useful tools for multistrip problems are 2D solvers that are integrated as models in linear and nonlinear circuit simulators. We can solve multistrip cases using stand alone solvers; however, the burden of transferring data makes this approach less attractive. The latest integrated 2D engines are also fast enough to be used inside an optimization loop.

VUSTLSn Model - LINMIC+/N (AC Microwave)

Numerical Method: Spectral Domain

Platforms: LINUX and Windows

Features: Handles up to 10 strips in six dielectric layers.

Second metal layer can be included in some cases.

Cover and sidewalls are included in the solution (closed box).

Look up table approach requires brief precomputation, but result is very fast analysis.

Comments: Very useful for Lange couplers, edge-coupled filters, interdigital filters, spiral inductors, etc.

MMICTL - LINMIC+/N (AC Microwave)

Numerical Method: Spectral Domain

Platforms: LINUX and Windows

Features: Handles up to 40 strips in six dielectric layers and two metal layers.

Cover and sidewalls are included in the solution (closed box).

Look up table approach requires brief precomputation, but result is very fast analysis.

Comments: Very useful for Lange couplers, edge-coupled filters, interdigital filters, spiral inductors, etc.

LINMIC+/N includes several other custom models built around this engine, including spiral inductors and spiral transformers.

MCPL Model - Ansoft Designer (Ansoft)

Numerical Method: Spectral Domain

Platforms: Windows and UNIX

Features: Handles up to 20 strips in microstrip or stripline.

Cover and sidewalls are included in the solution (closed box).

Up to four dielectric layers.

Option to speed up solution by computing only one specified frequency out of the entire sweep.

Comments: Very useful for Lange couplers, edge-coupled filters, interdigital filters, spiral inductors, etc.

PCLIN Model - ADS and Touchstone Series IV (Agilent EEsof EDA)

Numerical Method: Finite Difference

Platforms: UNIX

Features: Quasi-static solver for up to 10 strips in seven layers.

Strips not restricted to a single layer.

Comments: Useful for RF and high-speed digital applications.

Use with caution at microwave frequencies.

MSnCTL, SLnCTL, SSnCTL Models - MDS (Agilent EEsof EDA)

Numerical Method: Finite Difference

Platforms: UNIX

Features: Quasi-static solver for up for three to five strips in a single plane.

Covers microstrip, stripline, and suspended substrate.

Comments: Useful for RF and high-speed digital applications.

Use with caution at microwave frequencies.

MLnCTL_V and MLnCTL_C - ADS (Agilent EEsof EDA)

Numerical Method: Finite Difference?

Platforms: Windows and UNIX

Features: Quasi-static solver for two to 10 strips in multiple planes.

One special model for 16 strips, constant width and spacing.

Comments: Useful for RF and high-speed digital applications.

Use with caution at microwave frequencies.

A.1.5 Summary for Integrated 2D Field-Solvers

The Ansoft Designer and LINMIC+/N multistrip models have been used extensively to design distributed filters from a few GHz up to 40 GHz. The key feature is that they include the effects of the sidewalls and cover from first principles, not some tacked on correction factor. There are some second order effects they cannot predict that have to do with the waveguide channel. The 2.5D and 3D solvers do predict these second order effects but they are too slow to use for optimization.

A.2 2.5D PLANAR SOLVERS (3D MOSTLY PLANAR)

Let us continue our review of CAD tools with the 2.5D planar solvers. Remember that with this class of tools we are allowed multiple planes of metal, multiple homogeneous dielectric layers, and vias between layers. We have made quite a step up in geometric complexity compared to the 2D cross-section solvers, but we are not quite to the most general case of the 3D arbitrary geometry solvers. Also remember that this class of tools can be divided into two subclasses, laterally open and fully enclosed.

SFPMIC - LINMIC+/N (AC Microwave)

Numerical Method: Spectral Domain

Platforms: LINUX and Windows

Features: One of two solvers included in LINMIC+/N.

Up to six dielectric layers and two metal layers.

Irregular grid discretization with rectangles and triangles.

Vias between layers included.

Diakoptics technique accounts for coupling between blocks.

MLSIM - AC Microwave

Numerical Method: PEEC (Hybrid Quasi-TEM)

Platforms: LINUX and Windows

Features: One of two solvers included in LINMIC+/N.

Closed box formulation.

Up to 20 dielectric layers and 19 metal layers.

Vias between layers allowed.

A faster technique than 2.5D MoM.

Ensemble 8.0 - Ansoft

Numerical Method: MoM

Platforms: Windows and UNIX

Features: Laterally open and closed box formulations.

Additional features for antenna structures.

Allows finite ground planes or no ground plane.

Arbitrarily place one or two grounds with slots.

Box enclosure can be added.

Singular value decomposition (SVD) fast solve technology.

Optimization and parametrics options available.

Can insert black box elements for passive or active devices.

Bulk conductivity for semiconductor materials.

Adaptive meshing.

Synthesis of 1D arrays.

Estimate tool for transmission lines and patches is very handy.

Comments: The original Ensemble code was acquired from Boulder Microwave Technology.

At this writing, Ansoft Ensemble is being incorporated into Ansoft Designer with a new interface.

Momentum (ADS 2002) - Agilent EEsof EDA

Numerical Method: MoM

Platforms: Windows and UNIX

Features: Laterally open formulation.

Arbitrary number of layers and ports.

Dual formulation of voltage in a slot.

Can mix slot and strip formulations in different layers in the same problem.

Edge element meshing algorithm.

Box enclosure can be added.

Optional Empipe optimization module.

Optional visualization module.

Momentum RF (added in ADS 1.5):

Quasi-static solver;

Star-loop basis functions for low frequency stability;

Mesh reduction with polygonal cells.

Comments: Polygons are recombined before meshing; makes it hard to manually control the mesh.

Mesher recognizes standard discontinuities, applies predetermined meshing rules.

EMSight - Applied Wave Research

Numerical Method: MoM

Platforms: Windows

Features: Closed box formulation.

Current viewing module fully integrated.

Fast sweep option.

Integrated 3D model viewer.

Comments: Part of Microwave Office suite of tools.

Designed for Windows, optimized code for i86 processors.

em 8.0 - Sonnet Software

Numerical Method: MoM

Platforms: Windows and UNIX

Features: Closed box formulation.

Number of layers and ports limited only by memory and time.

Diagonal elements, calibrated internal ports, and dielectric bricks.

Bulk conductivity for semiconductor substrates.

Optimization and parametric analysis.

Optimization of netlist / geometry projects.

Fast sweep (Adaptive Band Synthesis)

Support modules for viewing currents and antenna patterns.

3rd party module for viewing geometry in 3D.

Comments: One of the first tools available in this class.

Free version with reduced capabilities is available.

IE3D 9.0 - Zeland Software

Numerical Method: MoM

Platforms: Windows and UNIX

Features: Laterally open and closed box formulations.

Arbitrary number of layers and ports.

Dual formulation of voltage in a slot.

Vias between layers not restricted to perpendicular orientation.

Metal patches can have arbitrary orientation.

Automatic edge meshing.

Optimization of netlist / geometry projects.

Comments: One of the first tools to focus on the Windows environment.

Translator from various CAD formats available.

Flexible display of currents and E-field.

A.3 3D ARBITRARY GEOMETRY SOLVERS

We conclude our CAD tool review with the 3D arbitrary geometry solvers. These are the most general tools available and can theoretically handle just about any problem. The price we pay for this generality is computation time. The numerical effort required is quite high because we have to mesh the entire problem space. It is probably easier to describe complicated geometries using FEM tools rather than time domain tools. However, time domain tools are quite efficient for generating broadband frequency data. Most of these tools are stand alone; however, there are a couple of special purpose integrated engines of note.

Ansoft HFSS 8.5 - Ansoft

Numerical Method: FEM

Platforms: Windows and UNIX

Features: Arbitrary geometry and resolution.

Tetrahedral edge elements; closed box formulation.

Second order absorbing boundary conditions (ABCs).

Perfectly matched layers (PMLs).

ACIS-based 3D modeler.

True-surface object modeling is an option.

Optimization capability added (V8.0).

Eigenmode-solver added (V8.0).

Dual processor support (V8.0).

Macro approach in editor, basic object info is lost.

Modes-to-nodes feature added (V8.0).

Comments: At this writing, Ansoft HFSS 9.0 is about to be released with a completely new interface.

Agilent HFSS 5.6 - Agilent EEsof EDA

Numerical Method: FEM

Platforms: Windows and UNIX

Features: Arbitrary geometry and resolution.

Tetrahedral edge elements; closed box formulation.

Second order absorbing boundary conditions.

ACIS-based 3D modeler.

New modes-to-nodes algorithm for multiconductor ports

Revised fast sweep algorithm.

Empipe3D optimization module is optional.

Basic objects remain editable, even after Boolean operations.

Comments: Discontinued by Agilent in May 2001.

CST Microwave Studio 4.0 - Computer Simulation Technology

Numerical Method: FIT (FDTD)

Platforms: Windows

Features: Conformal approximation in three dimensions.

Well integrated ACIS-based interface.

Large number of import and export formats supported.

Transient solver, eigenmode-solver, modal analysis simulator.

2D eigenmode-solver for port modes.

Built in parametric sweep and optimization.

Eigenmode-solver supports two processors on a PC.

Code is multithreaded.

Comments: Based on research begun in the late 1970s and early 1980s.

From the same research group that authored the MAFIA codes.

FIT starts from integral rather than differential formulation.

MEFiSTo-3D Pro - Faustus Scientific

Numerical Method: TLM

Platforms: Windows

Features: 2D and 3D geometry editor.

Automatic rectangular meshing of 2D structures.

Automatic cuboid meshing of 3D structures.

Smooth boundary fitting with local mesh modification.

Homogeneous and inhomogeneous electric and magnetic materials.

Frequency dispersive boundaries.

Lumped elements and active devices.

Real-time embedding of SPICE circuits into 2D and 3D field space.

Multi-threaded architecture.

Comments: Also offer MEFiSTo-2D Classic as shareware.

Micro-Stripes 6.0 - Flomerics Ltd.

Numerical Method: TLM

Platforms: Windows and UNIX

Features: ACIS-based modeler

Automatic mesh generation.

Multi-grid meshing.

Sub-cell models for wires, ports, circuits, slots, and thin sheets.

Support for frequency dependent materials.

Support for multiple processors.

Support for finite and infinite ground planes.

Can force some regions within solution space to be ignored.

Comments: Originally offered by KCC Ltd, which later merged with Flomerics.

Concerto - Vector Fields

Numerical Method: FDTD

Platforms: Windows

Features: Conformal approximations for irregular geometries in one plane.

Polygonal cells at curved metal surfaces.

Inhomogeneous cells at media interfaces.

Graded mesh with mesh snapping planes.

2D modal templates at ports (similar to FEM approach).

Differential decomposition / template filtering for *S*-parameter extraction.

Optional Prony module for convergence acceleration.

Optional optimization module.

This release supports multiple processors.

Macro language for building project geometry.

Comments: Was QuickWave3D from QWED.

Marketing agreement with Vector Fields.

Appendix B

List of Software Vendors

LINMIC+/N

AC Microwave GmbH
Kackertstr. 16-18
D-52072 Aachen, Germany

TEL: 49-241-879-3022
FAX: 49-241-879-3023
EMAIL: mail@linmic.com
WEB: www.linmic.com

Momentum and ADS

Agilent EEsof EDA
1400 Fountaingrove Parkway
Santa Rosa, CA 95401

TEL: 1-800-452-4844
FAX: 1-888-900-8921
EMAIL: eesof_support@agilent.com
WEB: www.eesof.tm.agilent.com

Ansoft HFSS and Designer

Ansoft Corporation
Four Station Square, Suite 660
Pittsburgh, PA 15219

TEL: 412-261-3200
FAX: 412-471-9427
EMAIL: info@ansoft.com
WEB: www.ansoft.com

EMSight and Microwave Office

Applied Wave Research, Inc.
2210A Graham Ave.
Redondo Beach, CA 90278

TEL: 310-370-2496
FAX: 310-793-6500
EMAIL: info@appwave.com
WEB: www.appwave.com

LINPAR, MULTLIN, and C_NL2

Artech House
685 Canton Street
Norwood, MA 02062

TEL: 781-769-9750
FAX: 781-769-6334
EMAIL: artech@artechhouse.com
WEB: www.artechhouse.com

IE3D and Fidelity

Bay Technology
1711 Trout Gulch Road
Aptos, CA 95003

TEL: 831-688-8919
FAX: 831-688-6435
EMAIL: sales@bay-technology.com
WEB: www.bay-technology.com

RealTime

CRC Press
2000 N.W. Corporate Blvd.
Boca Raton, FL 33431

TEL: 561-994-0555
FAX: 561-989-8732
EMAIL: orders@crcpress.com
WEB: www.crcpress.com

CST Microwave Studio (N. America)

CST of America
8 Grove Street, Suite 203
Wellesley, MA 02482

TEL: 781-416-2782
FAX: 781-416-2782
EMAIL: info@cst-america.com
WEB: www.cst.de

CST Microwave Studio (Europe)

CST GmbH
Buedinger Str. 2a
D-64289 Darmstadt, Germany

TEL: 49-(0)6151-7303-0
FAX: 49-(0)6151-7303-10
EMAIL: info@cst.de
WEB: www.cst.de

FEKO

EMSS-SA Ltd.
Technopark
Stellenbosch, South Africa

TEL: 27-21-8801880
FAX: 27-21-8801936
EMAIL: info@emss.co.za
WEB: www.feko.co.za

MEFiSTo-2D and MEFiSTo-3D

FAUSTUS Scientific Corporation
1256 Beach Drive
Victoria, BC V8S 2N3, Canada

TEL: 250-598-2834
FAX: 250-721-6230
EMAIL: marketing@faustcorp.com
WEB: www.faustcorp.com

Micro-Stripes

Flomerics Inc.
257 Turnpike Road, Suite 100
Southborough, MA 01772

TEL: 508-357-2012
FAX: 508-357-2013
EMAIL: info@flomerics.com
WEB: www.flomerics.com

ElecNet

Infolytica Corp.
300 Leo Pariseau, Suite 2222
Montreal, QC H2W 2P4, Canada

TEL: 514-849-8752 ext. 270
FAX: 514-849-4239
EMAIL: max@infolytica.com
WEB: www.infolytica.com

ELECTRO and SINGULA

Integrated Engineering Software
300 Cree Crescent
Winnipeg, MB R3J 3W9, Canada

TEL: 204-632-5636
FAX: 204-633-7780
EMAIL: info@integrated.ca
WEB: www.integrated.ca

FlexPDE

PDE Solutions Inc.
P.O. Box 4217
Antioch, CA 94531-4217

TEL: 925-776-2407
FAX: 925-776-2406
EMAIL: sales@pdesolutions.com
WEB: www.pdesolutions.com

XFDTD

REMCOM, Inc.
P.O. Box 10023
State College, PA 16805

TEL: 814-353-2986
FAX: 814-353-2986
EMAIL: xfdtd@remcominc.com
WEB: www.remcominc.com

Sonnet Suite

Sonnet Software
1020 Seventh North Street, Suite 210
Liverpool, NY 13088

TEL: 315-453-3096
FAX: 315-451-1694
EMAIL: info@sonnetusa.com
WEB: www.sonnetusa.com

QuickField (International)

Tera Analysis Ltd.
Knasterhovvej 21
DK-5700 Svendborg, Denmark

TEL: (+45) 6354 0080
FAX: (+45) 6254 2331
EMAIL: sales@quickfield.com
WEB: www.quickfield.com

QuickField (North America)

Tera Analysis Ltd.
Toronto, Ontario
Canada

TEL: 877-215-8688
FAX: 877-215-8688
EMAIL: ussales@quickfield.com
WEB: www.quickfield.com

Concerto (QuickWave-3D)

Vector Fields
24 Bankside
Kidlington
Oxford, OX5 1JE
United Kingdom

TEL: 44(0)1865-370151
FAX: 44(0)1865-370277
EMAIL: info@vectorfields.co.uk
WEB: www.vectorfields.co.uk

Appendix C

List of Internet Sites

The EMLIB site:

http://emlib.jpl.nasa.gov

This single site will point you to many other sites of interest. They maintain a list of commercial codes and have a small collection of shareware contributions. They also maintain an extensive list of university sites which will give you some sense of the research going on around the world.

The University of Missouri-Rolla Electromagnetic Compatibility Lab:

http://www.emclab.umr.edu

This is one of the few sites dedicated to EMC. They also maintain lists of shareware and commercial codes. There are some interesting technical reports that can be downloaded. The lab has some simple 3D finite element codes that may be of interest.

The Applied Computational Electromagnetics Society (ACES) site:

http://aces.ee.olemiss.edu/

The ACES group has been running a small conference in Monterrey, CA for many years. They also publish their own journal and newsletter. The bulk of their contributors are working with antennas or scatterers. There is an occasional article that is related to circuits.

The Los Alamos Accelerator Code Group (LAACG) site:

http://laacg1.lanl.gov/

Long before commercial codes were available to solve microwave circuits, specialized codes were written to design the hardware used in high energy physics experiments. Some of these structures are in fact microwave waveguides or resonant cavities. This site maintains a large database of EM modeling codes. Two codes developed at Los Alamos are POISSON and SUPERFISH.

The unofficial NEC site:

http://www.qsl.net/wb6tpu/

NEC-2 is a code developed at the Lawrence Livermore National Laboratory for modeling antennas and scattering from metallic structures. The software models antennas as a combination of thin wires and metal plates. ACES has a large group of NEC users.

FDTD.org:

http://www/fdtd.org/

A very comprehensive site for FDTD literature of all types. Maintained by Dr. John B. Schneider, Washington State University, Pullman, WA.

About the Authors

Daniel G. Swanson, Jr. received his BSEE degree from the University of Illinois in 1976 and his MSEE degree from the University of Michigan, Ann Arbor in 1978.

In 1978 he joined Narda Microwave, where he developed a 6 to 18-GHz low-noise amplifier, an 8 to 10-GHz low-noise amplifier, and a de-embedding system for *S*-parameter device characterization. He joined the Wiltron Company in 1980, where he designed YIG tuned oscillators for use in microwave sweepers. He also developed a broadband load-pull system for optimization of output power. In 1983, Mr. Swanson joined a startup company, Iridian Microwave, where he was responsible for the dielectric resonator oscillator product line. Iridian was closed by its investors in December, 1993. He joined Avantek Inc. in 1984, where he developed thin-film microwave filters, software for filter design, and a low-frequency broadband GaAs MMIC amplifier. In 1989, he joined Watkins-Johnson Company as a staff scientist. His work there included thin-film filter design for broadband surveillance receivers, high performance filters for wireless base stations, and the application of electromagnetic field-solvers to microwave component design. He developed and presented his first course on the practical application of electromagnetic field-solvers in 1995. Mr. Swanson joined AMP M/A-COM in 1997 where he was a senior principal engineer. As a member of the Central R&D group, he applied electromagnetic field-solvers to the design of multilayer PC boards, RF and digital connectors, couplers and other microwave components. Mr. Swanson joined Bartley R.F. Systems in 1999, where he designs high Q filters for wireless base stations. He is also a consultant on filter design and on the application of field-solvers to RF and high-speed digital problems. Bartley R.F. Systems became the Forem USA division of Allen Telecom in December, 2001.

Mr. Swanson is a Fellow of the IEEE. He has published numerous technical papers, given many workshop and short course presentations, and holds two patents. He servers on the MTT-1 Technical Committee on CAD and on the MTT-8 Technical Committee on Filters and Passive Components. He also serves on the editorial boards of the *IEEE-MTT Transactions*, the *IEEE Microwave and Wireless*

Component Letters, and the *International Journal of RF and Microwave Computer-Aided Engineering*.

Wolfgang J.R. Hoefer received a Dipl.-Ing. degree in electrical engineering from the Technische Hochschule Aachen, Germany, in 1965, and a D. Ing. degree from the University of Grenoble, France, in 1968.

During the academic year 1968/69 he was a lecturer at the Institut Universitaire de Technologie de Grenoble and a research fellow at the Institut National Polytechnique de Grenoble, France. In 1969 he joined the Department of Electrical Engineering, the University of Ottawa, Canada where he was a professor until March 1992. Since April 1992 he has held the NSERC/MPR Teltech Industrial Research Chair in RF Engineering in the Department of Electrical and Computer Engineering, the University of Victoria, Canada.

He held visiting appointments with the Space Division of AEG-Telefunken in Backnang, Germany, the Electromagnetics Laboratory of the Institut National Polytechnique de Grenoble, France, the Space Electronics Directorate of the Communications Research Centre in Ottawa, Canada, the University of Rome "Tor Vergata", Italy, the University of Nice - Sophia Antipolis, France, The Ferdinand Braun Institute in Berlin, and the Technical University of Munich, both in Germany.

His research interests include numerical techniques for modeling electromagnetic fields and waves, computer-aided design of microwave and millimeter-wave circuits, microwave measurement techniques, and engineering education. He serves regularly on the Technical Program Committees of IEEE-MTT and AP Symposia, is the chair of the MTT-15 Technical Committee on Field Theory, and the cofounder and managing editor of the *International Journal of Numerical Modelling*. He was associate editor of the *IEEE MTT Transactions*, and serves on the editorial boards of the *IEEE-MTT Transactions*, *Proceedings of the IEE*, *Electromagnetics*, the *International Journal of Microwave and Millimeter-Wave Computer Aided Engineering*, and the *Microwave and Optical Technology Letters*. He is a Fellow of the IEEE and of the Advanced System Institute of British Columbia. Dr. Hoefer is also the cofounder and president of Faustus Scientific Corporation.

Index

Recent Titles in the Artech House Microwave Library

Advanced Techniques in RF Power Amplifier Design, Steve C. Cripps

Automated Smith Chart, Version 4.0: Software and User's Manual, Leonard M. Schwab

Behavioral Modeling of Nonlinear RF and Microwave Devices, Thomas R. Turlington

Computer-Aided Analysis of Nonlinear Microwave Circuits, Paulo J. C. Rodrigues

Design of FET Frequency Multipliers and Harmonic Oscillators, Edmar Camargo

Design of Linear RF Outphasing Power Amplifiers, Xuejun Zhang, Lawrence E. Larson, and Peter M. Asbeck

Design of RF and Microwave Amplifiers and Oscillators, Pieter L. D. Abrie

Distortion in RF Power Amplifiers, Joel Vuolevi and Timo Rahkonen

EMPLAN: Electromagnetic Analysis of Printed Structures in Planarly Layered Media, Software and User's Manual, Noyan Kinayman and M. I. Aksun

Feedforward Linear Power Amplifiers, Nick Pothecary

Generalized Filter Design by Computer Optimization, Djuradj Budimir

High-Linearity RF Amplifier Design, Peter B. Kenington

Lumped Elements for RF and Microwave Circuits, Inder Bahl

Microwave Circuit Modeling Using Electromagnetic Field Simulation, Daniel G. Swanson, Jr. and Wolfgang J. R. Hoefer

Microwave Component Mechanics, Harri Eskelinen and Pekka Eskelinen

Microwave Engineers' Handbook, Two Volumes, Theodore Saad, editor

Microwave Filters, Impedance-Matching Networks, and Coupling Structures, George L. Matthaei, Leo Young, and E.M.T. Jones

Microwave Materials and Fabrication Techniques, Third Edition, Thomas S. Laverghetta

Microwave Mixers, Second Edition, Stephen A. Maas

Microwave Radio Transmission Design Guide, Trevor Manning

Microwaves and Wireless Simplified, Thomas S. Laverghetta

Neural Networks for RF and Microwave Design, Q. J. Zhang and K. C. Gupta

Nonlinear Microwave and RF Circuits, Second Edition, Stephen A. Maas

QMATCH: Lumped-Element Impedance Matching, Software and User's Guide, Pieter L. D. Abrie

Practical RF Circuit Design for Modern Wireless Systems, Volume I: Passive Circuits and Systems, Les Besser and Rowan Gilmore

Practical RF Circuit Design for Modern Wireless Systems, Volume II: Active Circuits and Systems, Rowan Gilmore and Les Besser

Radio Frequency Integrated Circuit Design, John Rogers and Calvin Plett

RF Design Guide: Systems, Circuits, and Equations, Peter Vizmuller

RF Measurements of Die and Packages, Scott A. Wartenberg

The RF and Microwave Circuit Design Handbook, Stephen A. Maas

RF and Microwave Coupled-Line Circuits, Rajesh Mongia, Inder Bahl, and Prakash Bhartia

RF and Microwave Oscillator Design, Michal Odyniec, editor

RF Power Amplifiers for Wireless Communications, Steve C. Cripps

RF Systems, Components, and Circuits Handbook, Ferril Losee

Stability Analysis of Nonlinear Microwave Circuits, Almudena Suárez and Raymond Quéré

TRAVIS 2.0: Transmission Line Visualization Software and User's Guide, Version 2.0, Robert G. Kaires and Barton T. Hickman

Understanding Microwave Heating Cavities, Tse V. Chow Ting Chan and Howard C. Reader

For further information on these and other Artech House titles, including previously considered out-of-print books now available through our In-Print-Forever® (IPF®) program, contact:

Artech House
685 Canton Street
Norwood, MA 02062
Phone: 781-769-9750
Fax: 781-769-6334
e-mail: artech@artechhouse.com

Artech House
46 Gillingham Street
London SW1V 1AH UK
Phone: +44 (0)20 7596-8750
Fax: +44 (0)20 7630 0166
e-mail: artech-uk@artechhouse.com

Find us on the World Wide Web at:
www.artechhouse.com